AF400536

The Maz'ya Anniversary Collection

Volume 1:
On Maz'ya's work in functional analysis,
partial differential equations and applications

Jürgen Rossmann
Peter Takáč
Günther Wildenhain
Editors

Springer Basel AG

Editors:
Jürgen Rossmann
Peter Takáč
Günther Wildenhain
University of Rostock
Department of Mathematics
D-18051 Rostock
Germany

1991 Mathematics Subject Classification 43-06; 35-06

A CIP catalogue record for this book is available from the
Library of Congress, Washington D.C., USA

Deutsche Bibliothek Cataloging-in-Publication Data

The **Maz'ya anniversary collection** / Jürgen Rossmann ... ed. – Basel ;
Boston ; Berlin : Birkhäuser
   ISBN 978-3-0348-9726-6   ISBN 978-3-0348-8675-8 (eBook)
   DOI 10.1007/978-3-0348-8675-8

Vol. 1. On Maz'ya's work in functional analysis, partial differential
equations and applications. – 1999
(Operator theory ; Vol. 109)
ISBN 978-3-0348-9726-6

© 1999 Springer Basel AG
Originally published by Birkhäuser Verlag in 1999
Softcover reprint of the hardcover 1st edition 1999

Printed on acid-free paper produced from chlorine-free pulp. TCF ∞
Cover design: Heinz Hiltbrunner, Basel

ISBN 978-3-0348-9726-6

Vladimir Maz'ya

Maz'ya's portrait is by N. Singer

Contents

Contents

Introduction

The contributions in this volume are dedicated to Vladimir G. Maz'ya and are partially based on talks given at the conference "Functional Analysis, Partial Differential Equations, and Applications", which took place at the University of Rostock from August 31 to September 4, 1998, to honour Prof. Maz'ya. This conference (a satellite meeting of the ICM) gave an opportunity to many friends and colleagues from all over the world to honour him. This academic community is very large. The scientific field of Prof. Maz'ya is impressively broad, which is reflected in the variety of contributions included in the volumes. Vladimir Maz'ya is the author and co-author of many publications (see the list of publications at the end of this volume), the topics of which extend from functional analysis, function theory and numerical analysis to partial differential equations and their broad applications. Vladimir G. Maz'ya provided significant contributions, among others to the theory of Sobolev spaces, the capacity theory, boundary integral methods, qualitative and asymptotic methods of analysis of linear and nonlinear elliptic differential equations, the Cauchy problem for elliptic and hyperbolic equations, the theory of multipliers in spaces of differentiable functions, maximum principles for elliptic and parabolic systems, and boundary value problems in domains with piecewise smooth boundaries. Surveys on Maz'ya's work in different fields of mathematics and areas, where he made essential contributions, form a major part of the present first volume of The Maz'ya Anniversary Collection. Other articles of this volume have their origin in the joint work with Maz'ya. (Most of the invited lectures of the Rostock conference are included in the second volume.)

V. G. Maz'ya's scientific commitment and creativity has influenced both his colleagues throughout the world and his research students. He is a master of creating a school of thought, and many young mathematicians owe their graduation to him. An active research group was created by him in Linköping where he works at present. In 1991 and 1994 he organized international conferences on applied and industrial mathematics in Linköping. The international reputation of Prof. Maz'ya is reflected in his membership in many editorial boards of scientific journals in Germany, Great Britain, Israel, Netherlands and Sweden.

In addition to the Rostock conference, a dedication to Vladimir Maz'ya on the occasion of his 60th birthday took place at the "International Symposium on Boundary Element Methods" organized by the Ecole Polytechnique in Palaiseau (Paris) in May 1998. The conference "Functional Analysis, Partial Differential Equations and Applications" which took place in Rostock a few months later was jointly organized by the Department of Mathematics at the University of Rostock and the Weierstrass Institute of Applied Analysis and Stochastics in Berlin and was motivated by many scientific contacts of both institutions with Vladimir Maz'ya for many decades. In particular, it was motivated by the friendly relations which have grown over this period between Vladimir Maz'ya and the organizers of the conference. As a result of this long lasting cooperation many joint publications appeared.

My personal contact with V. G. Maz'ya goes back to my visit in the former city of Leningrad in 1969, upon his invitation. Joint interests in potential theory and the theory of elliptic differential equations of higher order have deepened our personal friendship and understanding. I remember my talk in the Mikhlin seminar very well. I was fascinated with the mathematical atmosphere in Leningrad at that time. It was characterized by such names as Smirnov, Mikhlin, Ladyzhenskaya, Havin, Solonnikov, and of course Maz'ya. The extremely stimulating discussions with V. G. Maz'ya have decisively influenced my further scientific research as well as the work of our research group, which was formed later in Rostock. That was the beginning of a personal friendship between us, which has lasted and strengthened through the decades, and I am very proud of that. I was not only impressed by V. G. Maz'ya's mathematical ideas but also by his fascinating personality. His interests are not limited to mathematics, but also extend to many other fields of life, such as art, especially of music and literature. My visits to Leningrad were always associated with impressive and enjoyable musical experiences, and I thank V. G. Maz'ya for that. I give him special thanks for his mathematical stimulus. Mathematicians in Rostock owe him a debt of gratitude as well as mathematicians throughout the world. It was a special honour for Rostock University to give Prof. Maz'ya an honorary doctorate in 1990. He takes his place together with such eminent scientists as Max Planck and Albert Einstein. With this honour and the conference of 1998, the university wants to thank V. G. Maz'ya for his support of mathematicians in Rostock.

The contributions in this volume show the extent of V. G. Maz'ya's research in an impressive way, his creativity, influence and worldwide recognition. The editors of this volume wish V. G. Maz'ya continued creativity and success. We also wish him and his family good health, happiness and stimulating and fruitful contacts, combined with a discussion of challenging mathematical problems.

Prof. Dr. Günther Wildenhain,
Rector of the University of Rostock,
D-18051 Rostock, Germany
rektor@uni-rostock.de

Operator Theory:
Advances and Applications, Vol. 109

Vladimir Maz'ya: Friend and mathematician. Recollections

ISRAEL GOHBERG

It gives me great pleasure to write about Vladimir Maz'ya whom I have known for almost forty years. For me his name is closely connected with Leningrad, its mathematical school and, particularly, with the name of Solomon G. Mikhlin, and I would like to begin my recollections a decade before I met Maz'ya.

When I was in my fourth year at the Kishinev University (autumn 1949) a young D.Sc. from Leningrad came to work there. As he himself said, he taught us everything he knew. He was knowledgeable as well as a good lecturer and I listened to everything he had to say. That was I.A. Itscovitch, a very recent Ph.D. student of S.G. Mikhlin. I am grateful to him for introducing me to Mikhlin's school. This way I learned about singular integral equations, Fredholm operators, the theory of the operator index, and other topics in which I later started to work actively. In 1949, I obtained my first results in this area and the following year I met Mikhlin for the first time when visiting Leningrad. He made a very strong impression on me, a kind and wise man with a fatherly behavior. After that visit I made several trips to Leningrad. I lectured at the seminar of V.I. Smirnov and made many good friends in Leningrad. By the way, my wife and I got married in Leningrad and I also defended my Ph.D. thesis in Leningrad with Mikhlin as an external examiner (opponent). I am proud to consider Mikhlin as one of my teachers.

In one of my trips to Leningrad Mikhlin introduced to me Volodia Maz'ya (Volodia is the diminutive of Vladimir). I had heard from Mikhlin that the young man was very promising. He had already written several interesting papers discovering, in particular, the equivalence of Poincaré-Sobolev imbedding theorems and isoperimetric inequalities for measures and capacities. In the future these ideas of Maz'ya had a broad influence on the development of Sobolev spaces, potential theory, and partial differential equations on nonsmooth domains and manifolds. His monograph "Sobolev spaces" [1] played an important role in this, continuing to inspire numerous researchers.

After the first encounter, I seldom met Maz'ya, since we lived in different cities and later on even in different countries. Although our mathematical directions did not show much similarity, I followed his work with great interest. Thus in 1968, I learned of his counterexamples for the nineteenth and twentieth Hilbert problems for higher order analytic elliptic equations and regular variational problems. The amazing fact I remember was that these equations and problems can have nonanalytic and even nonsmooth solutions. For experts this created a sensation.

Mikhlin started to pay attention to the 18 year old Maz'ya when the latter became a freshman at the Faculty of Mathematics and Mechanics in Leningrad.

The student had solved all the problems for both first and second year while participating in the traditional olympiad of the Faculty. As Maz'ya did not make this a secret, his fellow students all decided not to submit their solutions. An unexpected result was that the jury deemed the contest a failure and no prizes were awarded to anyone, including the winner. Instead, Maz'ya got another, much greater prize, which came to shape his entire mathematical life. Professor Mikhlin invited him to his home, soon making this a habit. Vladimir's father died at the front in 1941, and as he once confessed to me, his deep longing for a father plagued him during his childhood and youth. Maz'ya never was a formal student of Mikhlin, but Mikhlin was for him more than a teacher. Maz'ya had found the topics of his dissertations by himself, while Mikhlin taught him mathematical ethics and rules of writing, referring and reviewing. He also expressed his opinions about the work of others and his views on relationships between people. Maz'ya's prevailing style of summarizing his work on a particular theme by writing a book is certainly inherited from Mikhlin. By the way, the professor had mixed feelings about Vladimir's early work on Sobolev spaces, being himself more application-oriented. He once expressed his opinion on Maz'ya's geometrical counterexamples in the following way: "Your domains are very interesting, but no mother would let her child play in such ravines."

Mikhlin considered the theory of singular integral equations as his favorite creation. Very soon his results led to pseudodifferential operators, and his notion of the symbol (1936) became a cornerstone of this new theory which revolutionized partial differential equations. I myself was attracted to Mikhlin's theory even in the 1950s and followed this theory with great interest.

Maz'ya was very enthusiastic about this theory and became interested in the case of the degenerate symbol, which emerged at that time thanks to Mikhlin. I would especially like to mention Maz'ya's later study of the nonelliptic boundary value problem with oblique derivative, which can be reduced to a singular integral equation with a degenerate symbol. He provided an ingenious proof of unique solvability for the so called generic case, when the behavior of the vector field in the boundary operator is quite complicated.

Another area of his research, that has lasted for many years, and that was close to my interests, is boundary integral equations on nonsmooth contours and surfaces. Certainly, Maz'ya is a leader in this rapidly developing field, which is highly important for applications.

In the new Springer monograph by Kozlov and Maz'ya [2], an asymptotic theory of ordinary differential equations originating from Poincaré and Birkhoff, is extended to general classes of differential equations with unbounded operator coefficients in a Banach space. These results build the foundation of a powerful theory of singularities of solutions to linear and nonlinear partial differential equations developed in recent years by the authors.

Having given a quick glance at the list of Maz'ya's publications, I am very far from giving a complete analysis of his entire work. In general this cannot be done by one expert. I know that other contributors to this volume fulfil the task of doing justice to the man, who as a result of four decades of incessant labour, has authoured and coauthoured over 300 papers and more than a dozen books.

Maz'ya's exceptional productivity is due to a rare combination of talent, working ability, and inner drive. The development of this drive was stimulated by the hardships of his childhood during and after World War II. The obstacles raised by Soviet antisemitism in his youth strengthened his will to survive. The stifling atmosphere did not make the mature years easier, hindering Maz'ya's international contacts, publications and trips abroad.

The enormous amount and variety of his work go hand in hand with excellent quality. Whatever he writes is beautiful, his love for art, music and literature seeming to feed his mathematical esthetic feeling. It appears as if Maz'ya intuitively aims to bring harmony to domains previously ruled by disorder. Degenerations and singularities form the sphere in which he feels at home. He either gives necessary and sufficient conditions or constructs striking counterexamples, as in the topics of his work mentioned above.

In fact, his work has many other facets. One example is concerned with the linear theory of time-harmonic water waves. In 1977, he proved a uniqueness theorem for the waves generated by a submerged body [3], thus solving an important problem posed by F. John in 1950.

Of quite another style are the Kresin and Maz'ya studies of the maximum modulus principle for elliptic and parabolic systems. They discovered necessary and sufficient conditions, giving an answer to a long-standing classical problem.

Coefficients of partial differential equations and even symbols of pseudodifferential operators can be considered as functions, which map one Sobolev space to another, by pointwise multiplication. Together with his wife Tatyana, Maz'ya described spaces of such multipliers and investigated their properties. Their Pitman monograph of 1985 [4] is a comprehensive account of this theory and its applications.

Maz'ya's long-standing interest in elasticity resulted in a number of excellent achievements. I mention only two of them here. The first is the Maz'ya and Plamenevskii method for calculating stress intensity factors in fracture mechanics [5, 1974]. The second, due to Maz'ya and Nazarov [6, 1986], is an asymptotic analysis of the well-known polygon-circle paradox: when a thin circular plate is approximated by regular polygons with freely supported edges, the limit solution does not satisfy the conditions of the free support on the circle.

Maz'ya's character leaves a stamp of perfectionism on his work. Definitive solutions are a must when he deals with problems of any kind. I know a couple of examples when he lost priority because of his reluctance to publish partial results. Some of the areas which he once explored either alone or with a few colleagues later became popular.

Finishing the short review of Maz'ya's mathematical results, I would like to name his other areas of work, following the list published in the article [7]:

- Theory of capacities and nonlinear potentials

- Boundary behaviour of solutions to elliptic equations

- Estimates for general diferential operators

- The Cauchy problem for the Laplace equation

- Characteristic Cauchy problem for hyperbolic equations

- Boundary value problems in domains with piecewise smooth boundaries

- Iterative procedures for solving ill-posed boundary value problems

- Asymptotic theory of singularly perturbed boundary value problems

- "Approximate approximations" and their applications

During the last 12 years, everything concerning the life and work of Jacques Hadamard has been a hobby of Vladimir and Tatyana. Their book on the subject, which was recently published by the American and London Mathematical Societies [8] is a masterpiece both from a biographical and a mathematical point of view and is definitely worth keeping on your bedside table.

Vladimir is a good friend and I would like to extend my most sincere congratulations to him on the occasion of his 60th birthday. May his creativity and energy never leave him.

References

[1] Maz'ya, V. G., *Sobolev spaces,* Springer, 1985.

[2] Kozlov, V., Maz'ya, V., *Differential equations with operator coefficients,* Springer, 1999.

[3] Maz'ya, V. G., *On the steady problem of small oscillations of a fluid in the presence of a submerged body,* Proc. Semin. Sobolev. No. 2, 57–79, Novosibirsk, Inst. of Math. 1977.

[4] Maz'ya, V. G., Shaposhnikova, T. O., *Theory of multipliers in spaces of differentiable functions,* Pitman, 1985.

[5] Maz'ya, V. G., Plamenevskii, B. A. *On the coefficients in the asymptotics of solutions of elliptic boundary value problems in domains with conical points,* In: Elliptic boundary value problems, American Mathematical Society Translations, Ser. 2, vol. 123, 1984.

[6] Maz'ya, V. G., Nazarov, S. A., *Paradoxes of limit passage in solutions of boundary value problems involving the approximation of smooth domains by polygonal domains,* Math. USSR Izvestia **29**, No. 3 (1987).

[7] Eidus, D., Khvoles, A., Kresin, G., Merzbach, E., Prössdorf, S., Shaposhnikova, T., Sobolevskii, P., Solomiak, M., *Mathematical work of Vladimir Maz'ya*, Functional Differential Equations **4**, No. 1–2 (1997), 3–11.

[8] Maz'ya, V. G., Shaposhnikova, T. O., *Jacques Hadamard, a universal mathematician*, American Mathematical Society and London Mathematical Society, 1998.

School of Mathematical Sciences,
Raymond and Beverly Sackler Faculty of Exact Sciences,
Tel Aviv University,
Ramat Aviv, Tel Aviv 69978, Israel,
gohberg@math.tau.ac.il

1991 Mathematics Subject Classification: Primary 01A70

Submitted: 03.02.1999

Operator Theory:
Advances and Applications, Vol. 109

On Maz'ya's work in potential theory and the theory of function spaces

LARS INGE HEDBERG

Abstract: A presentation of some of the highlights in Vladimir Maz'ya's remarkable early work on function spaces, potential theory, and partial differential operators.

1.   Introduction

I am happy and honored to have been asked to give this presentation of a portion of the work of Vladimir Maz'ya, whom I have known for over a quarter of a century, and who is now since eight years my colleague at Linköping University in Sweden.

It is a particular pleasure to give such a talk in Rostock, which although geographically so close, until quite recently was impossible to visit freely. The University of Rostock was founded before there was any university in Sweden, and in the 15th century it was the most popular place to go to for young Swedes who wanted a university education. The most famous of these students was Jakob Ulfsson, who was immatriculated here in 1457. He became the Archbishop of Sweden, and is remembered for having founded Uppsala University in 1477.

Everybody who has taken a look at a list of Maz'ya's publications, or only a part of it, will understand that it is impossible to do his achievements justice in a short lecture. I have chosen to concentrate on what I think are some of the highlights in his work in the 1960's and 1970's. In addition to the limitations of time and competence of the speaker, a reason for this choice is that many of his papers from this period were published in Soviet journals of limited circulation, and were translated into English only belatedly, or not at all.

2.   Embeddings and isoperimetric inequalities

Vladimir Maz'ya started publishing early. His first paper appeared in 1959, and in his second paper in 1960 he already set the tone for much of what was going to follow. In this paper [18], a short note in the Doklady without any proofs, he gave embedding theorems of Sobolev type for functions defined in domains with irregular boundaries. S. L. Sobolev had proved his inequality

$$\|u\|_{L^{p^*}(\Omega)} \le C\big(\|\nabla u\|_{L^p(\Omega)} + \|u\|_{L^p(\Omega)}\big), \quad 1 < p < n, \quad p^* = np/(n-p),$$

under the assumption that the domain $\Omega \subset \mathbb{R}^n$ satisfies an interior cone condition. For $p = 1$ an extension of this result had been proved in 1958 by E. Gagliardo.

Maz'ya set himself the task of finding necessary and sufficient conditions on the domain Ω for this type of inequality to be true, and also for the compactness of the embedding. In [18] he was able to do so in terms of a highly implicit kind of isoperimetric inequalities. However, in the case $p = 1$ one of these is nothing but the classical isoperimetric inequality

$$m_n(D)^{(n-1)/n} \leq n^{-1} v_n^{-1/n} m_{n-1}(\partial D),$$

where $v_n = \pi^{n/2}\Gamma(\frac{n}{2}+1)$ is the volume of the n-dimensional unit ball. As a consequence he obtained the best constant, $C = n^{-1} v_n^{-1/n}$, in Gagliardo's inequality,

$$\|u\|_{L^{n/(n-1)}(\mathbb{R}^n)} \leq C\|\nabla u\|_{L^1(\mathbb{R}^n)},$$

valid, for example, for all $u \in C_0^\infty(\mathbb{R}^n)$. The elegant proof (which appeared later) depends on the coarea formula,

$$\int_\Omega |\nabla u|\, dx = \int_0^\infty m_{n-1}(E_t)\, dt, \quad E_t = \{\, x : |u(x)| = t\,\}.$$

He had a bit of bad luck in that this inequality also appeared in a long paper on currents by H. Federer and W. H. Fleming [9], which was published in the same year. Also, the distinguished mathematician who reviewed Maz'ya's paper in the Mathematical Reviews did not mention this result. Instead, he noticed that there were formulas with as many as four different norms!

Maz'ya's discovery, that embedding theorems are equivalent to inequalities of isoperimetric type, was, of course, an important one. He pursued it in many subsequent papers, and it pervades much of his book "Sobolev Spaces" [34], which appeared in its definite form in 1985.

He soon published (among other things) two more short notes [19, 20] in the Doklady, which were followed by the longer [23] and [30]. The title of the second note, "The negative spectrum of the n-dimensional Schrödinger operator", indicates clearly at least one source of such problems.

Among the many results proved in these articles were conditions for the validity of the inequalities

$$\|u\|_{L^q(\Omega,\mu)} \leq C\big(\|\nabla u\|_{L^p(\Omega)} + \|u\|_{L^p(\Omega)}\big)$$

for $1 \leq p \leq q$. Here Ω is a domain in $\mathbb{R}^n$ whose boundary can be irregular, μ is a positive measure on Ω, and

$$\|u\|_{L^q(\Omega,\mu)} = \left(\int_\Omega |u|^q\, d\mu(x)\right)^{1/q}.$$

For functions vanishing near the boundary the results simplify, and for $p = 1$ the same proof as in the unweighted case gives the beautiful result that the inequality

$$\|u\|_{L^q(\Omega,\mu)} \leq \|\nabla u\|_{L^1(\Omega)}, \quad q \geq 1,$$

holds for all $u \in C_0^\infty(\Omega)$, if and only if the "isoperimetric inequality",

$$\mu(G)^{1/q} \le m_{n-1}(\partial G),$$

is valid for each open $G \Subset \Omega$ with smooth boundary ∂G. For $\Omega = \mathbb{R}^n$ the sets G can be replaced by balls, but with different constants in the necessary and sufficient conditions.

In order to prove similar conditions for $p > 1$, Maz'ya needed to introduce p-capacities. For any compact $K \subset \Omega$ and $p \ge 1$ the *condenser capacity* is defined by

$$C_{1,p}(K;\Omega) = \inf\left\{ \int_\Omega |\nabla\varphi|^p \, dx : \varphi \in C_0^\infty(\Omega), \ \varphi|_K \ge 1\right\} \ .$$

If $\Omega = \mathbb{R}^n$, the quantity is called *capacity*, and denoted $C_{1,p}(K)$.

The result is the following: For the inequality

$$\|u\|_{L^q(\Omega,\mu)} \le \|\nabla u\|_{L^p(\Omega)}, \quad q \ge p \ge 1,$$

to hold for all $u \in C_0^\infty(\Omega)$, it is sufficient that

$$\mu(K)^{p/q} \le (p-1)^{p-1} p^{-p} C_{1,p}(K;\Omega),$$

and necessary that

$$\mu(K)^{p/q} \le C_{1,p}(K;\Omega),$$

for all compact $K \subset \Omega$. Here the constant is interpreted as 1 for $p = 1$. For $\Omega = \mathbb{R}^n$ the sets K can again be replaced by balls if $q > p$, but not if $q = p$ (D. R. Adams [1]).

For $p = 2$ this result is directly related to properties of the Schrödinger operator $S = -\Delta - \mu$, where μ is a positive measure on a domain Ω. Indeed, this operator is associated to a quadratic form,

$$(Su, u) = \int_\Omega |\nabla u|^2 \, dx - \int_\Omega |u|^2 \, d\mu(x), \quad u \in C_0^\infty(\Omega),$$

and the above gives conditions for the positivity of this form.

A few years after first discovering this result, Maz'ya proved the following inequality ([30], Theorem 3): For any $p \ge 1$ and $u \in C_0^\infty(\Omega)$

$$\int_0^\infty C_{1,p}(\{\, x : u(x) \ge \lambda\,\};\Omega) \, d\lambda^p \le \frac{p^p}{(p-1)^{p-1}} \int_\Omega |\nabla u|^p \, dx,$$

where the constant is best possible. This result trivially implies the above embedding theorem. The inequality is the first of a type that D. R. Adams [2] has named *capacitary strong type* inequalities. There have been many further extensions of this result, that have to be omitted here.

I also have to omit any discussion of the applications to the theory of multipliers between function spaces, which were given a full treatment in the book by Maz'ya and T. O. Shaposhnikova [40]. Two joint papers [38, 41], with Yu. V. Netrusov and I. E. Verbitsky, respectively, testify to Maz'ya's continuing interest in the field.

For those desirous of further information, I would like to draw attention to Maz'ya's very readable encyclopedia article [35], and to his recent book with S. V. Poborchi [39], in addition to the very comprehensive book [34] already mentioned. The subject is also treated in the book [3] by D. R. Adams and the speaker.

3.   Regularity of solutions

According to a famous theorem of E. De Giorgi and J. Nash from 1957, any weak solution $u \in W^{1,2}$ of an elliptic partial differential equation

$$\sum_{i,j=1}^{n} \partial_i(a_{ij}(x)\partial_j u) = 0,$$

with bounded, measurable, real-valued coefficients is locally Hölder continuous. It was at first believed that this result could be extended to higher order equations and systems, but these hopes were crushed in 1968, when Maz'ya [27] (and independently E. De Giorgi [7] followed by E. Giusti and M. Miranda [11]) gave counterexamples, also providing a negative solution of Hilbert's nineteenth problem.

Maz'ya considered, in fact, the functional

$$\int_{|x|<1} \left(\nu(\Delta u)^2 + 2\kappa \frac{x_i x_j}{|x|^2} \partial_i \partial_j u \Delta u + \mu \left(\frac{x_i x_j}{|x|^2} \partial_i \partial_j u \right)^2 \right) dx,$$

where the summation convention is used, and $\nu > 0$, κ, μ are constants such that $\kappa^2 < \mu\nu$. Then its Euler equation is a strongly elliptic equation of order 4, and it has a solution in $W^{2,2}$ of the form $u = |x|^a$ for a certain number a. If $n > 4$ the constants can be chosen so that $a < 0$, and consequently the solution is unbounded.

He also expanded this example to give equations of any even order $2l$, $l \geq 2$, with solutions $u = |x|^a$ belonging to $W^{2,l}$. If $n > 2l$, a can again be made negative. It should be noted that for $n < 2l$ functions in $W^{2,l}$ are Hölder continuous by the Sobolev embedding theorem.

It is a consequence of the De Giorgi-Nash theorem, and earlier work beginning with S. N. Bernstein, that solutions of quasilinear elliptic equations of the type

$$\sum_{i,j=1}^{n} \partial_i(a_{ij}(x,u)\partial_j u) = 0,$$

with infinitely differentiable coefficients are infinitely differentiable. In the same paper, Maz'ya gave examples of equations of the form

$$\sum_{|\alpha|=|\beta|=l} D^\alpha(a_{\alpha\beta}(x,u,\nabla u,\dots,\nabla^k u)D^\beta u) = 0, \quad l \geq 2, \quad k \leq l-1,$$

with analytic coefficients, which have solutions of the form $u = c|x|^k$. This function does not belong to C^k, but it has Lipschitz continuous derivatives of order $k - 1$. On the other hand, it was known from the work of S. Agmon, A. Douglis and L. Nirenberg [4] that any solution which is à priori known to belong to C^k, also has to belong to C^∞, so the result is sharp.

For more information on regularity theory we refer to the books [10] by M. Giaquinta, and [17] by J. Malý and W. P. Ziemer, and the survey article of E. Bombieri [6].

4.   Boundary regularity

In 1924 N. Wiener published his criterion for the solvability of Dirichlet's problem. It can be formulated in the following way:

Let Ω be a domain in $\mathbb{R}^n$, $n \geq 2$. Then a point $a \in \partial\Omega$ is regular for the Dirichlet problem for the Laplace equation in Ω if and only if

$$\int_0^1 \frac{C_{1,2}(\Omega^c \cap B(a,r))}{r^{n-2}} \frac{dr}{r} = \infty.$$

(A slight modification of the definition of capacity is needed in the case $n = 2$, see below.)

There are many generalizations and extensions of this result, but the earliest Wiener condition for boundary regularity for non-linear equations was given by Maz'ya [29] in 1970. He considered a class of quasilinear equations of second order, including the so called p-Laplace equation,

$$\Delta_p u = \operatorname{div}\left(\nabla u |\nabla u|^{p-2}\right) = 0,$$

and defined regular boundary points for solutions in $W^{1,p}$. His main result contains the following:

For a point $a \in \partial D$ to be regular for the Dirichlet problem for the p-Laplace equation $\Delta_p u = 0$, $1 < p \leq n$, in a region $\Omega \subset \mathbb{R}^n$ it is sufficient that

$$\int_0^1 \left(\frac{C_{1,p}\left(\Omega^c \cap B(a,r)\right)}{r^{n-p}}\right)^{1/(p-1)} \frac{dr}{r} = \infty.$$

Whether this condition is also necessary was an open problem until 1993, when the necessity was proved for $p > 1$ by T. Kilpeläinen and J. Malý [16]; see also the book [17] mentioned above.

The rapidity of divergence of the above integrals is closely related to the modulus of continuity of solutions at the boundary. As early as in 1963 the first sharp such estimates were obtained by Maz'ya in the case of linear equations, see [22, 25, 26].

For solutions vanishing at a part of the boundary containing a point a he proved the following estimate:

$$\sup_{\Omega \cap B(a,r)} |u(x)| \leq A_1 \sup_{\Omega \cap B(a,r_0)} |u(x)| \exp\left(-A_2 \int_r^{r_0} \frac{C_{1,2}(\Omega^c \cap B(a,\rho))}{\rho^{n-2}} \frac{d\rho}{\rho}\right).$$

In [29] Maz'ya obtained similar estimates in terms of his integral for the quasilinear equations mentioned above.

It should be noted that these integrals are majorized by $A \log(r_0/r)$, so the estimates can never give more than polynomial decay. In the harmonic case Maz'ya [33] improved the estimate by adding a certain negative term in the exponential, measuring the narrowness of the complement, which for some domains gives superpolynomial decay. This result was extended to general second order elliptic equations in divergence form in a recent joint paper with his former student Jana Björn [5].

For equations of higher order very little is known, but in 1977 Maz'ya [32] gave a sufficient condition of Wiener type for boundary regularity for the biharmonic equation, $\Delta^2 u = 0$, in dimensions $4 \leq N \leq 7$, and Maz'ya and T. Donchev [37] extended this result to the equation $\Delta^k u = 0$, $k > 2$, for $2k \leq N \leq 2k+2$. Another extension was given recently by Maz'ya's student S. Eilertsen [8]. The proofs depend on inequalities that are not true in other dimensions, and the extension of the sufficient condition to higher dimensions, as well as the question of its necessity in any dimension, remain tantalizing open problems.

Maz'ya gave his own view of the problems discussed in this section in his address [36] at the Wiener Centennial Symposium in Cambridge, Massachusetts in 1994.

5.   Nonlinear potential theory

Much of what has been said above belongs to what is now often called nonlinear potential theory. This is not the place to try to describe the history of this extension of classical potential theory, the roots of which go back to the early 1950's. (An attempt was made in [3], see p. 49, where many references are given.) Suffice it to say that Maz'ya is one of the creators and main contributors to the theory. He introduced general (α, p)-capacities for integral α and $1 \leq p < \infty$ in [19, 21, 24, 28] and applied them in many subsequent papers.

The (α, p)-capacity can be defined for compact sets $K \subset \mathbb{R}^n$ in the following way:

$$C_{\alpha,p}(K) = \inf\{\|\varphi\|_{W^{\alpha,p}}^p : \varphi \in C_0^\infty, \varphi|_K \geq 1\}.$$

However, this definition has some drawbacks. The extremal function is the solution of a complicated nonlinear differential equation on K^c, and there is no simple representation corresponding to the Newtonian potential in the classical case.

It is a fact, which is a consequence of the theory of singular integral operators, that functions in $W^{\alpha,p}$ for $1 < p < \infty$ can be represented as Riesz (or Bessel)

α-potentials of L^p-functions, and conversely (if Bessel potentials are used). In the late 1960's it occurred to a number of people, in different countries, that by means of this representation one could get an equivalent definition of (α, p)-capacity, which is easier to handle, and which also has the advantage that it immediately extends to non-integer values of α. Among these were Maz'ya and V. P. Havin, who now worked jointly. They announced some of their results in a brief note [12] in 1970, which was followed by two long papers [13, 14].

Incidentally, this note was the reason for my first getting acquainted with Havin and Maz'ya in the fall of 1971. I wanted very much to have the proofs of their results, so I went to Leningrad, and was fortunate enough to meet them, which at that time was not entirely trivial, and even against the laws of the Soviet Union.

They defined (α, p)-capacity for compact sets (and $\alpha p < n$) in a way which can be reformulated as follows:

$$C_{\alpha,p}(K) = \inf \left\{ \|f\|_{L^p}^p : f \in L_+^p, \int_{\mathbb{R}^n} f(y)|x - y|^{n-\alpha}\, dy \geq 1 \text{ on } K \right\}.$$

They proved by a variational argument that the extremal f_K has the form

$$f_K(x) = \left(\int_{\mathbb{R}^n} |x - y|^{n-\alpha}\, d\mu_K(y) \right)^{1/(p-1)},$$

where μ_K is a certain positive measure on K whose total mass $\mu_K(K)$ is $C_{\alpha,p}(K)$.

Thus, the extremal potential, $\int_{\mathbb{R}^n} f_K(y)|x - y|^{n-\alpha}\, dy$ is nonlinear as a function of μ_K, and for this reason the authors coined the term *nonlinear potential theory* for the theory of such potentials. (Later, the term has come to be used in a somewhat more general sense.) For $p = 2$ the potential is linear, and by a classical convolution formula of M. Riesz it becomes a classical Riesz potential of order 2α, $\int_{\mathbb{R}^n} |x - y|^{n-2\alpha}\, d\mu_K(y)$.

They were able to extend a large part of classical potential theory, which corresponds to the Hilbert space case $p = 2$, to this nonlinear setting, and to give definite solutions to many problems in analysis, in particular questions concerning removable sets for various classes of solutions to elliptic equations; see also Maz'ya's paper [31].

Nonlinear potential theory is treated at length, from different points of view, in the books [3, 15, 17], and I have to refer to these for further information.

References

[1] ADAMS, D. R.: Traces of potentials arising from translation invariant operators, *Ann. Scuola Norm. Sup. Pisa Cl. Sci.* **25** (1971), 203–217

[2] ———: On the existence of capacitary strong type estimates in $\mathbf{R}^N$, *Ark. mat.* **14** (1976), 125–140

[3] ——— and HEDBERG, L. I.: *Function Spaces and Potential Theory*, Springer, Berlin Heidelberg, 1996

[4] AGMON, S., DOUGLIS, A. and NIRENBERG, L.: Estimates near the boundary for solutions of elliptic partial differential equations satisfying general boundary conditions, I-II, *Comm. Pure Appl. Math.* **12** (1959), 623–727, **17** (1964), 35–92.

[5] BJÖRN, J. and MAZ'YA, V. G.: Capacitary estimates for solutions of the Dirichlet problem for second order elliptic equations in divergence form, Preprint, Linköping University, 1998

[6] BOMBIERI, E.: Variational problems and elliptic equations, in *Mathematical developments arising from Hilbert problems* (F. E. Browder, ed.), 525–535, Proc. Symp. Pure Math. **28**:2, Amer. Math. Soc. Providence, R. I., 1976.

[7] DE GIORGI, E.: Un esempio di estremali discontinue per un problema variazionale di tipo ellittico, *Boll. Un. Mat. Ital.* (4) **1** (1968), 135–137

[8] EILERTSEN, S.: On weighted positivity and the Wiener regularity of a boundary point for the fractional Laplacian, Preprint, Linköping University, 1998

[9] FEDERER, H. and Fleming, W. H.: Normal and integral currents, *Ann. of Math.* **72** (1960), 458–520

[10] GIAQUINTA, M.: *Multiple Integrals in the Calculus of Variations and Nonlinear Elliptic Systems*, Annals of Mathematics Studies, 105, Princeton University Press, Princeton, NJ, 1983

[11] GIUSTI, E. and MIRANDA, M.: Un esempio di soluzioni discontinue per un problema di minimo relativo ad un integrale regolare del calcolo delle variazioni, *Boll. Un. Mat. Ital.* (4) **1** (1968), 219–226

[12] HAVIN, V. P. (KHAVIN, V. P.) and MAZ'YA, V. G.: A nonlinear analogue of the Newtonian potential and metric properties of the (p, l)-capacity (Russian), *Dokl. Akad. Nauk SSSR* **194** (1970), 770–773. English translation: *Soviet Math. Dokl.* **11** (1970), 1294–1298

[13] ______: Non-linear potential theory (Russian), *Uspekhi Mat. Nauk* **27**:6 (1972), 67–138. English translation: *Russian Math. Surveys* **27**:6 (1972), 71–148

[14] ______: Use of (p, l)-capacity in problems of the theory of exceptional sets (Russian), *Mat. Sb.* **90(132)** (1973), 558–591. English translation: *Math. USSR-Sb.* **19** (1973), 547–580.

[15] HEINONEN, J., KILPELÄINEN, T. and MARTIO, O.: *Nonlinear Potential Theory of Degenerate Elliptic Equations*, Oxford University Press, Oxford, 1993

[16] KILPELÄINEN, T. and MALÝ, J.: The Wiener test and potential estimates for quasilinear elliptic equations, *Acta Math.* **172**, 137–161, 1994

[17] MALÝ, J. and ZIEMER, W. P.: *Fine Regularity of Solutions of Elliptic Partial Differential Equations*, Amer. Math. Soc., Providence, R.I., 1997

[18] MAZ'YA, V. G.: Classes of domains and embedding theorems for function spaces (Russian), *Dokl. Akad. Nauk SSSR* **133** (1960), 527–530. English translation: *Soviet Math.* **1** (1961), 882–885

[19] ______: The p-conductivity and theorems on embedding certain function spaces into a C-space (Russian), *Dokl. Akad. Nauk SSSR* **140** (1961), 299–302. English translation: *Soviet Math.* **2** (1961), 1200–1203

[20] ———: The negative spectrum of the n-dimensional Schrödinger operator (Russian), *Dokl. Akad. Nauk SSSR* **144** (1962), 721–722. English translation: *Soviet Math.* **3** (1962), 808–810

[21] ———: The Dirichlet problem for elliptic equations of arbitrary order in unbounded regions (Russian), *Dokl. Akad. Nauk SSSR* **150** (1963), 1221–1224. English translation: *Soviet Math.* **4** (1963), 860–863

[22] ———: Regularity at the boundary of solutions of elliptic equations and conformal mapping (Russian), *Dokl. Akad. Nauk SSSR* **152** (1963), 1297–1300. English translation: *Soviet Math.* **4** (1963), 1547–1551

[23] ———: On the theory of the n-dimensional Schrödinger operator (Russian), *Izv. Akad. Nauk SSSR, Ser. Mat.* **28** (1964), 1145–1172

[24] ———: Polyharmonic capacity in the theory of the first boundary value problem (Russian), *Sibirsk. Mat. Zh.* **6** (1965), 127–148

[25] ———: On the modulus of continuity of a solution of the Dirichlet problem near an irregular boundary (Russian), *Problemy Mat. Anal.* **1**, Izdat. Leningrad. Univ., Leningrad, 1966, 45–58. English translation: *Problems in Math. Anal.* **1**, Plenum Press, New York, 1968, 41–54

[26] ———: The behavior near the boundary of the solution of the Dirichlet problem for a second order elliptic equation in divergence form (Russian), *Mat. Zametki* **2** (1967), 209–220. English translation: *Math. Notes* **2** (1967), 610–617

[27] ———: Examples of nonregular solutions of quasilinear elliptic equations with analytic coefficients (Russian), *Funkcional. Anal. i Prilozhen.* **2**:3 (1968), 53–57. English translation: *Funkt. Anal. Appl.* **2** (1968), 230–234

[28] ———: Classes of sets and measures connected with embedding theorems (Russian), in *Embedding Theorems and Their Applications* (Russian), Proc., Baku 1966 (L. D. Kudryavtsev, ed.), 142–159, Nauka, Moscow, 1970

[29] ———: On the continuity at a boundary point of solutions of quasilinear equations (Russian), *Vestnik Leningrad. Univ. Mat. Mekh. Astronom.* **25**:13 (1970), 42–55. Correction, *ibid.* **27**:1 (1972), 160. English translation: *Vestnik Leningrad Univ. Math.* **3** (1976), 225–242

[30] ———: On certain integral inequalities for functions of many variables (Russian), *Problemy Matematicheskogo Analiza, Leningrad. Univ.* **3** (1972), 33–68. English translation: *J. Soviet Math.* **1** (1973), 205–234

[31] ———: Removable singularities of bounded solutions of quasilinear elliptic equations of any order (Russian), *Zap. Nauchn. Sem. Leningrad. Otdel. Mat. Inst. Steklov. (LOMI)* **26** (1972), 116–130. English translation: *J. Soviet Math.* **3** (1975), 480–492

[32] ———: Behaviour of solutions to the Dirichlet problem for the biharmonic operator at a boundary point, in *Equadiff IV*, Proc., Prague, 1977 (J. Fábera, Ed.) *Lecture Notes in Math.* **703**, 250–262, Springer-Verlag, Berlin–Heidelberg, 1979

[33] ———: The modulus of continuity of a harmonic function at a boundary point (Russian), *Zap. Nauchn. Sem. Leningrad. Otdel. Mat. Inst. Steklov. (LOMI)* **135** (1984), 87–95. English translation: *J. Soviet Math.*

[34] ______ : *Sobolev Spaces*, Springer-Verlag, Berlin–New York, 1985. Russian edition, Izd. Leningrad. Univ., Leningrad, 1985

[35] ______ : Classes of domains, measures and capacities in the theory of differentiable functions, in *Encyclopaedia of Mathematical Sciences, Vol. 26, Analysis III*, S. M. Nikolskiĭ ed., Springer, Berlin Heidelberg, 1991, 141–211

[36] ______ : Unsolved problems connected with the Wiener criterion, in *The Legacy of Norbert Wiener: A Centennial Symposium (Proc., Cambridge, Massachusetts, 1994)* 199–208, Proc. Sympos. Pure Math., 60, Amer. Math. Soc., Providence, Rhode Island, 1997

[37] ______ and DONCHEV, T.: On Wiener regularity at a boundary point for a polyharmonic operator (Russian), *C. R. Acad. Bulgare Sci.* **36** (1983), 177–179. English translation: *Amer. Math. Soc. Transl.* (2) **137** (1987), 53–55

[38] ______ and NETRUSOV, YU. V.: Some counterexamples for the theory of Sobolev spaces on bad domains, *Potential Analysis* **4** (1995), 47–65

[39] ______ and POBORCHI, S. V.: *Differentiable Functions on Bad Domains*, World Scientific, Singapore, 1997

[40] ______ and SHAPOSHNIKOVA, T. O.: *The Theory of Multipliers in Spaces of Differentiable Functions*, Pitman, Boston–London, 1985. Russian edition (with additions), Izd. Leningrad. Univ., Leningrad, 1986

[41] ______ and VERBITSKY, I. E.: Capacitary inequalities for fractional integrals with applications to partial differential equations and Sobolev multipliers, *Ark. mat.* **33** (1995), 81–115

Department of Mathematics
Linköping University
S-58183 Linköping, Sweden,
lahed@mai.liu.se

1991 Mathematics Subject Classification: Primary 31-02, 35-02, 46-02

Submitted: 17.08.1998

Operator Theory:
Advances and Applications, Vol. 109

Maz'ya's works in the linear theory of water waves

N.G. KUZNETSOV, B.R. VAINBERG

Abstract. The paper surveys results of V. Maz'ya in the linear theory of water waves. All main topics of his work in this field are considered. At first, we describe Maz'ya's achievements concerning the tough question of the unique solvability of two steady-state problems, which are: (1) the problem of time-harmonic waves in a layer of variable depth, and above a totally submerged body; (2) the problem of wave patterns due to the uniform forward motion of a body in water of constant depth. The review ends with a description of asymptotic expansions for unsteady waves arising from brief and high-frequency disturbances. A complete list of Maz'ya's publications on water waves is given.

1. Introduction

As colleagues of Vladimir Maz'ya (the first author was also his Ph.D. student), and admirers of his work, we are very pleased to contribute to this Volume. We are going to discuss a part of his results in Mathematical Hydrodynamics. To be specific, we shall concentrate on certain problems in the extensive field of the linear water-wave theory. Vladimir had brought our attention to this area of Mathematical Physics circa 1970. At that time he was a part-time Professor at the Department of Applied and Computational Mathematics in the Leningrad Shipbuilding Institute, where he had learned that many mathematical questions arising in the sea-keeping theory and the theory of wave-making resistance are still open. Since then we worked upon these problems in collaboration with him (for the first author water waves became the main topic of research for the life-time).

We shall mainly discuss results related to two fundamental steady-state problems of the surface-wave theory: the problem of time-harmonic water waves and the problem of wave pattern due to a body in the uniform forward motion in a calm water. The former problem is known under several names: the floating-body problem, the sea-keeping problem, the water-wave problem etc. (in what follows we use the last one); the latter problem is often referred to as the Neumann-Kelvin problem. Also, we describe results on asymptotic expansions for transient velocity potentials for waves due to brief and high-frequency disturbances.

References are arranged in two lists: Maz'ya's works form the first one, and the other related papers are listed in the second one (with a few exceptions those papers which are published after 1973 contain references to Maz'ya's works). In either list papers are given in the chronological order.

2. The unique solvability of the water-wave problem

We begin with the statement of the water-wave problem in two and three dimensions. It describes the irrotational motion of an inviscid heavy fluid without surface tension (water is a standard example), and all motions are assumed to be of small amplitude with the radian frequency ω. Under these assumptions the velocity field can be expressed as the gradient of a scalar potential having the form $\mathrm{Re}\{u(x,y)\,e^{-i\omega t}\}$. Here (x,y) are rectangular Cartesian coordinates with origin in the mean free surface, and with the y-axis pointing vertically upwards; $x \in \mathbf{R}$ and $x = (x_1, x_2) \in \mathbf{R}^2$ for the two- and three-dimensional cases respectively.

Let W denote a domain occupied by water, outside any bodies present, and let S be the union of wetted surfaces of all the bodies in equilibrium. By F we denote the free surface at rest, which coincides with $\{y = 0\}$ outside all bodies. At last, the water bottom B is positioned below $F \cup S$, and coincides with $\{y = -d\}$ $(d > 0)$ outside some compact set, that is, as $|x| \to \infty$. The case of deep water, when there is no bottom, is included allowing $d = \infty$.

Now, the boundary value problem for determining u can be written as follows:

$$(2.1) \qquad\qquad \nabla^2 u = 0 \quad \text{in} \quad W,$$

$$(2.2) \qquad\qquad u_y - \nu u = 0 \quad \text{on} \quad F,$$

$$(2.3) \qquad\qquad \partial u/\partial n = f \quad \text{on} \quad S,$$

$$(2.4) \qquad\qquad \partial u/\partial n = 0 \quad \text{on} \quad B,$$

Here $\nu = \omega^2/g$, where g is the acceleration due to gravity, $\mathbf{n}$ is a unit normal to $S \cup B$ directed into W, and f is prescribed by the motion of S. For deep water (when $B = \emptyset$), condition (2.4) should be replaced by the following one:

$$(2.5) \qquad\qquad |\nabla u| \to 0 \quad \text{as} \quad y \to -\infty.$$

Furthermore, the appropriate condition at infinity ensuring the uniqueness of solution is the radiation condition. If the water domain is m-dimensional ($m = 2, 3$), and $d \leq \infty$, it has the form:

$$(2.6) \qquad\qquad u_{|x|} - ik_0 u = o\left(|x|^{(2-m)/2}\right) \quad \text{as} \quad |x| \to \infty.$$

Here k_0 is the unique positive root of the dispersion equation $\nu = k_0 \tanh k_0 d$ for $d < \infty$, and $k_0 = \nu$ for $d = \infty$. Moreover, the asymptotic behaviour in (2.6) is assumed to be uniform in y if $m = 2$, and uniform with respect to both y, and the polar angle in the plane $\{y = 0\}$ if $m = 3$.

To get a rough idea of intrinsic difficulties of the problem one has to take into account that (2.1)–(2.6) can be reduced to an equation with pseudodifferential operator on the free surface, involving ν as a spectral parameter. The essential spectrum of this operator is the half-axis $\nu \geq 0$. In order to prove the unique solvability of (2.1)–(2.6), one has to establish that there are no point eigenvalues embedded into the essential spectrum. Until now, there are no general approaches to problems of this type.

By 1950, Kochin and John laid the groundwork in studies of the fundamental question of unique solvability of (2.1)–(2.6). Kochin (1939, 1940) proved the unique solvability of the two- and three-dimensional water-wave problems for sufficiently small and large values of ν in the case of a body totally immersed in a fluid of infinite depth. These results are consequences of the unique solvability of the limit problems as $\nu \to 0, +\infty$, having the Neumann and Dirichlet conditions respectively on the free surface. All other known results are obtained under some restrictions on the body shape. In his famous paper John (1950) established the uniqueness for all ν, when a semi-immersed body is assumed to have no parts strictly below the free surface, that is, when any vertical straight line intersecting the body does not intersect the free surface. The existence of a solution was proved under further condition, that the free surface and body's surface are orthogonal along the water-line.

It is obvious that for proving the uniqueness theorem one needs only to show that any solution u of the homogenous problem, having the finite energy integral, is zero. To this end John invented the following ingenious trick which we describe only for the fluid of infinite depth. First, he notes that by (2.1) and (2.2) the function

$$I(x) = \int_{-\infty}^{0} u(x,y)\, e^{\nu y} dy \quad , \quad x \in F \subset \mathbf{R}^2$$

solves the Helmholtz equation $\nabla_x^2 I + \nu^2 I = 0$, where $\nabla_x = (\partial/\partial x_1, \, \partial/\partial x_2)$. Therefore, $I(x)$ vanishes by the uniqueness theorem of Rellich. This implies that

$$u(x,0) = \int_{-\infty}^{0} \frac{\partial}{\partial y}[u(x,y)\, e^{\nu y}]dy = \int_{-\infty}^{0} \frac{\partial u}{\partial y}(x,y)\, e^{\nu y} dy \, , \quad x \in F.$$

Hence, the Cauchy inequality yields

$$\int_{F} |u|^2\, dx \le \frac{1}{2\nu} \int_{W} |\nabla u|^2\, dx dy.$$

Comparing this with Green's formula

$$\int_{F} |u|^2\, dx = \frac{1}{\nu} \int_{W} |\nabla u|^2\, dx dy,$$

we see that $u \equiv 0$. Note, that $I(x)$ is defined on the whole F only under the above John's geometrical assumption.

In contrast to John's uniqueness result, Ursell (1951) and Jones (1953) demonstrated the existence of non-trivial solutions to the homogeneous problem for the fluid in a channel (trapping modes, in Ursell's terminology). It should be emphasized that these results rely heavily on channel's boundedness in one of horizontal directions.

Although the last decades brought a considerable progress to the linear theory of time-harmonic water waves, we are still far from the final solution of the fundamental question of unique solvability. In the list of unsolved problems in this theory given by Ursell (1992), the problem of uniqueness of the velocity potential is placed first.

Apparently, the water-wave problem did not attract the attention of mathematicians for twenty years or so. However, the 1970's brought new important results concerning this problem. Vainberg & Maz'ya (1973b) investigated this problem for a layer of variable depth in the absence of immersed bodies. They found two conditions having simple geometrical interpretations, under either of which the problem is uniquely solvable. The first condition means that the intersection of W with any horizontal plane $\{y = y_0\}$, where $y_0 < 0$, is starlike relative to the point $(0, 0, y_0)$. It imposes no restrictions on ν. The second condition is that W is starlike itself with respect to some point at a depth $h \geq 0$, but it ensures the unique solvability only if $0 \leq \nu h \leq 1$.

The uniqueness of the finite energy solution under the first condition was proved with the help of the following identity which plays the role of John's trick:

$$\operatorname{Re}\left[(r\overline{u}_r + \overline{u})\nabla^2 u\right] = \operatorname{Re} \nabla \cdot \left[(r\overline{u}_r + \overline{u})\nabla u\right] - |\nabla_x u|^2 - \frac{1}{2}\nabla_x \cdot \left(x|\nabla u|^2\right),$$

where $r = |x|$. In fact, if u is a finite energy solution, then by integrating this identity over W one obtains

$$2\int_W |\nabla_x u|^2 \, dx dz = \int_B r\mathbf{x} \cdot \mathbf{n} |\nabla u|^2 \, ds \leq 0, \quad \mathbf{x} = (x/r, 0),$$

where the last inequality follows from the geometrical assumption. This implies that $u \equiv 0$. Similar arguments were used to prove the uniqueness of the finite energy solution under the second condition. The solvability of the water-wave problem and the existence of Green's function (with the source placed in the water domain W, on the free surface F, or on the bottom B) are proved by reducing the problem to an operator equation for which the Fredholm alternative holds. The investigation of the latter equation requires rather sophisticated techniques.

In the same paper, Vainberg & Maz'ya showed that their first condition ensures that the solution of the problem in the channel is unique. This geometrical condition is necessary in some sense. In fact, it does not allow the bottom to rise above its level at infinity but allows it to deepen, whereas Jones (1953) has shown that a bottom protrusion generates non-trivial solutions of the homogeneous problem in a channel. At the time being, it is still unknown whether any geometrical condition is necessary for the unique solvability of (1)–(6) in a layer of variable depth infinite in both horizontal directions.

Kuznetsov & Maz'ya (1974) and Kuznetsov (1989) extended the above mentioned result of John (1950) on solvability. In particular, they admitted non-orthogonal intersections of the body with the free surface along the water-line.

Now, we turn to the papers Maz'ya (1977, 1978). Only the second brief note had been translated into English, and for a few years a great part of researchers was unaware of the method developed by Maz'ya (1977) for proving the uniqueness theorem, which was a significant breakthrough in the theory. Luckily, in Hulme (1984) one can find a treatment which is rather close to the original presentation of Maz'ya. Also, a number of examples illustrating the uniqueness theorem are given in this paper.

It should be mentioned that in the 1970's less was known about the water-wave problem for a totally submerged body than for a surface-piercing body. In the latter case the results obtained by John as early as in 1950 were widely known. However, his method requires that there is no water strictly above immersed bodies, and hence, leaves the question of uniqueness for submerged bodies open. John had mentioned that he

> 'has been unable to prove uniqueness' despite 'there is no physical reason why in those cases [...] the motion of the obstacle should not determine the motion of the liquid uniquely'

(see p. 49 in John 1950). For totally submerged bodies, only the two-dimensional problem of a circular cylinder was investigated in detail by means of multipole expansions in Ursell (1950), who proved the uniqueness theorem as well.

Here we outline the scheme of Maz'ya's work. He begins the proof of uniqueness with the derivation of an auxiliary integral identity, which now is usually referred to as Maz'ya's identity. Let $\mathbf{V} = (V_1, V_2, V_3)$ be a vector field on $\overline{W}$ (V_3 is its projection on the y-axis), whose components are real and uniformly Lipschitz functions on $\overline{W}$. By H we denote a real function on $\overline{W}$ with uniformly Lipschitz first derivatives. The following equality

$$2\mathrm{Re}\left\{(\mathbf{V} \cdot \nabla u + Hu)\nabla^2\overline{u}\right\} = 2\mathrm{Re}\,\nabla \cdot \left\{(\mathbf{V} \cdot \nabla u + Hu)\nabla\overline{u}\right\}$$

$$(2.7) \qquad +(Q\nabla\overline{u}) \cdot \nabla u - \nabla \cdot [|\nabla u|^2\mathbf{V} + |u|^2\nabla H] + |u|^2\nabla^2 H$$

can be verified directly. Here the matrix Q has the elements

$$Q_{ij} = (\nabla \cdot \mathbf{V} - 2H)\delta_{ij} - \left(\frac{\partial V_i}{\partial x_j} + \frac{\partial V_j}{\partial x_i}\right), \quad i,j = 1, 2, 3 \text{ and } x_3 = y,$$

δ_{ij} denotes the Kronecker delta.

Assuming that u satisfies the water-wave problem, let us integrate (2.7) over $W_a = W \cap \{|x| < a\}$, where a is large enough (so that S is contained within the cylinder $\{|x| < a\}$). Then, after using the Laplace equation and the free surface boundary condition, one can integrate by parts over the free surface, which is supposed to be the whole plane $\{y = 0\}$. Under assumption that u and its gradient

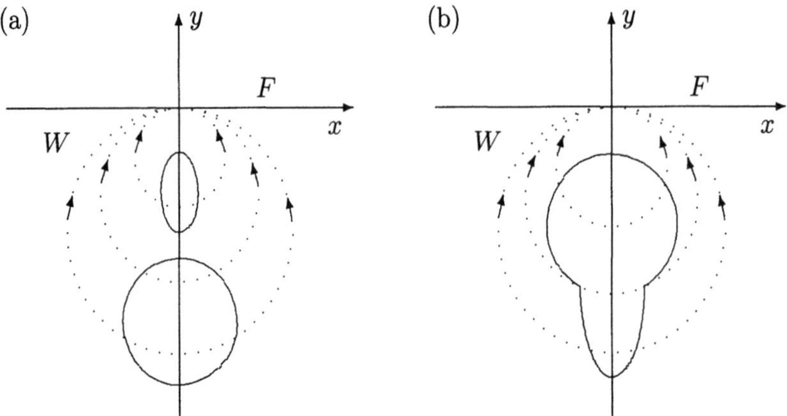

Figure 1. Curves satisfying Maz'ya's criterion are shown by solid lines; the integral curves (2.11) (dotted lines) have only points of entry into W.

are $O(|x|^{-3/2})$ as $|x| \to \infty$ (which is true for a solution to the homogeneous problem as is shown, for example, in Vainberg & Maz'ya 1973b), one can pass to the limit as $a \to \infty$. The result is *Maz'ya's identity*:

$$\int_W \left[(Q\nabla\bar{u}) \cdot \nabla u + |u|^2 \nabla^2 H \right] dx\, dy + \int_{S \cup B} \left[|\nabla u|^2 \mathbf{V} \cdot \mathbf{n} + |u|^2 \frac{\partial H}{\partial n} \right] dS$$

$$+ \int_F \left(\nu^2 V_3 + \nu[2H - \nabla_x \cdot \mathbf{V}] - H_y \right) |u|^2 \, dx - \int_F V_3 |\nabla_x u|^2 \, dx$$

$$(2.8) \quad = 2\mathrm{Re} \int_{S \cup B} [\mathbf{V} \cdot \nabla u + Hu] \, \partial\bar{u}/\partial n \, dS.$$

If W has infinite depth, one has to replace the integral over $S \cup B$ by the integral over S.

To obtain a criterion of uniqueness from (2.8), one has to choose such H and $\mathbf{V}$, that all terms in the left-hand side are non-negative, and at least one of them is strictly positive for non-trivial u. Maz'ya's choice was $H = -1$, and an axisymmetric field $\mathbf{V}$ with radial and vertical components

$$(2.9) \qquad\qquad V_r = r\frac{y^2 - r^2}{y^2 + r^2}, \quad V_y = \frac{-2yr^2}{y^2 + r^2}$$

respectively ($r = |x|$). Then, for u being a solution of the homogeneous water-wave problem, (2.8) takes the form

$$\int_W \left\{ \left| u_r \frac{2yr}{y^2 + r^2} + u_y \frac{y^2 - r^2}{y^2 + r^2} \right|^2 + |\nabla u|^2 \frac{y^2}{y^2 + r^2} \right\} dx\, dy$$

$$+ \frac{1}{2} \int_{S \cup B} |\nabla u|^2 \frac{r}{y^2 + r^2} \left\{ (y^2 - r^2)\frac{\partial r}{\partial n} - 2yr\frac{\partial y}{\partial n} \right\} dS = 0.$$

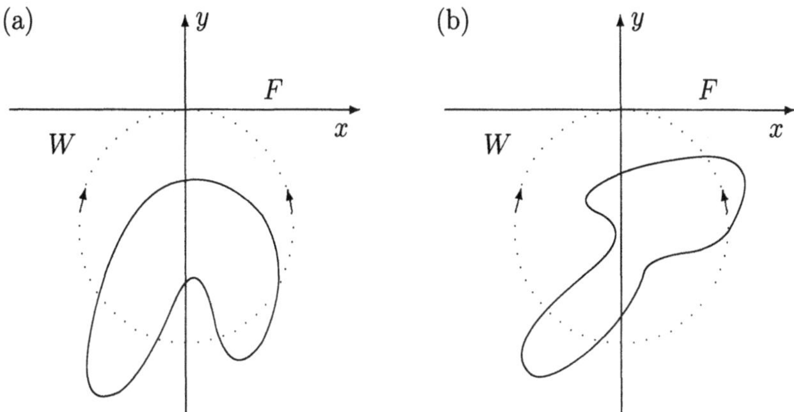

Figure 2. Curves violating Maz'ya's criterion are shown by solid lines; the integral curves (2.11) (dotted lines) have points of entry into W, as well as points of entry into the body.

This identity yields the following uniqueness theorem.

The homogeneous water-wave problem has only trivial solution when the inequality

(2.10) $$(y^2 - r^2)\frac{\partial r}{\partial n} - 2yr\frac{\partial y}{\partial n} \geq 0$$

holds on $S \cup B$.

Geometrically this condition means that the angle between $\mathbf{V}$ and $\mathbf{n}$ does not exceed $\pi/2$ on $S \cup B$. It is easy to see that integral curves of the field (2.9) are semicircles

(2.11) $$r^2 + (y + c)^2 = c^2, \quad \theta = \text{const}, \quad c > 0$$

beginning at $(0, -2c)$, and ending at the origin (θ is the polar angle in the plane $\{y = 0\}$). Hence, (2.10) is equivalent to the fact that *all transversal intersections of the curves* (2.11) *with* $S \cup B$ *are points of entry into* W.

It is easy to illustrate Maz'ya's criterion in the two-dimensional case, when r should be replaced by x in (2.10). Examples of cylinders for which this inequality holds are plotted in fig. 1, and fig. 2 demonstrates when the criterion fails. Sometimes, it is not clear whether (2.10) is satisfied or not. Even if S is an ellipse:

$$\frac{x^2}{(\lambda b)^2} + \frac{(y + h)^2}{b^2} = 1, \quad \lambda > 0, \quad h > b,$$

a calculation is needed to verify (2.10). It gives (see Hulme 1984) that Maz'ya's criterion holds when

(2.12) $$\frac{h}{b} \geq \max\{1, \ 2\lambda^2 - 1\},$$

which is obviously true when the major axis is vertical ($0 < \lambda \leq 1$). If the major axis is horizontal ($\lambda > 1$), then (2.12) holds when $h + b \geq 2b\lambda^2$. Both cases are shown in fig. 3.

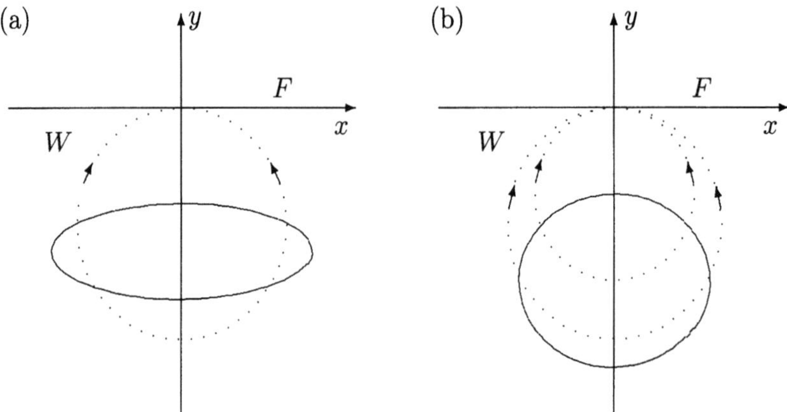

Figure 3. (a) A 'bad' ellipse with $\lambda = h/b = 3$. (b) A 'good' ellipse with $\lambda = 6/5$ and $h/b = 2$.

Note that two integral identities in Vainberg & Maz'ya (1973b) are particular cases of (2.8). They correspond to

$$(2.13) \qquad\qquad H = \text{const}, \qquad \mathbf{V} = (-x, -k(y + a)),$$

with special choices of non-negative parameters k, a. In the case of star-like W one has to put $a = h$, $k = 1$ and $H = 0$; the case of star-like horizontal cross-sections corresponds to $k = 0$ and $H = -1/2$. More examples of vector fields guaranteeing uniqueness under corresponding assumptions on S and B were proposed by Weck (1990) and Kuznetsov (1991). In the last paper, Maz'ya's identity is extended to geometries with surface-piercing bodies. Moreover, in Kuznetsov (1991) some parameter-dependent generalizations of (2.13) and (2.9) are found, which guarantee positivity of the left-hand side of (2.8). Conditions (2.13) are considered in detail, and the vector field (2.9) is generalized being a particular case of a one-parameter family of vector fields. This allows more geometries with the uniqueness property.

Other applications of Maz'ya's identity are possible. In Kuznetsov (1992, 1993b), it is shown that this identity is an appropriate tool for derivation of lower bounds to the first positive point eigenvalue in the sloshing problem, and in the problem of trapped modes over a bottom protrusion in a channel (see Evans & Kuznetsov (1997) for details of using Maz'ya's identity in the latter problem). Another use of this identity is given in Kuznetsov & Simon (1995), where it is applied to proving the uniqueness theorem in the case of two surface-piercing bodies.

The uniqueness theorem of Maz'ya has found application in the optimal design problem considered in the series of papers: Angell & Kleinman (1984, 1987, 1991) and Angell, Hsiao & Kleinman (1986). The importance of this theorem was amplified when McIver (1996) had discovered non-uniqueness examples in the water-wave problem.

3. The Neumann-Kelvin problem

There are two essential features which distinguish the Neumann-Kelvin problem from the water-wave problem. Since the Neumann-Kelvin problem describes a steady-state wave pattern due to *the uniform forward motion* of a body, there is one horizontal direction having special properties (apart from the vertical y-direction related to the gravity). We choose the x-axis directed along the path of body, and the z-axis orthogonal to the x- and y-axes in the three-dimensional case. In such a coordinate system attached to the body, that is, moving forward with it at the speed U, different conditions must be imposed on the velocity potential u as $x \to \pm\infty$ (they are different for two- and three-dimensional problems and will be stated shortly). Also, the free-surface boundary condition (2.2) must be replaced by the following one:

$$(3.1) \qquad u_{xx} + \nu u_y = 0 \quad \text{on} \quad F,$$

where $\nu = g/U^2$. The equation (2.1) and the Neumann condition (2.4) on the horizontal bottom remain valid in the Neumann-Kelvin problem as well as (2.3) on the wetted body's surface/contour, where the case when $f = U\mathbf{n} \cdot \mathbf{x}$ in (2.3) corresponds to S being rigid and impenetrable.

Thus in two dimensions, the solution of the Neumann-Kelvin problem must satisfy (2.1), (3.1), (2.3)–(2.5), and the following conditions at infinity:

$$(3.2) \qquad |\nabla u| \to 0 \text{ as } x \to +\infty, \quad \sup_W |\nabla u| < \infty.$$

Vainberg & Maz'ya (1973a) considered the two-dimensional Neumann-Kelvin problem for a smooth simply connected body totally immersed in a fluid having either infinite or finite depth d. First of all, we bring attention to the result which is given as a remark. It says that *in a fluid of infinite depth the Neumann-Kelvin problem is uniquely solvable for all values of U with the possible exception of a finite number of values.* It is a consequence of the solvability theorem proved by Kochin (1937) for sufficiently small and large values of ν combined with the well-known result on the invertibility of an operator-function, analytically depending on a parameter (see, for example, Gohberg & Krein 1969).

In the rest of their paper, Vainberg & Maz'ya (1973a) treat the Neumann-Kelvin problem under assumption that

$$(3.3) \qquad x\,\mathbf{n} \cdot \mathbf{x} \geq 0 \quad \text{on } S, \quad \mathbf{x} = (x/r, 0).$$

This assumption is opposite to the condition obtained in Vainberg & Maz'ya (1973b), which guarantees the uniqueness of the solution to the water-wave problem. Condition (3.3) means that the vertical line $y = 0$ divides the body into parts, each of which is convex in the x-direction. In particular, they demonstrate that the problem is uniquely solvable (up to a constant term, which is natural in the Neumann-Kelvin problem) when the flow of finite depth is supercritical, that is, when the forward velocity U exceeds $(gd)^{1/2}$. Ten years later, Lahalle (1984) proved this theorem without condition (3.3).

For a subcritical flow (this includes the case of infinite depth, when the flow is subcritical for any value of U), Vainberg & Maz'ya (1973a) demonstrated that the geometrical condition (3.3) guarantees the uniqueness of a finite energy solution. At that point, the main difference between the Neumann-Kelvin problem and the water-wave problem emerges. In fact, any solution to the homogeneous water-wave problem has finite energy. This follows from Green's formula and the radiation condition (2.6). For the subcritical Neumann-Kelvin problem the unconditional validity of this property is still an open question.

Vainberg & Maz'ya (1973a) found several equivalent forms of necessary and sufficient conditions on S, providing the unique solvability of the two-dimensional Neumann-Kelvin problem in the subcritical case. One form of this condition is expressed in terms of a special solution to an auxiliary scattering problem. This problem differs from the Neumann-Kelvin problem only by the condition at infinity, which is the radiation condition instead of (3.2). We describe Vainberg & Maz'ya's criterion for the flow of infinite depth. First, they prove that the solution has the following asymptotic behaviour at infinity:

$$u \sim D_\pm \, e^{\nu(y+i|x|)} \text{ as } x \to \pm\infty,$$

when

$$(3.4) \qquad\qquad \int_S f \, ds = 0.$$

Then they prove that the Neumann-Kelvin problem is uniquely solvable if and only if $|D_-| < 1$ for the solution of the scattering problem with

$$(3.5) \qquad\qquad f = \frac{\partial}{\partial n} e^{\nu(y+ix)}.$$

Note, that (3.5) satisfies (3.4), and that $|D_-| \leq 1$ for any contour S, when f is given by (3.5). Hence, the unique solvability fails only when a contour gives the maximum value (which is equal to one) to the functional $|D_-|$. Up to the present, it is not known whether such an extremal contour exists.

In the same paper, Vainberg & Maz'ya showed that the homogeneous Neumann-Kelvin problem has at most one non-trivial solution (up to multiplicative and additive constants) for any fixed ν. Livshits & Maz'ya (1997) proved that there is no such non-trivial solution when S is a circle in deep water. This is the only result, establishing the unique solvability of the Neumann-Kelvin problem for all values of the forward velocity U.

The three-dimensional Neumann-Kelvin problem is essentially more difficult in comparison with the two-dimensional one. First, the asymptotic behaviour of solutions at infinity is more complicated in this case. Secondly, the boundary condition (3.1) on the free surface is not elliptic which creates obstacles in the study of the problem for a surface-piercing body. In the discussed above paper, which is mostly devoted to the two-dimensional problem, Vainberg & Maz'ya (1973a) proved the uniqueness of a finite energy solution for the three-dimensional problem of a totally submerged body satisfying (3.3). Unfortunately, this is not sufficient for uniqueness of an arbitrary solution of the problem.

We mentioned above a remark in Vainberg & Maz'ya (1973a), that Kochin's results on the solvability of the two-dimensional Neumann-Kelvin problem for small and large ν imply the solvability of the problem for all ν, except possibly for a finite number of values. In his paper, Kochin (1937) proved the solvability of the three-dimensional Neumann-Kelvin problem in deep water only for sufficiently large values of ν. The question, whether the three-dimensional problem is solvable for small ν, remained open until 1993, when Maz'ya & Vainberg solved it positively. From their paper, it becomes clear why Kochin did not consider the three-dimensional problem for small values of ν. This case is really more difficult and requires rather sophisticated asymptotic evaluation of the corresponding Green's function. This result, as in two dimensions, implies that the unique solvability of the three-dimensional Neumann-Kelvin problem might fail for at most a finite number of values of the forward velocity U.

In the same paper, Maz'ya & Vainberg (1993) conclude a long series of studies of Green's function for the three-dimensional Neumann-Kelvin problem by Kelvin, Hogner, Havelock, Peters, Ursell, Euvrard, Newman. This paper deals with an asymptotic representation of Green's function at infinity. The form of this asymptotics was known when the field point goes to infinity along any fixed ray. The new feature of Maz'ya & Vainberg's work is the uniformity of their asymptotic representation with respect to the position of the field point and source's depth. In particular, it was known that Green's function is infinite, when the source is on the free surface and the field point is located on the track of the source, and that Green's function vanishes, when the field point is strictly below of the track of the source. Maz'ya & Vainberg (1993) singled out a wave which describes this singularity of Green's function. The remainder term in their expansion decays uniformly with respect to all of variables.

Kuznetsov & Maz'ya (1989) considered the Neumann-Kelvin problem of a surface-piercing two-dimensional body. The difficulty arising from the fact, that there are points of intersection of the body and the free surface, manifests itself in the existence of a family of solutions depending on two real parameters. Ursell (1981) demonstrated this by considering a semicircle by means of multipole expansions. Earlier, this fact was discovered numerically (see Suzuki (1982) and references cited therein). In particular, as early as in 1963, Bessho & Mizuno solved numerically the problem for a semi-submerged circular cylinder and found that there are infinitely many solutions and, hence, the calculated wave resistance can have any value. This fact gave rise to a discussion usually referred to as 'the problem of line integral' in hydrodynamic literature (there is a water-line in the three-dimensional case instead of two points of intersection with the free surface). One can find more about this discussion in Kuznetsov & Motygin (1997).

Thus, in order to define the solution uniquely, the Neumann-Kelvin problem must be augmented by two supplementary conditions. In Kuznetsov & Maz'ya (1989), two types of such conditions have been suggested. In one of the formulations, it is necessary to prescribe the values $u_x(a_\pm, 0)$, where $a_+ > a_-$ are the x-coordinates of points where the body intersects the free surface. The second formulation includes the prescription of the differences $u_x(a_+, 0) - u_x(a_-, 0)$ and $u(a_+, 0) - u(a_-, 0)$.

The quantities listed can be interpreted physically in the following manner. The derivative $u_x(a_\pm, 0)$ is proportional to the elevation of the free surface at the point $(a_\pm, 0)$, the difference $u_x(a_+, 0) - u_x(a_-, 0)$ is proportional to the rate of flow at infinity behind the cylinder, and, finally, the difference $u(a_+, 0) - u(a_-, 0)$ is proportional to the lifting force, acting on a unit length of the cylinder. In 1992, Kuznetsov & Maz'ya demonstrated that the above two types of conditions at the points $(a_\pm, 0)$ are particular cases of one general well-posed formulation, for which results similar to those in Vainberg & Maz'ya (1973a) hold. During the decade after publishing Kuznetsov & Maz'ya (1989), a number of other well-posed statements of the Neumann-Kelvin problem for a surface-piercing body appeared (see, for example, Kuznetsov & Motygin 1997, 1999).

4. Asymptotic expansions for transient water waves due to brief and high-frequency disturbances

A series of papers by Kuznetsov & Maz'ya (1985, 1986, 1987, 1997) is concerned with problems describing waves developing from the rest. The time-dependent velocity potential $u(x, y, z, t)$ satisfies relations (2.1), (2.3), (2.4) and the free surface boundary conditions containing the first order derivatives with respect to t:

$$(4.1) \qquad -u_t + \eta_t = 0, \quad u_t + g\eta = -p/\rho \quad \text{on } F \text{ for } t \geq 0.$$

Here $\eta(x, z, t)$ denotes the vertical elevation of the free surface which is another unknown function, $p(x, z, t)$ is the pressure prescribed on the free surface and ρ is fluid's density. Excluding η, one arrives at

$$u_{tt} + gu_y = -p_t/\rho \quad \text{on } F \text{ for } t \geq 0.$$

Also, for motions starting from the rest, the zero initial data for u and u_t should be given on the free surface.

In their paper written first, Kuznetsov & Maz'ya (1986) investigated problems describing waves due to brief disturbances. The motivation was to justify rigorously the widely-known heuristic consideration deriving the initial condition for the velocity potential from the linearized Bernoulli's equation. A typical pattern of the hydrodynamic approach to this condition is given by Stoker (1957), pp. 149–150 (see also Lamb 1932):

'In water wave problems it is of particular interest to consider cases in which the motion of the water is generated by applying an impulsive pressure to the surface when the water is initially at rest. To obtain the condition appropriate for an initial impulse we start from' the second equation (4.1) *'and integrate it over the small interval $0 \leq t \leq \epsilon$. The result is*

$$(4.2) \qquad \int_0^\epsilon p \, dt = -\rho u(x,0,z,\epsilon) - \rho g \int_0^\epsilon \eta \, dt,$$

since $u(x,y,z,0)$ can be assumed to vanish. One now imagines that $\epsilon \to +0$ while $p \to \infty$ in such a way that the integral on the left tends to a finite value – the impulse I per unit area. Since it is natural to assume that η is finite it follows that the integral on the right vanishes as $\epsilon \to +0$, and we have the formula

$$(4.3) \qquad I = -\rho u(x,0,z,+0)$$

for the initial impulse per unit area at the free surface in terms of the value of u there. If I is prescribed on the free surface (together with appropriate conditions at other boundaries), it follows that $u(x,y,z,+0)$ can be determined, or, in other words, the initial velocity of particles is known.'

Despite this argument looks very convincing one can pose some questions. Here is a couple of them.

- How should the passage to the limit be understood in (4.2) as $\epsilon \to 0$?
- Why the force

$$(4.4) \qquad \mathbf{F}(t) = \rho \int_S u_t \mathbf{n} \, dS,$$

acting on a submerged body bounded by a surface S, is finite when calculated by virtue of u determined with the help of (4.3), in spite of the fact that the pressure p is infinite at the initial moment?

These and other questions were answered by Kuznetsov & Maz'ya (1986), who used two-scale asymptotic expansions for this purpose. To give an idea of their results, we present a simple consequence concerning the problem (2.1), (2.3) with $f = 0$, (2.4) and (4.1) with zero initial data for η and η_t. The pressure p in the second condition (4.1) is prescribed as follows:

$$(4.5) \qquad p(x,z,t,\epsilon) = \epsilon^{-1} Q(t/\epsilon) \, I(x,z),$$

where $I(x,z)$ is a smooth function, and $Q(\tau)$ is a continuous function of $\tau \geq 0$, vanishing for $\tau = 0$ and $\tau \geq 1$, and such that

$$\int_0^1 Q(\tau) \, d\tau = 1.$$

Under these conditions, the infinite two-scale asymptotic expansion is derived for the velocity potential $u^{(\epsilon)}(x,y,z,t)$. The form of this expansion is similar to (4.6) below, but we restrict ourselves with the following initial terms

$$u^{(\epsilon)}(x,y,z,t) = u(x,y,z,t) + \rho^{-1}v(x,y,z)\int_{t/\epsilon}^{\infty} Q(\tau)\,d\tau + O(\epsilon),$$

which are of importance for the hydrodynamic corollaries. Here the velocity potential u satisfies (4.3) and its t-derivative is zero at the initial moment, and v is the unique solution to the following time-independent problem:

$$\nabla^2 v = 0 \text{ in } W, \quad v = I \text{ on } F, \quad \partial v/\partial n = 0 \text{ on } S \cup B.$$

Now, we get the asymptotic formula for the force in the form:

$$\mathbf{F}^{(\epsilon)}(t) = \mathbf{F}(t) - \frac{Q(t/\epsilon)}{\epsilon}\int_S v\mathbf{n}\,dS + O(\epsilon),$$

where $\mathbf{F}(t)$ must be determined from (4.4). Thus, for a small initial interval, (4.4) gives only a negligible part of the force, and the principal term in the force asymptotics, tending to infinity as $\epsilon \to 0$, requires the knowledge of v. However, the second term vanishes for $t \geq \epsilon$.

Along with an impulse of surface pressure, some other types of brief disturbances, such as a bottom shift and a submerged source, are treated. For simple geometries, explicit formulae are given for terms in asymptotic expansions.

Even before the publication of their first paper in this series, Kuznetsov & Maz'ya had recognized, that the same method works for high-frequency disturbances and rapidly accelerating perturbations. Disregarding the last type of perturbations considered in the paper published in 1987, we turn to the effect of high-frequency oscillations of surface and of submerged disturbances on hydrodynamic characteristics of the wave motion. Three papers published in 1985, 1987 and 1997 are devoted to this effect.

To show what kind of results is typical in this field, let us consider the pressure p in (4.5), given by $\kappa(t/\epsilon)\mathcal{P}(x,z)$, where κ is a one-periodic function. The frequency ϵ^{-1} is assumed to be high in comparison with the reciprocal of the characteristic time $(d/g)^{1/2}$. It is demonstrated that the potential has the asymptotic expansion

$$(4.6) \quad u^{(\epsilon)}(x,y,z,t) \sim \sum_{m=0}^{\infty} \epsilon^m \alpha_m w_m(x,y,z,t) + \sum_{m=1}^{\infty} \epsilon^{2m-1}\beta_m(t/\epsilon)v_m(x,y,z).$$

Here $\{\beta_m\}$ are certain one-periodic functions, $\{\alpha_m\}$ are constants expressed in terms of $\{\beta_m\}$, and $\{v_m, w_m\}$ are harmonic functions, which do not depend on ϵ. The functions $\{v_m\}$ are subject to the Dirichlet condition on the free surface, and the Neumann condition on the rigid surfaces, whereas $\{w_m\}$ are solutions of the Cauchy-Poisson problem. The sequences $\{v_m\}$ and $\{w_m\}$ can be found recurrently.

Analysis of the principal term in (4.6) demonstrates that up to $O(\epsilon)$ the waves from the high-frequency surface pressure are the same as those, resulting from the initial elevation of the free surface of the form

$$[\langle \kappa \rangle - \kappa(0)]\mathcal{P}(x, z),$$

where $\langle \kappa \rangle$ is the mean value of the periodic function κ. This term describes a slow wave motion. Additionally, there is a high-frequency motion of amplitude $O(\epsilon)$ with zero mean value. The latter gives a finite contribution to the force acting on a submerged body. This fact is of a substantial importance and may be used, for example, for reducing wave-making resistance as is shown by Kuznetsov (1993) later on.

References

PUBLICATIONS OF VLADIMIR MAZ'YA

VAINBERG, B.R. & MAZ'JA, V.G. 1972 On some stationary problems in the linear theory of surface waves. *Soviet Physics Dokl.* **17**, 640–643.

VAINBERG, B.R. & MAZ'JA, V.G. 1973*a* On the plane problem of the motion of a body immersed in a fluid. *Trans. Moscow Math. Soc.* **28**, 33–55.

VAINBERG, B.R. & MAZ'JA, V.G. 1973*b* On the problem of the steady state oscillations of a fluid layer of variable depth. *Trans. Moscow Math. Soc.* **28**, 56–73.

KUZNETSOV, N.G. & MAZ'YA, V.G. 1974 Problem concerning steady-state oscillations of a layer of fluid in the presence of an obstacle. *Soviet Physics Dokl.* **19**, 341–343.

MAZ'YA, V.G. 1977 On the steady problem of small oscillations of a fluid in the presence of a submerged body. *Proc. Sobolev's Semin.* No. 2, 57–79. Novosibirsk: Inst. of Maths, Siberian Branch, Acad. Sci. USSR (in Russian).

MAZ'JA, V.G. 1978 Solvability of the problem on the oscillations of a fluid containing a submerged body. *J. Soviet Math.* **10**, 86–89.

KUZNETSOV, N.G. & MAZ'YA, V.G. 1985 Asymptotic expansions for transient surface waves due to short-period oscillating disturbances. *Proc. Leningrad Shipbuild. Inst. / Math. Modelling and Automated Design in Shipbuilding*, 57–64 (in Russian).

KUZNETSOV, N.G. & MAZ'YA, V.G. 1986 Asymptotic expansions for surface waves caused by brief disturbances. *Asymptotic methods / Problems in Mechanics.* Novosibirsk: 'Nauka', pp. 103–138 (in Russian).

Where possible English translations are given instead of Russian originals.

KUZNETSOV, N.G. & MAZ'YA, V.G. 1987 Asymptotic expansions for surface waves caused by rapidly oscillating or accelerating disturbances. *Asymptotic methods / Problems and Models in Mechanics.* Novosibirsk: 'Nauka', pp. 136–175 (in Russian).

KUZNETSOV, N.G. & MAZ'YA, V.G. 1988 Unique solvability of a plane stationary problem related to the motion of a solid body submerged in a liquid. *Diff. Equat.* **24**, 1291–1301.

KUZNETSOV, N.G. & MAZ'YA, V.G. 1989 On unique solvability of the plane Neumann-Kelvin problem. *Math. USSR Sborn.* **63**, 425–446.

KUZNETSOV, N.G. & MAZ'YA, V.G. 1992 On a well-posed formulation of the two-dimensional Neumann-Kelvin problem for a surface-piercing body. *Preprint LiTH-MAT-R-92-42*, Dept. of Maths, University of Linköping, 34 p.

MAZ'YA, V. & VAINBERG, B. 1992 On uniqueness and asymptotic behavior of solutions of the Neumann-Kelvin problem. *Proc. of the 7th Int. Workshop on Water Waves & Floating Bodies*, France. Ed. R. Cointe, pp. 177–181.

MAZ'YA, V.G. & VAINBERG, B.R. 1993 On ship waves. *Wave Motion* **18**, 31–50.

LIVSHITS, M. & MAZ'YA, V. 1997 Solvability of the two-dimensional Kelvin-Neumann problem for a submerged circular cylinder. *Applicable Analysis* **64**, 1–5.

KUZNETSOV, N.G. & MAZ'YA, V.G. 1997 Asymptotic analysis of surface waves due to high-frequency disturbances. *Rend. Mat. Acc. Lincei, Ser. 9,* **8**, 5–29.

OTHER WORKS

LAMB, H. 1932 *Hydrodynamics.* Cambridge: Camb. Univ. Press.

KOCHIN, N.E. 1937 On the wave resistance and lift of bodies submerged in a fluid. *Proc. Conf. on the Wave Resistance Theory.* Moscow: TsAGI, pp. 65–134. (In Russian; English transl. in *SNAME Tech. & Res. Bull.* **1–8** (1951)).

KOCHIN, N.E. 1939 The two-dimensional problem of steady oscillations of bodies under the free surface of a heavy incompressible fluid. *Acad. Sci. USSR, Izvestia OTN*, No. 4, 37–62. (In Russian; English transl. in *SNAME Tech. & Res. Bull.* **1–10** (1952)).

KOCHIN, N.E. 1940 The theory of waves generated by oscillations of a body under the free surface of a heavy incompressible fluid. *Trans. Moscow Univ.* **46**, 85–106. (In Russian; English transl. in *SNAME Tech. & Res. Bull.* **1–10** (1952)).

JOHN, F. 1950 On the motion of floating bodies. II. *Comm. Pure Appl. Math.* **3**, 45–101.

URSELL, F. 1950 Surface waves on deep water in the presence of a submerged circular cylinder. I, II. *Proc. Camb. Phil. Soc.* **46**, 141–152, 153–158.

URSELL, F. 1951 Trapping modes in the theory of surface waves. *Proc. Camb. Phil. Soc.* **47**, 347–358.

JONES, D.S. 1953 The eigenvalus of $\nabla^2 u + \lambda u = 0$ when the boundary conditions are given on semiinfinite domains, *Proc. Camb. Phil. Soc.,* **49**, 668–684.

STOKER, J.J. 1957 *Water Waves. The Mathematical Theory with Applications.* New York: Intersci. Publ.

GOHBERG, I. & KREIN, M.G. 1969 *Introduction to the Theory of Linear Non-self-adjoint Operators in Hilbert Space.* Transl. Math. Mon. **18**. Providence, RI: Amer. Math. Soc.

URSELL, F. 1981 Mathematical notes on the two-dimensional Kelvin-Neumann problem. *Proceedings of the 13th Symposium on Naval Hydrodynamics.* Tokyo: Shipbuilding Research Association of Japan, pp. 245–251.

SUZUKI, K. 1982 Numerical studies of the Neumann-Kelvin problem for a two-dimensional semisubmerged body. *Proceedings of the 3d International Conference on Numerical Ship Hydrodynamics.* Paris: Bassin d'Essais des Carènes, pp. 83–95.

ANGELL, T.S. & KLEINMAN, R.E. 1984 A Galerkin procedure for optimization in radiation problems. *SIAM J. Appl. Math.* **44**, 1246–1257.

HULME, A. 1984 Some applications of Maz'ja's uniqueness theorem to a class of linear water wave problems. *Math. Proc. Camb. Phil. Soc.* **95**, 511–519.

LAHALLE, D. 1984 Calcul des efforts sur un profil portant d'hydroptere par couplage éléments finis – représentation intégrale. *ENSTA Rapport de Recherche* **187**.

ANGELL, T.S., HSIAO, G.C. & KLEINMAN, R.E. 1986 An optimal design problem for submerged bodies. *Math. Meth. Appl. Sci.* **8**, 50–76.

ANGELL, T.S. & KLEINMAN, R.E. 1987 On a domain optimization problem in hydrodynamics. *Optimal Control of Partial Differential Equations. II.* Basel *et al.*: Birkhäuser, pp. 9–27.

KUZNETSOV, N.G. 1989 Steady waves on the surface of fluid having variable depth and containing floating bodies. Part 4 in: N.G. Kuznetsov, Yu.F. Orlov, V.B. Cherepennikov, R.Yu. Shlaustas, *Regular Asymptotic Algorithms in Mechanics.* Novosibirsk: 'Nauka', pp. 200–270 (in Russian).

WECK, N. 1990 On a boundary value problem in the theory of linear water-waves. *Math. Meth. Appl. Sci.* **12**, 393–404.

ANGELL, T.S., KLEINMAN, R.E., 1991 A constructive method for shape optimization: a problem in hydrodynamics. *IMA J. Appl. Math.* **47**, 265–281.

KUZNETSOV, N.G. 1991 Uniqueness of a solution of a linear problem for stationary oscillations of a liquid. *Diff. Equat.* **27**, 187–194.

KUZNETSOV, N.G. 1992 The lower bound of the eigenfrequencies of plane oscillations of a fluid in a channel. *J. Appl. Math. Mech.* **56**, 293–297.

URSELL, F. 1992 Some unsolved and unfinished problems in the theory of waves. *Wave Asymptotics.* Cambridge: Camb. Univ. Press.

KUZNETSOV, N.G. 1993*a* Asymptotic analysis of wave resistance of a submerged body moving with an oscillating velocity. *J. Ship Res.* **37**, 119–125.

KUZNETSOV, N.G. 1993*b* The Maz'ya identity and lower estimates of eigenfrequencies of steady-state oscillations of a liquid in a channel. *Russian Math. Surveys* **48**(4), 222.

KUZNETSOV, N.G. & SIMON, M.J. 1995 On uniqueness in the two-dimensional water-wave problem for surface-piercing bodies in fluid of finite depth. *Appl. Math. Rep.* **95/4**. University of Manchester.

MCIVER, M. 1996 An example of non-uniqueness in the two-dimensional linear water wave problem. *J. Fluid Mech.* **315**, 257–266.

EVANS, D.V. & KUZNETSOV, N.G. 1997 Trapped modes. In: *Gravity Waves in Water of Finite Depth* (ed. J.N. Hant), pp. 127-168, Comp. Mech. Int., Southampton.

KUZNETSOV, N. & MOTYGIN, O. 1997 On waveless statement of the two-dimensional Neumann-Kelvin problem for a surface-piercing body. *IMA J. Appl. Math.* **59**, 25–42.

KUZNETSOV, N. & MOTYGIN, O. 1999 On the resistanceless statement of the two-dimensional Neumann-Kelvin problem for a surface-piercing tandem. *IMA J. Appl. Math.* **62**, 1–18.

Laboratory for Mathematical Modelling of Wave Phenomena,
Institute of Problems in Mechanical Engineering, Russian Academy of Sciences,
V.O., Bol'shoy pr. 61,
St. Petersburg, 199178, Russian Federation,
nikuz@snark.ipme.ru

University of North Carolina at Charlotte,
Charlotte, NC 28223, USA,
brvainbe@uncc.edu

1991 Mathematics Subject Classification: Primary 76B15; Secondary 35Q35

Submitted: 17.07.1998

Operator Theory:
Advances and Applications, Vol. 109
© 1999 Birkhäuser Verlag Basel/Switzerland

The work of Vladimir Maz'ya on integral and pseudodifferential operators

JOHANNES ELSCHNER

Since integral and pseudodifferential operators are one of the main themes of Maz'ya's vast mathematical work, it is a difficult task to describe his diverse results in this field in a short survey. In fact, this article was to have been written by Maz'ya's close friend Siegfried Prössdorf, who was my teacher at the Technical University of Chemnitz and my colleague at the Weierstrass Institute in Berlin. Siegfried's unexpected and untimely death was a tragic loss for everybody who knew him. Siegfried had followed Maz'ya's work for over thirty years. In this respect I would like to draw attention to their comprehensive joint monograph published as Volume 27 of the Encyclopaedia of Mathematical Sciences.

1. Non-elliptic operators

I learned about Vladimir's work for the first time in 1972 when two articles by Maz'ya and Plamenevskiĭ on multidimensional singular integral operators with degenerate symbol were reported in Prössdorf's seminar at the Technical University of Chemnitz. The ideas from these papers influenced the research done in Chemnitz in the same area, including some of my own work. This is why I have chosen to discuss Maz'ya's contributions to non-elliptic singular integral and pseudodifferential operators first.

I start with some definitions. An operator of the form

$$(1) \qquad Au(x) := a(x)u(x) + \int_{\mathbb{R}^n} \frac{f(x,\theta)}{|x-y|^n} u(y)dy\,,$$

where $x \in \mathbb{R}^n, \theta = (x-y)|x-y|^{-1} \in S^{n-1}$, is called *singular integral operator* in $\mathbb{R}^n$. The *symbol* of A, which was first introduced by Mikhlin for $n = 2$ and somewhat later by Giraud for $n > 2$ as a series in spherical harmonics, can equivalently be defined by

$$\sigma(x,\xi) := a(x) + F_{y\to\xi}\{|y|^{-n}f(x,y/|y|)\}\,, \quad x\,,\xi \in \mathbb{R}^n\,, \quad \xi \neq 0\,,$$

where F refers to the Fourier transform. Note that σ is positively homogeneous of degree 0 in ξ. It was proved by Calderón and Zygmund that (1) can be written in the form

$$(2) \qquad Au(x) = F^{-1}_{\xi\to x}\{\sigma(x,\xi)(Fu)(\xi)\}\,.$$

The operator (1) is called *elliptic* if $\sigma(x,\xi) \neq 0$ for all $x \in \mathbb{R}^n$ and $\xi \in S^{n-1}$, otherwise it is called non-elliptic or degenerate.

At the beginning of the sixties the solvability properties of elliptic multidimensional singular integral operators were well understood, due to the fundamental contributions by Tricomi, Mikhlin, Giraud, Calderón and Zygmund, and Gohberg, whereas nothing was known in the non-elliptic case. Influenced by S. Mikhlin, V. Maz'ya had already started working in this field in 1964. The short but illuminating paper [40] by Maz'ya and Plamenevskiĭ was the first dealing with non-elliptic pseudodifferential operators in higher dimensions. It was followed by another short note [41] and the longer paper [42]. Among other things, it was proved that the equation

$$Au = g\,, \quad g \in L_2(\mathbb{R}^n)\,,$$

is always solvable in an appropriate anisotropic Sobolev space provided the symbol of A does not depend on x and has zeroes of constant (finite) multiplicities on disjoint smooth submanifolds of S^{n-1}. Furthermore, a complete description of the infinite dimensional kernel (null space) of A and formulations of well-posed problems for the inhomogeneous equation were given. Maz'ya and Plamenevskiĭ were also able to treat some cases of symbols depending additionally on x.

Apparently, these pioneering works on non-elliptic operators remained completely unknown outside the Iron Curtain. However, the case of degenerating symbol became rather fashionable after the theory of pseudodifferential operators had emerged in the works by Eskin and Vishik, Kohn and Nirenberg, Bokobza and Unterberger, and Hörmander. Recall that a (classical) *pseudodifferential operator* in $\mathbb{R}^n$ is defined by relation (2), where the symbol σ admits an asymptotic expansion into positively homogeneous terms in ξ,

$$\sigma(x,\xi) \sim \sum_{k=0}^{\infty} \sigma_k(x,\xi)\,, \quad \sigma_k(x,t\xi) = t^{l-k}\sigma_k(x,\xi) \quad \forall\, t > 0\,.$$

Here σ_0 is called the principal symbol, l is the order of A, and A is said to be elliptic if σ_0 is nowhere vanishing.

After 1965 solvability and regularity theory for pseudodifferential equations with various types of degeneration became a vast area of study. This theory was also applied to non-elliptic boundary value problems. In the late sixties, Maz'ya and Paneyah made an important contribution to this field. In their papers [37], [38], [39] they studied a rather general class of pseudodifferential operators on a smooth manifold Γ without boundary, with symbol vanishing on a submanifold of codimension one. Assuming that the principal symbol σ_0 satisfies the condition

$$\mathrm{Im}\ \sigma_0(x,\xi) = 0 \quad \Longleftrightarrow \quad x \in \Gamma_0\,,$$

they introduced a classification of the types of degeneration (depending on the sign of $\mathrm{Im}\ \sigma_0$ near Γ_0) and developed a complete solvability theory for each of them. Moreover, sharp a priori estimates leading to precise regularity results for weak solutions were proved. These results have direct applications to the degenerate oblique derivative problem, which will be discussed now.

For simplicity we restrict ourselves to the formulation for the Laplace operator; all results hold of course for general elliptic operators of second order. Let $\Omega \subset \mathbb{R}^n, n \geq 3$, be a bounded domain with smooth boundary Γ, and denote the exterior unit normal to Γ by ν. The oblique derivative or Poincaré problem consists in determining a function u satisfying

$$(3) \qquad\qquad \Delta u = 0 \quad \text{in } \Omega, \quad \partial u/\partial \ell = f \quad \text{on } \Gamma,$$

where ℓ denotes a field of unit vectors on Γ. The problem (3) can be converted into a pseudodifferential equation of first order on Γ with the principal symbol

$$\sigma_0(x,\xi) = -\cos(\nu,\ell)|\xi| + i\cos(\xi,\ell)|\xi|, \ x \in \Gamma, \ \xi \in T_x\Gamma,$$

where T_x stands for the tangent space at the point x. Observe that this equation is elliptic if and only if the vector field ℓ is nowhere tangent to Γ.

In the elliptic case, the Fredholm property and regularity of problem (3) follow from standard elliptic theory of pseudodifferential operators, while its unique solvability is a consequence of Giraud's theorem on the sign of the oblique derivative at the extremum point.

Until the mid-sixties almost nothing was known about the degenerate problem (3). For *transversal degeneration* where the field ℓ is tangent to Γ on some $(n-2)$-dimensional submanifold Γ_0, but is not tangential to Γ_0, this situation changed when the first results on non-elliptic pseudodifferential operators became available.

As a by-product of his subelliptic estimates for pseudodifferential equations, Hörmander [18] proved that the dimension of the kernel of this problem may be infinite or the regularity of solutions may fail. Correct formulations leading to Fredholm operators were first studied by Malyutov [29] and by Egorov, Kontrat'ev [9], using entirely different methods from the theory of elliptic second order differential operators. In the above mentioned papers, Maz'ya and Paneyah [37]–[39] presented a unified pseudodifferential approach to all cases of transversal degeneration, proving complete unique solvability results and studying regularity properties of solutions.

2. Oblique derivative problem: breakthrough in the generic case of degeneration

Geometrically, the transversal degeneration leads to the following three types of components of the set Γ_0 (where the vector field ℓ is tangent to Γ): those consisting of the so-called "entrance" points (of ℓ into Ω), "exit" points, and "status quo" points where ℓ remains on the same side of Γ; see Figs. 1–3. In 1969, after the Malyutov, Egorov & Kondrat'ev and Maz'ya & Paneyah studies, the following properties of transversal degeneration became clear. The status quo components do not affect the unique solvability of the problem; they only generate some loss of regularity of solutions. In order to preserve unique solvability, one should allow discontinuities of solutions on the entrance components and prescribe additional boundary conditions on the exit components.

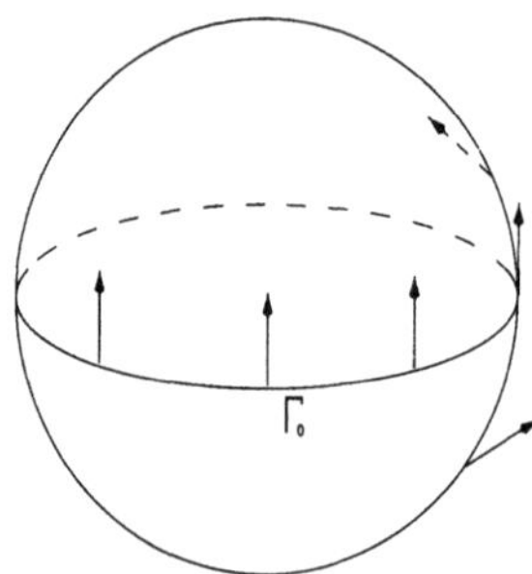

Fig. 1   Entrance points – under-
determined problem

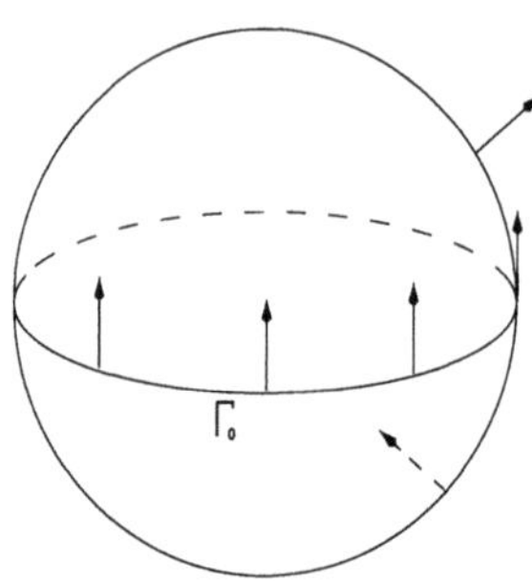

Fig. 2   Exit points – over-
determined problem

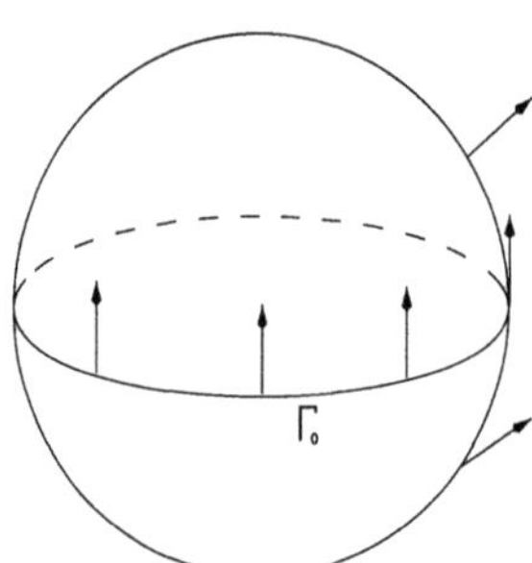

Fig. 3   Status quo –
well-posed problem

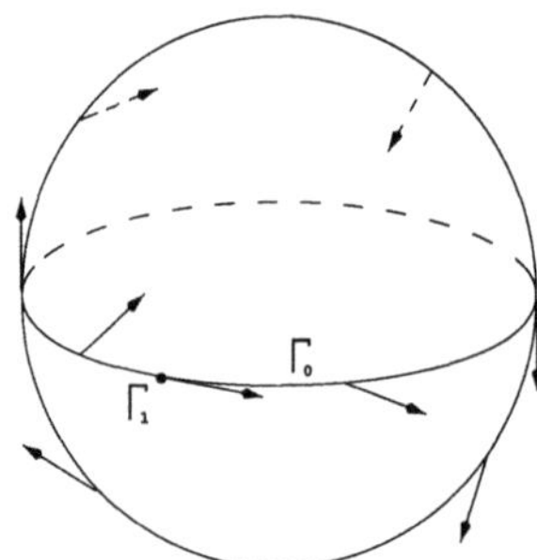

Fig. 4   Generic case of
degeneration

Around 1970 V. Arnold [1] stressed the importance of the so-called *generic case of degeneration*, where the vector field ℓ is no longer transversal to Γ_0; see also his well-known book [2], p. 203. More precisely, one assumes that there are smooth manifolds (without boundary) $\Gamma_0 \supset \Gamma_1 \ldots \supset \Gamma_s$ of dimensions $n-2, n-3, \ldots, n-2-s$ such that ℓ is tangent to Γ_j exactly at the points of Γ_{j+1}, whereas ℓ is nowhere tangent to Γ_s; see Fig. 4. A local model of this situation is given by the following:

$$\Omega = \{x \in \mathbb{R}^n : x_1 > 0\}, \quad \Gamma = \{x_1 = 0\},$$

$$\ell = x_2 \partial_1 + x_3 \partial_2 + \ldots + x_k \partial_{k-1} + \partial_k, \quad k \leq n,$$

$$\Gamma_j = \{x_1 = x_2 = \ldots = x_{2+j} = 0\}, \quad j = 0, \ldots, k-2.$$

The generic case is much more difficult from the analytical point of view than the transversal one, because entrance and exit points are permitted to belong to one and the same component of Γ_0 and the usual localization technique does not apply.

In 1972 V. Maz'ya [30] published a deep result related to the generic degeneration, still the only known one. He found function spaces of right-hand sides and solutions for the unique solvability of the problem. The success was achieved by an ingenious choice of weight functions in the derivation of a priori estimates for the solution. Additionally, Maz'ya proved that the inverse operator of the problem is always compact on $L_p(\Gamma), 1 < p \leq \infty$. It turned out that the manifolds Γ_j of codimension greater than one do not influence the correct statement of the problem, contrary to Arnold's expectations; see [2], p. 203. By the way, a description of the asymptotics of solutions near the points of tangency of the field ℓ to Γ_0 remains a difficult long-standing problem.

3. Estimates for differential operators in the half-space

At the beginning of the seventies Gelman and Maz'ya wrote a series of papers on the topic of this section, and the results were summarized in their monograph [13] published only in German. The fact that the book could not appear in the Soviet Union at that time sheds some light upon the antisemitic policy of the Soviet scientific administration. This is the right place to emphasize the role of Maz'ya's friends Siegfried Prössdorf and Günther Wildenhain. The first of them brought the manuscript illegally to East Germany, and the second became the editor.

The Gelman-Maz'ya book starts with the following epigraph by L. Gårding:

> "When a problem about partial differential operators has been fitted into the abstract theory, all that remains is usually to prove a suitable inequality and much of our knowledge is, in fact, essentially contained in such inequalities. But the abstract theory is not only a tool, it is also a guide to general and fruitful problems."

It contains indeed a great variety of inequalities for differential and pseudodifferential operators with constant coefficients. The authors obtain results of final character, without any restrictions on the type of the differential operators. They found necessary and sufficient conditions for the validity of the corresponding a priori estimates and presented more easily verifiable either necessary or sufficient conditions.

I will now describe a few typical results from this book. Let $\mathbb{R}^n_+ = \{(x,t) : x \in \mathbb{R}^{n-1}, t \geq 0\}$ and consider pseudodifferential operators $R(D), P_j(D), Q_s(D)$ with symbols $R(\xi,\tau), P_j(\xi,\tau), Q_s(\xi,\tau), \xi \in \mathbb{R}^{n-1}, \tau \in \mathbb{R}$, not depending on x and t, which are further assumed to be polynomials in τ with locally bounded measurable coefficients of polynomial growth in ξ. The book [13] presents a detailed and complete study of estimates in the half-space,

$$(4) \quad ||R(D)u||^2_{L_2(\mathbb{R}^n_+)} \le c \left\{ \sum_{j=1}^{m} ||P_j(D)u||^2_{L_2(\mathbb{R}^n_+)} + \sum_{s=1}^{r} ||Q_s(D)u||^2_{L_2(\partial\mathbb{R}^n_+)} \right\},$$

$$u \in C_0^\infty(\mathbb{R}^n_+),$$

and of trace estimates of the form

$$(5) \quad ||R(D)u||^2_{H^\mu(\partial\mathbb{R}^n_+)} \le c \left\{ \sum_{j=1}^{m} ||P_j(D)u||^2_{L_2(\mathbb{R}^n_+)} + \sum_{s=1}^{r} ||Q_s(D)u||^2_{L_2(\partial\mathbb{R}^n_+)} \right\},$$

$$u \in C_0^\infty(\mathbb{R}^n_+),$$

where H^μ denotes the Sobolev space of order μ. Gelman and Maz'ya found necessary and sufficient conditions stated in algebraic terms for these inequalities in full generality. Well-known results by Aronszajn, Agmon, Douglis and Nirenberg, and Schechter became part of the general theory developed in [13].

To give an idea of the results, I consider the example

$$r = 0, \quad m = 2, \quad P_1(\xi,\tau) = P(\xi,\tau), \quad P_2(\xi,\tau) = 1$$

and assume that the leading coefficient of P is equal to one. Now (5) takes the form

$$(6) \qquad ||R(D)u||^2_{H^\mu(\partial\mathbb{R}^n_+)} \le c \left\{ ||P(D)u||^2_{L_2(\mathbb{R}^n_+)} + ||u||^2_{L_2(\mathbb{R}^n_+)} \right\}.$$

Let $H(\xi,\tau)$ be a polynomial in τ with roots in the half-plane Im $\zeta > 0$, $\zeta = \tau + i\sigma$, and such that

$$|P(\xi,\tau)|^2 + 1 = |H(\xi,\tau)|^2.$$

The authors show that estimate (6) holds if and only if

$$\int_{\mathbb{R}} \frac{|T_1(\xi,\tau)|^2 + |T_2(\xi,\tau)|^2}{|P(\xi,\tau)|^2 + 1} d\tau \le c(1 + |\xi|^2)^{-\mu},$$

where $T_1(\xi,\tau), T_2(\xi,\tau)$ denote the quotient and the remainder obtained when the polynomial (in τ) $R(\xi,\tau)\overline{H(\xi,\tau)}$ is divided by $P(\xi,\tau)$; see [13], p. 170.

It is a pity that this Gelman-Maz'ya book did not attract much attention despite the beauty and the completeness of the results obtained. Clearly they have a great potential of generalization to partial differential and pseudodifferential equations with variable coefficients, both in the half-space and on domains.

4. The characteristic Cauchy problem for hyperbolic equations

If one thinks of well-posed problems for hyperbolic differential equations, the first that comes to one's mind is the Cauchy problem with initial data given on a space-like initial surface. On the other hand, it is known (see the book [8] by Courant and Hilbert) that the solution of the wave equation is already uniquely determined if its values are prescribed on the characteristic cone.

This is the simplest example of a characteristic Cauchy problem, which was the topic of an important paper by Vainberg and Maz'ya [52]. They studied general hyperbolic operators of arbitrary even order $2m$. In contrast to previous work by Gårding, Kotake, Leray [12] and Kondrat'ev [25], they were able to avoid any assumption regarding the set where the initial surface S is characteristic. This set, which will be denoted by V, may even have positive measure. In this case derivatives of order $2m - 1$ need not be prescribed on V in the formulation of the problem. The existence and uniqueness theorems and energy estimates obtained in [52] show that this formulation leads to a well-posed problem in appropriately chosen function spaces.

To illustrate this, I will present the Vainberg-Maz'ya energy estimate in the special case of the homogeneous differential equation. Let $\{S_\tau\}_{0\leq\tau\leq T}$ be a one-parameter family of surfaces $S_\tau = \{(t,x) : t = s(\tau,x),\ x \in \mathbb{R}^n\}$, where $S_0 = S$. On some compact sets V_τ the surfaces S_τ have characteristic directions, while at the remaining points they are space-like. Consider the weight function

$$\rho(\tau, x) = P_0(s(\tau,x), x, 1, -s'_x(\tau,x)),\ 0 \leq \tau \leq T,$$

where P_0 is the principal symbol of the hyperbolic differential operator P under consideration. The function ρ is non-negative and satisfies $\rho(\tau,x) = 0$ if and only if $(t,x) \in V_\tau$. Finally, let E^ρ denote the "weighted" energy

$$E^\rho(\tau, u) = \|\sqrt{\rho}\, \partial^{2m-1}u/\partial t^{2m-1}\|^2_{L_2(S_\tau)} + \sum_{j=0}^{2m-2} \|\partial^j u/\partial t^j\|^2_{H^{2m-1-j}(S_\tau)},$$

and define $E(\tau, u)$ by setting $\rho = 1$ in the first term. Then, for the characteristic Cauchy problem

$$P(t, x, \partial/\partial t, \partial/\partial x)u = 0,$$

$$\partial^j u/\partial t^j|_S = \varphi_j,\quad 0 \leq j \leq 2m - 2,\quad \partial^{2m-1}u/\partial t^{2m-1}|_{S\setminus V} = \varphi_{2m-1},$$

the energy estimate

$$\max_{0\leq\tau\leq T} E^\rho(\tau, u) + \int_0^T E(\tau, u)d\tau \leq c \left\{ \|\sqrt{\rho}\, \varphi_{2m-1}\|^2_{L_2(S\setminus V)} + \sum_{j=0}^{2m-2} \|\varphi_j\|^2_{H^{2m-1-j}(S)} \right\}$$

holds. Hörmander, who did not know about the paper by Vainberg and Maz'ya, obtained the same results in [19] for hyperbolic equations of second order. It should be noted that the proofs for higher order equations are much more complicated.

5. New methods for solving ill-posed boundary value problems

In 1986–1990, when running a mathematical division of the Leningrad Engineering Research Institute, Maz'ya organized a consultation center for engineers. The center proved to be useful not only for engineers, because it was also a source of many real-world problems of great mathematical interest.

In this way Kozlov and Maz'ya learned about an engineering approach to compute thermoelastic fields in a Tokamak fusion reactor. Starting with the rigorous foundation of this algorithm in [23], they developed a mathematical theory of new iterative methods for solving ill-posed boundary value problems for elliptic, parabolic and hyperbolic equations. Previously known and widely used numerical methods to solve those problems are based on changing the type or the order of the equation by means of small singular perturbations. The Kozlov-Maz'ya procedures are of different type. They are iterative, and in every step a well-posed problem for the *original* differential equation is solved.

The regularizing character of these methods is provided solely by an appropriate choice of boundary conditions for each iteration. For example, the solution of the Cauchy problem for the Laplace equation reduces to an alternating sequence of mixed boundary value problems which can easily be solved with standard software. Similarly, for the wave equation in a cylinder with Dirichlet boundary conditions, the iterative procedure consists in obtaining solutions of Cauchy's problem for the same equation with data alternatively prescribed either on the lower or upper base; see [22].

6. Applications of multiplier theory to integral operators

In 1979–1983, together with his wife Tatyana Shaposhnikova, Maz'ya developed a new theory of multipliers in spaces of differentiable functions, the results of which were published in their book [43]. One should not mix up these multipliers with the Fourier multipliers already studied in the thirties. Maz'ya and Shaposhnikova showed that their multipliers provide a natural language for various questions of analysis and the theory of differential and pseudodifferential operators.

Before their work, only a few isolated facts concerning multipliers in Sobolev spaces were known. In [43] one can find, in particular, a complete description of the multiplier spaces $M(W_p^m \to W_p^l)$ as well as of the spaces $M^\circ(W_p^m \to W_p^l)$ of compact multipliers. Here W_p^m stands for the Sobolev-Slobodetskiĭ space on $\mathbb{R}^n$, and $M(X \to Y)$ denotes the set of functions for which the corresponding multiplication operator maps the Banach space X into another Banach space Y. For $X = Y$, we simply write $M(X)$.

I dwell upon some of the many results in [43] which are close to the topic of the present article. As an application of their multiplier results for $p = 2$, Maz'ya and Shaposhnikova proved two-sided estimates for the essential norm and necessary and sufficient conditions for the compactness of convolution operators $u \to k * u$, acting from the weighted L_2 space $L_2(\mathbb{R}^n; (1 + |x|^2)^{m/2})$ into $L_2(\mathbb{R}^n; (1 + |x|^2)^{l/2})$. Note that the Fourier transform is an isomorphic map of W_2^l onto the latter weighted L_2 space.

The authors also study singular integral operators in Sobolev spaces. It turns out that the basic properties of these operators are retained under minimal smoothness assumptions on the symbol concerning the first variable. I present a typical result in this direction.

Consider singular integral operators A, B of the form (1) in $\mathbb{R}^n$, which have the symbols $a(x, \xi)$, $b(x, \xi)$. Let $A \circ B$ denote the singular integral operator with the product symbol ab. It was proved in [43], Chap. 4.5, that if a and b satisfy the smoothness conditions

$$a \in C^\infty(M(W_p^{l+1}), S^{n-1}), \quad \nabla_x b \in C^\infty(M(W_p^l), S^{n-1})$$

then the operator $C := AB - A \circ B$ is a continuous map of W_p^l into W_p^{l+1}. If, in addition, there exists $b_\infty \in C^\infty(S^{n-1})$ such that

$$b - b_\infty \in C^\infty(M^\circ(W_p^l), S^{n-1}),$$

then C is compact on W_p^l.

In conclusion I mention that the book [43] contains very interesting applications of multipliers to the theory of elliptic boundary value problems on domains with non-smooth boundaries.

7. Integral equations of harmonic potential theory on general non-regular surfaces

In the sixties Maz'ya and his colleagues made a major breakthrough in the theory of boundary integral equations on very general surfaces, generalizing the classical Radon theory to higher dimensions. In order to explain this, I need some notation. Let $\Omega \subset \mathbb{R}^n$ $(n \geq 2)$ be a domain with boundary Γ. The harmonic double layer potential is given by

$$Wu(x) = \int_\Gamma u(\xi) \frac{\partial}{\partial \nu_\xi} G(x, \xi) d_\xi \Gamma, \quad x \in \mathbb{R}^n \backslash \Gamma,$$

where G denotes the fundamental solution of the Laplacian. Let $T = 2W_0$, with W_0 being the value of W on Γ. The interior and exterior Dirichlet problem for the Laplacian in Ω can be reduced to the classical second kind integral equation with the boundary integral operators $I + T$ and $I - T$, respectively, whereas the corresponding Neumann problems lead to the adjoint operators $I \mp T^*$. Here I denotes the identity operator.

To formulate Radon's classical results on these integral equations, I need some further notation. Let A be a bounded linear operator acting on a Banach space X. The Fredholm radius $R(A)$ of A, introduced by Radon [47], is the radius of the largest disk centered at the origin of the complex λ-plane such that the operators $I - \lambda A$ are Fredholm for all λ in its interior. The quantity $|A| = \inf \|A - K\|_X$, where the infimum is taken over all compact operators K on X, is referred to as the essential norm of A and also appeared first in [47].

Let Γ be a planar curve of bounded rotation, i.e., Γ is rectifiable and the angle $\vartheta(s)$ between the positively oriented tangent and the abscissa is of bounded variation on, say, $0 \leq s \leq l$. In 1919 Radon [48] proved that, for the harmonic double layer potential operator T acting on the space $C(\Gamma)$ of continuous functions, the equality

$$R(T) = |T|^{-1} = \pi/\alpha$$

is satisfied with $\alpha = \sup\{|\vartheta(s+0) - \vartheta(s-0)| : 0 \leq s \leq l\}$.

As a corollary of this result, one obtains that if Γ has no cusps then $R(T) > 1$, so that in this case the Fredholm theory applies to the operators $I \pm T$ in $C(\Gamma)$. Moreover, Radon then obtained the basic solvability results for these operators as well, e.g., the invertibility of $I + T$.

In their famous course of functional analysis [50] F. Riesz and B. Sz.-Nagy noted that "in the case of the spatial problem an analogue of curves with bounded rotation has not yet been found." This inspired the work by Burago, Maz'ya, Sapozhnikova [4], [5] and Burago, Maz'ya [3], who not only extended Radon's theory to higher dimensions but also improved Radon's result for contours in the plane.

The basic solvability results (e.g., invertibility of $I + T$ in $C(\Gamma)$ and of $I + T^*$ in $C(\Gamma^*)$) were obtained for domains subject to the following two conditions:

(A) $\sup\{\operatorname{var} \omega(\xi, \Gamma \backslash \xi) : \xi \in \Gamma\} < \infty,$

(B) $\lim_{r \longrightarrow 0} \sup\{\operatorname{var} \omega(\xi, \Gamma \cap B_r(\xi)) : \xi \in \Gamma\} < \sigma_n/2.$

Here $\omega(\xi, B)$ denotes the solid angle at which the set B is seen from the point ξ, or rather its generalization to a certain set function. Furthermore, var denotes the variation of the charge, $B_r(\xi)$ is the ball with center ξ and radius r, and σ_n is the area of the unit sphere S^{n-1}.

Condition (A) is necessary to apply the potential method in the spaces under consideration, whereas (B) is solely needed to prove the Fredholm alternative. In fact, independently of each other, Král [25] and Burago and Maz'ya [3] proved that (B) is equivalent to the inequality $|T| < 1$. Condition (A) is of course valid for any curve of bounded rotation, while (B) is satisfied if it has no cusps. However, there exist plane curves satisfying (A) and (B) which are not of bounded rotation. Before the works of Král and Burago, Maz'ya and Sapozhnikova it was a common belief that Radon generalized the theory of potentials in spaces of continuous functions to its natural limit; see [50], Sect. 91.

We now discuss some deep results by Kresin and Maz'ya [26], [27] on the essential norm of the general vector-valued double layer potentials

$$Tu(x) = 2 \int_\Gamma k(e_{xy})u(y)\omega(x,dy) \, .$$

Here k is a continuous even $(m \times m)$-matrix-valued function on the unit sphere, which is homogeneous of degree 0 and normalized by the condition $\int_{S^{n-1}} k = I$, and e_{xy} denotes $(y-x)/|y-x|$. The authors succeeded in proving general formulas for the norm and the essential norm in the space $C(\Gamma)^m$ of continuous vector-valued functions on Γ with m components, provided the surface Γ satisfies condition (A).

The case where $k(e)$ is the matrix with elements

$$\sigma_n^{-1}\{(1-\kappa)\delta_{ij} + n\kappa(e,e_i)\}, \quad i,j = 1,\ldots,n, \quad 0 \le \kappa \le 1,$$

e_j being the unit vector directed along the jth coordinate axis, is of particular interest. Note that $\kappa = 0$ yields a diagonal matrix of harmonic potentials, and putting $\kappa = 1$ and $\kappa = (\lambda+\mu)(\lambda+3\mu)^{-1}$ with the Lamé constants λ, μ, we obtain the hydrodynamic and elastic potentials, respectively.

For $n = 2$ and Γ a polygon with interior angles $\alpha_1,\ldots,\alpha_N$, Kresin and Maz'ya deduced the following beautiful formula:

$$|T| = \frac{2}{\pi}(1+\kappa)E\left(\frac{\pi - \alpha_{\min}}{2}, \frac{2\sqrt{\kappa}}{1+\kappa}\right), \quad \alpha_{\min} = \min\{\alpha_i : i = 1,\ldots,N\},$$

where E denotes the elliptic integral of the second kind. For the hydrodynamic potential, i.e. for $\kappa = 1$, this gives $|T| = 4\pi^{-1}\cos(\alpha_{\min}/2)$. On the other hand, Shelepov [51] found the Fredholm radius,

$$R(T) = \pi\{|\pi - \alpha_{\min}| + \kappa \sin|\pi - \alpha_{\min}|\}^{-1} \, .$$

These two formulae imply the unexpected inequality $R(T) > |T|^{-1}$ whenever $\kappa > 0$. Explicit formulae for $|T|$ were also given in [27] in the three-dimensional case where Γ has an edge or a conical point.

8. Boundary integral equations on piecewise smooth surfaces

Despite the great generality of surfaces which Burago, Maz'ya and Král dealt with in the sixties, their theory did not apply to quite natural geometries because of the rather restrictive condition (B) on the local variation of the solid angle, and of course they only studied the Dirichlet and Neumann problem for harmonic potentials. Neither specific problems of mathematical elasticity and hydrodynamics nor general elliptic boundary value problems were touched. The main obstacle was that the theories of Fredholm and singular integral equations were not sufficiently developed to deal with irregular boundaries.

Around 1980 Maz'ya arrived at a simple but bold idea which brought a new light to the whole domain of boundary integral equations. He understood that such equations could be exhaustively studied without using general theories of integral equations. His talk [31] presented at the Petrovskiĭ Conference in 1981 became a breakthrough in the solvability theory of boundary integral equations on piecewise smooth surfaces.

Maz'ya's approach is based on the fact that solutions of boundary integral equations can be expressed in terms of solutions to certain auxiliary interior and exterior boundary value problems. Then the full force of the newly developed theory of boundary problems in non-smooth domains, with its theorems on solvability and Fredholm property in various function spaces as well as the results on asymptotics of solutions near boundary singularities, becomes available. In this theory Maz'ya also did major and pioneering work; see J. Rossmann's survey in the present Volume.

The new approach was first exemplified in [31] by the classical boundary value problems of potential theory, although it was clear from the very beginning that it has universal character. In [32] and [33] it was applied to the two fundamental boundary value problems of elasticity. As a result of fifteen years' work, Maz'ya and his collaborators developed an extensive solvability theory of boundary integral equations on surfaces with conical points, edges, vertices, and also cusps. A detailed account of the results obtained before 1990 can be found in Maz'ya's survey article [35], which has become a standard reference in the field.

Another important development during the last two decades was the theory of boundary equations in Lipschitz domains and with data from L_p, originating from the result of Calderón and Coifman, McIntosh, Meyer on the boundedness of the Cauchy singular integral operator over Lipschitz curves. Based on these results and on the Rellich-Nečas identities, Verchota [53], Kenig [20], Fabes [11], and others proved the solvability in L_p of various boundary integral equations on a Lipschitz surface. However, it is worth mentioning that there are simple polyhedra that are not Lipschitz (in the sense that they can in local Cartesian coordinates be described by Lipschitz functions).

To give an impression of how Maz'ya's method works, I will consider the special case of the second kind boundary integral equation

$$(7) \qquad\qquad (I + T)u = f, \quad T = 2W_0$$

for the Laplacian, which has already been discussed in the previous section. Let $\mathcal{D} : f \to v$ be the inverse operator of the interior Dirichlet problem

$$\Delta v = 0 \quad \text{in } \Omega, \quad \mathcal{T}_0 v := v|_\Gamma = f,$$

and let $\mathcal{N}$ be the inverse of the corresponding exterior Neumann problem. Then the inverse of $I + T$ can be represented as

$$(8) \qquad (I + T)^{-1} = \frac{1}{2}(I - \mathcal{T}_0 \mathcal{N} \mathcal{T}_1 \mathcal{D}),$$

where $\mathcal{T}_1$ stands for the trace operator $\mathcal{T}_1 v := (\partial/\partial \nu)v|_\Gamma$. This also applies to the systems of integral equations of elasticity and hydrodynamics if the role of $\mathcal{T}_1$ is played by the stress operator. In [33] Maz'ya studied the solvability of these equations in weighted Hölder spaces in the case of a surface with conical points, edges and polyhedral angles, using the representation (8) and his results with Plamenevskiĭ on boundary value problems. Corresponding results for the mixed problem of 3D elasticity in domains with edges were obtained in his paper [34]. Moreover, Maz'ya's approach turned out to be useful to calculate the Fredholm radius of boundary integral operators. In the paper [15] with his student Grachev, an explicit formula for the harmonic double layer potential operator in weighted Hölder spaces on surfaces with edges was proved.

If one wants to obtain solvability in the space C or in weighted L_p and Sobolev spaces, then the representation (8) is not directly applicable (as long as there are no solvability results in Hardy spaces for the auxiliary boundary value problems). In 1989 Grachev and Maz'ya developed an approach which also works in this case. They established sharp estimates for the kernel of the inverse operator $(I + T)^{-1}$, implying, in particular, the solvability of the integral equation (7) in C without any assumption on the essential norm of T. For a smooth cone Γ of vertex 0 in $\mathbb{R}^n$, it was shown in their paper [16] that $(I + T)^{-1}$ decomposes into a sum $I + M_1 + M_2$ with certain integral operators M_1 and M_2 on Γ whose kernels admit the estimates

$$|M_1(x, y)| \leq c(1 + |y|^{\mu - 1}),$$

$$|M_2(x, y)| \leq \begin{cases} c|y|^{1-n}(|x|/|y|)^\mu & \text{for} \quad |x| < |y|/2, \\ c|y|^{-1}|x - y|^{2-n} & \text{for} \quad |y|/2 < |x| < 2|y|, \\ c|y|^{-1}|x|^{2-n}(|y|/|x|)^\mu & \text{for} \quad |x| > 2|y|, \end{cases}$$

where the number $\mu, 0 < \mu \leq 1$, depends on the shape of the cone.

As a consequence of these estimates, general solvability results for equation (7) in C and weighted L_p spaces on a boundary with a finite number of conical points were obtained. Applying the same method, Maz'ya and Grachev [17] settled the long-standing classical question on solvability of (7) in the space C for an arbitrary polyhedron; see also [14]. At the same time, this problem was solved by Rathsfeld [49] using Mellin transformation and Banach algebra techniques. In addition [49] establishes the inequality $R(T) > 1$ for the Fredholm radius in C, which also holds for L_2 and certain weighted Sobolev spaces (see [10]).

As another application of Maz'ya's method, the asymptotics of solutions to boundary integral equations near singular points of the boundary can be derived, including the computation of the coefficients appearing in the asymptotic formulas. These results were set up by Zargaryan and Maz'ya [54], [55] for the integral equations of harmonic potential theory on a polygonal boundary. A different approach to solvability and asymptotics of solutions of integral equations over curves with corners was worked out by Costabel and Stephan [6], [7], who used the Mellin transform to solve the model equation on the legs of an infinite angle. The first result on the asymptotics of solutions to boundary integral equations over three-dimensional regions is due to Maz'ya and Levin [28] and refers to the case of conical points on the boundary. Recently Kozlov, Wendland and Goldberg [24] combined Maz'ya's approach with Mellin techniques to find the asymptotics of solutions of the first-kind boundary integral equation for the Lamé system in a polyhedral cone.

During the last decade, Maz'ya and Solov'ev developed a theory of boundary integral equations on plane contours with cusps, including the Dirichlet and Neumann problem for the Laplace and the Lamé operator. Their extensive research on this topic started in 1988 with [44] and is still going on; see e.g. the recent papers [45], [46]. They proved theorems on the unique solvability in appropriate weighted L_p, Sobolev and Hölder spaces and on asymptotic representations for solutions near peaks, which are the only known results in the area. I refer the interested reader to Maz'ya's survey [36] for further information. It is a tempting perspective to generalize the Maz'ya-Solov'ev results to multidimensional domains with cusps as well as to other boundary value problems.

I hope what I have said in this section shows convincingly that Maz'ya's approach opened new horizons in the theory of boundary integral equations on non-regular surfaces, and it will definitely inspire fruitful research in this field in the future.

References

[1] V. I. Arnold, On local problems of mathematical analysis (Russian), *Vestnik Moscov. Univ.* **2** (1970), 52–56.

[2] ______ , *Additional chapters of the theory of ordinary differential equations* (Russian), Nauka, Moscow, 1978.

[3] Yu. D. Burago and V. G. Maz'ya, Potential theory and function theory for irregular regions, *Sem. Math. V. A. Steklov Inst. Leningr.* **3** (1969), 1–68, translation from *Zap. Nauchn. Semin. Leningr. Otd. Mat. Inst. Steklova* **3** (1967), 1–152.

[4] Yu. D. Burago, V. G. Maz'ya, and V. D. Sapozhnikova, On the double layer potential for nonregular regions, *Sov. Math. Dokl.* **3** (1962), 1640–1642, translation from *Dokl. Akad. Nauk SSSR* **147** (1962), 523–525.

[5] ______ , On the theory of simple and double-layer potentials for domains with irregular boundaries, *Probl. Mat. Anal.* **1** (1968), 1–30, translation from *Probl. Mat. Analiza, Kraevye Zadachi, Integral. Uravn. Leningrad* (1966), 3–34.

[6] M. Costabel and E. Stephan, Curvature terms in the asymptotic expansions for solutions of boundary integral equations on curved polygons, *J. Integral Equations* **5** (1983), 353–371.

[7] ______ , Boundary integral equations for mixed boundary value problems in polygonal domains and Galerkin approximation, *Mathematical Models in Mechanics*, Banach Center Publication, vol. 15, PWN–Polish Scientific Publishers, Warsaw, 1985, pp. 175–251.

[8] R. Courant and D. Hilbert, *Methods of mathematical physics*, vol. 2, Interscience Publishers, New York, 1962.

[9] Yu. V. Egorov and V. A. Kondrat'ev, The oblique derivative problem, *Math. USSR-Sb.* **7** (1969), 139–169, translation from *Mat. Sb.* **78** (1969), 148–176.

[10] J. Elschner, The double layer potential operator over polyhedral domains. I: Solvability in weighted Sobolev spaces, *Appl. Anal.* **45** (1992), 117–134.

[11] E. B. Fabes, Boundary value problems of linear elastostatics and hydrostatics on Lipschitz domains, *Proc. Centre Math. Anal. Austral. Nat. Univ.* **9** (1985), 27–45.

[12] L. Gårding, T. Kotake, and J. Leray, Uniformisation et development asymptotique de la solution du problème de Cauchy lineaire, *Bull. Soc. Math. France* **92** (1964), 263–361.

[13] I. W. Gelman and V. G. Maz'ya, *Abschätzungen für Differentialoperatoren im Halbraum*, Akademie-Verlag, Berlin, 1981.

[14] N. V. Grachev, Representations and estimates for inverse operators of the potential theory integral equations in a polyhedron, *Potential Theory* (M. Kishi, ed.), Walter de Gruyter, Berlin, 1991, pp. 201–206.

[15] N. V. Grachev and V. G. Maz'ya, On the Fredholm radius of operators of the double layer potential type on piecewise smooth surfaces, *Vestnik Leningrad Univ. Math.* **19** (1986), no. 4, 20–25, translation from *Vestnik Leningr. Univ.*, Ser. I, no. 4 (1986), 60–64.

[16] ______ , Representations and estimates of inverse operators of integral equations of potential theory for surfaces with conic points (Russian), *Soobshch. Akad. Nauk Gruzin. SSR* **132**, no. 1 (1988), 21–24.

[17] ______ , Solvability of a boundary integral equation on a polyhedron, Preprint LITH-MAT-R91-50, Linköping University, Department of Mathematics, 1992.

[18] L. Hörmander, Pseudo-differential operators and non-elliptic boundary value problems, *Ann. Math.* **83** (1966), 129–209.

[19] ______ , A remark on the characteristic Cauchy problem, *J. Funct. Anal.* **93** (1990), 270–277.

[20] C. E. Kenig, Boundary value problems of linear elastostatics and hydrostatics on Lipschitz domains, *Seminar Goulaouic-Meyer-Schwartz, Equat. Deriv. Partielles* 1983–1984, Exp. no. 21, 1984, pp. 1–12.

[21] V. A. Kondrat'ev, The Cauchy problem with characteristic points on the initial surface, *Moscow Univ. Math. Bull.* **29**, no. 1/2 (1974), 68–74, translation from *Vestnik Moskov. Univ.*, Ser. I, **29**, no. 1 (1974), 84–92.

[22] V. A. Kozlov and V. G. Maz'ya, On iterative procedures for solving ill-posed boundary value problems that preserve differential equations, *Leningrad Math. J.* **1**, no. 5 (1990), 1207–1228, translation from *Algebra i Analiz* **1**, No. 5 (1989), 144–170.

[23] V. A. Kozlov, V. G. Maz'ya, and A. V. Fomin, An iterative method for solving the Cauchy problem for elliptic equations, *Comput. Maths. Math. Phys.* **31**, no. 1 (1991), 45–52, translation from *Zh. Vychisl. Mat. i Mat. Fiz.* **31**, no. 1 (1991), 64–74.

[24] V. A. Kozlov, W. L. Wendland, and H. Goldberg, The behaviour of elastic fields and boundary integral Mellin techniques near conical points, *Math. Nachr.* **180** (1996), 95–133.

[25] J. Král, The Fredholm radius of an operator in potential theory, *Czechoslovak Math. J.* **15** (1965), 454–473 and 565–588.

[26] G. I. Kresin and V. G. Maz'ya, On the essential norm of an operator of double layer potential type in the space C_m, *Sov. Math. Dokl.* **20** (1979), 459–462, translation from *Dokl. Akad. Nauk SSSR* **246** (1979), 272–275.

[27] ———, The norm and the essential norm of the double layer elastic and hydrodynamic potentials in the space of continuous functions, *Math. Methods Appl. Sci.* **18** (1995), 1095–1131.

[28] A. V. Levin and V. G. Maz'ya, On the asymptotics of the density of harmonic potentials close to the vertex of a cone (Russian), *Z. Anal. Anwendungen* **8** (1989), 501–514.

[29] M. B. Malyutov, On the boundary value problem of Poincaré (Russian), *Trudy Moscov. Mat. Obshch.* **20** (1969), 173–204.

[30] V. G. Maz'ya, On the degenerate oblique derivative problem, *Math. USSR-Sb.* **16** (1972), 429–469, translation from *Mat. Sb.* **87** (1972), 417–454.

[31] ———, The integral equations of potential theory in domains with piecewise smooth boundary (Russian), *Uspekhi Mat. Nauk* **36**, no. 4 (1981), 229–230.

[32] ———, On the solvability of the integral equations of classical elasticity theory in domains with piecewise smooth boundary, *Proceedings Conference on General Mechanics and Elasticity Theory*, Telavi, Metsniereba, Tbilisi, 1981, pp. 55–56.

[33] ———, Boundary integral equations of elasticity in domains with piecewise smooth boundaries, *Differential equations and their applications*, Equadiff 6, Proceedings 6th Int. Conf., Brno/Czech., 1985, *Lecture Notes Math.* **1192**, Springer, Berlin, 1986, pp. 235–242.

[34] ———, Potential theory for the Lamé equations in domains with piecewise smooth boundary, *Proceedings All-Union Symposium*, Tbilisi, April 21–23, 1982, Metsniereba, Tbilisi, 1986, pp. 123–129.

[35] ———, Boundary integral equations, *Analysis IV* (V. G. Maz'ya and S. M. Nikol'skiĭ, eds.), *Encyclopaedia of Mathematical Sciences*, vol. 27, Springer-Verlag, Berlin, 1991, pp. 127–222.

[36] ———, Boundary integral equations on a contour with peaks, *Boundary integral methods for nonlinear problems* (L. Morino et al., ed.), Proceedings of the

IABEM symposium, Pontignano, Italy, May 28–June 3, 1995, Kluwer Academic Publishers, Dortrecht, 1997, pp. 145–153.

[37] V. G. Maz'ya and B. P. Paneyah, Degenerate elliptic pseudo-differential operators on a smooth manifold without boundary, *Functional Anal. Appl.* **3** (1969), 159–160, translation from *Funktional. Anal. Prilozhen.* **3**, no. 2 (1969), 91–92.

[38] ———, Coercive estimates and regularity of the solution of degenerate elliptic pseudodifferential equations, *Functional Anal. Appl.* **4** (1970), 299–311, translation from *Funktional. Anal. Prilozhen.* **4**, no. 4 (1970), 41–56.

[39] ———, Degenerate elliptic pseudodifferential operators and the oblique derivative problem, *Trans. Moscow Math. Soc.* **31** (1974), 247–305, translation from *Trudy Moscov. Mat. Obshch.* **31** (1974), 237–295.

[40] V. G. Maz'ya and B. A. Plamenevskiĭ, Singular equations with a vanishing symbol, *Sov. Math. Dokl.* **6** (1965), 294–297, translation from *Dokl. Akad. Nauk SSSR* **160** (1965), 1250–1253.

[41] ———, The Cauchy problem for hyperbolic singular integral equations of convolution type (Russian), *Vestnik Leningr. Univ.* **20**, no. 19 (1965), 161–163.

[42] V. G. Maz'ya, B. A. Plamenevskiĭ, and Yu. E. Khaikin, A correctly posed problem for singular integral equations with vanishing symbols, *Differential Equations* **13** (1977), 1028–1033, translation from *Differencial'nye Uravneniya* **13** (1977), 1479–1486.

[43] V. G. Maz'ya and T. O. Shaposhnikova, *Theory of multipliers in spaces of differentiable functions*, Pitman, Boston, 1985.

[44] V. G. Maz'ya and A. A. Solov'ev, Solvability of an integral equation for the Dirichlet problem in a plane domain with cusps on the boundary, *Sov. Math. Dokl.* **37** (1988), 255–258, translation from *Dokl. Akad. Nauk SSSR* **298**, no. 6 (1988), 1312–1315.

[45] ———, L_p-theory of a boundary integral equation on a cuspidal contour, *Appl. Anal.* **65** (1997), 289–305.

[46] ———, L_p-theory of a boundary integral equation on a contour with outward peak, *Integral Equation Operator Theory* **32** (1998), 75–100.

[47] J. Radon, Über lineare Funktionaltransformationen und Funktionalgleichungen, *Sitzungsberichte Akad. Wiss.*, Abt. 2a, Wien **128** (1919), 1083–1121.

[48] ———, Über die Randwertaufgaben beim logarithmischen Potential, *Sitzungsberichte Akad. Wiss.*, Abt. 2a, Wien **128** (1919), 1123–1167.

[49] A. Rathsfeld, The invertibility of the double layer potential operator in the space of continuous functions defined on a polyhedron. The panel method, *Appl. Anal.* **45** (1992), 135–177.

[50] F. Riesz and B. Sz.-Nagy, *Functional analysis*, Ungar Publishing Co., New York, 1955, French original: Akad. Kiado, Budapest, 1952.

[51] V. Yu. Shelepov, On investigations by Ya. B. Lopatinskiĭ's method of matrix integral equations in the space of continuous functions (Russian), *General theory of boundary value problems*, Collect. Sci. Works, Naukova Dumka, Kiev, 1983, pp. 220–226.

[52] B. R. Vainberg and V. G. Maz'ya, Characteristic Cauchy problem for a hyperbolic equation, *J. Soviet Math.* **31** (1985), 3135–3147, translation from *Trudy Sem. Petrovsk.* **7** (1981), 101–117.

[53] G. Verchota, Layer potentials and regularity for the Dirichlet problem for Laplace's equation in Lipschitz domains, *J. Funct. Anal.* **59** (1984), 572–611.

[54] S. S. Zargaryan and V. G. Maz'ya, On singularities of solutions of a system of integral equations in potential theory for the Zaremba problem, *Vestnik Leningrad Univ. Math.* **16** (1984), 49–55, translation from *Vestnik Leningr. Univ.*, Ser. I, no. 1 (1983), 43–48.

[55] ———, The asymptotic form of solutions of integral equations of potential theory in the neighbourhood of the corner points of a contour, *J. Appl. Math. Mech.* **48** (1985), 120–124, translation from *Prikl. Mat. Mekh.* **48** (1984), 169–174.

Weierstrass Institute
for Applied Analysis and Stochastics,
Mohrenstrasse 39,
D–10117 Berlin, Germany,
elschner@wias-berlin.de

1991 Mathematics Subject Classification: Primary 31–02, 35–02, 45–02, 58–02

Submitted: 22.12.1998

Operator Theory:
Advances and Applications, Vol. 109
© 1999 Birkhäuser Verlag Basel/Switzerland

Contributions of V. Maz'ya to the theory of boundary value problems in nonsmooth domains

JÜRGEN ROSSMANN

0. Introduction

A large part of the scientific work of Vladimir Maz'ya is concerned with the theory of boundary value problems in nonsmooth domains. Different variations of this theme are constantly heard in his mathematical symphony. The present article is aimed to give a review of his results in this field.

Elliptic boundary value problems constitute an essential part in the modern theory of partial differential equations. The question of the behaviour of solutions near the boundary and, in particular, near its singular points plays an essential role when studying these problems. This question is of great importance for many applications in aero- and hydrodynamics, elasticity, fracture mechanics etc. During the last fifty years a vast number of mathematical works dealt with the boundary behaviour of solutions, and Maz'ya is one of a few pioneers in the development of this theory. He contributed both to special problems of mathematical physics and general boundary value problems for elliptic equations and systems.

Maz'ya's works influenced the mathematical research in Rostock in various areas such as capacity theory, Wiener criterion, boundary value problems in piecewise smooth or singularly perturbed domains. My cooperation with V. G. Maz'ya started in 1983, when I was a Ph.D. student and spent eight months at the Leningrad University. Our joint work at that time and during the following years greatly affected my scientific career. This is one of the reasons why I am pleased to describe a part of his enormous work in this volume honouring him.

Here is a rough plan of my survey:

- Maz'ya's early works on elliptic equations in nonsmooth domains (Section 1)

- Theory of general elliptic boundary value problems in domains having point singularities, edges and polyhedral vertices (Sections 2 and 3)

- Spectral theory of operator pencils generated by elliptic problems in a cone (Section 4)

- Applications to particular problems of mathematical physics (Section 5)

- Conic singularities of solutions to nonlinear elliptic equations (Section 6)

Since the whole set of publications on boundary value problems in nonsmooth domains is unobservable, I have included (with a small number of exceptions) only Maz'ya's works in the bibliography. References to the results of other authors can be found, e.g., in the books [2], [24] and in the survey paper [4].

1. Maz'ya's early work on boundary value problems in nonsmooth domains

By 1960 numerous results on boundary value problems were established by functional analytic methods under restrictive requirements on regularity of the boundary, and the question whether these assumptions are necessary was raised very seldom. Necessary and sufficient conditions by Wiener and Molchanov stated in terms of the harmonic capacity were rare exceptions of this rule. In general, such questions were considered as subtle and difficult.

Since 1960 Maz'ya obtained a large number of fundamental results concerning the influence of the quality of boundary on the solvability in various function spaces and spectral properties of the classical boundary value problems for linear elliptic operators of the second or higher order. In many cases, using his own approaches, he was able to make crucial advances in the field. Thus, he was able to include well-known properties of boundary value problems obtained for smooth domains as particular cases in scales of properties changing gradually under deterioration of the domain.

1.1. Existence of generalized solutions

Criteria for integral inequalities. In his paper [31], published in 1960 when Maz'ya was still a student at the Leningrad University, he found necessary and sufficient conditions, under which the integral inequalities of Poincaré and Sobolev type

$$(1.1) \qquad \inf_{\gamma \in \mathbb{R}} \|v - \gamma\|_{L_q(\Omega)} \le c \, \|\nabla v\|_{L_p(\Omega)}$$

are valid. He proved that not only the class of sets Ω determines the parameters p and q but, conversely, (1.1) characterizes the class of sets Ω, for which this inequality holds. For $p = 1$ Maz'ya's criterion is stated as the so-called relative isoperimetric inequality connecting the volume and the interior (with repect to Ω) part of the surface area of subsets of Ω. However, for $p > 1$ this geometric characterization is not sufficient, and (1.1) is equivalent to an isoperimetric inequality for the volume and the p-conductivity. When $p = 2$, the last set function is closely related to the classical Wiener capacity, which is a mathematical expression for the electrostatic capacity. A subset $K \subset \Omega$ is called 'conductor' if there are an open subset $G \subset \Omega$ and a relatively closed subset $F \subset G$ such that $K = G \backslash F$.

The *conductivity* of K is defined as

$$\operatorname{cond}(K) = \inf\left\{ \int_K |\nabla f|^2 \, dx \ : \ f \in C^\infty(\Omega), \ f = 1 \text{ on } F, \ f = 0 \text{ outside } G\right\}.$$

Maz'ya [34, 48] proved that inequality (1.1) with $p = 2$ and $q \geq 2$ is equivalent to the isoperimetric inequality

$$(1.2) \qquad\qquad \sup_K \frac{\left(\operatorname{mes}_n(F)\right)^{2/q}}{\operatorname{cond}(K)} < \infty.$$

Here the supremum is taken over the set of all conductors $K = G\backslash F$, such that $F \subset G \subset \Omega$ and $2\operatorname{mes}_n(G) \leq \operatorname{mes}_n(\Omega)$.

As early as in 1966, Maz'ya had understood that the methods of proof of the above and similar results do not rely upon the specific character of the Euclidean space. In [45], he wrote that these methods can be applied to the case of nonregular Riemannian manifolds. In this case the results depend additionally on the singularities of the metric of the manifold.

The Neumann problem in the energy space. The importance of inequality (1.1) with $p = 2$ is explained by its close relation to the Neumann problem for the second order elliptic equation. Let

$$(1.3) \qquad\qquad \mathcal{L}u = -\sum_{i,j=1}^{n} \frac{\partial}{\partial x_j}\left(a_{i,j}(x)\frac{\partial u}{\partial x_i}\right),$$

where $a_{i,j}$ are measurable and bounded functions such that

$$\sum_{i,j=1}^{n} a_{i,j}(x)\,\xi_i\,\xi_j \geq c|\xi|^2 \quad \text{for all} \quad \xi \in \mathbb{R}^n, \quad c > 0.$$

The Neumann problem for the equation $\mathcal{L}u = f$ can be formulated in the following way. Let $q' = q/(q-1)$. For any $f \in L_{q'}(\Omega)$ orthogonal to a constant we have to find a function $u \in L_2^1(\Omega)$ [1] satisfying the equality

$$\int_\Omega \sum_{i,j=1}^{n} a_{i,j}\frac{\partial u}{\partial x_i}\frac{\partial v}{\partial x_j}\,dx = \int_\Omega f\,v\,dx$$

for all $v \in L_2^1(\Omega) \cap L_\infty(\Omega)$. By Riesz' representation theorem, this problem is solvable for every f if inequality (1.1) with $p = 2$ is satisfied for all $v \in L_2^1(\Omega)$. The Hahn-Banach theorem guarantees that the converse assertion is also true. So, Maz'ya's result gives a necessary and sufficient condition on the domain Ω,

---

[1] $L_p^1(\Omega)$ denotes the set of all distributions u on Ω such that $\nabla u \in L_p(\Omega)^n$.

ensuring the solvability of the Neumann problem. For the validity of condition
(1.2) it is sufficient to verify a more simple inequality

$$(1.4) \qquad \sup_{G} \frac{\left(\mathrm{mes}_n(G)\right)^{(2+q)/2q}}{\mathrm{mes}_{n-1}(\Omega \cap \partial G)} < \infty,$$

where the supremum is taken over all open sets $G \subset \Omega$, such that $\Omega \cap \partial G$ is a
manifold of class C^∞ and $2\,\mathrm{mes}_n(G) \le \mathrm{mes}_n(\Omega)$.

Using the well-known Rellich's lemma, one reduces the question whether the
spectrum of the Neumann problem for $-\Delta$ is discrete, to another one of the compactness of the embedding operator: $W_2^1(\Omega) \mapsto L_2(\Omega)$ [2]. The condition obtained
in [48] for the compactness of this embedding has the form

$$(1.5) \qquad \sup_{\{\mathrm{mes}_n(G) \le \varepsilon\}} \frac{\mathrm{mes}_n(F)}{\mathrm{cond}(K)} \to 0 \quad \text{as } \varepsilon \to 0.$$

In a particular case, conditions of the form (1.2)–(1.5) are effectively verifiable.

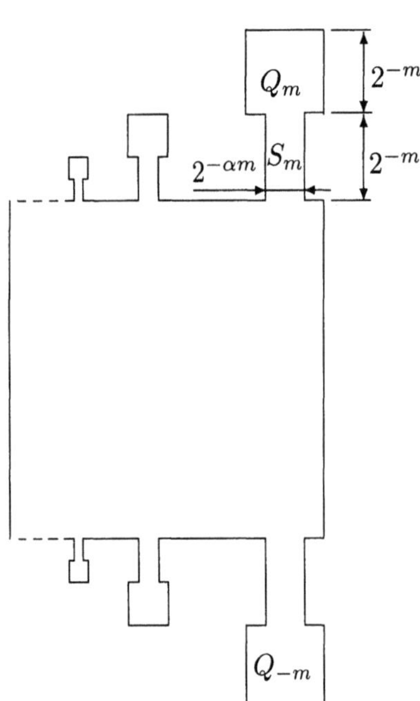

Let, for example, the plane domain Ω be the union of the square
$(0,2) \times (-1,+1)$ with squares Q_m,
Q_{-m} and rectangles S_m, S_{-m},
$m = 1,2,\dots$ arranged symmetri-
cally. Here the length of the sides
of Q_m, Q_{-m} and the height of S_m,
S_{-m} are equal to 2^{-m}, whilst the
width of S_m, S_{-m} is equal to $2^{-\alpha m}$.
This domain was introduced in the
book of Courant and Hilbert [1] for
$\alpha = 4$, as an example of a domain
for which the Poincaré inequality
(1.1) is not valid for $p = q = 2$.
Maz'ya's criterion allows to show
that Poincaré's inequality is valid
(and, consequently, the Neumann
problem for $-\Delta u = f$ is solvable in
$L_2^1(\Omega)$ for any $f \in L_2(\Omega)$ orthogonal to unity) if and only if $\alpha \le 3$.

The discreteness of the spectrum for the operator $-\Delta$ holds if and only if $\alpha < 3$
(see [34, 48]).

---

[2] $W_2^l(\Omega)$ denotes the Sobolev space of all functions, which are square integrable together with
all derivatives up to order l over the domain Ω.

Spectral theory of the Schrödinger operator. Of special mention is Maz'ya's paper [40] of 1964 concerning the multidimensional Schrödinger operator with zero Dirichlet data generated by the quadratic form

$$\int_\Omega |\nabla u|^2\, dx - \int_\Omega |u|^2\, P(dx),$$

where Ω is an arbitrary open set in $\mathbb{R}^n$, $n > 2$, and P is a non-negative measure. A short announcement appeared in 1962 [35]. Unfortunately, this work, rich with ideas, was not translated into English and, therefore, is never referred to by experts in the spectral theory of the Schrödinger operator.

To my knowledge, it was for the first time that a measure P, not necessarily continuous with respect to the n-dimensional Lebesgue measure, had been used as a potential. However, this is useful for many applications. Maz'ya introduced the notion of the absolute continuity of a measure with respect to a capacity, and proved that the validity of this property is a necessary and sufficient condition for the closability of the quadratic form

$$\int_\Omega |u|^2\, P(dx)$$

with respect to the Dirichlet integral. In [35, 40], final results on the positivity, semiboundedness and the structure of the negative spectrum of the Schrödinger operator were given.

Uniqueness of solution to the Dirichlet problem. In his first works, Maz'ya was also interested in conditions ensuring the unique solvability of the Dirichlet problem. Let A be a matrix such that $\xi A\xi \geq \kappa|\xi|^2$ with a positive constant κ for all $\xi \in \mathbb{R}^n$. One of the conditions obtained by Maz'ya [50] for the uniqueness in $L_2^1(\Omega)$ of the solutions of the Dirichlet problem for the equation

$$(1.6) \qquad \mathcal{L}u = -\nabla(A\nabla)u + \mathbf{a}\nabla u + bu = f,$$

$f \in L_{2n/(n+2)}(\Omega)$, in a domain Ω with finite measure is the inequality

$$\sup\mathrm{vrai}\, b^- < \lambda \exp\left(-\frac{c}{\kappa}\left(\mathrm{mes}_n(\Omega)\right)^{n^{-1}-r^{-1}} \|\mathbf{a}\|_r\right).$$

Here b^- denotes the negative part of b, λ is the first eigenvalue of the Dirichlet problem for the operator $-\nabla(A\nabla)$, $\|\cdot\|_r$ is the norm in $L_r(\Omega)$, $r > n$, and c is a constant depending only on n and r.

Solutions with the infinite energy integral. In 1961, Maz'ya [32] considered solutions whose gradient is not square integrable. If, for example, f belongs to the space $L_p(\Omega)$, $1 < p < 2n/(n+2)$, then, in general, the Dirichlet problem for (1.6) does not have a solution with finite energy integral. For this case Maz'ya

[50] proved the unique solvability of the Dirichlet problem in the space $L_q^1(\Omega)$ with $q = pn/(n-p) < 2$, which in the agreement with Sobolev's embedding theorem.

While these properties of the Dirichlet problem are valid for any Ω of finite volume, the Neumann problem is more sensitive with respect to the boundary regularity. If the boundary is too bad, the Neumann problem may have no solution in $L_q^1(\Omega)$ for arbitrarily small $q > 0$. It is proved in Maz'ya [50], that for any domain Ω having finite volume, the existence and uniqueness of a generalized solution such that

$$\int_\Omega M(|\nabla u|)\,dx < \infty,$$

where $M(x)$ is a convex function depending only on the domain Ω and tending to ∞ as $x \to +\infty$. The boundary $\partial\Omega$ is characterized by the asymptotic behaviour of the function

$$\lambda(t) = \inf_{\{E\}} \mathrm{mes}_{n-1}(\Omega \cap \partial E)$$

as $t \to +0$. Here E is any subset of Ω, such that the part of its boundary interior to Ω is smooth, and $t \le \mathrm{mes}_n(E) \le \frac{1}{2}\,\mathrm{mes}_n(\Omega)$. Now, $\lambda(t)$ is called 'isoperimetric function' and is often used in the theory of Riemannian manifolds. In a similar way, Maz'ya treated the solvability of the Neumann problem in $L_q^1(\Omega)$, $0 < q \le 2$, and obtained $\lambda(t)$-dependent conditions for estimates of the maximum modulus norm and $L_s(\Omega)$-norms to hold. He also demonstrated that his conditions on the boundary were precise in a certain sense.

Higher-order elliptic equations. In the article [37] of 1963, Maz'ya studied the Dirichlet problem for the polyharmonic equation in the energy space. He derived necessary and sufficient conditions on the domain ensuring the unique solvability for all right-hand sides in $L_p(\Omega)$. In [37, 42], he described all domains where this problem has the discrete spectrum, which generalizes Molchanov's criterion for the Laplace operator. These results were obtained using a new important notion of polyharmonic capacity defined in the same article. This capacity, as was shown by Maz'ya later on, is equivalent to a potential theoretic M. Riesz capacity (see Hedberg's article in this volume). Generally speaking, this area of Maz'ya's studies was a part of his new approach to the theory of Sobolev spaces presented in his famous book [58]. I would like to cite Maz'ya's words from the introduction to his Encyclopaedia paper [59]:

> "In the past thirty years there has accumulated a large amount of information about conditions which are necessary and sufficient for various properties of spaces of Sobolev type to hold true. It is question of boundedness and compactness criteria for embedding operators characterizing the domain or the weight functions, of tests for the possibility of extending functions from the domain to $\mathbb{R}^n$, of conditions asserting the density of one space of differentiable functions in another etc. An adequate description of the properties of function spaces has

made it necessary to introduce new classes of domains of definition for the functions, or classes of measures entering in the norms. In this connection, the universal importance of the notion of capacity of a set became manifest."

Studying Maz'ya's papers of 1960-1970s one arrives at a conclusion that this 'large amount of information' as well as understanding of 'universal importance of the notion of capacity' results, to a great extent, from his work.

1.2.　Strong solutions of the Dirichlet problem

Now I turn from weak solutions to basic properties of strong ones which also became a subject of Maz'ya's work in the 1960s. I mean the questions of the L_2-coercivity, of boundedness of the first derivatives of solution, and selfadjointness of the Dirichlet problem.

Solvability of the Dirichlet problem in the Sobolev space $W_2^2(\Omega)$. The boundedness in L_2 of the second derivatives of solutions to the Dirichlet problem for second order elliptic equations in domains with regular boundary were proved already in the first half of the 20th century by S. Bernstein and J. Schauder. This result does not hold if the boundary is not smooth enough. In his D.Sc. thesis of 1965 and in the papers [45, 46], Maz'ya considered domains, whose boundaries belong to the class C^1 but are not in C^2. He derived conditions on $\partial\Omega$, ensuring that the solution $u \in \overset{\circ}{W}_2^1(\Omega)$ [3] is in $W_2^2(\Omega)$. Here, I restrict myself to the Dirichlet problem in a plane domain. The multidimensional case was investigated by Maz'ya in [53].

Let Ω be a plane domain with boundary $\partial\Omega$ of class C^1, and having the finite variation of rotation and continuous curvature outside of a point O, where the origin of the coordinate system is placed. We consider the Dirichlet problem for the elliptic equation

$$(1.7) \qquad \mathcal{L}u = a\,\frac{\partial^2 u}{\partial x_1^2} + 2b\,\frac{\partial^2 u}{\partial x_1 \partial x_2} + c\,\frac{\partial^2 u}{\partial x_2^2} = f \quad \text{in } \Omega,$$

In order to formulate Maz'ya's result, I need the following notation. Let $\partial\Omega$ be parametrized by the arc length l, and $l = 0$ at O; let $t(l)$ denote the tangent vector to $\partial\Omega$. If F is a closed subset of $\partial\Omega$, then l_F denotes the length of F, τ_F the rotation of $\partial\Omega$ on F, and τ_F^- the negative part of the rotation. Furthermore, let U_ε be the arc $\{x \in \partial\Omega : |x| < \varepsilon\}$.

It is shown by Maz'ya in [46], that the condition

$$(1.8) \qquad \lim_{\varepsilon \to 0}\sup_{F \subset U_\varepsilon} \tau_F^- \log\frac{1}{l} < \frac{\pi}{2}\,\frac{(\det A(O))^{1/2}}{(t(0),\,A(O)\,t(0))}$$

---

[3] $\overset{\circ}{W}_2^1(\Omega)$ denotes the closure in $W_2^1(\Omega)$ of the set of all smooth functions having compact support in Ω

(here A denotes the matrix of the coefficients of $\mathcal{L}$, which, in this statement, are assumed to be continuous) ensures the existence of a unique solution $u \in W_2^2(\Omega) \cap \overset{\circ}{W}{}_2^1(\Omega)$ to equation (1.7) for arbitrary $f \in L_2(\Omega)$. This solution satisfies the estimate

$$(1.9) \qquad \|u\|_{W_2^2(\Omega)} \le c \, \|f\|_{L_2(\Omega)}$$

with a constant c independent of f. In particular, if the left side in (1.8) is zero, then the above result is true for every elliptic operator $\mathcal{L}$ with continuous coefficients. Furthermore, Maz'ya showed that estimate (1.8) is sharp in the following sense. For an arbitrary elliptic operator $\mathcal{L}$ of the second order with constant coefficients one can find a domain Ω satisfying condition (1.8), where '$<$' is replaced by '$=$', and a function $f \in C^\infty(\overline{\Omega})$, such that the solution $u \in \overset{\circ}{W}{}_2^1(\Omega)$ of (1.7) does not belong to $W_2^2(\Omega)$.

Boundedness of the gradient. Maz'ya's article [49] on the boundedness of the gradient of the solution to the Dirichlet problem for the elliptic equation (1.7) with measurable bounded coefficients is related to the paper [46] just cited. As in [46], Maz'ya considered a contour of class C^1, having continuous curvature outside the point O, and proved that for every solution $u \in W_2^2(\Omega) \cap \overset{\circ}{W}{}_2^1(\Omega)$ of (1.7), where $f \in L_p(\Omega)$, $p > 2$, the estimate

$$\operatorname*{ess\,sup}_{\Omega} |\nabla u| \le C \, \|f\|_{L_p(\Omega)}$$

holds with a constant C independent of u, if

$$\int_0 \alpha(l) \, \frac{dl}{l} < \infty,$$

where $\alpha(l) = \sup\{\tau_F^- : F \subset U_\varepsilon,\ l_F \le l\}$.

The index of the closure of the Dirichlet Laplacian. Let $\bar{\Delta}$ be the closure of the operator $\Delta : W_2^2(\Omega) \times \overset{\circ}{W}{}_2^1(\Omega)$ in $L_2(\Omega)$, and let $\tilde{\Delta}$ be its Friedrichs' extension. If $\partial\Omega$ is a sufficiently smooth curve, then the operators Δ, $\bar{\Delta}$ and $\tilde{\Delta}$ do coincide. Generally speaking, this is not true, if $\partial\Omega \notin C^2$.

Using the notion of *conductivity* (see Subsection 1.1), Maz'ya established conditions sufficient for selfadjointness of the closure of the Laplace operator, i.e., for the equality $\bar{\Delta} = \tilde{\Delta}$ in the case $\partial\Omega \in C^1$, $\partial\Omega \backslash \{O\} \in C^2$. He proved (see [45, Th.2]) that the condition

$$(1.10) \qquad \int_0^\delta \exp\left(\frac{2\pi}{\operatorname{cond}(K_\rho)}\right) \rho \, d\rho = \infty$$

(here K_ρ is the 'conductor' $\Omega_\delta \backslash \Omega_\rho$, $\rho < \delta$, where δ is small, and $\Omega_r = \{x \in \Omega : |x| < r\}$) is necessary and sufficient for the selfadjointness of the operator $\bar{\Delta}$. If integral (1.10) converges, then $\operatorname{ind} \bar{\Delta} = 1$. The following geometric condition

$$\int_0^\delta \exp\left(\frac{2}{\pi} \int_\rho^\delta \frac{\pi - \omega(r)}{r}\right) \frac{d\rho}{\rho} = \infty,$$

where $\omega(r)$ is the angle measure of $\{x \in \Omega : |x| = r\}$, is sufficient for the selfadjointness.

Furthermore, Maz'ya proved in [55] that the latter condition for the self-adjointness of $\bar{\Delta}$ is also sufficient for vanishing of the index of the closure of an arbitrary second order elliptic operator $\mathcal{L}$ in a domain, having the C^2-boundary outside the origin (without assuming that $\partial\Omega \in C^1$). Under a certain additional assumption on the boundary, this condition is also necessary. If the condition is violated, then $\operatorname{ind} \bar{\mathcal{L}} = 1$.

The results mentioned above have some surprising consequences (see [41, 45, 46]). There exists a domain, having C^1-boundary, such that the closure of Δ is self-adjoint, but the estimate (1.9) does not hold, i.e., $\bar{\Delta} = \tilde{\Delta}$ but $\Delta \neq \bar{\Delta}$. The closure of the operator $(1 + \varepsilon)\partial_{x_1}^2 + \partial_{x_2}^2$, $\varepsilon = \text{const} > 0$, can be non-selfadjoint, whilst for the same domain the generalized solution $u \in \overset{\circ}{W}_2^1(\Omega)$ of the equation $\Delta u = f$ satisfies (1.9) and, consequently, $\Delta = \bar{\Delta} = \tilde{\Delta}$. These facts follow from the above given necessary and sufficient condition for the self-adjointness of the operator Δ.

To complete Section 1.2 I would like to note for the reader familiar with the modern theory of elliptic boundary value problems for domains with singularities, that Maz'ya's results mentioned here may open a new rich field of research in this rapidly developing area.

1.3. Boundary behaviour of solutions to the Dirichlet problem

Maz'ya's interest in the behaviour of solutions near a boundary point had manifested itself in 1963, when he published his widely known 'capacitary' estimate for the continuity modulus of a second order elliptic equation. This estimate can be reduced to that for the function

$$I(r) = \left(\int_{S_r} |u(r, \omega)|^2 \, d\omega\right)^{1/2},$$

where $r = |x|$, $\omega = x/|x|$, and S_r denotes the projection of the set $\{x \in \Omega : |x| = r\}$ from the point $O \in \partial\Omega$ onto the unit sphere S^{n-1}. In [38], Maz'ya proved that the bounded function u, harmonic in the set $\Omega \subset \mathbb{R}^n$, $n > 2$, near the boundary point O and equal to zero on the boundary near O, satisfies the inequality

$$I(r) \leq I(R) \exp\left(-\frac{n-2}{n-1} \int_r^R \frac{\operatorname{cap}(B_\rho \backslash \Omega)}{\rho^{n-1}} \, d\rho\right),$$

where R is sufficiently small, $r < R$, B_ρ is the ball of radius ρ centered at O, and cap denotes the Wiener capacity. Maz'ya's inequality gives a new proof of sufficiency in the classical Wiener regularity criterion for the boundary point and has other important applications. It was extended by Maz'ya and others to various classes of linear and quasilinear equations, in particular, to the p-Laplace equation (see Hedberg's article in this book).

In 1967, Maz'ya and Verzhbinskiĭ [91] arrived at another and in some cases more precise way of estimating the function $I(r)$, which enabled them to develop an extensive asymptotic theory of the Dirichlet problem for the n-dimensional Laplacian in domains with various types of point boundary singularities. (Note that the Kondrat'ev's fundamental work on elliptic boundary value problems in domains with conical points also appeared in 1967.) Since the methods applied in [91] are of importance for later works of Maz'ya, I describe some results of this article in more detail. (The complete proofs appeared in [92].) First, Maz'ya and Verzhbinskiĭ considered the bounded and unbounded solutions $y_1(r)$ and $y_2(r)$ of the ordinary differential equation

$$(1.11) \qquad y'' + \frac{n-1}{r} y' - \frac{\lambda(r)}{r^2} y = 0,$$

where $\lambda(r) \geq 0$ such that $\lambda(r)(\lambda(r) + n - 2)$ is the first eigenvalue of the Dirichlet problem for the Beltrami operator in S_r. They demonstrated that, if the above harmonic function u is bounded near O, then

$$(1.12) \qquad I(r) \leq \frac{y_1(r)}{y_1(R)} I(R)$$

for small R and $r < R$. The function u, unbounded in any neighborhood of O, harmonic in Ω, and subject to the zero Dirichlet condition, satisfies the estimate

$$I(r) \geq \frac{y_2(r)}{y_2(R)} I(R).$$

The comparison principle (1.12) leads directly to estimates of Green's function $G(x, \xi)$, and the harmonic measure $H(x, E)$, $E \subset \partial\Omega$:

$$(1.13) \qquad G(x, \xi) \leq c \, \frac{y_1(\alpha|x|)}{y_1(\beta|\xi|) \, |\xi|^{n-2}} \quad \text{for } \alpha|x| \leq \beta|\xi|,$$

$$H(x, \partial\Omega \backslash B_\rho) \leq c \, \frac{y_1(\alpha|x|)}{y_1(\beta\rho)} \quad \text{for } \alpha|x| \leq \beta\rho,$$

where c, α and β are positive constants. These estimates give much interesting information about solutions of the Dirichlet problem, which is formulated in terms of the first eigenvalue of the Beltrami operator on the spherical set $\Omega \cap \partial B_r$. In particular, for special singular boundary points it is possible to describe the asymptotics of solutions explicitly using (1.13). Maz'ya and Verzhbinskiĭ considered, for

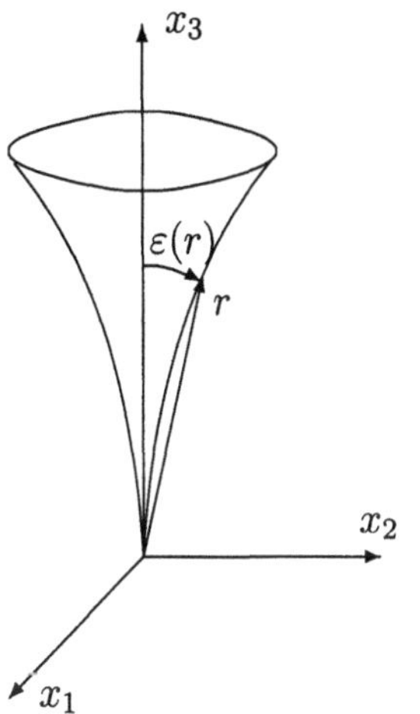

example, domains which are rotationally symmetric in a neighborhood of O, and have interior or exterior cusp, or are quasiconical. Here, I give only one result for the three-dimensional interior cusp $\{x : \varepsilon(r) < \theta \leq \pi\}$, where $\cos\theta = x_3/|x|$, and $\varepsilon(r)$ tends to 0 as $r \to 0$. If $r^{-1}(\log\varepsilon(r))^{-1} \notin L(0,1)$ and $r^{-1}(\log\varepsilon(r))^{-2} \in L(0,1)$, then there exists a solution of $\Delta u = f$, $f = 0$ near the point $x = 0$, which has the asymptotic representation

$$u(x) = r^{-1}\exp\left(\frac{1}{2}\int_r^1 \frac{d\rho}{\rho\,|\log\varepsilon(\rho)|}\right)\left(1 - \frac{\log\sin\theta/2}{\log\sin\varepsilon(r)/2} + o(1)\right).$$

Another interesting corollary of (1.13) is the first L_p coercivity result for the Dirichlet Laplacian in a domain, having a conical vertex at the boundary, which was formulated in the 1967 article by Maz'ya and Verzhbinskiĭ and proved (in the framework of more general context) in their paper [93]. Note that for a cone Ω with vertex O estimate (1.13) takes the simple form

$$(1.14) \qquad G(x,\xi) \leq c\,\frac{|x|^\lambda}{|\xi|^{n-2+\lambda}} \quad \text{for } |x| \leq |\xi|/2,$$

where λ is a positive number such that $\lambda(\lambda + n - 2)$ is the first eigenvalue of the Dirichlet problem for the Beltrami operator δ on $\Omega \cap S^{n-1}$.

An estimate similar to (1.12) was used by Maz'ya [52] in 1972 to generalize Beurling's minimum principle for positive harmonic functions in a plane domain. This extension is concerned with a bounded domain $\Omega \subset \mathbb{R}^n$ having $\partial\Omega \in C^{1,\alpha}$, $0 < \alpha < 1$, and with an elliptic operator $\mathcal{L}$ defined in (1.3), where $a_{i,j} \in C^{1,\alpha}(\overline{\Omega})$. To formulate the result we need the following definition. Let $O \in \partial\Omega$, and ν is the interior normal on $\partial\Omega$. A sequence $\{x_m\}_{m\geq 1}$, belonging to Ω and tending to O, is called *determining* if for every positive solution to $\mathcal{L}u = 0$ in Ω the inequality

$$u(x_m) \geq c\frac{\partial G}{\partial \nu}(x_m, O), \quad c > 0,$$

implies that

$$u(x) \geq c\frac{\partial G}{\partial \nu}(x, O) \quad \text{for all} \quad x \in \Omega.$$

Maz'ya proved that a necessary and sufficient condition for $\{x_m\}$ to be a determining sequence is that it contains a subsequence $\{x_{m_k}\}$, subject to

$$\inf_{k \neq l} \frac{|x_{m_k} - x_{m_l}|}{\mathrm{dist}(x_{m_k}, \partial\Omega)} > 0,$$

and such that

$$\sum_{k \geq 1} \left[\frac{\mathrm{dist}(x_{m_k}, \partial\Omega)}{|x_{m_k}|} \right]^n = \infty.$$

In a short article [57] of 1977, Maz'ya formulated deep results on the Martin boundary and the minimal positive harmonic functions for a class of n-dimensional domains. Roughly speaking, the question is related to the uniqueness of a positive harmonic function equal to zero on the boundary everywhere, except for one point. The classical example of Bouligand (1931) implies that there are infinitely many such functions in a domain bounded by two tangent spheres. This example became the starting point for Maz'ya who considered solids of revolution bounded by two surfaces tangent at O. He stated that the non-uniqueness (in other words, a difference in the Martin and Euclidian topologies) arises if and only if the domain is thin near the point 0 in the following sense:

$$\int_0 \frac{\Phi(\rho)\,d\rho}{\left(\pi - 2|\Psi(\rho)|\right)^2 \rho} < \infty,$$

where Φ and Ψ are sufficiently regular functions in the inequality

$$2\,|\theta - \Psi(\rho)| < \Phi(\rho)$$

describing the domain near O. By ρ and θ the spherical coordinates are denoted: $\rho = |x|$, $\cos\theta = x_n/|x|$, $0 < \theta < \pi$.

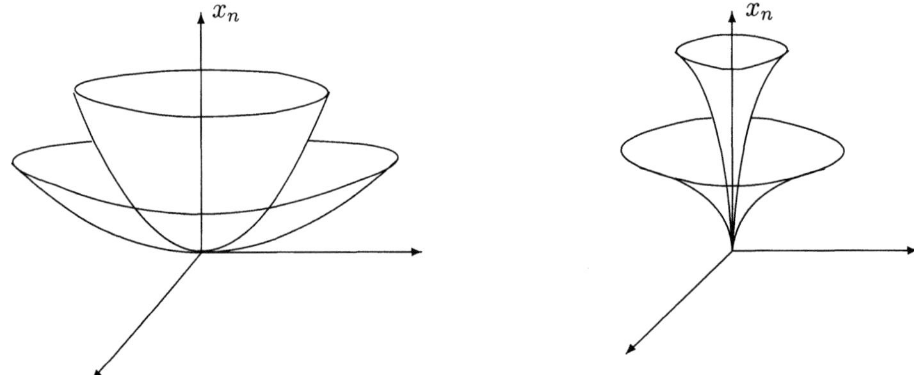

Although the article [57] contains a rather detailed description of main steps of the argument, the complete proof of the results has not been published yet.

2. General elliptic boundary value problems in domains with point singularities

Maz'ya's works have played an outstanding role in the development of the theory of general elliptic boundary value problems in domains with singularities (angular, conical and cuspidal points, edges, polyhedral vertices). It started as follows. In 1967, a comprehensive paper [3] of V. A. Kondrat'ev was published, where the Fredholm property in weighted L_2 Sobolev spaces, and the asymptotics of solutions for elliptic problems in domains with angular or conical points were obtained. By that time, Maz'ya had much experience in treating the behaviour of solutions to the second order elliptic equations near 'bad' boundaries. So he immediately recognized perspectives of further research opened by Kondrat'ev's theory, which, despite being rich with interesting results, was not mature yet for solving particular applied problems, especially nonlinear ones.

In 1970s, he together with B. A. Plamenevskiĭ launched a massive attack on the general elliptic theory for domains with piecewise smooth boundaries. I restrict myself in this section to their results and some subsequent work to elliptic problems in domains with isolated singularities. One of the goals of Maz'ya and Plamenevskiĭ was to extend Kondrat'ev's results to other function spaces (L_p Sobolev spaces, Hölder classes, spaces with inhomogeneous norms). They also obtained the Fredholm and regularity theorems for general elliptic problems in domains with other kinds of boundary singularities (quasiconical points, cuspidal points, quasicylindrical domains). Their formulas for the coefficients in the asymptotics of solutions near conical points are of great importance for applications. Moreover, they studied Green's functions and Poisson's kernels near conical points.

Their estimates of Green's functions became a useful tool for the subsequent study of the Miranda-Agmon maximum principle by Maz'ya and Roßmann (1992), who gave a complete answer to the question of its validity for domains with conical points. Finally, I mention a comparison principle by Kozlov and Maz'ya (1985-1988), which represents a new and original approach to the elliptic theory for nonsmooth domains.

2.1. Solvability of elliptic problems in weighted Sobolev and Hölder spaces

Elliptic problems in domains with conical points. The solvability question for various function spaces is one of the central questions in the theory of elliptic boundary value problems. I start with the case of a domain with conical points. In the previous section, I already mentioned the L_p coercivity result for the Dirichlet Laplacian obtained by Maz'ya and Verzhbinskiĭ in 1967. In the paper [75] of 1978, Maz'ya and Plamenevskiĭ obtained weighted L_p and Hölder estimates as well as Fredholm theorems for general elliptic boundary value problems in domains with conical points.

Let, for simplicity, $\mathcal{G}$ be a bounded domain in $\mathbb{R}^n$, having one conical point at the origin, i.e., $\mathcal{G}$ coincides with the cone

$$(2.1) \qquad \mathcal{K} = \{x \in \mathbb{R}^n : 0 < r = |x| < \infty, \; \omega = x/|x| \in \Omega\}$$

in a neighborhood of the origin. Here Ω is a subdomain of the unit sphere S^{n-1} with a smooth boundary.

By $V^l_{p,\beta}(\mathcal{G})$ and $\Lambda^{l,\alpha}_\beta(\mathcal{G})$ one means the weighted Sobolev and Hölder spaces respectively supplied with the norms

$$(2.2) \qquad \|u\|_{V^l_{p,\beta}(\mathcal{G})} = \left(\int_{\mathcal{G}} \sum_{|\gamma| \le l} r^{p(\beta - l + |\gamma|)} |D^\gamma_x u|^p \, dx \right)^{1/p},$$

$$(2.3) \qquad \|u\|_{\Lambda^{l,\alpha}_\beta(\mathcal{G})} = \sup_{x \in \mathcal{G}} \sum_{|\gamma| \le l} |x|^{\beta - l - \alpha + |\gamma|} \, |D^\gamma_x u(x)|$$

$$+ \sup_{x,y \in \mathcal{G}} |x - y|^{-\alpha} \sum_{|\gamma| = l} \left| |x|^\beta D^\gamma_x u(x) - |y|^\beta D^\gamma_y u(y) \right|.$$

Here l is a nonnegative integer and p, α, β are real numbers, $1 < p < \infty, 0 < \alpha < 1$. Let $V^{l-1/p}_{p,\beta}(\mathcal{G})$ and $\Lambda^{l,\alpha}_\beta(\partial\mathcal{G})$ be the corresponding trace spaces on $\partial\mathcal{G} \setminus \{0\}$.

We consider the boundary value problem

$$(2.4) \qquad \mathcal{L}u = f \quad \text{in } \mathcal{G}, \qquad \mathcal{B}u = g \quad \text{on } \partial\mathcal{G} \setminus \{0\},$$

where $\mathcal{L}$ is an elliptic differential operator of order $2m$ and $\mathcal{B}$ is a vector of differential operators $\mathcal{B}_j$ of orders m_j, $j = 1, \dots, m$, which cover $\mathcal{L}$ on $\partial\mathcal{G} \setminus \{0\}$. Suppose, for simplicity, that the coefficients of $\mathcal{L}$ and $\mathcal{B}_j$ are smooth on $\overline{\mathcal{G}}$. (In fact, this assumption can be essentially weakened. In [75], Maz'ya and Plamenevskiĭ considered boundary value problems for general elliptic systems, where all differential operators are so-called admissible operators.) Extending Kondrat'ev's result for $p = 2$ [3], they showed that the operator

$$(2.5) \qquad V^l_{p,\beta}(\mathcal{G}) \ni u \to (\mathcal{L}u, \mathcal{B}u) \in V^{l-2m}_{p,\beta}(\mathcal{G}) \times \prod_{j=1}^{m} V^{l-m_j-1/p}_{p,\beta}(\partial\mathcal{G})$$

is Fredholm for $l \ge 2m$, $l > \max m_j$, $p \in (1, \infty)$, if the line $\operatorname{Im} \lambda = \beta - l + n/p$ does not contain eigenvalues of a certain operator pencil $\mathfrak{A}(\lambda)$. This operator pencil appears if one represents the principal parts of the operators $\mathcal{L}$ and $\mathcal{B}_j$ with coefficients frozen at $x = 0$, in spherical coordinates:

$$\mathcal{L}^\circ = r^{-2m} L(\omega, D_\omega, rD_r), \qquad \mathcal{B}^\circ_j = r^{-m_j} B_j(\omega, D_\omega, rD_r),$$

and then applies the Mellin transform $r \to \lambda$ to L and B_j. This leads to the parameter depending boundary value problem

$$L(\omega, D_\omega, \lambda) u = f \quad \text{in } \Omega, \qquad B_j(\omega, D_\omega, \lambda) u = g_j \quad \text{on } \partial\Omega, \; j = 1, \dots, m.$$

The operator of this problem is denoted by $\mathfrak{A}(\lambda)$.

An analogous result is true for the weighted Hölder spaces. More precisely, the operator

$$\Lambda_\beta^{l,\alpha}(\mathcal{G}) \ni u \to (\mathcal{L}u, \mathcal{B}u) \in \Lambda_\beta^{l-2m,\alpha}(\mathcal{G}) \times \prod_{j=1}^{m} \Lambda_\beta^{l-m_j,\alpha}(\partial\mathcal{G})$$

is Fredholm if and only if the line $\operatorname{Im}\lambda = \beta - l - \alpha$ is free of eigenvalues of the pencil $\mathfrak{A}$. Furthermore, Maz'ya and Plamenevskiĭ obtained the following index formula for the operator (2.5). If $\mathcal{A}_\beta$ denotes the operator (2.5), where l is a given integer, $l \geq 2m$, $l > \max m_j$, and $p \in (1,\infty)$ is a given real number, and if the lines $\operatorname{Im}\lambda = \beta - l + n/p$ and $\operatorname{Im}\lambda = \gamma - l + n/p$, $\beta > \gamma$, do not contain eigenvalues of the pencil $\mathfrak{A}$, then

$$\operatorname{ind}\mathcal{A}_\beta = \operatorname{ind}\mathcal{A}_\gamma + \kappa,$$

where κ is the sum of the algebraic multiplicities of all eigenvalues in the strip $\gamma - l + n/p < \operatorname{Im}\lambda < \beta - l + n/p$.

In the recent book [24] of Kozlov, Maz'ya and Roßmann, these results were extended to weighted L_2 Sobolev spaces of an arbitrary integer (not necessarily positive) order.

Solvability in weighted spaces with inhomogeneous norms. The introduced above classes of the weighted spaces $V_{p,\beta}^l$ and $\Lambda_\beta^{l,\alpha}$ do not contain the usual Sobolev and Hölder spaces. One way to obtain Fredholm theorems in the usual Sobolev spaces W_2^k is to study relations between these spaces and $V_2^0(\mathcal{G})$. This way was adopted, for example, by Kondrat'ev [3]. Maz'ya and Plamenevskiĭ [71] studied the weighted Sobolev and Hölder spaces $W_{p,\beta}^l$ and $C_\beta^{l,\alpha}$ with so-called inhomogeneous norms

$$\|u\|_{W_{p,\beta}^l(\mathcal{G})} = \left(\int_{\mathcal{G}} \sum_{|\alpha|\leq l} |x|^{p\beta} |D^\alpha u|^p dx \right)^{1/p},$$

$$\|u\|_{C_\beta^{l,\alpha}(\mathcal{G})} = \sup_{x\in\mathcal{G}} \sum_{|\gamma|\leq l} |x|^{\beta} |D_x^\gamma u(x)|$$
$$+ \sup_{x,y\in\mathcal{G}} |x-y|^{-\alpha} \sum_{|\gamma|=l} \left| |x|^{\beta} D_x^\gamma u(x) - |y|^{\beta} D_y^\gamma u(y) \right|$$

which coincide with the usual Sobolev and Hölder spaces for $\beta = 0$. They proved that, for noninteger $\beta + n/p$ and $\beta - \alpha$, these spaces are direct sums of $V_{p,\beta}^l$ and $\Lambda_\beta^{l,\alpha}$, respectively, and a set of homogeneous polynomials. For $\beta < -n/p$ or $\beta > l - n/p$ the spaces $W_{p,\beta}^l$ and $V_{p,\beta}^l$ coincide. The same is true for the weighted Hölder spaces $C_\beta^{l,\alpha}$ and $\Lambda_\beta^{l,\alpha}$, if $\beta < 0$ or $\beta > l + \alpha$. Using these relations, Maz'ya and Plamenevskiĭ obtained Fredholm theorems for general elliptic problems in the classes of the spaces $W_{p,\beta}^l$ and $C_\beta^{l,\alpha}$.

Quasiconical and cuspidal points. The above mentioned paper [75] of Maz'ya and Plamenevskiĭ is not restricted to elliptic boundary value problems in domains with conical points. It contains also boundary value problems in domains with other singular boundary points which can be transformed by suitable coordinate changes to boundary value problems on cylindrical domains. I give here only two examples of domains which are included there.

1) *Quasicylindrical domains.* This means that $\mathcal{G}$ is a domain in $\mathbb{R}^n$ with smooth
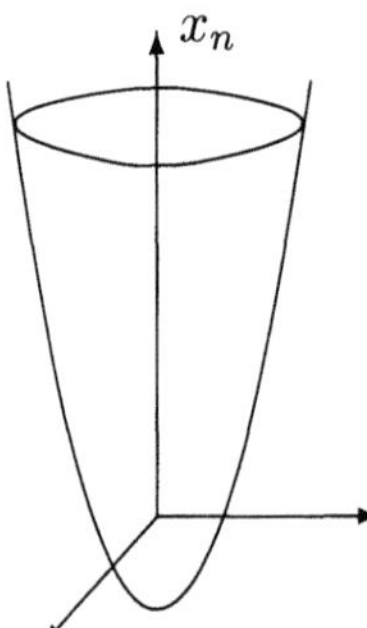
boundary $\partial\mathcal{G}$, such that $\{x \in \mathcal{G} : x_n < 1\}$ is bounded, and $\{x \in \mathcal{G} : x_n > 1\}$ can be transformed by the mapping

$$y_j = \frac{h_j(x)}{\varphi(x_n)}, \quad j = 1,\dots,n-1, \quad y_n = \int_1^{x_n} \frac{dt}{\varphi(t)}$$

into the half-cylinder

$$\{y = (y', y_n) : y' = (y_1,\dots,y_{n-1}) \in \Omega, \ y_n > 0\}.$$

Here φ is a positive function on $[1, \infty)$, satisfying the conditions

$$\lim_{t\to\infty} \varphi(t)^{k-1}\, \varphi^{(k)}(t) < \infty \quad \text{and} \quad \int_1^\infty \frac{dt}{\varphi(t)} = \infty,$$

and h_j are functions, such that $\det \left(\dfrac{\partial h_j}{\partial x_k} \right)_{j,k=1}^{n-1} \geq$ const > 0, and there exist the finite limits (uniform with respect to $y' \in \Omega$) $\lim\limits_{x_n\to\infty} \varphi(x_n)^{|\alpha|-1}\, D_x^\alpha h_j(x)$ for all multi-indices α.

2) *Domains of exterior cusp type.* This means that $\mathcal{G}$ is a bounded domain in
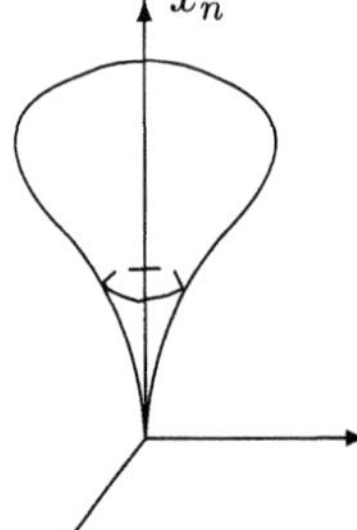
$\mathbb{R}^n$ with boundary $\partial\mathcal{G}$ containing the origin, and such that $\partial G\backslash\{0\}$ is smooth, and $\{x \in \mathcal{G} : x_n < 1\}$ can be transformed by the mapping

$$y_j = \frac{h_j(x)}{\varphi(x_n)}, \quad j = 1,\dots,n-1, \quad y_n = \int_{x_n}^1 \frac{dt}{\varphi(t)}$$

into the half-cylinder

$$\{y = (y', y_n) : y' = (y_1,\dots,y_{n-1}) \in \Omega, \ y_n > 0\}.$$

Here φ is a positive function on $(0, 1]$, satisfying the conditions

$$\lim_{t \to 0} \varphi(t)^{k-1} \, \varphi^{(k)}(t) < \infty \quad \text{and} \quad \int_0^1 \frac{dt}{\varphi(t)} = \infty,$$

and the functions h_j satisfy the same assumptions as in the previous example (here $\lim_{x_n \to \infty}$ has to be replaced by $\lim_{x_n \to 0}$).

For such domains Maz'ya and Plamenevskiĭ introduced weighted Sobolev spaces $V^l_{p,\beta,\gamma}(\mathcal{G})$ with the norm

$$\|u\|_{V^l_{p,\beta,\gamma}(\mathcal{G})} = \left(\int_{\mathcal{G}} e^{p\beta y_n(x_n)} \sum_{|\alpha| \le l} \varphi(x_n)^{p(\gamma-l+|\alpha|)} \, |D^\alpha u(x)|^p \, dx \right)^{1/p},$$

and the corresponding trace spaces $V^{l-1/p}_{p,\beta,\gamma}(\partial\mathcal{G})$. They proved that (under certain conditions on the differential operators and on the eigenvalues of a certain operator pencil) the operator

$$V^l_{p,\beta,\gamma}(\mathcal{G}) \ni u \to (\mathcal{L}u, \mathcal{B}u) \in V^{l-2m}_{p,\beta,\gamma}(\mathcal{G}) \times \prod_{j=1}^m V^{l-m_j-1/p}_{p,\beta,\gamma}(\partial\mathcal{G})$$

of problem (2.4) is a Fredholm operator. Moreover, the paper [75] contains regularity assertions for the solutions. Analogous results were obtained for weighted Hölder spaces.

Elliptic problems in domains having inward cusps. The situation of a multidimensional inward cusp is more complicated and, up to now, there are only few works concerning elliptic problems in such domains. I have mentioned above the papers [91, 92] of Maz'ya and Verzhbinskiĭ, where the Dirichlet problem for the second order elliptic equations in domains with singular boundary points, including vertices of interior cusps, was considered. A breakthrough in the theory of general elliptic problems in domains with inward cusps was the result by Maz'ya, Nazarov and Plamenevskiĭ of 1984. Their paper [61] contains Fredholm theorems in weighted Sobolev spaces and regularity assertions for the solutions.

Here, I formulate the main results of this paper for the particular case of the Dirichlet problem for an elliptic differential operator $\mathcal{L}$ of the order $2m$ in a bounded domain $\mathcal{G}$ with the following properties:

(i) $\partial\mathcal{G}\backslash\{0\}$ is smooth,

(ii) the set $\{x \in \mathcal{G} : |x| = \rho\}$ coincides with $\{x = (x', x_n) : |x| = \rho, \, x'/\rho\delta(\rho) \in \tilde{\Omega}\}$ for small ρ.

Here $\tilde{\Omega}$ is a domain in $\mathbb{R}^{n-1}$, and δ is a positive function on $\mathbb{R}_+$ such that $\rho^j \delta^{(j)}(\rho) = o(\delta(\rho))$, $j \geq 1$, and $\delta(\rho) \leq \delta_0$, where δ_0 is a sufficiently small number. Let $\mathcal{V}_{2,\beta,\gamma}^l(\mathcal{G})$ be the weighted Sobolev space with the norm

$$\|u\|_{\mathcal{V}_{2,\beta,\gamma}^l(\mathcal{G})} = \Big(\sum_{|\alpha| \leq l} \int_{\mathcal{G}} |x|^{2(\gamma-l+|\alpha|)} \, \theta^{2(\beta-l+|\alpha|)} \, |D_x^\alpha u|^2 \, dx \Big)^{1/2},$$

where θ denotes the angle between x and the x_n-axis. Let $\mathcal{V}_{2,\beta,\gamma}^{l-1/2}(\partial\mathcal{G})$ be the corresponding trace space.

By [61], the operator of the Dirichlet problem for $\mathcal{L}$ realizes an isomorphism

$$\mathcal{V}_{2,\beta,\gamma}^{2m}(\mathcal{G}) \to \mathcal{V}_{2,\beta,\gamma}^0(\mathcal{G}) \times \prod_{j=1}^{m} \mathcal{V}_{2,\beta,\gamma}^{2m-j+1/2}(\partial\mathcal{G}),$$

if the coefficients of $\mathcal{L}$ are infinitely differentiable, Gårding's inequality

$$\mathrm{Re}\, (\mathcal{L}u, u)_{L_2(\mathcal{G})} \geq c \|u\|_{W_2^m(\mathcal{G})}^2$$

is satisfied for all $u \in \overset{\circ}{W}{}_2^m(\mathcal{G})$, $2m < n-1$, $\beta \in \big(2m - (n-1)/2, (n-1)/2\big)$, and $\gamma \in (2m - n/2, n/2)$. Furthermore, the following regularity result for the generalized solution $u \in \overset{\circ}{W}{}_2^m(\mathcal{G})$ of $\mathcal{L}u = f$ holds as a consequence of [61]: If $f \in \mathcal{V}_{2,\beta,\gamma}^0(\mathcal{G})$, $2m < n-1$, $\beta \in \big(2m - (n-1)/2, (n-1)/2\big)$, $\gamma \in (2m - n/2, n/2)$, then $u \in \mathcal{V}_{2,\beta,\gamma}^{2m}(\mathcal{G})$.

2.2. Asymptotics of solutions

Asymptotics near conical points. In [3], Kondrat'ev obtained the following asymptotic formula for the solution $u \in V_{2,\beta}^l(\mathcal{G})$ of (2.4) near the conical point O:

$$(2.6) \quad u = \sum_{\mu=1}^{N} \sum_{\rho=1}^{I_\mu} \sum_{\sigma=0}^{\kappa_{\rho,\mu}-1} c_\mu^{\sigma,\rho} \, r^{i\lambda_\mu} \sum_{s=0}^{\sigma} \frac{1}{s!} (i \log r)^s \, \varphi_\mu^{\sigma-s,\rho}(\omega) + v, \quad v \in V_{2,\gamma}^l(\mathcal{G}),$$

provided $f \in V_{2,\gamma}^{l-2m}(\mathcal{G})$, $g_j \in V_{2,\gamma}^{l-m_j-1/2}(\partial\mathcal{G})$, $\beta - 1 < \gamma < \beta$, and the lines $\mathrm{Im}\,\lambda = \beta - l + n/2$, $\mathrm{Im}\,\lambda = \gamma - l + n/2$ do not contain eigenvalues of the pencil $\mathfrak{A}$. Here $c_\mu^{\sigma,\rho}$ are certain constants, $\lambda_1, \ldots, \lambda_N$ are the eigenvalues of $\mathfrak{A}$ within the strip $\gamma - l + n/p < \mathrm{Im}\,\lambda < \beta - l + n/p$, and $\{\varphi_\mu^{s,\rho} : \rho = 1, \ldots, I_\mu, \; s = 0, \ldots, \kappa_{\rho,\mu}\}$ is a canonical system of Jordan chains corresponding to λ_μ. If $\gamma < \beta - 1$, then the asymptotics of u contains additional terms of a more complicated form.

Maz'ya and Plamenevskiĭ [72] proved that the same asymptotic representation holds for solutions in the weighted L_p Sobolev and Hölder spaces. Moreover, they derived formulas for $c_\mu^{\rho,\sigma}$, which are of importance for applications. (In fracture mechanics, the coefficients $c_\mu^{\rho,\sigma}$ are called stress intensity factors.) These formulas contain special solutions of the equation $\mathcal{A}^*(v, \vec{w}) = 0$, where $\mathcal{A}^*$ denotes the adjoint operator to $\mathcal{A} = (\mathcal{L}, \mathcal{B})$, or of the homogeneous formally adjoint problem. I give here only one of their formulas which is valid if $m_j < 2m$ for $j = 1, \ldots, m$, the system of the boundary operators $\mathcal{B}_j$ is normal on $\partial\mathcal{G}\backslash\{0\}$, and the Green formula

$$\int_{\mathcal{G}} \mathcal{L}u \cdot \bar{v} \, dx + \sum_{j=1}^{m} \int_{\partial\mathcal{G}} \mathcal{B}_j u \cdot \overline{\mathcal{T}_j v} \, dx = \int_{\mathcal{G}} u \cdot \overline{\mathcal{L}^+ v} \, dx + \sum_{j=1}^{m} \int_{\partial\mathcal{G}} \mathcal{S}_j u \cdot \overline{\mathcal{C}_j v} \, dx$$

is valid for all smooth functions u, v equal to zero near the point $x = 0$. Let $\mathfrak{A}^+(\lambda)$ denote the operator pencil generated by the formally adjoint boundary value problem

$$\mathcal{L}^+ u = f \text{ in } \mathcal{G}, \quad \mathcal{C}_j u = g_j \text{ on } \partial\mathcal{G}\backslash\{0\}, j = 1, \ldots, m,$$

and let $\{\psi_\mu^{\rho,\sigma}, \ \rho = 1, \ldots, I_\mu, \ s = 0, \ldots, \kappa_{\rho,\mu}\}$ be a canonical system of Jordan chains of the pencil $\mathfrak{A}^+$ corresponding to the eigenvalue $\bar{\lambda}_\mu + i(n - 2m)$ satisfying the so-called *biorthonormality condition*. Then for every μ, σ, ρ there exists a solution v of the homogeneous formally adjoint problem such that

$$v - r^{i\bar{\lambda}_\mu + 2m - n} \sum_{s=0}^{\kappa_{\mu,\rho} - \sigma - 1} \frac{(i \log r)^s}{s!} \psi_\mu^{\kappa_{\mu,\rho} - \sigma - s - 1, \rho}(\omega) \in V_{p', -\beta + l}^{2m}(\mathcal{G}), \ p' = p/(p - 1).$$

In the case when the kernel of the operator (2.5) is trivial, the coefficients $c_\mu^{\sigma,\rho}$ in the asymptotics of the solution $u \in V_{p,\beta}^l(\mathcal{G})$ can be directly expressed in terms of the above introduced function v and the right-hand side terms f and g in the boundary value problem (2.4). For this particular case, the coefficients formula in [72] (see also [68, 69]) takes the form

$$(2.7) \qquad c_\mu^{\sigma,\rho} = \left(f, iv\right)_{\mathcal{G}} + \sum_{j=1}^{m} \left(g_j, i\mathcal{T}_j v\right)_{\partial\mathcal{G}}.$$

In the book [24], similar coefficients formulas were obtained which are based on a modification of the classical Green formula and are applicable to all elliptic problems.

Stable asymptotics. Often the asymptotic representation of solutions near angular or conical points obtained by means of Kondrat'ev's theory is unstable under variation of the coefficients of the differential operators or the angle at the singular boundary point. Let us consider, for example, the Dirichlet problem

$$\Delta u = f \text{ in } \mathcal{G}, \qquad u|_{\partial\mathcal{G}} = 0$$

in a two-dimensional domain $\mathcal{G}$, which coincides with the angle

$$\mathcal{K} = \{x = (x_1, x_2) : 0 < \theta = \arg(x_1 + ix_2) < \alpha\}, \ \pi/3 < \alpha < 2\pi/3$$

in a neighborhood of the origin. If f is smooth and $\alpha \neq \pi/2$, then Kondrat'ev's theory yields the asymptotics

$$u = c_\alpha \, |x|^{\pi/\alpha} \, \sin \frac{\pi\theta}{\alpha} + f(0) \, |x|^2 \, \frac{1 - \cos 2\theta}{4} + f(0) \, \frac{\cos 2\alpha - 1}{4 \sin 2\alpha} \, |x|^2 \, \sin 2\theta + O(r^3)$$

near the vertex. This asymptotics is unstable for α close to $\pi/2$, since the coefficients c_α and $f(0)\,(\cos 2\alpha - 1)/(4\sin 2\alpha)$ tend to infinity as $\alpha \to \pi/2$. (When $\alpha = \pi/2$, there is an additional logarithmic term in the asymptotics.) During Maz'ya's stay in Maryland in 1988, I. Babuška posed the question whether it is possible to get a stable asymptotic representation for u, if α belongs to a neighborhood of $\pi/2$. There are not only numerical reasons to consider the question of stability of the asymptotics. This question is also of importance for the description of the behaviour of solutions near edges when the angle along the edge varies and assumes a critical value at a certain point on the edge. Asymptotic representations of solutions for this case were obtained by V. Maz'ya and J. Roßmann in [88] and, in a more abstract form, in papers of B.-W. Schulze, M. Costabel and M. Dauge.

An answer to the above question of Babuška was given in the paper [87] of Maz'ya and Roßmann. Here stable asymptotic formulas for solutions of the Dirichlet problem for second order differential equations were found by replacing the power-logarithmic terms in Kondrat'ev's asymptotics by divided differences of power functions. For the above example, the following asymptotic formula holds:

$$u = c'_\alpha \, r^{\pi/\alpha} \, \sin \frac{\pi\theta}{\alpha} + f(0) \, r^2 \, \frac{1 - \cos 2\theta}{4}$$
$$+ f(0) \, \frac{(2\alpha - \pi)\,(\cos 2\alpha - 1)}{4 \sin 2\alpha} \, \frac{r^2 \sin 2\theta - r^{\pi/\alpha} \sin(\pi\theta/\alpha)}{2\alpha - \pi} + u_1.$$

This representation of u contains the singular functions $r^{\pi/\alpha} \, \sin \dfrac{\pi\theta}{\alpha}$ and $\dfrac{r^2 \sin 2\theta - r^{\pi/\alpha} \sin(\pi\theta/\alpha)}{2\alpha - \pi}$. The coefficients c'_α and $f(0)\,\dfrac{(2\alpha - \pi)\,(\cos 2\alpha - 1)}{4 \sin 2\alpha}$ are bounded.

Asymptotics in quasicylindrical and cuspidal domains. Besides asymptotic formulas for solutions near conical points, Maz'ya was also interested in analogous formulas for other singular boundary points. I have already mentioned the papers [91, 92] of Maz'ya and Verzhbinskiĭ on the Dirchlet problem for the second order elliptic equations. In 1977, Maz'ya and Plamenevskiĭ [73] obtained asymptotic formulas for the solutions of the Dirichlet problem for an elliptic operator of the order $2m$ in quasicylindrical domain and near cuspidal points. Let $\mathcal{G}$ be a quasicylindrical domain coinciding with $\{x = (x', x_n) : x'/x_n^\alpha \in \Omega\}$, $\alpha < 1$, for $x_n > 1$. If u is a solution of the Dirichlet problem

$$\mathcal{L}(D_{x'}, D_{x_n})u = f \;\; \text{in } \mathcal{G}, \quad D_\nu^{j-1} u = 0 \;\; \text{on } \partial\mathcal{G}, \; j = 1, \ldots, m,$$

such that $u = o\big(e^{\beta_1 x_n^{1-\alpha}/(\alpha-1)}\big)$, where $f = 0$ for large x_n, the line $\operatorname{Im}\lambda = \beta_0$ contains only a simple eigenvalue λ_0 of the pencil $\mathcal{L}(D_{x'}, \lambda)$, and the strip $\beta_0 < \operatorname{Im}\lambda \leq \beta_1$ is free of eigenvalues, then

$$u(x', x_n) \sim C\, x_n^{\gamma}\, e^{i\lambda_0 x_n^{1-\alpha}/(1-\alpha)} \sum_{j=0}^{\infty} \varphi_j(x'/x_n^{\alpha})\, x_n^{(\alpha-1)j}.$$

Here γ is a certain constant, φ_0 is the eigenfunction of the pencil $\mathcal{L}(D_{x'}, \lambda)$ corresponding to λ_0, and φ_j are smooth functions on Ω. An asymptotic formula of the same type holds for the solution of the Dirichlet problem near a cuspidal point.

2.3. Estimates of Green's function

Green's functions play an important role in the elliptic theory both for smooth and nonsmooth domains. In several of his works, Maz'ya dealt with the singularities of these functions for special and general elliptic problems in domains with conical points and edges. To this set of works the papers [91, 92] of Maz'ya and Verzhbinskiĭ on the Dirichlet problem for second order elliptic equations belong, which I have already discussed in Section 1. In 1979, Maz'ya and Plamenevskiĭ obtained asymptotic formulas for Green's function of general elliptic boundary value problems near conical boundary points. These formulas are essentially based on Maz'ya's and Plamenevskiĭ's formulas for coefficients mentioned above and allow to estimate Green's function similarly to (1.13).

I consider here, for simplicity, only the case when problem (2.4) is uniquely solvable in $V_{p,\beta}^{2m}(\mathcal{G})$ for arbitrary $f \in V_{p,\beta}^{0}(\mathcal{G})$, $g_j \in V_{p,\beta}^{2m-m_j-1/p}(\partial\mathcal{G})$ and for certain p and β. These assumptions ensure (see [79, Th.3.5]) the existence of the unique solution $G(x, \xi)$ of the boundary value problem

$$\mathcal{L}(x, D_x)\, G(x, \xi) = \delta(x - \xi), \quad x, \xi \in \mathcal{G},$$
$$\mathcal{B}(x, D_x)\, G(x, \xi) = 0, \quad x \in \partial\mathcal{G}\backslash\{0\}, \ \xi \in \mathcal{G},$$

such that the function $x \to \eta(x, \xi)\, G(x, \xi)$ belongs to $V_{p,\beta+k}^{2m+k}(\mathcal{G})$ for each fixed $\xi \in \mathcal{G}$ and each integer $k \geq 0$. Here η is an arbitrary smooth function on $\mathcal{G} \times \mathcal{G}$ equal to zero in a neighborhood of the diagonal. The function G is called Green's function for problem (2.4). It is infinitely differentiable in $x, \xi \in \overline{\mathcal{G}}\backslash\{0\}$, $x \neq \xi$. Furthermore, the function $x \to \overline{G(\xi, x)}$ is a solution of the formally adjoint problem.

As a consequence of the asymptotic formula for Green's function obtained in [79], the following estimates of Green's function $G(x, \xi)$ hold for x, ξ in a neighborhood of the origin:

$$|D_x^{\alpha} D_{\xi}^{\gamma} G(x, \xi)| \leq c\, |x|^{-\Lambda_- - |\alpha| - \varepsilon}\, |\xi|^{\Lambda_- - n + 2m - |\gamma| + \varepsilon} \qquad \text{for } 2|x| < |\xi|,$$
$$|D_x^{\alpha} D_{\xi}^{\gamma} G(x, \xi)| \leq c\, |x|^{-\Lambda_+ - |\alpha| + \varepsilon}\, |\xi|^{\Lambda_+ - n + 2m - |\gamma| - \varepsilon} \qquad \text{for } |x| > 2|\xi|.$$

Here ε is an arbitrarily small positive number and $\Lambda_- < \operatorname{Im}\lambda < \Lambda_+$ is the widest strip in the complex plane which contains the line $\operatorname{Im}\lambda = \beta - 2m + n/2$ and is free of eigenvalues of the pencil $\mathfrak{A}$. For $|x|/2 < |\xi| < 2|x|$, there is the estimate

$$|D_x^\alpha D_\xi^\gamma G(x,\xi)| \le c\left(|x-\xi|^{2m-n-|\alpha|-|\gamma|} + |\xi|^{2m-n-|\alpha|-|\gamma|}\right)$$

if $|\alpha| + |\gamma| \ne 2m - n$, and the other one

$$|D_x^\alpha D_\xi^\gamma G(x,\xi)| \le c\left(\left|\log\frac{|x-\xi|}{|\xi|}\right| + 1\right)$$

if $|\alpha| + |\gamma| = 2m - n$.

2.4.   The Miranda-Agmon maximum principle

In 1958, C. Miranda (in the two-dimensional case) and in 1960 S. Agmon (in
the higher-dimensional case) proved the following generalization of the classical
maximum principle for harmonic functions: Every solution of a strongly elliptic
differential equation $\mathcal{L}u = 0$ of the order $2m$ in a smooth domain $\mathcal{G}$ satisfies the
estimate

$$(2.8) \qquad \|u\|_{W_\infty^{m-1}(\mathcal{G})} \le c\left(\sum_{j=0}^{m-1} \|D_\nu^j u\|_{W_\infty^{m-1-j}(\partial\mathcal{G})} + \|u\|_{L(S)}\right),$$

where c is a positive constant independent of u and S is a compact subset of
$\mathcal{G}$. (The last term in the right-hand side in (2.8) can be omitted if the Dirichlet
problem for the operator L is uniquely solvable.)

The estimates for Green's functions given in the previous subsection allow to
prove the Miranda-Agmon maximum principle for domains with conical points.
However, this requires additional knowledge of the eigenvalue distribution for the
corresponding operator pencil. In particular, one has to estimate the width of
the strip $\Lambda_- < \operatorname{Im}\lambda < \Lambda_+$, which is free of eigenvalues and contains the line
$\operatorname{Im}\lambda = -m + n/2$. For the biharmonic equation such estimates were obtained
in [81]. In this paper, Maz'ya and Plamenevskiĭ also showed that biharmonic
functions in two- and three-dimensional domains with conical points satisfy (2.8)
for $m = 2$. For $n \ge 4$ the Miranda-Agmon maximum principle holds if the first
eigenvalue of the Dirichlet problem for the operator $-\delta$ is not less than $n - 1$.
(Here δ denotes the Beltrami operator in the domain Ω which is cut out on the
unit sphere by the cone $\mathcal{K}$.) The last assumption on the first eigenvalue of $-\delta$ is
satisfied, in particular, if $\mathcal{K}$ is confined within a half-space.

In [89], Maz'ya and Roßmann studied two versions of the Miranda-Agmon maximum principle for $2m$ order strongly elliptic differential equations in a domain
with conical points. The first one is concerned with the validity of (2.8) for the
generalized solution $u \in W_2^m(\mathcal{G})$ of the Dirichlet problem, whilst the second one
consists in the validity of (2.8) for smooth (C^∞) solutions of the equation $\mathcal{L}u = 0$.
The authors proved that (2.8) is valid for all generalized solutions if and only if
the strip $1 - m \le \operatorname{Im}\lambda \le n/2 - m$ does not contain eigenvalues of the pencil $\mathfrak{A}$.
Due to a result of Kozlov and Maz'ya [9], this is always true if $n = 2, 3$ and $\mathcal{K}$ has

an explicit representation in Cartesian coordinates. By [89], the validity of (2.8) holds for all smooth solutions of $\mathcal{L}u = f$ if and only if the line $\operatorname{Im}\lambda = 1 - m$ is free of eigenvalues. Moreover, in [89] it is shown that for $n \geq 4$ and for every strongly elliptic differential operator L of the order $2m \geq 4$ there exists a domain $\mathcal{G} \subset \mathbb{R}^n$, such that (2.8) is not satisfied for all smooth solutions of $\mathcal{L}u = 0$.

2.5. The comparison principle for solutions of elliptic problems in a cone

In 1986, Kozlov and Maz'ya [5, 8, 19] developed another powerful tool for study of solutions of elliptic problems near conical points. Their approach is based on comparison of solutions to a boundary value problem and to an ordinary differential equation. A similar comparison principle was developed in 1967 by Verzhbinskiĭ and Maz'ya for second order elliptic operators.

I give a short description of Kozlov's and Maz'ya's approach in the case of a model problem in a cone (2.1). Let $\mathcal{L}$ be an elliptic model operator of the order $2m$, which means that $\mathcal{L}$ has the representation

$$\mathcal{L} = r^{-2m} L(\omega, D_\omega, rD_r)$$

where $r = |x|$, $\omega = x/|x|$, and $L(\omega, \mu, \lambda)$ is a polynomial in μ and λ of the degree $2m$ with smooth coefficients on $\overline{\Omega}$. Furthermore, let $\mathcal{B}$ be a system of model operators $\mathcal{B}_j$ of the order m_j, $j = 1, \ldots, m$, which covers $\mathcal{L}$ on $\partial\mathcal{K}\backslash\{0\}$. We consider the problem

$$(2.9) \qquad \mathcal{L}u = f \text{ in } \mathcal{K}, \quad \mathcal{B}u = g_j \text{ on } \partial\mathcal{K}\backslash\{0\}, \ j = 1, \ldots, m.$$

For arbitrary numbers: an integer $l \geq 0$ and a real number $p > 1$, we define the space $W^l_{p,loc}(\overline{\mathcal{K}}\backslash\{0\})$ as the set of all functions u, such that

$$M^l_{p;\mathcal{K}}(u;\rho) \overset{\text{def}}{=} \left(\sum_{|\alpha|\leq l} \rho^{p|\alpha|} \int_\rho^{2\rho} \int_\Omega |D_x^\alpha u|^p \, d\omega \, \frac{dr}{r} \right)^{1/p} < \infty$$

for every $\rho > 0$. Furthermore, let $W^{l-1/p}_{p,loc}(\partial\mathcal{K}\backslash\{0\})$, $l \geq 1$, be the corresponding trace space on $\partial\mathcal{K}\backslash\{0\}$. On the last space we introduce the family of seminorms

$$M^{l-1/p}_{p;\partial\mathcal{K}}(v;\rho) = \inf M^l_{p;\mathcal{K}}(u;\rho), \quad \rho > 0,$$

where the infimum is taken over all $u \in W^l_{p,loc}(\overline{\mathcal{K}}\backslash\{0\})$, such that $u = v$ on $\partial\Omega \times (\rho, 2\rho)$. For the sake of brevity, we set

$$F^l_p(\rho) = M^{l-2m}_{p;\mathcal{K}}(f;\rho) + \sum_{j=1}^m M^{l-m_j-1/p}_{p;\partial\mathcal{K}}(g_j;\rho)$$

for arbitrary functions $f \in W^{l-2m}_{p,loc}(\overline{\mathcal{K}}\backslash\{0\})$ and $g_j \in W^{l-m_j-1/p}_{p,loc}(\partial\mathcal{K}\backslash\{0\})$. We fix real numbers λ_-, λ_+, $\lambda_- < \lambda_+$, such that the strip $\lambda_- < \operatorname{Im}\lambda < \lambda_+$ does not contain eigenvalues of the pencil $\mathfrak{A}$. By $m_\pm$ we denote an arbitrary integer upper bound for the maximal lengths of the Jordan chains corresponding to the eigenvalues of $\mathfrak{A}$ on the lines $\operatorname{Im}\lambda = \lambda_\pm$ (if there are no eigenvalues on the line $\operatorname{Im}\lambda = \lambda_\pm$, we set $m_\pm = 1$).

Kozlov and Maz'ya [5] proved that the model problem (2.9) has a solution $u \in W^l_{p,loc}(\overline{\mathcal{K}}\backslash\{0\})$ satisfying the inequality

$$
(2.10) \qquad M^l_{p;\mathcal{K}}(u;r) \;\leq\; c\Bigg(\int\limits_0^r \Big(\frac{\rho}{r}\Big)^{\lambda_+} \Big(1+\log\frac{r}{\rho}\Big)^{m_+ -1} F^l_p(\rho)\,\frac{d\rho}{\rho}
$$
$$
+ \int\limits_r^\infty \Big(\frac{\rho}{r}\Big)^{\lambda_-} \Big(1+\log\frac{\rho}{r}\Big)^{m_- -1} F^l_p(\rho)\,\frac{d\rho}{\rho} \Bigg)
$$

if f, g_j are functions in $W^{l-2m}_{p,loc}(\overline{\mathcal{K}}\backslash\{0\})$ and $W^{l-m_j-1/p}_{p,loc}(\partial\mathcal{K}\backslash\{0\})$, respectively, such that

$$
\int\limits_0^1 \rho^{\lambda_+}\big(1+|\log\rho|\big)^{m_+ -1} F^l_p(\rho)\,\frac{d\rho}{\rho} + \int\limits_1^\infty \rho^{\lambda_-}\big(1+|\log\rho|\big)^{m_- -1} F^l_p(\rho)\,\frac{d\rho}{\rho} < \infty .
$$

From (2.10) it follows that the function $\mathbb{R}_+ \ni r \to M^l_{p;\mathcal{K}}(u;r)$ satisfies the comparison principle
$$
(2.11) \qquad M^l_{p;\mathcal{K}}(u;r) \leq c\,w(\log r),
$$

where M denotes the ordinary differential operator

$$
M(\partial_t) = (-\partial_t - \lambda_-)^{m_-}\,(\partial_t + \lambda_+)^{m_+}.
$$

and w is the solution of the equation $M(\partial_t)\,w(t) = F^l_p(e^t)$ on $\mathbb{R}$ satisfying $w(t) = o(e^{-\lambda_\mp t})$ as $t \to \pm\infty$. (Note that the conditions on f and g_j in the above given result are necessary and sufficient for the existence of w.) A similar result without the restriction that $\mathcal{L}$ and $\mathcal{B}_j$ are model operators was established in [8].

2.6. The heat equation in a cone

Kozlov and Maz'ya contributed to the theory of parabolic differential equations in domains with singular boundary points. In 1987, they published two papers [6, 7] on the first boundary value problem for the heat equation in a cone and in bounded domains with angular and conical points, where solvability theorems were proved. Also, complete asymptotic expansions for the solutions in the zones $|x|^2 \ll t$ and $|x|^2 \gg t$ as well as representations for the coefficients in the asymptotics were obtained. The results in [6] were derived without the compatibility condition at $t = 0$, which is important when considering thermal stresses developing under instantaneous cooling or heating of a body. Later on, Kozlov extended these results to general higher-order parabolic equations, and obtained asymptotic formulas for Green's functions and Poisson's kernels.

Using the results just mentioned, Kozlov and Maz'ya proved an interesting asymptotic formula for the eigenfunctions of the Dirichlet problem for the Laplace operator in a domain $\mathcal{G}$ with a conical point. Let $\{\lambda_j\}$ be the sequence of eigenvalues of this problem and let $\{u_j\}$ be an orthonormal system of the corresponding eigenfunctions. It is assumed that $\mathcal{G}$ coincides with the cone $\mathcal{K} = \{x \in \mathbb{R}^n : \omega = x/|x| \in \Omega\}$ in a neighborhood of the origin. Then u_j behaves asymptotically like $C_j\,|x|^\mu\,\varphi(\omega)$ near the vertex, where μ is a positive number, such that $\mu(\mu + n - 2)$ is the first eigenvalue of the operator $-\delta$ (with Dirichlet boundary condition) on Ω and φ is the first eigenfunction (positive and normed). According to [10], the following formula:

$$\sum_{\lambda_j < \lambda} C_j^2 = 4^{1-\mu-n/2}\,(n + 2\mu)^{-1}\,\left(\Gamma\!\left(\mu + \frac{n}{2}\right)\right)^{-2} \lambda^{\mu+n/2}\,(1 + o(1)).$$

holds for the coefficients C_j.

3. Boundary value problems in domains with edges

Simultaneously with elliptic problems in domains with point singularities Maz'ya and Plamenevskiĭ dealt with the more complicated situation when the boundary of the domain contains edges. Their papers in this direction appeared in 1971 and 1975 and were concerned with boundary value problems for the second order elliptic equations. In these papers, the authors obtained explicit conditions ensuring the Fredholm property for the boundary value problem. More abstract conditions for the solvability of general elliptic boundary value problems in domains with edges were published by Maz'ya and Plamenevskiĭ in 1973, 1977 and 1978. They proved Fredholm theorems in weighted Sobolev spaces not only for a domain with smooth disjoint edges, but also for domains whose boundaries contain intersecting edges of various dimensions. By means of estimates for Green's function which were proved in 1978, they obtained Fredholm theorems in weighted Hölder spaces. Furthermore, Maz'ya together with Plamenevskiĭ and later with Roßmann studied the asymptotics of solutions to general elliptic boundary value problems near edges.

3.1. Solvability of boundary value problems in weighted Sobolev spaces

Elliptic problems for second order differential operators. I start with the question of the solvability of boundary value problems for the second order elliptic equations which was studied in the papers [63, 70] of Maz'ya and Plamenevskiĭ. Here the authors obtained necessary and sufficient conditions ensuring the Fredholm property for the boundary value problem in weighted L_2 Sobolev spaces with

homogeneous and nonhomogeneous norms. I give one of their results from these
papers.

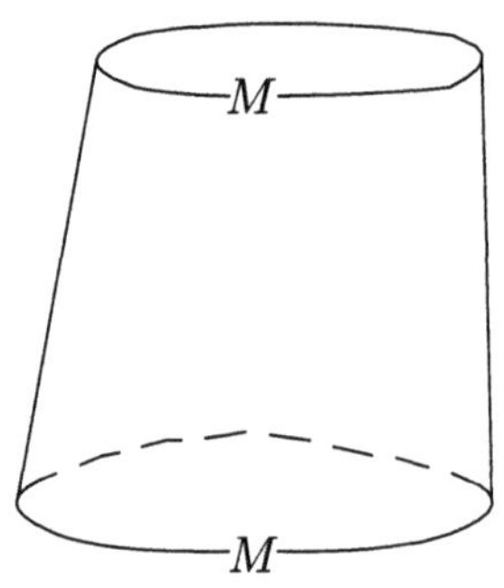

Let $\mathcal{G}$ be a domain in $\mathbb{R}^n$ with compact closure
bounded by a $(n-1)$-dimensional manifold $\partial\mathcal{G}$.
We suppose that there exists a $(n-2)$-dimensional
smooth subset M of $\partial\mathcal{G}$ such that

   (i) $\partial\mathcal{G}\backslash M$ is smooth,

  (ii) for every $x^{(0)} \in M$ there is a neighborhood $\mathcal{U}$
in $\mathbb{R}^n$ and a two-dimensional cone $\mathcal{K}$, such that
$\mathcal{G} \cap \mathcal{U}$ is diffeomorphic to $\mathcal{K} \times \mathbb{R}^{n-2} \cap B_1(0)$,
where $B_1(0)$ is the unit ball in $\mathbb{R}^n$ centered at
the origin.

The connected components of $\partial\mathcal{G}\backslash M$ are denoted by $\Gamma_1,\ldots,\Gamma_T$. On $\mathcal{G}$ the
weighted Sobolev spaces $V^l_{p,\beta}(\mathcal{G})$ and $W^l_{p,\beta}(\mathcal{G})$ are defined as follows. For functions
with support in a neighborhood of M, the norms in these spaces are defined as
follows:

$$\|u\|_{V^l_{p,\beta}(\mathcal{G})} = \left(\int_{\mathcal{G}} \sum_{|\alpha|\leq l} r^{p(\beta(z)-l+|\alpha|)} |D^\alpha u(x)|^p \, dx \right)^{1/p},$$

$$\|u\|_{W^l_{p,\beta}(\mathcal{G})} = \left(\int_{\mathcal{G}} \sum_{|\alpha|\leq l} r^{p\beta} |D^\alpha u(x)|^p \, dx \right)^{1/p},$$

where $r = r(x)$ denotes the distance from the point x to M and z is the nearest
to x point on M. If u vanishes near M, then its norm in $V^l_{p,\beta}(\mathcal{G})$ and $W^l_{p,\beta}(\mathcal{G})$ is
equivalent to the W^l_p-norm.

Maz'ya and Plamenevskiĭ considered, in particular, the boundary value problem

$$\mathcal{L}u = f \ \text{in} \ \mathcal{G}, \qquad \frac{\partial u}{\partial \ell_k} = g_k \ \text{on} \ \Gamma_k, \ k = 1,\ldots,T,$$

for a second order differential operator $\mathcal{L}$ with smooth coefficients such that
$\mathrm{ord}\,(\mathcal{L} - \Delta) \leq 1$. Here ℓ_k is a nondegenerate vector field on Γ_k which is directed outside of $\mathcal{G}$ and transversal to Γ_k. Let z be an arbitrary fixed point on
M. We denote the connected components of $\partial\mathcal{G}\backslash M$ adjoint to z by Γ_{k+} and Γ_{k-}.
There exist $(n-1)$-dimensional half-spaces $\Gamma^\pm$ tangent to $\Gamma_{k\pm}$ in z, such that
$M_0 = \overline{\Gamma^+} \cap \overline{\Gamma^-}$ is an $(n-2)$-dimensional space tangent to M at z. The dihedron
bounded by Γ^+, Γ^- and the edge M_0 is denoted by $\mathcal{D}$, and the angle between its
sides $\Gamma^\pm$ is denoted by $\alpha(z)$. We assume that the limits $\ell^\pm(z) = \lim_{\zeta\to z} \ell_{k\pm}(\zeta)$
exist and the angles between the vectors $\ell^\pm(z)$ and the exterior normals $\nu^\pm(z)$
to $\Gamma^\pm$ are acute. By $\ell_0^\pm(z)$ we denote the projections of $\ell^\pm(z)$ to the the plane
$\Pi(z)$ orthogonal to M at z, and by $\gamma^\pm(z)$ the angles between $\nu^\pm(z)$ and $\ell_0^\pm(z)$,

$-\pi/2 < \gamma^{\pm}(z) < \pi/2$. Finally, we put $\sigma(z) = \gamma^+(z) + \gamma^-(z)$. Then, according to [70], the operator

$$V_{2,\beta}^2(\mathcal{G}) \ni u \to \left(\mathcal{L}u, \frac{\partial u}{\partial \ell_k}\Big|_{\Gamma_k}\right) \in V_{2,\beta}^0(\mathcal{G}) \times \prod_{k=1}^{T} V_{2,\beta}^{1/2}(\Gamma_k)$$

is a Fredholm operator, if and only if for all $z \in M$ one of the conditions

$$1 - \frac{\sigma(z)}{\alpha(z)} < \beta(z) < 1, \qquad 1 < \beta(z) < 1 - \frac{\sigma(z)}{\alpha(z)}$$

is satisfied.

Solvability of general elliptic problems in weighted L_p Sobolev spaces. In the papers [66] of 1973 and [76] of 1978, Maz'ya and Plamenevskiĭ developed a theory of general elliptic problems in domains with edges including Fredholm theorems in weighted L_2 and L_p Sobolev spaces, a priori estimates and regularity assertions for the solutions. They considered the boundary value problem

$$(3.1) \qquad \mathcal{L}(x, D_x)u = f \quad \text{in } \mathcal{G},$$

$$(3.2) \qquad \mathcal{B}_{j,k}(x, D_x)u = g_{j,k} \quad \text{on } \Gamma_k, \; j = 1, \ldots, m, \; k = 1, \ldots, T,$$

where $\mathcal{L}$ is an elliptic differential operator of the order $2m$, and $\mathcal{B}_{j,k}$ are differential operators of order $m_{j,k} < 2m$, which cover $\mathcal{L}$ on Γ_k. The coefficients of $\mathcal{L}$ and $\mathcal{B}_{j,k}$ were assumed to be smooth on $\overline{\mathcal{G}}$. Maz'ya and Plamenevskiĭ reduced the question whether the Fredholm property holds (in weighted L_p Sobolev spaces) for the boundary value problem, to the question of the unique solvability of a "model problem" in weighted L_2 Sobolev spaces. Let z be an arbitrary point on M, and let $\mathcal{L}^\circ(z, D_x)$, $\mathcal{B}_{j,k\pm}^\circ(z, D_x)$ be the principal parts of $\mathcal{L}$ and $\mathcal{B}_{j,k\pm}$, respectively, with coefficients frozen at z. Furthermore, let $\mathcal{D}$ be the above dihedron with the sides Γ^+ and Γ^-. It is shown in [76] that the operator

$$(3.3) \qquad V_{p,\beta}^l(\mathcal{G}) \ni u \to \left(\mathcal{L}u, \{\mathcal{B}_{j,k}u|_{\Gamma_k}\}\right) \in V_{p,\beta}^{l-2m}(\mathcal{G}) \times \prod_{j,k} V_{p,\beta}^{l-m_{j,k}-1/p}(\Gamma_k)$$

of the boundary value problem (3.1), (3.2) is a Fredholm operator, if and only if for every $z \in M$ the model problem

$$(3.4) \quad \mathcal{L}^\circ(z, D_x)u = f \quad \text{in } \mathcal{G}, \qquad \mathcal{B}_{j,k\pm}^\circ(z, D_x)u = 0 \quad \text{on } \Gamma_{k\pm}, \; j = 1, \ldots, m,$$

is uniquely solvable in $V_{2,\gamma}^{2m}(\mathcal{D})$ for every $f \in V_{2,\gamma}^0(\mathcal{D})$, $\gamma = \beta(z) - l + 2m - 1 + 2/p$. The last condition with $\gamma = m$ can be easily verified for the Dirichlet problem and other V-elliptic problems. Furthermore, the paper [76] contains necessary and sufficient conditions for the unique solvability of (3.4), stated in terms of another model problem in an angle on the plane.

Elliptic problems in domains with intersecting edges. Also, Maz'ya and Plamenevskiĭ [74] dealt with the situation when the boundary of the domain contains edges of different dimensions which may intersect at nonzero angles. For example, polyhedra are such domains. Analogously to the case of disjoint edges discussed above, they defined a collection of model problems and proved that the operator of the boundary value problem is a Fredholm operator, if and only if the model problems are uniquely solvable.

3.2.   Asymptotics of solutions

The asymptotic formulas for solutions of elliptic boundary value problems near angular and conical points given in the previous section made it possible to describe the behaviour of solutions of general elliptic problems near edges as well. When the right-hand side of the differential equation and the boundary data are sufficiently smooth, asymptotic representations near edges were given first by Maz'ya and Plamenevskiĭ [66] in 1973. Kondrat'ev (in 1977) and Nikishkin (in 1979) proved similar asymptotic formulas for solutions of the Dirichlet problem for the second order elliptic equations without additional conditions on the right-hand sides. In the paper [85] of Maz'ya and Roßmann, these formulas were generalized to elliptic boundary value problems for elliptic equations of arbitrary order. Here I present the asymptotic representation obtained in [85].

For the sake of simplicity, let the domain $\mathcal{G}$ coincide with the dihedron $\mathcal{D} = \{x \in \mathbb{R}^n : (x_1, x_2) \in \mathcal{K}\}$ in a neighborhood of a certain edge point z_0, where $\mathcal{K} = \{(x_1, x_2) : \theta_- < \theta = \arg(x_1 + ix_2) < \theta_+\}$ is a plane angle. Omitting all terms which contain the derivatives $D_{x_3}, \ldots, D_{x_n}$ in the model problem (3.4), we get the problem

$$(3.5) \qquad \mathcal{L}^\circ(z_0, D_{x_1}, D_{x_2}, 0)u = f \quad \text{in } \mathcal{K},$$

$$(3.6) \qquad \mathcal{B}^\circ_{j,k\pm}(z_0, D_{x_1}, D_{x_2}, 0)u = g_j^\pm \quad \text{for } \theta = \theta_\pm,\ j = 1, \ldots, m,$$

in the plane angle $\mathcal{K}$. By one of the results in [76], the unique solvability of this problem in the weighted space $V_{2,\gamma}^{2m}(\mathcal{K})$, where $\gamma = \beta(z_0) - l + 2m - 1 + 2/p$, is a necessary condition for the Fredholm property for the operator (3.3).

We denote by $\mathfrak{A}(z_0; \lambda)$ the operator

$$u(\theta) \to \left(L(z_0, D_\theta, \lambda)u(\theta),\ B_j^\pm(z_0, D_\theta, \lambda)\, u(\theta)\big|_{\theta=\theta\pm}, j = 1, \ldots, m \right),$$

where $L(z_0, D_\theta, \lambda)u(\theta) = r^{2m-i\lambda}\mathcal{L}^\circ(z_0, D_{x_1}, D_{x_2}, 0)r^{i\lambda}u$, and $B_j^\pm(z_0, D_\theta, \lambda)\, u(\theta) = r^{m_{j,k\pm}-i\lambda}\mathcal{B}^\circ_{j,k\pm}(z_0, D_{x_1}, D_{x_2}, 0)r^{i\lambda}u$. As a result of the theory of elliptic boundary value problems in domains with angular or conical points, one obtains that problem (3.5), (3.6) is uniquely solvable in $V_{2,\gamma}^{2m}(\mathcal{K})$ for arbitrary $f \in V_{2,\gamma}^0(\mathcal{K})$ and $g_j^\pm = 0$, $j = 1, \ldots, m$, if there are no eigenvalues of the pencil $\mathfrak{A}(z; \lambda)$ on the line $\operatorname{Im}\lambda = \gamma - 2m + 1$. Furthermore, the eigenvalues of this problem determine the behaviour of the solutions both to problem (3.5), (3.6) near the origin and to problem (3.1), (3.2) near the edge point z_0. If, for example, $u \in V^l_{p,\beta}(\mathcal{G})$ is a solution of problem (3.1), (3.2), the functions f, $g_{j,k}$ belong to the spaces $V^{l'-2m}_{p',\beta'}(\mathcal{G})$ and $V^{l'-m_{j,k}-1/p'}_{p',\beta'}(\partial\mathcal{G})$, respectively, the model problem (3.4) is uniquely solvable in $V^{2m}_{2,\gamma}(\mathcal{K})$, $\gamma = \beta(z) - l + 2m - 1 + 2/p$, and certain additional conditions on the eigenvalues of the pencil $\mathfrak{A}(z;\lambda)$ are satisfied, then u admits the following representation in a neighborhood of z_0:

$$u = \sum_{1 \le \mu < N} \sum_{0 \le \nu < \operatorname{Im} \lambda_\mu - \beta' + l' - 2/p'} C_{\mu,\nu}(x)\, P_{\mu,\nu}(\theta, z, \log r) + v, \quad v \in V^{l'}_{p',\beta'}(\mathcal{G}).$$

Here z denotes the nearest edge point to x, $\lambda_j(z)$ are the eigenvalues of the pencil $\mathfrak{A}(z;\lambda)$ in the strip $\beta'(z) - l' + 2/p' < \operatorname{Im}\lambda < \beta(z) - l + 2/p$, $P_{\mu,\nu}(\theta, z, \cdot)$ are polynomials with coefficients, which are infinitely differentiable with respect to θ and z, and $C_{\mu,\nu}$ are extensions of functions from $W^{\operatorname{Im}\lambda_\mu - \beta' + l' - \nu - 2/p' - \varepsilon}_{p'}(M)$, ε is an arbitrarily small positive number. If f and $g_{j,k}$ are sufficiently smooth, then (see [66]) the coefficients $C_{\mu,\nu}$ are also smooth and can be replaced by their traces on M.

3.3. Estimates of Green's functions and Schauder estimates

Estimates of Green's functions and Poisson's kernels. In the previous section, I presented estimates of Green's functions for elliptic boundary value problems in domains with conical points which follow from Maz'ya's and Plamenevskiĭ's asymptotic formulas proven in [79]. Shortly after that, they obtained estimates of the same type for Green's functions and Poisson's kernels of elliptic boundary value problems in an n-dimensional dihedral angle.

Let $\mathcal{D} = \{x \in \mathbb{R}^n : \theta_- < \theta < \theta_+\}$, where $r = (x_1^2 + x_2^2)^{1/2}$ and $\theta = \arg(x_1 + ix_2)$ are the polar coordinates in the (x_1, x_2)-plane. In [77], Maz'ya and Plamenevskiĭ studied Green's function of the boundary value problem

$$(3.7) \qquad \mathcal{L}(D_x)u = f \text{ in } \mathcal{D}, \qquad \mathcal{B}^\pm_j u = g^\pm_j \text{ on } \Gamma^\pm, \ j = 1, \ldots, m,$$

where $\mathcal{L}$ is a homogeneous elliptic differential operator of the order $2m$ with constant coefficients, $\mathcal{B}^\pm_j$ are homogeneous differential operators of order $m^\pm_j$ such that every of the systems $\{\mathcal{B}^\pm_j\}^m_{j=1}$ is normal and covers $\mathcal{L}$ on $\Gamma^\pm = \{x \in \mathbb{R}^n : \theta = \theta^\pm\}$. If this problem is uniquely solvable in $V^{2m}_{2,\beta}(\mathcal{D})$ for every $f \in V^0_{2,\beta}(\mathcal{D})$ and for $g^\pm_j = 0$, then, by [77, Th.1.3], there exists a unique solution $G(x,y)$ of the problem

$$\mathcal{L}(D_x)\,G(x,y) = \delta(x-y), \quad x, y \in \mathcal{D},$$
$$\mathcal{B}^\pm_j(D_x)\,G(x,y) = 0, \quad x \in \gamma^\pm, \ y \in \mathcal{D}, \ j = 1, \ldots, m,$$

such that the function $x \to \varphi\big(\frac{r(x)}{r(y)}\big)\, G(x,y)$ belongs to $V^{l+2m}_{2,\beta+l}(\mathcal{D})$, $l = 0, 1, \ldots$, for every fixed $y \in \mathcal{D}$. Here φ is an infinitely differentiable function on the positive half-axis equal to one outside of the interval $(1/2, 2)$, and to zero on the interval

$(3/4, 1)$. When $|x - y| < \min(r(x), r(y))$, Maz'ya and Plamenevskiĭ obtained the estimate

$$|D_x^\alpha D_y^\gamma G(x,y)| \le c_{\alpha,\gamma} \left(|x-y|^{2m-n-|\alpha|-|\gamma|} + [r(y)]^{2m-n-|\alpha|-|\gamma|} \right)$$

for $2m - n \ne |\alpha| + |\gamma|$, and

$$|D_x^\alpha D_y^\gamma G(x,y)| \le c_{\alpha,\gamma} \left(\left| \log \frac{|x-y|}{r(y)} \right| + 1 \right)$$

for $2m - n = |\alpha| + |\gamma|$.

In order to estimate Green's function in the case $|x - y| < \min(r(x), r(y))$, one has to consider the eigenvalues of $\mathfrak{A}$. From our assumption on solvability of the boundary value problem it follows that the line $\operatorname{Im} \lambda = \beta - 2m + 1$ does not contain eigenvalues of the corresponding operator pencil $\mathfrak{A}$. Let $\Lambda_- < \operatorname{Im} \lambda < \Lambda_+$ be the widest strip in the complex plane which contains the line $\operatorname{Im} \lambda = \beta - 2m + 1$ and is free of eigenvalues, and let $\alpha = (\alpha', \alpha'')$, $\gamma = (\gamma', \gamma'')$ be arbitrary multi-indices (here α', γ' contain the first two components and α'', γ'' the last $n - 2$ components of the multi-indices α and γ, respectively). Then, by [77, Th.2.1], for $|x - y| \ge \min(r(x), r(y))$ there is the estimate

$$|D_x^\alpha D_y^\gamma G(x,y)| \le c_{\alpha,\gamma} \, \frac{r(x)^{-\Lambda_- - \varepsilon - |\alpha'|} \, r(y)^{\Lambda_+ + 2m - 2 - \varepsilon - |\gamma'|}}{|x-y|^{\Lambda_+ - \Lambda_- + n - 2 - 2\varepsilon + |\alpha''| + |\gamma''|}} \, .$$

Similar estimates hold for Poisson's kernels.

Schauder estimates. The estimates for Green's functions and Poisson's kernels in [77] are used in the same paper to get estimates in weighted Hölder spaces $\Lambda_\delta^{l,\alpha}$ with the norm

$$\|u\|_{\Lambda_\delta^{l,\alpha}(\mathcal{D})} = \sup_{\substack{x \in \mathcal{D} \\ |\gamma| \le l}} \sum [r(x)]^{\delta - l - \alpha + |\gamma|} |D_x^\gamma u(x)|$$

$$+ \sup_{x,y \in \mathcal{D}} |x-y|^{-\alpha} \sum_{|\gamma|=l} \left| [r(x)]^\delta D_x^\gamma u(x) - [r(y)]^\delta D_y^\gamma u(y) \right|.$$

Maz'ya and Plamenevskiĭ proved (see [77, Th.3.1]) that the solution $u \in V_{2,\beta}^{2m}(\mathcal{D})$ of the boundary value problem (3.7) belongs to $\Lambda_\delta^{l,\alpha}(\mathcal{D})$ with arbitrary integer $l \ge 2m$ and real $\delta \in (\Lambda_- + l + \alpha, \Lambda_+ + l + \alpha)$ if $f \in \Lambda_\delta^{l-2m,\alpha}(\mathcal{D})$ and the functions $g_j^\pm$ belong to the corresponding trace spaces $\Lambda_\delta^{l-m_j^\pm,\alpha}(\Gamma^\pm)$. The solution u satisfies the estimate

$$\|u\|_{\Lambda_\delta^{l,\alpha}(\mathcal{D})} \le c \left(\|f\|_{\Lambda_\delta^{l-2m,\alpha}(\mathcal{D})} + \sum_\pm \sum_{j=1}^m \|g_j^\pm\|_{\Lambda_\delta^{l-m_j^\pm,\alpha}(\Gamma^\pm)} \right)$$

with a constant c independent of f and $g_j^{\pm}$. As a consequence of this result, they obtained that the operator

$$\Lambda_\delta^{l,\alpha}(\mathcal{D}) \ni u \to (\mathcal{L}u, \mathcal{B}_j^{\pm}u) \in \Lambda_\delta^{l-2m,\alpha}(\mathcal{D}) \times \prod_{\pm}\prod_{j=1}^{m} \Lambda_\delta^{l-m_j^{\pm},\alpha}(\Gamma^{\pm})$$

is a bijection for $\delta \in (\Lambda_- + l + \alpha, \Lambda_+ + l + \alpha)$.

In [78], Maz'ya and Plamenevskiĭ applied the results above to boundary value problems on manifolds with edges. They proved the Fredholm property for the operator of the boundary value problem in weighted Hölder spaces and obtained a priori estimates for the solutions.

Green's functions of the Dirichlet problem in polyhedral domains. Estimates for Green's function of the Dirichlet problem for the Stokes and the Lamé systems in three-dimensional domains of polyhedral type were obtained by Maz'ya and Plamenevskiĭ in [83]. Using these estimates, they proved theorems on the unique solvability in weighted Sobolev and Hölder spaces, regularity assertions for the solutions, and maximum principles.

In [86], Maz'ya and Roßmann proved estimates for Green's functions of the Dirichlet problem for the $2m$-order elliptic equations in three-dimensional poly-hedral domains. These estimates and an estimate for the eigenvalues of the cor-responding operator pencil obtained by Kozlov and Maz'ya [9] were used for the proof of the Miranda-Agmon maximum principle.

4. Spectral properties of operator pencils generated by elliptic boundary value problems in a cone

Almost all results in Sections 2 and 3 involve an a priori knowledge about the eigen-values, eigenfunctions and generalized eigenfunctions of certain operator pencils. Without this information the general theory cannot answer even simple questions. For example, the differentiability properties of solutions in a neighborhood of con-ical points or edges depend on the presence or absence of eigenvalues in a certain strip $\alpha \leq \operatorname{Im}\lambda \leq \beta$ which belong to nonpolynomial singular terms. In this sec-tion, I have collected results by Kozlov and Maz'ya on the spectral properties of operator pencils generated by boundary value problems for general elliptic equa-tions and systems in a cone. Besides, there are some results of the same nature concerning boundary value problems of elasticity and hydrodynamics, which will be discussed in the next section.

4.1. The operator pencil generated by the Dirichlet problem in a cone

I start with the Dirichlet problem for the elliptic system

$$(4.1) \qquad \mathcal{L}(D_x)u = \sum_{|\alpha|=|\beta|=m} A_{\alpha,\beta}\, D_x^{\alpha+\beta} u = f$$

in a cone $\mathcal{K} = \{x \in \mathbb{R}^n : \omega = x/|x| \in \Omega\}$. Here $A_{\alpha,\beta}$ are constant $\ell \times \ell$-matrices, such that $A_{\alpha,\beta} = A_{\beta,\alpha}^*$. The operator $\mathcal{L}$ is assumed to be strongly elliptic, i.e.,

$$\sum_{|\alpha|=|\beta|=m} \left(A_{\alpha,\beta}\xi^{\alpha+\beta}z, z \right)_{\mathbb{C}^\ell} \geq c\,|\xi|_{\mathbb{R}^n}^{2m}\,|z|_{\mathbb{C}^\ell}^2$$

for all $\xi \in \mathbb{R}^n$, $z \in \mathbb{C}^\ell$. The Dirichlet problem for system (4.1) generates the operator pencil $\mathfrak{A}$ on $\overset{\circ}{W}{}_2^m(\Omega)$ which is defined by the equality

$$\mathfrak{A}(\lambda)\Phi = |x|^{-i\lambda+2m}\,\mathcal{L}(D_x)\,|x|^{i\lambda}\,\Phi(\omega), \qquad \Phi \in \overset{\circ}{W}{}_2^m(\Omega).$$

By results discussed in Section 2, the singularities of the solutions of (4.1) can be expressed in terms of the eigenvalues, eigenfunctions and generalized eigenfunctions of $\mathfrak{A}$. In particular, to describe the singularities of solutions which are square integrable together with all their derivatives up to order m in a neighborhood of the vertex O one has to study the spectrum of this pencil in the half-plane $\operatorname{Im}\lambda \leq (n - 2m)/2$. First, I want to present the following important result of Kozlov and Maz'ya. They proved in [9] that the strip

$$(4.2) \qquad\qquad |\operatorname{Im}\lambda - (n - 2m)/2| \leq 1/2$$

does not contain eigenvalues if the cone $\mathcal{K}$ has an explicit representation in Cartesian coordinates. As I mentioned in Sections 2.4 and 3.3, this result played an essential role in the proof of the Miranda-Agmon maximum principle for solutions of the Dirichlet problem in three-dimensional domains of polyhedral type.

A result of Kozlov and Maz'ya in [11], demonstrates that estimate (4.2) is sharp. Their paper [11] is concerned with the Sobolev problem in the exterior of a ray and with the Dirichlet problem in the exterior of a thin cone of dimension $n \leq 2m$. It is shown there that the spectrum of the operator pencil generated by the Sobolev problem coincides with the set $\{m - (n - 1)/2 + k\}_{k=0,\pm1,\dots}$ for odd n is and with $\{m - n/2 + k/2\}_{k=\pm1,\pm2,\dots}$ for even n. Furthermore, Kozlov and Maz'ya proved that the spectrum of the pencil generated by the Dirichlet problem in the exterior of a thin cone is close to the spectrum of the operator pencil generated by the Sobolev problem. This implies that for any positive ε there exists a cone $\mathcal{K} = \{x = (x', x_n) : x_n > -N\,|x'|\}$, such that the strip $(n - 2m - 1)/2 - \varepsilon < \operatorname{Im}\lambda < (n - 2m)/2$ contains at least one eigenvalue of the pencil $\mathfrak{A}$ and, therefore, estimate (4.2) is sharp. For $n = 2m + 1$ this follows from results in the book [60] of Maz'ya, Nazarov and Plamenevskiĭ.

For a second order system in a three-dimensional polyhedral angle Kozlov and Maz'ya [15, 16] showed that even the wider strip $|\operatorname{Im}\lambda - 1/2| \leq 1$ is free of eigenvalues of $\mathfrak{A}$ provided there is at most one edge where the interior angle is greater than π.

The results of Kozlov and Maz'ya [14] on the singularities of solutions to the Dirichlet problem in a cone of the form $\mathcal{K} = K_d \times \mathbb{R}^{n-d}$, where K_d is a d-dimensional cone, are of great importance for the study of edge asymptotics. In [14], these singularities were expressed in terms of the eigenvalues, eigenfunctions and generalized eigenfunctions of the pencil generated by the Dirichlet problem in the d-dimensional cone $\mathcal{K}_d$.

4.2. The operator pencil generated by the Neumann problem

In [13], Kozlov and Maz'ya dealt with the spectral properties of operator pencils generated by the Neumann problem for the elliptic system

$$\mathcal{L}(D_x)\, u = (-1)^m \sum_{|\alpha|=|\beta|=m} A_{\alpha,\beta}\, D_x^{\alpha+\beta} u = f,$$

in a cone $\mathcal{K}$. Here $A_{\alpha,\beta}$ are constant $\ell \times \ell$ matrices, such that $A_{\alpha,\beta} = A_{\beta,\alpha}^*$. In order to describe the singularities of the solutions to the Neumann problem, one has to find all power-logarithmic solutions u of the equation

$$b(u,v) = \int_{\mathcal{K}} \sum_{|\alpha|=|\beta|=m} \left(A_{\alpha,\beta}\, D_x^\beta u,\, D_x^\alpha v \right)_{\mathbb{C}^\ell} dx = 0,$$

where v is an arbitrary function in $W_2^m(\mathcal{K})^\ell$ vanishing in a neighborhoods of $x = 0$ and infinity. This leads to a spectral problem for the operator pencil $\mathfrak{A}(\lambda)$ generated by the parameter-depending sesquilinear form

$$a(\varphi,\psi;\lambda) = \frac{1}{2\log 2} \int_{\{x\in\mathcal{K},\, 1/2<|x|<2\}} \sum_{|\alpha|=|\beta|=m} \left(A_{\alpha,\beta}\, D_x^\beta u,\, D_x^\alpha v \right)_{\mathbb{C}^\ell} dx,$$

where $u = |x|^{i\lambda}\varphi(x/|x|)$, $V = |x|^{2m-n+i\bar\lambda}\psi(x/|x|)$, $\varphi,\psi \in W_2^m(\Omega)^\ell$. Kozlov and Maz'ya obtained a complete description of the eigenvalues (and the corresponding eigenvectors and generalized eigenvectors) in the strip

(4.3) $$|\operatorname{Im}\lambda - (n-2m)/2| \leq 1/2.$$

Under the conditions that $\mathcal{K}$ is given by the inequality $x_n > h(x_1,\ldots,x_{n-1})$, where h is a smooth function on $\mathbb{R}^{n-1}\backslash\{0\}$ positively homogeneous of degree 1, and the inequality

$$\sum_{|\alpha|=|\beta|=m} \left(A_{\alpha,\beta} f_\beta,\, f_\alpha \right)_{\mathbb{C}^\ell} \geq c \sum_{|\alpha|=m} |f_\alpha|_{\mathbb{C}^\ell}^2$$

with a positive constant c is satisfied for all $f_\alpha \in \mathbb{C}^\ell$, they obtained, in particular, the following results:

(i) If $2m < n - 1$, then the strip (4.3) contains no points of the spectrum of $\mathfrak{A}$.

(ii) For $2m \geq n - 1$ and even n the strip (4.3) contains only one eigenvalue $\lambda_0 = (n - 2m)/2$.

(iii) For $2m \geq n - 1$ and odd n the strip (4.3) contains exactly two eigenvalues $\lambda_\pm = (n - 2m)/2 \pm 1/2$.

In case (ii), the eigenfunctions corresponding to λ_0 are the homogeneous vector-polynomials of degree $m - n/2$ restricted to Ω and there is exactly one generalized eigenfunction for every eigenfunction. The eigenfunctions corresponding to λ_- in case (iii) are restrictions to Ω of homogeneous vector-polynomials of degree $(2m - n + 1)/2$. To each of these eigenvectors there corresponds at most one generalized eigenvector.

5. Applications to elastostatics and hydrodynamics

While studying singularities of solutions to elliptic problems near singular boundary points, Maz'ya was always interested in applications to problems of mathematical physics. Many of his papers are concerned, in particular, with boundary value problems of elasticity and hydrodynamics in nonsmooth domains and contain more detailed information about the spectra of the operator pencils which are generated by the boundary value problems in a cone. This was the basis to get, for example, theorems on the solvability in various function spaces, regularity results and a priori estimates for the solutions.

5.1. Problems of linear elasticity

I start with the description of results by Maz'ya and Plamenevskiĭ published in the paper [82]. This article along with [81] initiated in 1981 a series of works dealing with spectral properties of the above mentioned operator pencils. In [82], the Dirichlet problems for the Lamé system

$$(5.1) \qquad \Delta u + \gamma^{-1} \operatorname{grad}\, \operatorname{div} u = f$$

and for the Stokes system in a three-dimensional domain with conical points were considered. The authors proved that there are no eigenvalues of the operator pencil generated by the Lamé system in a cone (cf. definition of the pencil $\mathfrak{A}$ in Section 4.1) in the strip

$$(5.2) \qquad \left| \operatorname{Im} \lambda - \frac{1}{2} \right| \leq \alpha + \frac{1}{2}, \qquad \text{where } \alpha = \frac{(2\gamma + 1)\mu}{\mu + 2\gamma + 4}$$

and $\mu(\mu + 1)$ is the first eigenvalue of the Dirichlet problem for the operator $-\delta$ on the subdomain Ω, which is cut out by the cone on the sphere S^2. This result enabled Maz'ya and Plamenevskiĭ to estimate Green's function and to prove the maximum principle

$$(5.3) \qquad \|r^\beta u\|_{L_\infty(\mathcal{G})} \le c\|r^\beta u\|_{L_\infty(\partial\mathcal{G})} ,$$

for solutions of the homogeneous system (5.1). Here r is the distance from the set of conical points, and β is an arbitrary number in $(-\alpha, 1 - \alpha)$.

Furthermore, the paper [82] includes theorems on the solvability and on the regularity of solutions to the Dirichlet problem for the Lamé system in quasicylinders and cuspidal domains.

A more complicated situation when a domain has polyhedral vertices was studied by the same authors in [83]. Here the authors proved theorems on the solvability in weighted Sobolev spaces $V^{l,p}_{\vec{\beta},\vec{\delta}}$, where both the distances to the vertices and the distances to the edges appear in the weight function. If a function has its support in a neighborhood of the j-th vertex q, the norm in $V^{l,p}_{\vec{\beta},\vec{\delta}}(\mathcal{G})$ is equivalent to

$$\left(\rho^{p\beta_j} \prod_k r_k^{p\delta_k} \sum_{|\alpha|\le l} r^{p(|\alpha|-l)} \, |D^\alpha u|^p \, dx\right)^{1/p},$$

where ρ denotes the distance to q, r is the distance to the set of boundary singularities, r_k is the distance to the k-th edge e_k, and the product is taken over all k, such that $q \in \bar{e}_k$. Furthermore, Maz'ya and Plamenevskiĭ studied the solvability in weighted Hölder classes, proved regularity assertions for the solutions, obtained estimates for Green's matrices and a weighted maximum principle analogous to (5.3).

More detailed information on the eigenvalues, eigenvectors and generalized eigenvectors of the operator pencil generated by the Dirichlet problem for the Lamé system was obtained in [27]. In this paper, Kozlov, Maz'ya and Schwab showed, in particular, that all eigenvalues of this pencil in the strip

$$(5.4) \qquad \left|\operatorname{Im}\lambda - 1/2\right| \le F_\gamma(\Omega),$$

where F_γ is a certain set-function, $F_\gamma(\Omega)^2 \ge \gamma^2 + (\gamma + 3/2)^2$, are purely imaginary and there are no generalized eigenvectors corresponding to the eigenvalues in the open strip (5.4). Furthermore, they established a variational principle for the eigenvalues in the strip $- \min\{F_\gamma - 1/2, 1 + 2\gamma\} < \operatorname{Im}\lambda \le 1/2$, and showed that these eigenvalues depend monotonically on Ω. The monotonicity of these eigenvalues allows to get more precise estimates for the width of the strip which is free of eigenvalues of the above mentioned operator pencil. For example, it was shown in [27], that there are no eigenvalues in $|\operatorname{Im}\lambda - 1/2| \le 3/2$, if Ω is a subdomain of the unit hemi-sphere S_+^2. In particular, it follows from the last result that the solution of the Dirichlet problem for (5.1) in a domain of polyhedral type belongs to $C^{1,\alpha}(\mathcal{G})^3$ for a certain positive α, if $f \in W_2^1(\mathcal{G})^3$, all tangent cones at vertices of $\mathcal{G}$ are contained in a half-space and all interior edge angles are not greater than π.

In [25], Kozlov, Maz'ya and Roßmann studied the spectral properties of the operator pencil generated by the Lamé system in a three-dimensional cone $\mathcal{K}$ with the boundary conditions

$$
\begin{array}{ll}
\text{(i)} & u = 0, \\
\text{(ii)} & u_n = 0, \ \sigma_{n,\tau}(u) = 0, \\
\text{(iii)} & u_\tau = 0, \ \sigma_{n,n}(u) = 0
\end{array}
$$

on different parts of $\partial\mathcal{K}\backslash\{0\}$. Here $u_n = u \cdot n$ denotes the normal component of the displacement vector u, u_τ is the tangential component of u on the boundary and $\sigma_{n,n}$, $\sigma_{n,\tau}$ are the normal and tangential, respectively, components of $\sigma_n(u) = \sigma(u) \cdot n$, and $\sigma(u)$ denotes the stress tensor. The main results of this paper are the absence of eigenvalues with nonzero real parts in the strip $|\mathrm{Im}\,\lambda - 1/2| \leq \left[\gamma^2 + (\gamma + 3/2)^2\right]^{1/2}$ and the nonexistence of generalized eigenvectors corresponding to eigenvalues in the interior of this strip. As in the case of Dirichlet boundary conditions, a variational principle for the eigenvalues in this strip was established.

Furthermore, the following result by Kozlov and Maz'ya concerning the Neumann problem for the Lamé system is of importance. In [9], they proved that the strip $0 \leq \mathrm{Im}\,\lambda \leq 1$ contains exactly two eigenvalues $\lambda_0 = 0$ and $\lambda_1 = i$ of the operator pencil generated by the Neumann problem if the cone $\mathcal{K}$ is given by the inequality $x_3 > \varphi(x_1, x_2)$, where φ is positively homogeneous function of degree one, and smooth on $\mathbb{R}^2\backslash\{0\}$. Both eigenvalues have geometric and algebraic multiplicities equal to three, and the eigenspace corresponding to $\lambda = 0$ consists of the constant vectors. The same result was proved by Kozlov and Maz'ya in [12] for a two-dimensional polygonal crack.

Finally, I mention the papers [21]–[23] by Kozlov, Maz'ya and Parton which are concerned with the quasistatic problem of thermoelasticity in a two-dimensional domain with a crack. Here the authors investigated the asymptotic form of the stresses in a neighborhood of the crack tip. The paper [21] treats the instantaneous change of the temperature at the boundary and provides the asymptotics of stress intensity factors for small times. In [23], the authors considered point sources for the heat equation with three types of boundary conditions on the crack sides (ideal contact, zero temperature, and zero heat transfer). For each of these types they obtained the principal term in the asymptotics of the tensile and shear stress intensity factors as $t \to 0$. The case of a thermal shock, i.e. the action of stresses caused by an instantaneous jump of temperature, is investigated in [22].

5.2. Problems of hydrodynamics

The above mentioned papers [82, 83] of Maz'ya and Plamenevskiĭ are concerned with the Dirichlet problem for the Stokes system

$$
(5.5) \qquad\qquad -\Delta u + \nabla p = f, \quad \mathrm{div}\, u = g
$$

as well. In [82], the Stokes system in a three-dimensional domain with conical points is considered. It is shown that the strip (5.2) with $\gamma = 0$ does not contain eigenvalues of the corresponding operator pencil. Using this result, the authors derived estimates of Green's matrix and of solutions to the homogeneous system (5.5). For example, these solutions satisfy the inequality (5.3) for $\beta \in (-\alpha, 1-\alpha)$, $\alpha = \mu/(\mu+4)$. A similar result was obtained for the generalized solution $(u,p) \in W_2^1(\mathcal{G})^3 \times L_2(\mathcal{G})$ of the Navier-Stokes system

$$(5.6) \qquad -\Delta u + \nabla p + (u \cdot \nabla)u = 0, \quad \operatorname{div} u = 0 \quad \text{in } \mathcal{G}.$$

Maz'ya and Plamenevskiĭ proved that $r^\beta u \in L_\infty(\mathcal{G})$ if $r^\beta u|_{\partial\mathcal{G}} \in L_\infty(\partial\mathcal{G})$ and $\beta \in (-\alpha, 1-\alpha)$. Furthermore, they studied the solvability of the Dirichlet problem for the Stokes system in quasicylinders and cuspidal domains.

The Dirichlet problem for the Stokes and Navier-Stokes systems in three-dimensional domains of polyhedral type was investigated by Maz'ya and Plamenevskiĭ in [80, 83]. The comprehensive work [83] gives a detailed presentation of the material announced in [80], and is very rich with new techniques and results. Unfortunately, there is no English translation of this important paper, and since the results obtained in it are unknown, their weaker versions are rediscovered from time to time. For the Stokes system, basic theorems in [83] are concerned with solvability of the Dirichlet problem in weighted Sobolev and Hölder spaces, regularity assertions for the weak and strong solutions, estimates for Green's matrix in a neighborhood of singular boundary points, and maximum principles. In particular, it is proved that certain conditions on the weight parameters $\vec{\beta}$ and $\vec{\delta}$ guarantee that every generalized solution (u,p) of the homogeneous Stokes system satisfies the estimate

$$\|u\|_{V_{\vec{\beta},\vec{\delta}}^{0,\infty}(\mathcal{G})} \leq c \|u|_{\partial\mathcal{G}}\|_{V_{\vec{\beta},\vec{\delta}}^{0,\infty}(\partial\mathcal{G})}.$$

For $\vec{\beta} = 0$, $\vec{\delta} = 0$, the maximum principle

$$\|u\|_{L_\infty(\mathcal{G})} \leq c\|u\|_{L_\infty(\partial\mathcal{G})}$$

holds.

Also, the authors of [83] obtained regularity assertions for the solution $(u,p) \in W_2^1(\mathcal{G})^3 \times L_2(\mathcal{G})$ of the Dirichlet problem for the Navier-Stokes system. They proved that it belongs to $V_{\vec{\beta},\vec{\delta}}^{l,p}(\mathcal{G}) \times V_{\vec{\beta},\vec{\delta}}^{l-1,p}(\mathcal{G})$, $l = 1$ or $l = 2$, if $\vec{\beta}$ and $\vec{\delta}$ satisfy certain inequalities. A similar result was obtained for a class of weighted Hölder spaces. Furthermore, it is shown that the vector u of an arbitrary solution $(u,p) \in W_2^1(\mathcal{G})^3 \times L_2(\mathcal{G})$ of (5.6) with the boundary condition $u = \varphi$ on $\partial\mathcal{G}$ belongs to the space $C^1(\overline{\mathcal{G}})$ if $\varphi \in C^1(\partial\mathcal{G})$.

In [84], Maz'ya, Plamenevskiĭ and Stupyalis studied the boundary value problem of steady-state motion of a viscous incompressible fluid with a free surface in a cylindrical vessel, i.e., in a vessel which coincides with a cylinder $\Omega \times \mathbb{R}$ up to a certain depth. In this case the boundary of the domain $\mathcal{G}$ consists of two components F and Σ. The Dirichlet condition is imposed on Σ (the walls of the vessel), whilst on the free surface $F : x_3 = \varphi(x_1, x_2)$ (the function φ is unknown a priori and is to be determined) the normal component of u, and the tangential component $s_{n,\tau}(u)$ of the vector $s_n(u) = s(u) \cdot n$, where $s(u)$ is the matrix with the components $s_{i,j} = \partial u_i/\partial x_j + \partial u_j/\partial x_i$, vanish. So we have the boundary value problem

$$(5.7) \qquad -\Delta u + \nabla p + (u \cdot \nabla)u = 0, \quad \operatorname{div} u = 0 \quad \text{in } \mathcal{G},$$

$$(5.8) \qquad u = Ra \quad \text{on } \Sigma, \qquad s_{n,\tau}(u) = 0 \quad \text{on } F,$$

where φ, p and u are subjected to relations

$$(5.9) \qquad \operatorname{div}\left(\frac{\nabla\varphi}{\sqrt{1 + |\nabla\varphi|^2}}\right) - \beta\varphi = W(-p + s_n(u))(x_1, x_2, \varphi(x_1, x_2)) \quad \text{in } \Omega,$$

$$(5.10) \qquad (1 + |\nabla\varphi|^2)^{-1/2}\frac{\partial\varphi}{\partial\nu} = \cos\theta \quad \text{on } \partial\Omega.$$

Here R, W, β are positive parameters, and R is assumed to be small. The exterior normal to $\partial\Omega$ is denoted by ν, and $\theta \in (0, \pi)$ is a given angle. A vector-function a is equal to zero on the cylindrical part of Σ, and satisfies

$$\int_\Sigma a \cdot n d\sigma = 0.$$

The main theorem in [84] proves the unique solvability of (5.7)–(5.10). A similar result was independently obtained by V.A. Solonnikov.

A detailed analysis of the operator pencil generated by the Dirichlet problem for the Stokes system in a three-dimensional cone $\mathcal{K}$ is given by Kozlov, Maz'ya and Schwab in [28]. In particular, it is shown there that the strip

$$(5.11) \qquad \left|\operatorname{Im}\lambda - 1/2\right| < F(\Omega),$$

contains only purely imaginary eigenvalues. Here the set-function $F(\Omega)$ depends monotonically on the domain $\Omega = \mathcal{K} \cap S^2$ and $F(\Omega) \geq 3/2$. Moreover, there is no generalized eigenvectors corresponding to eigenvalues $\lambda \neq -i$, $\lambda \neq 2i$. At last, estimates for the widest strip, centered at the line $\operatorname{Im}\lambda = 1/2$ and free of eigenvalues, are obtained by means of a non-standard variational principle. In the case when $\mathcal{K}$ is a dihedron, it is possible to express the eigenvalues, eigenvectors and generalized eigenvectors of the above mentioned operator pencil in terms of the same quantities for the pencil generated by the two-dimensional Stokes problem. This is done in the paper [17] of Kozlov and Maz'ya.

In [9], Kozlov and Maz'ya proved that there are exactly two eigenvalues $\lambda_0 = 0$ and $\lambda_1 = i$ of the operator pencil generated by the Neumann problem for the Stokes system in the strip $0 \leq \operatorname{Im}\lambda \leq 1$, if the cone $\mathcal{K}$ is given by the inequality $x_3 > \varphi(x_1, x_2)$ and φ is positively homogeneous function of the degree one and smooth on $\mathbb{R}^2 \backslash \{0\}$. Geometric and algebraic multiplicities of both eigenvalues are equal to three, and the eigenspace corresponding to $\lambda = 0$ consists of the vectors $(c, 0)$, where $c \in \mathbb{C}^3$.

Kozlov, Maz'ya and Roßmann [26] studied the singularities of solutions to the Stokes system in a three-dimensional cone $\mathcal{K}$, when the boundary conditions

$$\begin{aligned}
&\text{(i)} && u = 0, \\
&\text{(ii)} && u_n = 0, \ s_{n,\tau}(u) = 0, \\
&\text{(iii)} && u_\tau = 0, \ s_{n,n}(u) = 0
\end{aligned}$$

are imposed on different parts of $\partial \mathcal{K} \backslash \{0\}$. Here u_n, u_τ denote the normal and tangential components of u, and $s_{n,n}$, $s_{n,\tau}$ are the normal and tangential components of the vector $s_n(u) = s(u) \cdot n$ introduced above. It is shown in [26], that the strip $0 \le \operatorname{Im} \lambda \le 1$ is free of eigenvalues of the corresponding operator pencil, the eigenvalues in the strip $-1 \le \operatorname{Im} \lambda \le 2$ are purely imaginary, and there are no generalized eigenvectors corresponding to eigenvalues inside the last strip.

All spectral results formulated in this section have immediate corollaries concerning the asymptotic formulas for solutions of the Stokes and Navier-Stokes systems near a cone vertex.

6. Singularities of solutions to nonlinear elliptic equations at a cone vertex

This survey has become already very lengthy, but approaching its end, I see that it is still incomplete. I either said too little about some themes or just missed a relevant material. However, the topic of the present section could not be omitted because the boundary singularities of solutions to quasilinear elliptic equations attracted Maz'ya's attention as early as the late 1960s, and he still works in this rapidly developing field.

In 1969, Maz'ya and his post-graduate student I.Krol' constructed special solutions to the p-Laplace equation:

$$\operatorname{div}(|\operatorname{grad}|^{p-2} \operatorname{grad} u) = 0,$$

$1 < p \le n - 1$, in a rotational cone (see [29, 30]). To be more precise, they proved the existence of a nonnegative solution

$$u(x) = |x|^{\lambda(\theta)} \Phi(x_n |x|^{-1}),$$

which vanishes on the boundary of the cone $\{x : x_n |x|^{-1} \ge \cos \theta\}$, where θ is sufficiently close to π and $\lambda(\theta) > 0$. They showed that the asymptotic behavior of $\lambda(\theta)$ as $\theta \to \pi$ has the form

$$\frac{1}{\lambda(\theta)} = c \int_{\pi/2}^{\theta} \frac{d\phi}{(\sin \phi)^{\frac{n-2}{p-1}}} + O(1),$$

where c is a constant depending on n and p. The function Φ and the exponent $\lambda(\theta)$ are determined by a spectral problem for a very complicated ordinary differential equation. In [90], Maz'ya, Slutskiĭ and Fomin showed that the same problem in the plane with a cut along a half-axis can be solved explicitly. The result is useful for fracture mechanics, since the problem describes the steady-state creep in a rod with a crack under torsion,

With the development of a linear theory of general elliptic problems in domains with conical points its applications to quasilinear boundary value problems became an issue. Description of strong singularities of solutions to these problems is a difficult question. However, it seems natural to construct a complete asymptotic expansion of the weakly singular solution by linearization. An explicit statement of such a result was given by Maz'ya and Plamenevskiĭ in the paper [67] of 1973. They considered the elliptic boundary value problem

$$\sum_{|\alpha|\leq 2m} a_\alpha(x,u,\ldots,D_{2m-1}u)\,D^\alpha u = f(x,u,\ldots,D_{2m-1}u) \ \text{ in } \mathcal{G},$$

$$\sum_{|\alpha|\leq m_j} b_{j,\alpha}(x,u,\ldots,D_{m_j-1}u)\,D^\alpha u = g_j(x,u,\ldots,D_{m_j-1}u) \ \text{ on } \partial\mathcal{G}\backslash\{0\}$$

(here D_k is the gradient of order k) in a domain $\mathcal{G}$ of $\mathbb{R}^n$ with the conical point $x=0$ on the boundary. Under the condition that the solution u belongs both to $W^{2m}_{p,loc}(\mathcal{G})$ and to the space $C^l_\beta(\mathcal{G})$ with the norm

$$\|u\|_{C^l_\beta(\mathcal{G})} = \|r^\beta u\|_{C^l(\overline{\mathcal{G}})} + \|r^{\beta-l}u\|_{C(\overline{\mathcal{G}})},$$

where $\beta < 0$ and l is the highest order of derivatives of u appearing in the coefficients a_α, $b_{j,\alpha}$ and in the functions f, g_j, they obtained an asymptotic representation for u

In the recent paper [20], Kozlov and Maz'ya described the asymptotics of solutions to the Neumann problem for the nonlinear elliptic equation

$$\Delta u + \alpha(x)\left(\frac{\partial u}{\partial x_1}\right)^2 + 2\beta(x)\,\frac{\partial u}{\partial x_1}\,\frac{\partial u}{\partial x_2} + \gamma(x)\left(\frac{\partial u}{\partial x_2}\right)^2 = 0$$

in the sector $K_\delta = \{x \in \mathbb{R}^2 : 0 < r < \delta, 0 < \theta < \varphi\}$ without a priori restrictions on the growth of u. (Here r,θ are the polar coordinates of x and $\varphi \in (0,2\pi]$.) They showed that there are two possibilities: either u is unbounded and then

$$u(x) = \varphi \int_r^{\delta/e} \frac{ds}{s}\left(\int_{x\in K_\delta\backslash K_s} \frac{\alpha x_1^2 + 2\beta x_1 x_2 + \gamma x_2^2}{|x|^4}\,dx\right)^{-1} + c_* + o(1)$$

or u is bounded and has the same asymptotics

$$u = c_0 + c_1\,r^{\pi/\varphi}\cos(\pi\theta/\varphi) + o\!\left(r^{\pi/\varphi}\right)$$

as in the case of the Neumann problem for the Laplace equation. Here c_*, c_0, c_1 are real constants.

References

[1] Courant, R., Hilbert, D., *Methoden der mathematischen Physik*, Bd. 2, Springer, Berlin – Heidelberg – New York 1968.

[2] Grisvard, P., *Elliptic problems in nonsmooth domains*, Monographs and Studies in Mathematics **21** Pitman, Boston 1985.

[3] Kondrat'ev, V. A., *Boundary value problems for elliptic equations in domains with conical or angular points*, Trudy Moskov. Mat. Obshch. **16** (1967) 209–292.

[4] Kondrat'ev, V. A., Oleĭnik, O. A., *Boundary value problems for partial differential equations in nonsmooth domains*, Uspekhi Mat. Nauk **38** (1983) 2, 3–76.

[5] Kozlov, V. A., Maz'ya, V. G., *Estimates of the L_p means and asymptotics of the solutions of elliptic boundary value problems in a cone, I*, Semin. Anal., Oper. Equ. and Numer. Anal. 1985/86, Inst. Math. Berlin, 55–91.

[6] Kozlov, V. A., Maz'ya, V. G., *Singularities of solutions to first boundary value problem for thermal conductivity equation in domains with conical points. I*, Izv. Vyssh. Uchebn. Zaved. Mat. **31** (1987) 2, 38–46.

[7] Kozlov, V. A., Maz'ya, V. G., *Singularities of solutions to first boundary value problem for thermal conductivity equation in domains with conical points. II*, Izv. Vyssh. Uchebn. Zaved. Mat. **31** (1987) 3, 37–44.

[8] Kozlov, V. A., Maz'ya, V. G., *Estimates of the L_p means and asymptotics of the solutions of elliptic boundary value problems in a cone, II: Operators with variable coefficients*, Math. Nachr. **137** (1988) 113–139.

[9] Kozlov, V. A., Maz'ya, V. G., *Spectral properties of operator pencils generated by elliptic boundary value problems in a cone*, Funktsional. Anal. i Prilozh. **22** (1988), 38–46; English transl. in: Functional Anal. Appl. **22** (1988) No. 2, 114–121.

[10] Kozlov, V. A., Maz'ya, V. G., *An asymptotic formula for eigenfunctions of the Dirichlet problem in a domain with a conical point*, Vestnik Leningrad. Univ. Mat. **21** (1988) 4, 1988.

[11] Kozlov, V. A., Maz'ya, V. G., *On the spectrum of the operator pencil generated by the Dirichlet problem in a cone*, Mat. Sb. **182** (1991) 5; English transl. in: Math. USSR Sbornik **73** (1992) 1, 27–48.

[12] Kozlov, V. A., Maz'ya, V. G., *On stress singularities near the boundary of a polygonal crack*, Proc. Roy. Soc. Edinburgh **117A** (1991), 31–37.

[13] Kozlov, V. A., Maz'ya, V. G., *On the spectrum of an operator pencil generated by the Neumann problem in a cone*, Algebra i Analiz **3** (1991), 111–131; English transl. in: St. Petersburg Math. J. **3** (1992) No. 2, 333–353.

[14] Kozlov, V. A., Maz'ya, V. G, *On quasihomogeneous solutions of the Dirichlet problem for elliptic systems in $K_d \times \mathbb{R}^{n-d}$, where K_d is a d-dimensional cone*, Preprint Lith-Math-R-91-12, Univ. Linköping 1991.

[15] Kozlov, V. A., Maz'ya, V. G, *On eigenvalues of the operator pencil generated by the Dirichlet problem for a strongly elliptic system in a polyhedral angle*, Preprint Lith-Math-R-91-14, Univ. Linköping 1991.

[16] Kozlov, V. A., Maz'ya, V. G, *Singularities of solutions to problems of mathematical physics in nonsmooth domains*, Partial Differential Equations and Functional Analysis, In Memory of Pierre Grisvard, Birkhäuser, Boston-Basel-Berlin 1996.

[17] Kozlov, V. A., Maz'ya, V. G, *On power-logarithmic solutions to the Dirichlet problem for the Stokes system in a dihedral angle*, Mathematical Methods in the Applied Sciences **20** (1997) 315–346.

[18] Kozlov, V. A., Maz'ya, V. G, *Theory of a higher order Sturm-Liouville equation*, Lecture Notes in Mathematics **1659**, Springer-Verlag Berlin 1997.

[19] Kozlov, V. A., Maz'ya, V. G., *Differential equations with operator coefficients*, Springer 1999.

[20] Kozlov, V. A., Maz'ya, V. G., *Angle singularities of solutions to the Neumann problem for the two-dimensional Ricatti's equation*, to appear in Asymptotic Analysis

[21] Kozlov, V. A., Maz'ya, V. G., Parton, V. Z., *Asymptotic form of the stress intensity coefficients in quasistatic temperature problems for a domain with a cut*, Prikl. Mat. Mekh. **49** (1985) 4, 627–636.

[22] Kozlov, V. A., Maz'ya, V. G., Parton, V. Z., *Thermal shock in a domain with a crack*, Prikl. Mat. Mekh. **52** (1988) 2, 318–326.

[23] Kozlov, V. A., Maz'ya, V. G., Parton, V. Z., *Asymptotics of the intensity factors for stresses induced by heat sources*, Journal of Thermal Stresses **17** (1994) 309–320.

[24] Kozlov, V. A., Maz'ya, V. G., Rossmann, J. *Elliptic boundary value problems in domains with point singularities*, Math. Surveys and Monogr. **52**, Amer. Math. Soc., Providence, Rhode Island 1997.

[25] Kozlov, V. A., Maz'ya, V. G., Rossmann, J. *Spectral properties of operator pencils generated by elliptic boundary value problems for the Lamé system*, Rostock. Math. Kolloq. **51** (1997) 5–24.

[26] Kozlov, V. A., Maz'ya, V. G., Rossmann, J. *Conic singularities of solutions to problems in hydrodynamics of a viscous fluid with a free surface*, to appear in Math. Scand.

[27] Kozlov, V. A., Maz'ya, V. G., Schwab, C., *On singularities of solutions of the displacement problem of linear elasticity near the vertex of a cone*, Arch. Rational Mech. Anal. **119** (1992) 197–227.

[28] Kozlov, V. A., Maz'ya, V. G., Schwab, C., *On singularities of solutions to the Dirichlet problem of hydrodynamics near the vertex of a cone*, J. Reine Angew. Math. **456** (1994) 65–97.

[29] Krol', I. N., Maz'ya, V. G., *The lack of the continuity and Hölder continuity of solutions of a certain quasilinear elliptic equations*, Zap. Nauchn. Sem. Leningrad. Otdel. Mat. Inst. Steklov. (LOMI) **14** (1969) 89–91.

[30] Krol', I. N., Maz'ya, V. G., *The absence of the continuity and Hölder continuity of solutions to quasilinear elliptic equations near a nonregular boundary*, Trudy Moskov. Mat. Obshch. **26** (1972) 75–94.

[31] Maz'ya, V. G., *Classes of domains and embedding theorems for function spaces*, Dokl. Akad. Nauk SSSR **133** (1960) 3; English transl. in: Soviet Math. Dokl. **1** (1960) 882–885.

[32] Maz'ya, V. G., *Some estimates of solutions to elliptic equations of second order*, Dokl. Akad. Nauk SSSR **137** (1961) 5, 1057–1059.

[33] Maz'ya, V. G., *Classes of sets and embedding theorems of function spaces. Some questions of the theory of elliptic equations*, Candidate's thesis, Moscow University, 1962.

[34] Maz'ya, V. G., *On the solvability of the Neumann problem*, Doklady Akad. Nauk SSSR **147** (1962) 2, 294–296.

[35] Maz'ya, V. G., *The negative spectrum of the higher-dimensional Schrödinger operator*, Dokl. Akad. Nauk SSSR **144** (1962) 721–722.

[36] Maz'ya, V. G., *Embedding theorems for arbitrary sets*, Uspekhi Mat. Nauk **17** (1962) 1, 247–248.

[37] Maz'ya, V. G., *On the Dirichlet problem for elliptic equations of arbitrary order in unbounded domains*, Dokl. Akad. Nauk SSSR **150** (1963) 6, 1221–1224.

[38] Maz'ya, V. G., *Regularity on the boundary of elliptic equations and conformal mappings*, Dokl. Akad. Nauk SSSR **152** (1963) 1297–1300.

[39] Maz'ya, V. G., *The solvability in $\overset{\circ}{W}{}^2_2$ of the Dirichlet problem for a region with a smooth irregular boundary*, Vestnik Leningrad. Univ. **19** (1963) 7, 163–165.

[40] Maz'ya, V. G., *On the theory of the higher-dimensional Schrödinger operator*, Izv. Akad. Nauk SSSR **28** (1964) 5, 1145–1172.

[41] Maz'ya, V. G., *The Dirichlet and Neumann problems in domains with nonregular boundaries*, D.Sc. thesis, Leningrad University, 1965.

[42] Maz'ya, V. G., *Polyharmonic capacity in the theory of the first boundary value problem*, Sibirsk. Mat. J. **6** (1965) 1, 127–148.

[43] Maz'ya, V. G., *On the theory of the multidimensional Schrödinger operator*, Vestnik Leningrad. Univ. **20** (1965) 1, 135–137.

[44] Maz'ya, V. G., *On the modulus of continuity of a solution of the Dirichlet problem near an irregular boundary*, Problems Math. Anal. Boundary Value Problems Integr. Equations, Izd. Leningrad. Univ., pp. 45–58, 1966.

[45] Maz'ya, V. G., *On the self-adjointness of the Laplace operator*, Embedding Theorems and their Applications, Proceedings of the Symposium on Embedding Theorems, Baku 1966.

[46] Maz'ya, V. G., *On the solvability in $\overset{\circ}{W}{}^2_2$ of the Dirichlet problem in a domain with smooth nonregular boundary*, Vestnik Leningrad. Univ., Ser. Mat., Mekh., Astron. **22** (1967) 7, 87–95.

[47] Maz'ya, V. G., *The behavior near the boundary of the solution of the Dirichlet problem for an elliptic equation of second order in divergence form*, Mat. Zametki **2** (1967) 209–220.

[48] Maz'ya, V. G., *On Neumann's problem in domains with nonregular boundaries*, Sibirsk. Matem. Zh. **9** (1968) 6, 1322–1350.

[49] Maz'ya, V. G., *On the boundedness of the first derivatives of a solution of the Dirichlet problem in a domain with smooth nonregular boundary*, Vestnik Leningrad. Univ., Ser. Mat., Mekh., Astron. **24** (1969) 1; English transl. in: Vestnik Leningrad Univ., Math. **2** (1975) 2, 87–95.

[50] Maz'ya, V. G., *On weak solutions of the Dirichlet and Neumann problems*, Trans. Moscow Math. Soc. **20** (1969) 135–172.

[51] Maz'ya, V. G., *On the Neumann problem for elliptic operators of arbitrary order in domains with nonregular boundaries*, Vestnik Leningrad. Univ., Ser. Mat. Mekh. Astron., No. 1 (1972) 26–33.

[52] Maz'ya, V. G., *On Beurling's theorem on the minimum principle for positive harmonic functions*, Investigations on linear operators and the theory of functions, III. Zap. Nauchn. Sem. Leningrad. Otdel. Mat. Inst. Steklov. (LOMI) **30** (1972) 76–90.

[53] Maz'ya, V. G., *On the coercivity of the Dirichlet problem in a domain with nonregular boundary*, Izv. Vyssh. Ucheb. Zaved., Matem., **4** (1973) 64–76.

[54] Maz'ya, V. G., *On a problem with oblique derivative for a domain of polyhedral type*, Dokl. Akad. Nauk SSSR **211** (1973) 1; English transl. in: Soviet Math Dokl. **14** (1973) 4.

[55] Maz'ya, V. G., *On the index of the closure of the operator to the Dirichlet problem in a domain with nonregular boundary*, Probl. Mat. Anal. **5** (1975) 98–121.

[56] Maz'ya, V. G., *On the behavior near the boundary of solutions to the Dirichlet problem for the biharmonic operator*, Dokl. Akad. Nauk SSSR **235** (1977) 6, 1263–1266.

[57] Maz'ya, V. G., *The connection between Martin's and Euclid's topologies*, Dokl. Akad. Nauk SSSR **233** (1977) 1, 27–30.

[58] Maz'ya, V. G., *Sobolev spaces*, Izdat. Leningrad Univ., Leningrad, 1985; English transl.: Springer-Verlag, Berlin – New York, 1985.

[59] Maz'ya, V. G., *Classes of domains, measures and capacities in the theory of differentiable functions*, Analysis, III, 141–211, Encyclopaedia Math. Sci. **26**, Springer, Berlin 1991.

[60] Maz'ya, V. G., Nazarov, S. A., Plamenevskiĭ, B. A., *On the singularities of solutions of the Dirichlet problem in the exterior of a slender cone*, Mat. Sb. **122** (1983) 4, 435–457; English transl. in: Math. USSR Sb. **50** (1985) 2, 415–437.

[61] Maz'ya, V. G., Nazarov, S. A., Plamenevskiĭ, B. A. *Elliptic boundary value problems in domains of the exterior of a cusp type*, Probl. Mat. Anal. **9** (1984) 105–148.

[62] Maz'ya, V. G., Plamenevskiĭ, B. A., *On the asymptotic behaviour of solutions of differential equations with operator coefficients*, Dokl. Akad. Nauk SSSR **196** (1971) 512–515; English transl. in: Soviet Math. Dokl. **12** (1971).

[63] Maz'ya, V. G., Plamenevskiĭ, B. A., *Problem with oblique derivative in a domain with piecewise smooth boundary*, Funktsional. Anal. i Prilozh. **5** (1971) 3, 102–103; English transl. in: Functional Anal. Appl. **5** (1971).

[64] Maz'ya, V. G., Plamenevskiĭ, B. A., *On a class of manifolds with singularities*, Izv. Vyssh. Uchebn. Zaved. Mat. **11** (1972) 46–52.

[65] Maz'ya, V. G., Plamenevskiĭ, B. A., *The asymptotic behavior of solutions of differential equations in Hilbert space*, Izv. Akad. Nauk SSSR, Ser. Mat. **36** (1972) 1080–1133,; erratum, ibid. **37** (1973) 709–710.

[66] Maz'ya, V. G., Plamenevskiĭ, B. A., *On elliptic boundary value problems in domains with piecewise smooth boundary*, Trudy Simp. Mekh. Sploshnoi Sredy 1971, Metsniereba, Tbilissi 1973, 171–182.

[67] Maz'ya, V. G., Plamenevskiĭ, B. A., *On the behavior of solutions to quasilinear elliptic boundary value problems in a neighborhood of a conical points*, Zap. Nauchn. Sem. Leningrad. Otdel. Mat. Inst. Steklov. (LOMI) **38** (1973) 94–97.

[68] Maz'ya, V. G., Plamenevskiĭ, B. A., *On the coefficients in the asymptotics of solutions of elliptic boundary value problems near conical points*, Dokl. Akad. Nauk SSSR **219** (1974) 286–289.

[69] Maz'ya, V. G., Plamenevskiĭ, B. A., *On the coefficients in the asymptotics of solutions of elliptic boundary value problems in a cone*, Zap. Nauchn. Sem. Leningrad. Otdel. Mat. Inst. Steklov. (LOMI) **52** (1975) 110–127.

[70] Maz'ya, V. G., Plamenevskiĭ, B. A., *On boundary value problems for a second order elliptic equation in a domain with edges*, Vestnik Leningrad. Univ. Mat. **1** (1975) 102–108; English transl. in: Vestnik Leningrad. Univ. Mat. **8** (1980) 99–106.

[71] Maz'ya, V. G., Plamenevskiĭ, B. A., *Weighted spaces with nonhomogeneous norms and boundary value problems in domains with conical points*, Ellipt. Differentialgleichungen (Meeting, Rostock, 1977) Univ. Rostock, 1978, 161–189; English transl. in: Amer. Math. Soc. Transl. **123** (1984) 89–107.

[72] Maz'ya, V. G., Plamenevskiĭ, B. A., *On the coefficients in the asymptotics of solutions of elliptic boundary value problems in domains with conical points*, Math. Nachr. **76** (1977) 29–60; English transl. in: Amer. Math. Soc. Transl. **123** (1984) 57–88.

[73] Maz'ya, V. G., Plamenevskiĭ, B. A., *On the asymptotics of the solution of the Dirichlet problem near an isolated singularity of the boundary*, Vestnik Leningrad. Univ. Mat. Mekh. Astron. **13** (1977) 59–66; English transl. in: Vestnik Leningrad Univ. Math. **10** (1982) 295–302.

[74] Maz'ya, V. G., Plamenevskiĭ, B. A., *Elliptic boundary value problems on manifolds with singularities*, Probl. Mat. Anal. **6** (1977) 85–142.

[75] Maz'ya, V. G., Plamenevskiĭ, B. A. *Estimates in L_p and Hölder classes and the Miranda-Agmon maximum principle for solutions of elliptic bvp in domains with singular points on the boundary*, Math. Nachrichten **81** (1978) 25–82; English transl. in: Amer. Math. Soc. Transl., Vol. 123 (1984) 1–56.

[76] Maz'ya, V. G., Plamenevskiĭ, B. A. *L_p estimates for solutions of elliptic boundary value problems in domains with edges*, Trudy Moskov. Matem. Obshch. **37** (1978) 49–93; English transl. in: Trans. Moscow. Math. Soc. **1** (1980) 49–97.

[77] Maz'ya, V. G., Plamenevskiĭ, B. A. *Estimates of Green functions and Schauder estimates for solutions of elliptic boundary value problems in a dihedral angle*, Sibirsk. Mat. Zh. **19** (1978) 5, 1065–1082.

[78] Maz'ya, V. G., Plamenevskiĭ, B. A. *Schauder estimates of solutions of elliptic boundary value problems in domains with edges on the boundary*, Partial Differential Equations (Proc. Sem. S. L. Sobolev, 1978, No. 2) Inst. Mat. Sibirsk. Otdel. Akad. Nauk SSSR, Novosibirsk, 1978, 69–102; English transl. in: Amer. Math. Soc. Transl. **123** (1984) 141–169.

[79] Maz'ya, V. G., Plamenevskiĭ, B. A. *On the asymptotics of the fundamental solutions of elliptic boundary value problems in regions with conical points*, Probl. Mat. Anal. **7** (1979) 100–145; English transl. in: Sel. Math. Sov. **4** (1985) 363–397.

[80] Maz'ya, V. G., Plamenevskiĭ, B. A. *On the first boundary value problem for equations of hydrodynamics in a domain with piecewise smooth boundary*, Zap. Nauchn. Sem. Leningrad. Otdel. Mat. Inst. Steklov. (LOMI) **98** (1980) 179–184.

[81] Maz'ya, V. G., Plamenevskiĭ, B. A. *On the maximum principle for the biharmonic equation in domains with conical points*, Izv. Vyssh. Uchebn. Zaved. Mat. **2** (1981) 52–59.

[82] Maz'ya, V. G., Plamenevskiĭ, B. A. *On properties of solutions of three-dimensional problems of elasticity theory and hydrodynamics in domains with isolated singular points*, Dinamika Sploshnoi Sredy **50** (1981) 99–120; English transl. in: Amer. Math. Soc. Transl. (2) **123** (1984) 109–123.

[83] Maz'ya, V. G., Plamenevskiĭ, B. A. *The first boundary value problem for classical equations of mathematical physics in domains with piecewise smooth boundaries I, II*, Zeitschr. Anal. Anw. **2** (1983) 335–359, 523–551.

[84] Maz'ya, V. G., Plamenevskiĭ, B. A., Stupyalis, L. I. *The three-dimensional problem of steady state motion of a fluid with a free surface*, Differentsial'nye Uravneniya i Primenen. – Trudy Sem. Protsessy Optimal. Upravleniya **23** (1979); English transl. in: Amer. Math. Soc. Transl. (2) **123** (1984) 171–268.

[85] Maz'ya, V. G., Roßmann, J., *Über die Asymptotik der Lösungen elliptischer Randwertaufgaben in der Umgebung von Kanten*, Math. Nachr. **138** (1988) 27–53.

[86] Maz'ya, V. G., Roßmann, J., *On the Agmon-Miranda maximum principle for solutions of elliptic equations in polyhedral and polygonal domains*, Ann. Global Anal. Geom. **9** (1991) 253–303.

[87] Maz'ya, V. G., Roßmann, J., *On a problem of Babuška (Stable asymptotics of the solution to the Dirichlet problem for elliptic equations of second order in domains with angular points*. Math. Nachr. **155** (1992) 199–220.

[88] Maz'ya, V. G., Roßmann, J., *On the behaviour of solutions to the Dirichlet problem for second order elliptic equations near edges and polyhedral vertices with critical angles*, Z. Anal. Anw. **13** (1994) 1, 19–47.

[89] Maz'ya, V. G., Roßmann, J., *On the Agmon-Miranda maximum principle for solutions of strongly elliptic equations in domains of $\mathbb{R}^n$ with conical points*, Ann. Global Anal. Geom. **10** (1992) 125–150.

[90] Maz'ya, V. G., Slutskiĭ, A. S., Fomin, V. A. *Asymptotic behavior of the stress function near the vertex of a crack in the problem of torsion under steady state creep*, Mekh. Tverd. Tela **4** (1986) 170–176.

[91] Verzhbinskiĭ, G. M., Maz'ya, V. G., *On the asymptotics of solutions of the Dirichlet problem near a nonregular boundary*, Dokl. Akad. Nauk SSSR **176** (1967) 3, 498–501.

[92] Verzhbinskiĭ, G. M., Maz'ya, V. G., *Asymptotic behaviour of solutions of elliptic equations of second order near the boundary I, II*, Sibirsk. Mat. Zh. **12** (1971) 6, 1217–1249, **13** (1972) 6, 1239–1271.

[93] Verzhbinskiĭ, G. M., Maz'ya, V. G., *Closure in L_p of the Dirichlet problem operator in a region with conical points*, Izv. Vyssh. Uchebn. Zaved. Mat. **18** (1974) 6, 8–19.

University of Rostock,
Department of Mathematics,
D-18051 Rostock, Germany,
juergen.rossmann@mathematik.uni-rostock.de

1991 Mathematics Subject Classification: Primary 35-03; Secondary 35B40, 35J40

Submitted: 01.02.1999

Operator Theory:
Advances and Applications, Vol. 109

On some potential theoretic themes in function theory

VICTOR P. HAVIN

On the 60th anniversary of my friend Vladimir Maz'ya I am happy to remember a fortunate time long ago when we worked on some problems in function theory (1965–1972). Our themes were

1) L^p-approximation by analytic and harmonic functions,

2) Uniqueness properties of analytic functions,

3) The Cauchy problem for the Laplace equation,

4) Non-linear potential theory.

A characteristic feature of everything we did was the heavy use of potential theoretic methods and ideas. Even if a problem didn't contain anything potential theoretic in its statement (as was the case with the first and to some extent with the second theme), then potential theory would emerge by itself in the solution. Working on 1) and 2) we were compelled to invent a "non-linear potential theory" (among the initial works on the subject are [MH1], [MH2], but this is another story, not to be discussed here; see [AH]). As to theme 3), the problem was potential theoretic from the outset, but it is related to and suggested by traditional themes of pure function theory (quasianalyticity, moment problem, and weighted polynomial approximation in the spirit of S. N. Bernstein) and so corresponds completely to the title of this article.

I am going to summarize the joint work V. Maz'ya and I did on themes 1)–3).

Taking into account the character of this volume I hope the reader will excuse a certain amount of personal digressions.

1. Approximation theory

I became a student of the department of mathematics and mechanics of the Leningrad State University in 1950. One of the major events in analysis of that time was Mergelyan's proof of his theorem about uniform polynomial approximation on plane compacta. The statement (not the proof!) of the result is simple, so I was able to understand it and be impressed by it rather early. My further interest in approximation properties of analytic functions was sparked by my father's 1953 New Year present to me, the one year "Uspekhi" subscription. I had already started taking my first course of complex analysis. One of the rare articles I was educated enough to understand was [Me1]. It was about weighted L^2-approximation by polynomials. Approximation in the mean (with its subtle phenomena making it so different from the uniform approximation) puzzled and attracted me for several years after.

At that time approximation in the complex plane meant "polynomial" or "rational" approximation and was generally perceived as the development and outgrowth of the Weierstrass theorem on approximation by polynomials on an interval. But the $\bar{\partial}$-ideology was in the air, and the time was ripe to understand (at last) that analytic functions are solutions of the Cauchy-Riemann system (*generalized* solutions at that). By the Runge theorem, rational uniform approximation on a compact set $K \subset \mathbb{C}$ is equivalent to the uniform approximation by functions analytic n e a r K, that is by solutions of a concrete system of differential equations.

An event which prepared me (without my realization) for this point of view and future cooperation with Vladimir Maz'ya was my candidate (=Ph.D preliminary) exam 1956. L. V. Kantorovich, my adviser, made me study a pile of books including Sobolev's book [So] published by our university in 1950. This great book was extremely hard to read as the density of misprints and small mistakes was exceedingly high. The subject had nothing to do with the theme of my Ph.D thesis, my permanent concern at the time. So I considered studying this book as a great nuisance, but had to obey my adviser, and so I spent one semester struggling with the text. At the beginning of the sixties the two themes (polynomial approximation in the mean and Sobolev spaces) somehow crossed in my head and led me to the following simple observation:

Let G be a plane domain, $1 < p < +\infty$; a function $f \in L^p(G)$ is analytic iff $\bar{\partial} f = 0$ (distributionally), that is $\int_G f \bar{\partial}\varphi = 0$ for any $\varphi \in C_0^\infty$ with $\operatorname{supp}\varphi \subset G$ ($\varphi \in \mathcal{D}(G)$ for short). This fact makes it possible to apply duality and reduce the L^p-approximation by analytic functions on subsets of G to some uniqueness questions addressed to the closure in $L^q(G)$, $q = p/(p-1)$, of functions $\bar{\partial}\varphi$, $\varphi \in \mathcal{D}(G)$. But (using the Calderón-Zygmund estimates of the simplest singular integrals) this closure can be shown to coincide with $\bar{\partial}\ \overset{\circ}{W}{}_q^1(G)$, $\overset{\circ}{W}{}_q^1(G)$ being the Sobolev space of funtions φ with $\operatorname{grad}\varphi \in L^q(G)$ vanishing (in a due sense) on the bundary of G. This scheme looked especially promising for $p = 2$, since the spaces $W_2^1(G)$ are within the reach of classical potential theory whose terms and methods were ready to provide simple proofs of Vitushkin-like approximation theorems by rational functions in L^2. The case $1 < p < 2$ is even simpler, since W_q^1 consists of continuous functions, but for $p \in (2, +\infty)$ an adequate "potential theory" was needed. From this point of view the classical L^∞-setting (uniform rational approximation) turns out to be the hardest, an extreme case crowning the easier L^p-scale of problems accessible to purely "real" means: the capacities used in L^p-approximation theorems are much more "real" and explicit than the analytic capacity involved into the uniform approximation (very "complex" in both senses of this word and resisting reduction to more palpable characteristics).

I was lucky to know V. Maz'ya. He was in his twenties when we met, but a true expert and master of potential theory and Sobolev spaces. He helped me a lot having taught me the "fine" W_p^l-theory ("fonctions précisées" de Deny-Lions [DL]). Very soon I was able to give the above vague ideas a definite form and describe the sets $E \subset \mathbb{C}$ such that $L^2(E)$-closure of functions analytic near E coincides with the set of all $L^2(E)$-functions analytic in the interior of E [H1]. The description was given in terms of the Cartan fine topology. This result is now a very particular case of subsequent results due to Bagby and Hedberg (see[AH]).

Then we started working together on problems of polynomial approximation in L^p-spaces. To describe these problems let us first look at a compact set $K \subset \mathbb{C}$ dividing the plane, so that $\mathbb{C}\backslash K$ has a nonempty bounded component g. Then any sequence of polynomials convergent u n i f o r m l y on K converges uniformly on $K \cup g$ as well. This fact is an obstacle for the uniform polynomial approximation on K (e.g. $(z - a)^{-1}$ cannot be uniformly approximated by polynomials on K if $a \in g$), and according to classical results it is the only obstacle if the approximated function is continuous on K and analytic in its interior. Suppose now our set K is "thick" enough; then the same can be said on sequences of polynomials converging in $L^p(K)$ for finite values of p (e.g. if K is a non-degenerate circular annulus). But this time not only topological, but some quantitative characteristics of K (of its "thickness") come into play. This phenomenon discovered by M. V. Keldyš can be illustrated by the so-called "crescent domains" K (see Figure 1); we do not suppose K to be compact anymore, this restriction being not necessary for the L^p-approximation: $K = G\backslash(g \cup \gamma)$, where G, g are Jordan domains with the boundaries Γ, γ such that $g \subset G$, $\Gamma \cap \gamma = \{p\}$. Keldyš showed that for the circular crescent ($\Gamma = \{|z| - 1\}$, $\gamma = \{|z - a| = 1 - a\}$, $0 < a < 1$) the $L^2(K)$-convergence of a sequence of polynomials implies their convergence in g, and $\mathcal{P}$, the set of all polynomials, is not dense in $L_a^2(K)$, the set of all $L^2(K)$-functions analytic in K; thus the circular crescent is "thick". On the other hand $\mathcal{P}$ is dense in $L_a^2(K)$ if the arclengths $l_1(r)$, $l_2(r)$ (see Figure 1) decrease very rapidly as $r \to 0$.

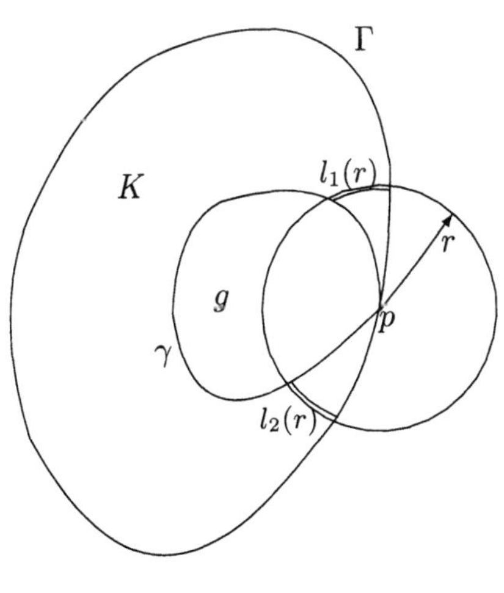

$$G = (K \cup \gamma \cup g)\backslash\{p\}$$

Figure 1

M. M. Dzhrbashyan and A. A. Shaginyan have proved that $\mathrm{Clos}_{L^2(K)}\mathcal{P} = L_a^2(K)$ iff

$$\int_0 \log l(r)\, dr = -\infty, \quad l(r) = l_1(r) + l_2(r),$$

provided γ and Γ satisfy some (rather strong) qualitative regularity conditions (see [Me1] for the historical information and references).

In [MH3], [MH4] the approach based on Sobolev spaces and sketched above was used. The density problem of $\mathcal{P}$ in L_a^p, $1 < p < +\infty$, was reduced to a real variable property of functions in the Sobolev class $\overset{\circ}{W}_q^1(G)$. This approach resulted in a new insight into the problem. We were able to replace G on Figure 1 by an a r b i t r a r y bounded domain and g by its a r b i t r a r y Jordan subdomain with γ satisfying some smoothness conditions; all restrictions imposed on $\gamma \cap \Gamma$ (where Γ is the boundary of G) were dropped, this set is not necessarily a singleton anymore. We obtained a lot of information on the closure of $L_a^p(G)$ in $L^p(K)$. In the case when $\mathrm{Clos}_{L^p(G)}\mathcal{P} = L_a^p(G)$ this information yields sharp conditions ensuring the equality $\mathrm{Clos}_{L^p(K)}\mathcal{P} = L_a^p(K)$. Here I only want to state the above mentioned real variable property of $\overset{\circ}{W}_q^1(G)$ (with respect to K), the most essential point of [MH3] (see also §3 of [MH4]):

Let μ be the harmonic measure of g on γ (w.r. to a point of g); any precised $\varphi \in \overset{\circ}{W}_q^1(G)$ satisfies

$$\int\limits_{\gamma} \log|\varphi|\, d\mu = -\infty.$$

(For $p \in (1,2)$ "precised" means just "continuous".) This property of the triple (G,g,p) expresses certain "thinness" of K in a rather implicit form: elements of $\overset{\circ}{W}_q^1(G)$ are "continuous" (this is almost literally true if $q > 2$, and can be made precise for $q \le 2$ using q-capacities, see [AH], [MH2], [M]); but $\varphi \in \overset{\circ}{W}_q^1$ is zero on Γ and, thanks to the "continuity" remains small on parts of γ which are close to $\gamma \cap \Gamma$ so that the geometric mean of φ along γ w.r. to the probability measure μ is zero. It is just this degree of thinness of K which ensures the desired approximation properties of $L_a^p(G)$ in $L^p(K)$. This very indirect characterization of the thinness of K can be made much more explicit in concrete situations using techniques of the Sobolev spaces and p-capacities (see [MH3], [MH4]).

In [H3] our approach was generalized to the L^p-approximation by solutions of some elliptic systems in $\mathbb{R}^n$. In [H2] it was used to obtain a quite explicit description of $\mathrm{Clos}_{L^p(K)}\mathcal{P}$ for a crescent K with regular γ and Γ (see Figure 1); this description is especially telling for the Keldyš circular cresent. An important progress in L^p-approximation by polynomials was made in [B1]–[B3]. And in [MH5], [MH6] an approach similar to that of [MH3], [MH4] was applied to the L^p-approximation by harmonic functions in $\mathbb{R}^n$. But this time the main difficulty was caused by the lack of uniqueness theorems for harmonic vector fields in $\mathbb{R}^n$, $n \ge 3$, and we had to use our uniqueness results from [MH7], see Section 3 below.

2. Uniqueness properties of analytic functions

In 1970 the publishing house "Mir" decided to publish the Russian translation of Carleson's book [Ca1]. I was the translator, V. Maz'ya was the editor. This episode of our cooperation (and our work described in Section 1) gave us new themes and incentives. The book dealt with all kinds of "small" and "exceptional" sets emerging in classical analysis (divergence sets for various classes of Fourier series, removable singularities of analytic and harmonic functions etc.). The smallness of exceptional sets was measured using Hausdorff measures and classical capacities adjusted to harmonic functions and Sobolev space W_2^1. Our article [MH4] was (in part) motivated by the desire to test methods of the non-linear potential theory in a new context including exceptional sets related to the Sobolev W_p^l-norms, $p \neq 2$. For example we generalized the classical Beurling theorem on the quasi everywhere convergence of the Fourier series $\sum\limits_{n=-\infty}^{+\infty} c_n e^{in\theta}$, $\sum\limits_{n=-\infty}^{+\infty} |n| \, |c_n|^2 < +\infty$, replacing the last finiteness condition by the L^p-summability of the function $\sum\limits_{n=-\infty}^{+\infty} |n|^\alpha c_n e^{in\theta}$, $\alpha > 0$; the set function whose zero-sets are divergence sets of such series is $c_{p,\alpha}$, a capacity from the non-linear potential theory.

But in this section I am going to discuss another theme suggested by [Ca1] and also by [Ca2]: the uniqueness sets for analytic functions with a finite Sobolev norm.

Suppose X is a class of functions analytic in the unit disc $\mathbb{D} := \{|z| < 1\}$ and having finite radial limits $\lim\limits_{r \uparrow 1} f(re^{i\theta}) =: f(e^{i\theta})$ for m-almost all $e^{i\theta} \in \mathbb{T} := \{|z| = 1\}$, where m denotes the normalized Lebesgue measure on $\mathbb{T}$. A set E is called *a uniqueness set* for X if

$$f \in X, \ f(\zeta) = 0 \ \text{ for all } \zeta \in E \Longrightarrow f = 0.$$

It is well known that for $X = H^p$ (= the Hardy space) the class of uniqueness sets coincides with the class of all sets of positive length on $\mathbb{T}$; if $X = A =$ the disc algebra (i.e. the set of all functions continuous on $\mathbb{D} \cup \mathbb{T}$ and analytic in $\mathbb{D}$), then closed uniqueness sets are precisely the closed sets $E \subset \mathbb{T}$ with $m(E) > 0$ (see [Ho]). If $X = A_\alpha := \{f \in A : f$ satisfies the Lipschitz condition of order $\alpha\}$, then uniqueness sets may be of zero length (or even countable). Namely a *closed set $E \subset \mathbb{T}$ is a uniqueness set for A_α iff one of the following conditions holds*: (a) $m(E) > 0$; (b) $\sum m(l) \log m(l) = -\infty$, *the sum being taken over all complementary arcs of E (the components of $\mathbb{T}\backslash E$)*; see [Ca2].

Let us now turn to the class $W_{2,a}^1$ of all functions f analytic in $\mathbb{D}$ with $\iint\limits_{\mathbb{D}} |f'(x + iy)|^2 \, dx \, dy < +\infty$ (analytic functions with finite Dirichlet integral). The uniqueness sets for this important class can hardly be given a complete and satisfactory description, the problem is too delicate and no one of the popular methods to measure "smallness" seems to be adequate. It is obvious that any

$E \subset \mathbb{T}$ with $m(E) > 0$ is a uniqueness set for $W^1_{2,a}$, since this class is contained in H^p for any $p > 0$. But at the same time functions $f \in W^1_{2,a}$ preserve certain "smoothness" at the boundary which can be expressed by the finiteness of the integral

$$\iint_{\mathbb{T} \times \mathbb{T}} \left| \frac{f(\zeta) - f(\eta)}{\zeta - \eta} \right|^2 \, dm(\zeta) \, dm(\eta).$$

Therefore it is not surprising that some uniqueness sets for $W^1_{2,a}$ look very much as the uniqueness sets for A_α. Carleson described a class of such sets $E \subset \mathbb{T}$ of zero length in [Ca2]; one of the conditions imposed on E in his theorem is the positivity of $c_\alpha(E)$, the Riesz capacity of a positive order α.

In [MH4] we described another class of uniqueness sets for $W^1_{2,a}$ which is not included into Carleson's. To state our theorem denote by E a Borel subset of $\mathbb{T}$ and let Δ be a set of non-overlapping open arcs $\delta \subset \mathbb{T}$. Put $|\delta| := m(\delta), E_\delta := E \cap \delta$. *Suppose $p \in (1,2)$ and*

$$(2.1) \qquad \sum_{\delta \in \Delta} |\delta| \log \frac{|\delta|}{c(\delta)} = -\infty$$

where $c(\delta) := \mathrm{cap}_{p,1}(E_\delta)$ ($= a$ "W^1_p-capacity of E" defined in the non-linear potential theory, see [MH4], [AH], [M]). Then E is a uniqueness set for $W^1_{p,a}$, the class of all functions f analytic in $\mathbb{D}$ with $f' \in L^p(\mathbb{D})$.

This theorem is also valid for $p = 2$ with a slightly different meaning of $c(\delta)$ due to the peculiarity of dimension two: it is now the $c_{2,1}$-capacity of E_δ with respect to the disc $2d_\delta$ where d_δ is a disc containing E_δ; this capacity can be expressed by the logarithmic capacity of E_δ or by its transfinite diameter. The case $p \in (2, +\infty)$ can also be included, but it is not interesting since $W^1_{p,a} \subset A_\alpha$, $\alpha = \alpha(p)$. It is not hard to construct a set E satisfying (2.1) whose Riesz capacities $c_\alpha(E)$ of any order $\alpha > 0$ vanish. Thus our theorem enlarges the class of uniqueness sets for $W^1_{2,a}$ found in [Ca2].

Our proof is based on an upper estimate of the geometric mean

$$\exp \left((m(O))^{-1} \int_O \log |\varphi| \, dm \right)$$

of a precised W^1_p-function φ vanishing on a set $E \subset O \subset \mathbb{T}$. This estimate involves the L^p-norm of $\nabla \varphi$ over $\mathbb{D}$ and a W^1_p-capacitary characteristic of E. It is a corollary of an inequality applicable to the $W^1_p(\mathbb{R}^n)$-functions for any n. Similar inequalities also apply to the approximation problems discussed in Section 1 (or rather to their dual counterparts dealing with the uniqueness properties of analytic functions).

3.   The Cauchy problem for the Laplace equation

This problem is mainly known as an example of ill posedness: given functions f, g defined on $\mathbb{R}^n$ find a function u harmonic in the upper half-space $\mathbb{R}^{n+1}_+ := \{(x, t) \in \mathbb{R}^{n+1} : x \in \mathbb{R}^n, t > 0\}$ satisfying

$$(3.1) \qquad\qquad u|\mathbb{R}^n = f, \ \frac{\partial u}{\partial t}\bigg|\mathbb{R}^n = g$$

(these equalities should be duly interpreted; for the time being we assume $u \in C^1(\mathbb{R}^{n+1}_+ \cup \mathbb{R}^n)$). Each of the Cauchy data $u|\mathbb{R}^n$, $\frac{\partial u}{\partial t}|\mathbb{R}^n$ determines u completely (under some reasonable restrictions imposed on u at infinity) so that conditions (3.1) are, as a rule, contradictory, and the problem is in general unsolvable. If it happens to be solvable (for a choice of f and g), then the solution is unique in a very strong sense: $u \equiv 0$ in $\mathbb{R}^{n+1}_+$ if $f \equiv g \equiv 0$ on an n-dimensional ball in $\mathbb{R}^n$.

A generic function u harmonic in $\mathbb{R}^{n+1}_+$ can be written as the Newton potential

$$u(z) = U^\varphi(z) := \int_{\mathbb{R}^n} \frac{\varphi(y)}{|z - y|^{n-2}} \, dy, \quad z = (x, t) \in \mathbb{R}^{n+1}_+,$$

where, under suitable conditions, $\varphi = c \cdot \frac{\partial u}{\partial t}|\mathbb{R}^n$. Thus the uniqueness of the solution of the Cauchy problem (3.1) can be interpreted as the uniqueness of the solution f of the following inverse problem of potential theory: determine the density φ by its restriction to an open set $E \subset \mathbb{R}^n$ and by $U^\varphi|E$. Note that if we drop the openness assumption, then the uniqueness becomes a problematic and very delicate property.

This theme is closely related to the uniqueness properties of harmonic vector fields, i.e. solution $\vec{v} = (v_1, \ldots, v_{n+1})$ of the multidimensional Cauchy-Riemann system

$$\sum_{k=1}^{n+1} \frac{\partial v_k}{\partial x_k} = 0, \quad \frac{\partial v_k}{\partial x_l} = \frac{\partial v_l}{\partial x_k}, \quad k, l = 1, 2, \ldots, n + 1; \ n = 1, 2, \ldots .$$

If $n = 1$, then these fields can be viewed (up to the complex conjugation) as analytic functions of a complex variable. In general they can also be defined (in the case of a simply connected domain) as gradients of harmonic functions, a very natural multidimensional generalization of analytic functions in $\mathbb{C}$.

In dimension one the following uniqueness result is well known: *any vector field continuous in $\mathbb{R}^2_+ \cup \mathbb{R}$, harmonic in $\mathbb{R}^2_+$, and vanishing on a subset of $\mathbb{R}$ of positive length is identically zero.* The continuity assumption can be essentially relaxed, but the positivity of length is necessary (see Section 2). This classical uniqueness theorem motivated the following question known as the Bers-Lavrentiev problem: *is it true that a function $u \in C^1(\mathbb{R}^{n+1}_+ \cup \mathbb{R}^n)$ harmonic in $\mathbb{R}^{n+1}_+$ is constant if its gradient vanishes on a subset of $\mathbb{R}^n$ of positive n-dimensional Lebesgue measure?*

This is actually only one of many similar questions suggested by the case $n = 1$ and addressed to the (gradients of) harmonic functions in $\mathbb{R}^{n+1}$, $n \geq 2$. One of them was posed by S. N. Mergelyan in [Me2]: *suppose a function $u \in C^1(\mathbb{R}_+^{n+1} \cup \mathbb{R}^n)$ is harmonic in $\mathbb{R}_+^{n+1}$ and its Cauchy data (see (3.1)) decrease rapidly enough at the origin along $\mathbb{R}^n$:*

$$(3.2) \qquad |u(p)| + \left| \frac{\partial u}{\partial t}(p) \right| \leq h(|p|), \quad p \in \mathbb{R}^n,$$

where h is a function on $[0, +\infty)$ positive on $(0, +\infty)$ and such that $h(r) \downarrow 0$ as $r \downarrow 0$; for which h is it true that (3.2) *implies $u \equiv 0$?*

The Moscow conference on functional analysis in January of 1956 was the first mathematical conference I attended (I was a first year graduate student at the time). The opening session included a speech by M. V. Keldyš. I remember one phrase from his speech: "It is necessary to study the Cauchy problem for the Laplace equation." I didn't understand what this actually means, but I memorized this recommendation. Much later, in 1962, during a conference on complex analysis in Rostov I met B. V. Shabat (former student and collaborator of M. A. Lavrentiev) who introduced me to this circle of problems which I also discussed with V. Matsaev. Later on I had several stimulating conversations with S. N. Mergelyan on this subject which became one of my favourites, although I didn't do anything in this direction until we have started working with V. Maz'ya. Our motivation was the desire to generalize the approximation results described in Section 1 above to higher dimensions. And, of course, the uniqueness properties of harmonic functions and vector fields are very attractive by themselves. We also turned to some neighbouring themes: normal families of harmonic functions and weighted approximation by Cauchy data.

Our results (published im [MH7]) give a complete answer to the Mergelyan question:

Suppose h satisfies certain regularity conditions (not to be specified here); (3.2) *implies $u \equiv 0$ iff*

$$(3.3) \qquad \int\limits_0 \log h(r)\, dr = -\infty.$$

This is a very rough and approximate version of our uniqueness result from [MH7]. Here I only add the following:

1. Our results are applicable to functions harmonic in a domain (not necessarily $\mathbb{R}_+^{n+1}$) whose boundary contains a flat n-dimensional neighbourhood of the origin,

2. In fact we replaced (3.2) by a pair of separate conditions $|u(p)| \leq h_1(p)$, $\left| \frac{\partial u}{\partial t}(p) \right| \leq h(|p|)$, where h satisfies (3.3), and $h_1(p) = o(|p|^n)$, $|p| \to 0$, $n = 1, 2, \ldots$; the last condition can be made still weaker (see[MH7]), especially for $n = 2$,

3. the uniqueness theorem proved in [MH7] applies to a much wider class of harmonic functions u (not necessarily C^1 up to the boundary). This is important for applications to the dual approximation problems (the larger is the class of admissible functions u, the stronger is the norm to be used in the dual approximation problem).

We also have solved two more problems posed in [Me2].

Suppose $G \subset \mathbb{R}^{n+1}$ is a bounded domain with a nice boundary bG of the type mentioned in Remark 1 above. Let v be a positive continuous function defined on $bG\backslash\{0\}$, $v(p) \equiv 1/h(|p|)$ for small $p \in \mathbb{R}^n$, $p \neq 0$. Denote by $\mathcal{F}_v$ the set of all functions $u \in C^1(G \cup bG)$ harmonic in G and satisfying

$$(3.4) \qquad |u(p)| + \left|\frac{\partial u}{\partial \nu}(p)\right| \leq v(p) \quad (p \in bG)$$

(ν is the unit inner normal vector on bG). Mergegylan asked in [Me2] for which v the set $\mathcal{F}_v$ is uniformly bounded on compact parts of G. In [MH7] we proved that *this is the case iff* (3.3) *holds.*

For the sake of brevity I omit some important details: in fact our result applies to essentially wider sets of harmonic functions.

The next Mergelyan's problem was about the stability of the Cauchy data for harmonic functions. Let S be a compact smooth n-dimensional surface (with border) in $\mathbb{R}^{n+1}$ diffeomorphic to the ball $\{x \in \mathbb{R}^n : |x| \leq 1\}$. Let ν be a side of S (one of two continuous normal unit fields on S). It is proved in [Me2] that the set of all pairs $\left(u, \frac{\partial u}{\partial \nu}\right)$ where u is a harmonic polynomial is dense in $C(S) \times C(S)$. This is no longer true if S is the border of a $(n+1)$-dimensional compact manifold in $\mathbb{R}^{n+1}$ (e.g., if $S = \mathbb{S}^n \subset \mathbb{R}^{n+1}$), but this is true for $S = \mathbb{S}^n\backslash k$ where k is a tiny relatively open spherical cap of $\mathbb{S}^n$. Answering a question from [Me2] we proved the following theorem:

Let G be a domain as in Remark 1 above. Then (3.3) *is equivalent to the following assertion: for any pair* $(f_1, f_2) \in C(bG) \times C(bG)$ *and for any* $\varepsilon > 0$ *there exists a harmonic polynomial u such that*

$$\left(|f_1 - u| + \left|f_2 - \frac{\partial u}{\partial \nu}\right|\right)h < \varepsilon \quad on \ bG\backslash\{0\}.$$

A result similar to our uniqueness theorem above was obtained by N. V. Rao in [R] using a different method. His result does not imply ours, nor does ours imply his (see a discussion in [MH7]).

The proofs in [MH7] are technically rather difficult and long. Using spherical harmonics we reduced our problem to some one-dimensional quasianalyticity problems (some of them had yet to be solved). Our results from [MH7] imply the following uniqueness property of harmonic gradients. *Suppose $E \subset \mathbb{R}^n$ is closed, $0 \in E$, and E is superthick at 0:*

$$\mathrm{mes}_n\left(\{|x| < r\}\backslash E\right) = O\big(\exp(-1/r)\big), \quad r \to 0;$$

then any $u \in C^1(\mathbb{R}^{n+1}_+ \cup \mathbb{R}^n)$ harmonic in $\mathbb{R}^{n+1}_+$ with $\operatorname{grad} u|E = 0$ *is constant* (see [HJ, Chap.5]; this is also proved in [W]).

Here is a sketch of the proof: suppose $u(0) = 0$; for $\omega \in \mathbb{S}^{n-1}$ and small $r > 0$

$$u(r\omega) = \int\limits_{\substack{0 \le \rho \le r \\ \rho\omega \notin E}} \frac{\partial u}{\partial \rho}(\rho\omega)\, d\rho$$

whence $\displaystyle\int\limits_{|p| \le r,\, p \in \mathbb{R}^n} |u(p)|\, dp = O\big(\exp(-r^{-1})\big);$

but $\displaystyle\int\limits_{|p| \le r,\, p \in \mathbb{R}^n} \left|\frac{\partial u}{\partial t}(p)\right| dp \le \int\limits_{|p| \le r,\, p \in \mathbb{R}^n} |\nabla u(p)|\, dp = O\big(\exp(-r^{-1})\big),$

and these estimates imply $u \equiv 0$ by [MH7], see also [HJ].

In [Ar1], [Ar2] some of our results are generalized to the Cauchy problem for the polyharmonic functions.

In [W] a remarkable counterexample has been contructed which solved the Bers-Lavrentiev problem in the negative; see also [AK]. A similar counterexample pertaining to the Cauchy problem for the Laplace equation is in [BW]. But some natural questions related to the theme of this section remain open.

References

[AH]   Adams, D.R., Hedberg, L.-I., *Function Spaces and Potential Theory*, Springer 1996.

[AK]   Aleksandrov, A.B., Kargaev, P.P., *Hardy classes of functions harmonic in a half-space*, Algebra i Analiz **5** (1993) 1–73 (Russian); Engl. transl. in: Saint Petersburg Math. J. **5** (1994) 229–286.

[Ar1]   Arushanyan, Z.A., *A boundary uniqueness theorem for the solutions of the Cauchy problem for the polyharmonic equation*, Izv. Akad. Nauk. Armjan. SSR, Ser. Mat. **11** (1976) 514–547 (Russian).

[Ar2]   Arushanyan, Z.A., *Solutions of the Cauchy problem for the polyharmonic equation (uniqueness, approximation)*, Zap. Nauchn. Sem. LOMI **65** (1976) 164–171 (Russian).

[B1]   Brennan, J.E., *Invariant subspaces and weighted polynomial approximation*. Ark. Mat. **11** (1973) 167–189.

[B2]   Brennan, J.E., *Weighted polymial approximation, quasianalyticity, analytic continuation*, J. reine angew. Math **357** (1985) 23–50.

[B3]   Brennan, J.E., *Weighted polynomial approximation and quasianalyticity on general sets*, Algebra i Analiz **6** (1994) 69–89.

[BW]   Bourgain, J., Wolff, T., *A remark on gradients of harmonic functions in dimension ≥ 3*, Colloq. Math. **50/51** (1990) 253–260.

[Ca1] Carleson, L., *Problems on Exceptional Sets*, Van Nostrand 1967; Russian transl. Mir 1971.

[Ca2] Carleson, L., *Sets of uniqueness for functions regular in the unit disc*, Acta. Math. **87** (1952) 325–345.

[DL] Deny, J., Lions, J.-L., *Les espaces du type de Beppo Levi*, Ann. Inst. Fourier **5** (1953/54) 305–370.

[H1] Havin, V. P., *Approximation in the mean by analytic functions*, Dokl. Akad. Nauk SSSR **178** (1968) 1025–1028; English transl. Soviet Math. Dokl. **9** (1968) 245–248.

[H2] Havin, V. P., *Polynomial approximation in the mean in certain non-Carathéodory regions*, I, II, Izv. Vyssh. Uchebn. Zaved. Mat. **9** (1968) 86–93; **10** (1968) 87–94 (Russian).

[H3] Havin, V. P., *On approximation in L^p by solutions of certain systems of linear differential equations*, Vestnik Leningrad. Univ. **1** (1975) 35–44 (Russian); English transl. Vestnik Leningrad Univ. Math. **8** (1980) 159–168.

[HJ] Havin, V. P., Jöricke, B., *The Uncertainty Principle in Harmonic Analysis*, Springer 1994.

[Ho] Hoffmann, K., *Banach Spaces of Analytic Functions*, Dover Publ. 1988.

[M] Maz'ya, V. G., *Sobolev Spaces*, Springer 1985.

[Me1] Mergelyan, S. N., *On the completeness of systems of analytic functions*, Uspekhi Mat. Nauk **8** (1953) 3–63 (Russian); English transl. in AMS Transl. **19** (1962) 109–166.

[Me2] Mergelyan, S. N., *Harmonic approximation and approximate solutions of the Cauchy problem for the Laplace equation*, Uspekhi Mat. Nauk **11** (1956) 3–26 (Russian).

[MH1] Maz'ya, V. G., Havin, V. P., *A non-linear analog of the Newtonian potential and metric properties of the (p,l)-capacity*. Dokl. Akad. Nauk SSSR **194** (1970) 770–773 (Russian); English transl. Soviet Math. Dokl. **11** (1970) 1294–1298.

[MH2] Maz'ya, V. G., Havin, V. P., *Nonlinear potential theory*. Uspekhi Mat. Nauk **27** (1972) 6, 67–138 (Russian); English transl. Russian Math. Surveys **27** (1972) 6.

[MH3] Maz'ya, V. G., Havin, V. P., *Approximation in the mean by analytic functions*, Vestnik Leningrad. Univ. **13** (1968) 62–74 (Russian); English transl. Vestnik Leningrad Univ. Math. **1** (1974) 3, 231–245.

[MH4] Maz'ya, V. G., Havin, V. P., *Use of (p, l)-capacity in problems of the theory of exceptional sets*, Mat. Sbornik, **90** (1973) 4, 558–591; English transl. Math. USSR Sb. **19** (1973) 547–580.

[MH5] Maz'ya, V. G., Havin, V. P.: *Approximation in the mean by harmonic functions*, Zap. Nauchn. Sem. LOMI **5** (1967) 196–200 (Russian); English transl. Sem. Math. V. A. Steklov Math. Inst. **5** (1967) 72–74.

[MH6] Maz'ya, V. G., Havin, V. P.: *Approximation in the mean by harmonic functions*, Zap. Nauchn. Sem. LOMI **30** (1972) 91–105 (Russian); English transl. J. Soviet Math. **4** (1975) 57–70.

[MH7] Maz'ya, V. G., Havin, V. P., *On the solutions of the Cauchy problem for the Laplace equation (uniqueness, normality, approximations)*, Trudy Moskov. Mat. Obshch. **30** (1974) 61–114 (Russian); English transl. Trans. Moscow. Math. Soc. **30** (1974) 65–117.

[R]   Rao, N. V., *Uniqueness theorem for harmonic functions*, Mat. Zametki **3** (1968) 247–252 (Russian); English transl. Math. Notes **3** (1968) 159–162.

[So]   Sobolev, S. L., *Applications of Functional Analysis in Mathematical Physics*, Providence, AMS, 1963.

[W]   Wolff, T., *Counterexamples with harmonic gradients in* $\mathbb{R}^3$, In the book: Essays on Fourier Analysis in Honor of Elias M. Stein (Princeton Mathematical, No 42) by Ch. Fefferman, R. Feffermann, St. Wainger (Eds.) 1995.

Faculty of Mathematics and Mechanics
St. Petersburg University
Bibliotechnaya Pl. 2
198904 Petrodvorets, Russia
havin@havin.usr.pu.ru

Submitted: 08.02.1999

Operator Theory:
Advances and Applications, Vol. 109

Approximate Approximations and their Applications

GUNTHER SCHMIDT

This paper gives a survey of an approximation method which was proposed by V. Maz'ya as underlying procedure for numerical algorithms to solve initial and boundary value problems of mathematical physics. Due to a greater flexibility in the choice of approximating functions it allows efficient approximations of multi-dimensional integral operators often occuring in applied problems. Its application especially in connection with integral equation methods is very promising, which has been proved already for different classes of evolution equations. The survey describes some basic results concerning error estimates for quasi-interpolation and cubature of integral operators with singular kernels as well as a multiscale and wavelet approach to approximate those operators over bounded domains. Finally a general numerical method for solving nonlocal nonlinear evolution equations is presented.

1. Introduction

V. Maz'ya proposed the method of "Approximate Approximations" in the late 80s when dealing with applied problems as chair of the division "Mathematical models in mechanics" of the Leningrad Research Engineering Institute of the USSR Academy of Sciences (see some of its applications in [8], [20], [21], and the first announcement in English [9]). The main idea of this method is based on the fact that the numerical solution of applied problems is always required only within some prescribed accuracy. Therefore it is possible to use numerical algorithms which provide good approximations only up to some prescribed error level, but do not converge in rigorous mathematical sense. By this way one can

- enlarge the set of approximating functions significantly

- obtain numerically cheap and accurate approximation formulas

- deduce effective formulas for the approximation of various integral and other pseudodifferential operators of mathematical physics

For the first time I heard about this concept at a lecture of V. Maz'ya in 1991 and I became interested in the rather unusual idea to use non-converging approximation methods in numerical computations together with the wide area of possible applications. At the end of 1992 we began our joint work on different aspects of approximate approximations which is still in progress. The present report gives an outline of some of the obtained results.

The starting point is the error analysis of approximate quasi-interpolation on uniform lattices. The quasi-interpolants are linear combinations of scaled translates of a sufficiently smooth and rapidly decaying basis function η and depend on two parameters, the "small" mesh width h and the "large" parameter $\mathcal{D}$ to scale the Fourier image. Under quite general assumptions on η the approximants converge if $\mathcal{D} \to \infty$ and the product $\sqrt{\mathcal{D}}h \to 0$. However, in practical computations it is advantageous to keep $\mathcal{D}$ fixed, which results in an approximation of some order N up to a saturation error which can be made arbitrarily small if $\mathcal{D}$ is chosen large enough. The order N is determined by the moments of the function η and can be increased to any integer value by some analytic or algebraic transformations of the given basis function. Thus one derives a new class of simple multi-variate formulas which behave in numerical computations like high order approximations. This quasi-interpolation procedure was recently extended to the approximation of functions on domains and manifolds with nonuniformly distributed nodes.

The great flexibility in the choice of generating functions η makes it easier to find approximations for which the action of a given operator can be effectively determined. For example, suppose one has to evaluate the convolution with a singular radial kernel as in the case of many potentials in mathematical physics. If the density is replaced by a quasi-interpolant with radial η then after passing to spherical coordinates the convolution is approximated by one-dimensional integrals. For many important integral operators $\mathcal{K}$ one can choose η even such that $\mathcal{K}\eta$ is analytically known, which results in semi-analytic cubature formulas for these operators. The special structure of the quasi-interpolation error gives rise to an interesting effect. Since the saturation error is a fast oscillating function and converges weakly to zero, the cubature formulas for potentials, for example, converge even in rigorous sense.

The presence of the two scaling parameters h and $\mathcal{D}$ is also useful to develop a multiscale approach within the concept of approximate approximations. This approach enables the construction of high order approximations to discontinuous functions and of accurate cubature formulas for integral operators over bounded domains. Here a two scale relation for generating functions from the Schwartz class is used which is valid within any prescribed tolerance for sufficiently large $\mathcal{D}$. This relation allows also to perform an approximate multiresolution analysis of spaces generated by those functions. Therefore a wavelet basis can be constructed in which elements of fine scale spaces are representable within a given tolerance. These approximate wavelets provide most of the properties utilized in wavelet based numerical methods and possess additionally simple analytic representations. Therefore the sparse approximation of important integral operators in the new basis can be computed using special functions or simple quadrature.

The capability of approximate approximations to treat multi-dimensional integral operators very efficiently is very promising for new integral equation based numerical methods. In the last section I present some examples of these methods which were developed by V. Maz'ya together with V. Karlin to solve linear and nonlinear problems. In particular, numerical results for evolution equations with nonlocal operators are given which cannot be solved by standard finite-difference or finite-element methods.

2. Quasi-interpolation

2.1. Example

To illustrate the idea of approximate approximations let us consider the following example of a simple quasi-interpolation formula in $\mathbf{R}^1$

$$(2.1) \qquad \mathcal{M}_h u(x) = \frac{1}{\pi\sqrt{\mathcal{D}}} \sum_{m=-\infty}^{\infty} u(mh)\, \mathrm{sech}\, \frac{x-mh}{\sqrt{\mathcal{D}}h}$$

If u is 2-times continuously differentiable then the Taylor expansion

$$u(mh) = u(x) + u'(x)(mh-x) + u''(x_m)\frac{(mh-x)^2}{2}$$

with some x_m between x and mh leads to

$$\mathcal{M}_h u(x) = \frac{u(x)}{\pi\sqrt{\mathcal{D}}} \sum_{m=-\infty}^{\infty} \mathrm{sech}\, \frac{x-mh}{\sqrt{\mathcal{D}}h} + \frac{u'(x)}{\pi\sqrt{\mathcal{D}}} \sum_{m=-\infty}^{\infty} (mh-x)\, \mathrm{sech}\, \frac{x-mh}{\sqrt{\mathcal{D}}h}$$
$$+ \frac{1}{2\pi\sqrt{\mathcal{D}}} \sum_{m=-\infty}^{\infty} u''(x_m)(mh-x)^2\, \mathrm{sech}\, \frac{x-mh}{\sqrt{\mathcal{D}}h}\,.$$

The Fourier transform of $\mathrm{sech}(x)$ is given as

$$(\mathcal{F}\mathrm{sech})(\lambda) := \int_{\mathbf{R}^1} \mathrm{sech}\,x\; e^{-2\pi i x\lambda}\, dx = \pi\, \mathrm{sech}\,\pi^2\lambda\,,$$

therefore Poisson's summation formula results in the relations

$$\frac{1}{\pi\sqrt{\mathcal{D}}} \sum_{m=-\infty}^{\infty} \mathrm{sech}\, \frac{x-m}{\sqrt{\mathcal{D}}} = 1 + 2\sum_{\nu=1}^{\infty} \mathrm{sech}(\pi^2\sqrt{\mathcal{D}}\nu)\, \cos(2\pi\nu x)\,,$$

$$\frac{1}{\pi\sqrt{\mathcal{D}}} \sum_{m=-\infty}^{\infty} \frac{x-m}{\sqrt{\mathcal{D}}}\, \mathrm{sech}\, \frac{x-m}{\sqrt{\mathcal{D}}} = \pi\sum_{\nu=1}^{\infty} \mathrm{sech}(\pi^2\sqrt{\mathcal{D}}\nu)\, \tanh(\pi^2\sqrt{\mathcal{D}}\nu)\, \sin(2\pi\nu x)\,.$$

It is clear that $\mathcal{M}_h u(x)$ does not converge to $u(x)$. However, one has

$$\left| \frac{1}{\pi\sqrt{\mathcal{D}}} \sum_{m=-\infty}^{\infty} \mathrm{sech}\, \frac{x-m}{\sqrt{\mathcal{D}}} - 1 \right| \le 2\,\varepsilon(\sqrt{\mathcal{D}})\,,$$

$$\left| \frac{1}{\pi\sqrt{\mathcal{D}}} \sum_{m=-\infty}^{\infty} \frac{x-m}{\sqrt{\mathcal{D}}}\, \mathrm{sech}\, \frac{x-m}{\sqrt{\mathcal{D}}} \right| \le \pi\,\varepsilon(\sqrt{\mathcal{D}})\,,$$

where the number

$$\varepsilon(\sqrt{\mathcal{D}}) := \sum_{\nu=1}^{\infty} \operatorname{sech}(\pi^2 \sqrt{\mathcal{D}} \nu)$$

can be made arbitrarily small by choosing $\mathcal{D}$ large enough. For example, if $\mathcal{D} = 4$ then $\varepsilon(\sqrt{\mathcal{D}}) = 0.000000005351$. Since

$$\frac{1}{2\pi\sqrt{\mathcal{D}}} \left| \sum_{m=-\infty}^{\infty} u''(x_m) \frac{(mh-x)^2}{\mathcal{D}h^2} \operatorname{sech} \frac{x-mh}{\sqrt{\mathcal{D}}h} \right| \le \frac{5}{4} \sup_{t\in\mathbf{R}} |u''(t)|$$

the difference between u and $\mathcal{M}_h u$ can be estimated by

$$|\mathcal{M}_h u(x) - u(x)| \le \frac{5}{4} \mathcal{D}h^2 \sup_{\mathbf{R}} |u''| + \varepsilon(\sqrt{\mathcal{D}})\big(2\,|u(x)| + \pi\sqrt{\mathcal{D}}h\,|u'(x)|\big) .$$

This means, above the tolerance $\varepsilon(\sqrt{\mathcal{D}})\big(2\,|u(x)| + \pi\sqrt{\mathcal{D}}h\,|u'(x)|\big)$ the quasi-interpolant $\mathcal{M}_h u$ approximates any 2-times continuously differentiable function like usual second order approximations and any prescribed accuracy can be reached if $\mathcal{D}$ is chosen appropriately.

2.2.   Quasi-interpolation with general basis functions

A similar approximation behavior as in the previous example can be obtained for quite arbitrary basis functions in arbitrary space dimension n and approximation order N. Consider the quasi-interpolation formula

$$(2.2) \qquad \mathcal{M}_h u(\mathbf{m}) = \mathcal{D}^{-\mathbf{n}/2} \sum_{\mathbf{m}\in\mathbf{Z}^n} \mathbf{u}(\mathbf{hm})\,\eta\Big(\frac{\mathbf{x}-\mathbf{hm}}{\sqrt{\mathcal{D}}\mathbf{h}}\Big)$$

with a continuous generating function η satisfying
A.1   Decay condition

$$(2.3) \qquad |\eta(\mathbf{x})| \le A_K\,(1+|\mathbf{x}|^2)^{-K/2}\,, \quad \mathbf{x}\in\mathbf{R}^{\mathbf{n}}\,,$$

for some constant A_K, $K > N + n$ and $N \ge 1$ is a given positive integer.
   Denoting by

$$\sigma_\alpha(\mathbf{x},\mathcal{D}) := \mathcal{D}^{-n/2} \sum_{\mathbf{m}\in\mathbf{Z}^n} \Big(\frac{\mathbf{x}-\mathbf{m}}{\sqrt{\mathcal{D}}}\Big)^\alpha \eta\Big(\frac{\mathbf{x}-\mathbf{m}}{\sqrt{\mathcal{D}}}\Big) ,$$

$$\rho_\alpha(\mathbf{x},\mathcal{D}) := \mathcal{D}^{-n/2} \sum_{\mathbf{m}\in\mathbf{Z}^n} \Big|\Big(\frac{\mathbf{x}-\mathbf{m}}{\sqrt{\mathcal{D}}}\Big)^\alpha \eta\Big(\frac{\mathbf{x}-\mathbf{m}}{\sqrt{\mathcal{D}}}\Big)\Big| ,$$

one gets after substituting Taylors expansion of u into (2.2)

$$(2.4) \qquad \mathcal{M}_h u(\mathbf{x}) = \sum_{|\alpha|=0}^{N-1} \frac{\partial^\alpha u(\mathbf{x})}{\alpha!} (-\sqrt{\mathcal{D}}h)^{|\alpha|} \sigma_\alpha(\mathbf{x}/h,\mathcal{D}) + R_N(\mathbf{x},\sqrt{\mathcal{D}}h) ,$$

where

$$|R_N(\mathbf{x}, \sqrt{\mathcal{D}}h)| \le (\sqrt{\mathcal{D}}h)^N \sum_{|\alpha|=N} \frac{\|\partial^\alpha u\|_{L_\infty(\mathbf{R}^n)}}{\alpha!} \rho_\alpha(\mathbf{x}/h, \mathcal{D})$$

The application of Poisson's summation formula to σ_α yields the equality

$$\sigma_\alpha(\mathbf{x}, \mathcal{D}) = \left(\frac{i}{2\pi}\right)^{|\alpha|} \sum_{\nu \in \mathbf{Z}^n} \partial^\alpha \mathcal{F}\eta(\sqrt{\mathcal{D}}\nu)\, e^{2\pi i(\nu, \mathbf{x})},$$

which holds if the sequence $\{\partial^\alpha \mathcal{F}\eta(\sqrt{\mathcal{D}}\,\cdot)\} \in l_1(\mathbf{Z}^n)$ (see e.g. [18]).

Thus $\mathcal{M}_h u$ would approximate a sufficiently smooth functions u with the order $\mathcal{O}((\sqrt{\mathcal{D}}h)^N)$ if $\sigma_\alpha(\mathbf{x}, \mathcal{D}) = \delta_{|\alpha|0}$ for all $\mathbf{x} \in \mathbf{R}^n$, δ_{ik} is the Kronecker symbol. These equalities imply the well-known Strang-Fix condition

$$\partial^\alpha \mathcal{F}\eta(\mathbf{0}) = \delta_{|\alpha|0}, \quad \partial^\alpha \mathcal{F}\eta(\sqrt{\mathcal{D}}\nu) = 0, \quad \nu \in \mathbf{Z}^n \setminus \{\mathbf{0}\}, \ 0 \le |\alpha| \le N-1.$$

The idea of approximate quasi-interpolation is to use generating functions η for which $\sigma_\alpha(\mathbf{x}, \mathcal{D})$ can be made arbitrarily close to $\delta_{|\alpha|0}$ by choosing appropriate values of $\mathcal{D}$. Therefore η and its Fourier transform $\mathcal{F}\eta$ have to satisfy

A.2   Moment condition

$$(2.5) \qquad \int_{\mathbf{R}^n} \eta(\mathbf{y})\, d\mathbf{y} = 1, \ \int_{\mathbf{R}^n} \mathbf{y}^\alpha \eta(\mathbf{y})\, d\mathbf{y} = 0, \quad \forall\, \alpha, \ 1 \le |\alpha| < N,$$

A.3   For any $\varepsilon > 0$ there exists $\mathcal{D} > 0$ so that

$$(2.6) \qquad \sum_{\nu \in \mathbf{Z}^n \setminus \{\mathbf{0}\}} |\partial^\alpha \mathcal{F}\eta(\sqrt{\mathcal{D}}\nu)| < \varepsilon, \quad \forall\, \alpha, \ 0 \le |\alpha| \le N.$$

In [17] it is shown that for M-times differentiable functions η (M the smallest integer greater than $n/2$), which satisfy together with all derivatives $\partial^\beta \eta$, $|\beta| \le M$, the decay condition A.1, assumption A.2 is true.

Consequently, if the generating function η satisfies the assumptions A.1–3 one obtains from (2.4) the error estimate

Theorem 2.1. ([14]) *If $u \in C^N(\mathbf{R}^n) \cap W_\infty^N(\mathbf{R}^n)$ then for any $\varepsilon > 0$ there exists $\mathcal{D} > 0$ such that*

$$|\mathcal{M}_h u(\mathbf{x}) - u(\mathbf{x})| \le c(\sqrt{\mathcal{D}}h)^N \sum_{|\alpha|=N} \frac{\|\partial^\alpha u\|_{L_\infty(\mathbf{R}^n)}}{\alpha!} + \varepsilon \sum_{|\alpha|=0}^{N-1} (\sqrt{\mathcal{D}}h)^{|\alpha|} \frac{|\partial^\alpha u(\mathbf{x})|}{\alpha!},$$

where the constant c depends only on η.

Similar error estimates for quasi-interpolation in the L_p-norm are proved in [15]. Note that for u from the Hölder class $C^{L,\mu}$, $0 \le L < N$, $0 < \mu < 1$, the error $|\mathcal{M}_h u(\mathbf{x}) - u(\mathbf{x})|$ is of the order $\mathcal{O}((\sqrt{\mathcal{D}}h)^{L+\mu})$ plus small remainder terms.

The local nature of the quasi-interpolant $\mathcal{M}_h$ should be emphasised, its behaviour at a given point $\mathbf{x}$ depends within some error level ε only on the values of u at grid points in some neighborhood of this point. Since in practice one has to truncate the summation in (2.2) it is important that depending on $\varepsilon > 0$ there exist $N_s > 0$ ensuring that the truncated quasi-interpolant

$$(2.7) \qquad \mathcal{M}_h^s u(\mathbf{x}) = \mathcal{D}^{-n/2} \sum_{|\mathbf{x}-h\mathbf{m}| \leq N_s h} u(h\mathbf{m})\, \eta\left(\frac{\mathbf{x} - h\mathbf{m}}{\sqrt{\mathcal{D}}\mathbf{h}}\right),$$

shows the same approximation behaviour as $\mathcal{M}_h u$. Thus, the number of summands in (2.7) does not depend on the mesh size h.

One interesting consequence of Theorem 2.1 is that the quasi-interpolant $\mathcal{M}_h u$ converges to u if the parameter $\mathcal{D}$ is chosen depending on h with $\mathcal{D}(h) \to \infty$ as $h \to 0$. For any generating functions η satisfying the assumptions A.1–3 one can specify this dependence and determine the optimal convergence order, which is of course less than N. For special one-dimensional or tensor-product generating functions this problem has been addressed in the literature ([19],[2],[1]).

We prefer to have $\mathcal{D}$ as additional parameter because this may be advantageous in some numerical applications, where the quasi-interpolation is used, as well as for further developments of approximate approximations mentioned in Sections 7 and 8.

2.3.   Quasi-interpolation with variable nodes

The quasi-interpolation formula (2.2) is not restricted to the case of uniform spacing of the nodes or grid points. In [17] an extension of this formula to the form

$$(2.8) \qquad \mathcal{D}^{-n/2} \sum_{\mathbf{x_m} \in \Omega} u(\mathbf{x_m})\, \eta\left(\frac{\mathbf{x} - \mathbf{x_m}}{\sqrt{\mathcal{D}}\, V_{\mathbf{m}}}\right), \qquad \mathbf{x} \in \Omega,$$

is discussed, where Ω is some domain in $\mathbf{R}^n$ or an n-dimensional manifold in $\mathbf{R}^s$ and the nodes $\mathbf{x_m} \in \Omega$ are the images of a lattice of width h under smooth parametrization of Ω. To retain the local character of the quasi-interpolant the scaling $V_{\mathbf{m}}$ should be proportional to h.

After applying a partition of unity it can be supposed that $\Omega = \phi(\omega)$, where $\omega \subset \mathbf{R}^n$ is a bounded domain and $\phi = (\varphi_1, \dots, \varphi_s) : \mathbf{R}^n \to \mathbf{R}^s$, $n \leq s$ is a sufficiently smooth and non-singular one-to-one mapping. In particular,

$$|\phi'(\mathbf{y})| = \left(\sum_{(i)} (\kappa_{(i)}(\mathbf{y}))^2\right)^{1/2} > 0, \qquad \mathbf{y} \in \omega,$$

where $\kappa^{(i)}$ denotes the minor of order n of the matrix $\phi'(\mathbf{y}) = (\partial \varphi_j / \partial y_k)_{j,k=1}^{s,n}$, corresponding to indices $i_1 < \dots < i_n$. The sum is extended over all distinct tuples $(i) = (i_1, \dots, i_n)$, $1 \leq i_p \leq s$, of this kind.

It turns out that if in (2.8) the nodes are given by $\mathbf{x_m} = \phi(h\mathbf{m})$, $h\mathbf{m} \in \omega$, then this quasi-interpolant has similar approximation properties as in the case of uniformly distributed nodes. For the sake of simplicity we assume here that η belongs to the Schwartz class, $\eta \in \mathcal{S}(\mathbf{R}^s)$. Further one has suppose that in the case $s > n$ the generating function η is radial.

Theorem 2.2. ([17]) *Assume that η satisfies the assumption A.2, $\phi : \omega \to \Omega$ is in the class C^{N+1} with $|\phi'(\mathbf{y}| > 0$, $\mathbf{y} \in \overline{\omega}$, and the numbers $V_{\mathbf{m}}$ are chosen such that*

$$|(h^{-1}V_{\mathbf{m}})^n - |\phi'(h\mathbf{m})|| \le c\,h^N .$$

If $u \in C_0^N(\Omega)$, then for any $\varepsilon > 0$ there exists $\mathcal{D} > 0$ such that at any point $\mathbf{x} \in \Omega$

$$(2.9) \qquad |u_h(\mathbf{x}) - u(\mathbf{x})| \le c_\eta \left(\sqrt{\mathcal{D}}h\right)^N \|u\|_{C^N(\overline{\Omega})} + \varepsilon \sum_{k=0}^{N-1} c_k \left(\sqrt{\mathcal{D}}h\right)^k ,$$

where c_η does not depend on u, h and $\mathcal{D}$ and the constants c_k can be obtained from the values $\partial^\alpha u(\mathbf{x})$, $|\alpha| \le k$.

Consider as simple example a fourth order quasi-interpolation formula on a surface Γ in $\mathbf{R}^3$. To define the numbers $V_{\mathbf{m}}$ denote by $Q_{\mathbf{m}}^j = h\mathbf{m} + hQ^j$, $j = 1, 2$, where $Q^1 = [-1,1]^2$ and Q^2 is the sqaure with corners at the points $(\pm 1, 0)$ and $(0, \pm 1)$. Then formula (2.8) with $\eta(|\mathbf{x}|) = \pi^{-1}(2 - |\mathbf{x}|^2)\exp(-|\mathbf{x}|^2)$ provides approximate approximations with the order $\mathcal{O}(\mathcal{D}^2 h^4)$ on Γ if one chooses

$$V_{\mathbf{m}} = \sqrt{\mathrm{meas}(\phi(Q_{\mathbf{m}}^2)) - \mathrm{meas}(\phi(Q_{\mathbf{m}}^1))/4} .$$

3. Generating functions for quasi-interpolation of high order

In fact any sufficiently smooth and decaying function η with $\mathcal{F}\eta(0) \ne 0$ can be used as generating function for quasi-interpolation processes. One interesting feature of approximate approximations is the possibility to construct from a given basis function η new generating functions η_N satisfying the assumptions A.2 on the moments for arbitrary given large N using different analytic and algebraic methods. Some of these methods are described in this section.

3.1. Examples of generating functions

The following table lists some examples of useful functions for which the Fourier transform is known. The first number in each row indicates the space dimension. In the next section it is shown how these functions can be modified to generate high order quasi-interpolants.

$\eta(x)$	$\mathcal{F}\eta(\lambda)$

	$\eta(x)$	$\mathcal{F}\eta(\lambda)$
1	$\pi^{-1}\operatorname{sech}(x)$	$\operatorname{sech}(\pi^2\lambda)$
1	$2\pi^{-2}x\operatorname{cosech}(x)$	$\operatorname{sech}^2(\pi^2\lambda)$
n	$\pi^{-n/2}\exp(-|\mathbf{x}|^2)$	$\exp(-\pi^2|\lambda|^2)$
1	$\sqrt{e/\pi}\,\exp(-x^2)\cos\sqrt{2}x$	$\exp(-\pi^2\lambda^2)\cosh\sqrt{2}\pi\lambda$
2	$\dfrac{\exp(-2\pi a\sqrt{|\mathbf{x}|^2+b^2})}{\sqrt{|\mathbf{x}|^2+b^2}}$	$\dfrac{\exp(-2\pi b\sqrt{|\lambda|^2+a^2})}{\sqrt{|\lambda|^2+a^2}}$
3	$\pi^{-4}\operatorname{sech}(|\mathbf{x}|)$	$\dfrac{\sinh(\pi^2|\lambda|)\operatorname{sech}^2(\pi^2|\lambda|)}{\pi^2|\lambda|}$
n	$\dfrac{4}{3\pi^{n+1/2}}\dfrac{\Gamma(\frac{n+5}{2})}{(1+|\mathbf{x}|^2)^{(n+5)/2}}$	$\left(1+2\pi|\lambda|+\dfrac{4}{3}\pi^2|\lambda|^2\right)e^{-2\pi|\lambda|}$
n	$(-1)^k\dfrac{\pi^{(n+1)/2}}{\Gamma(k+\frac{n+1}{2})}\dfrac{\partial^k}{\partial\tau^k}\dfrac{1}{\sqrt{\tau}}e^{-2\pi\sqrt{\tau}|\mathbf{x}|}\Big|_{\tau=1}$	$(1+|\lambda|^2)^{-k-(n+1)/2}$
n	$\dfrac{\Gamma(k+1+\frac{n}{2})}{\pi^{n/2}\Gamma(k+1)}(1-|\mathbf{x}|^2)^k\,\chi(\mathbf{x})$	$\Gamma(k+1+\frac{n}{2})\dfrac{J_{k+n/2}(2\pi|\lambda|)}{(\pi|\lambda|)^{k+n/2}}$

Here χ denotes the characteristic function of the unit ball $B(\mathbf{0},1)$ and J_ν the Bessel function of the first kind.

3.2.　A general formula

If $\mathcal{F}\eta(\mathbf{0})\neq 0$ then the function

$$(3.10)\qquad \eta_N(\mathbf{x})=\sum_{|\alpha|=0}^{N-1}\frac{\partial^\alpha(\mathcal{F}\eta)^{-1}(0)}{\alpha!\,(2\pi i)^{|\alpha|}}\,\partial^\alpha\eta(\mathbf{x})$$

satisfies the moment condition (2.5), where $\partial^\alpha(\mathcal{F}\eta)^{-1}(0):=\partial^\alpha\big(1/\mathcal{F}\eta(\lambda)\big)\big|_{\lambda=0}$.

For radial basis functions $\eta(\mathbf{x})=\eta(|\mathbf{x}|)$ this construction leads to the function

$$(3.11)\qquad \eta_{2M}(\mathbf{x})=\Gamma\Big(\frac{n}{2}\Big)\sum_{j=0}^{M-1}\frac{\Delta^j(\mathcal{F}\eta)^{-1}(0)}{j!\,(4\pi)^{2j}\,\Gamma(j+\frac{n}{2})}\,(-\Delta)^j\eta(\mathbf{x})$$

satisfying the moment condition (2.5) with $N=2M$ and possessing the Fourier transform

$$\mathcal{F}\eta_{2M}(\lambda)=\mathcal{F}\eta(\lambda)\,\Gamma\Big(\frac{n}{2}\Big)\sum_{j=0}^{M-1}\frac{\Delta^j(\mathcal{F}\eta)^{-1}(0)}{j!\,4^j\,\Gamma(j+\frac{n}{2})}\,|\lambda|^{2j}\,.$$

The additive structure of formula (3.11) allows to increase the order of a given quasi-interpolant by adding a new formula of the form (2.2) with the next term of (3.11) as generating function.

An interesting example is provided by the Gaussian function $\eta(\mathbf{x}) = \exp(-|\mathbf{x}|^2)$, where the application of (3.11) leads to the generating function

$$(3.12) \qquad \eta_{2M}(\mathbf{x}) = \frac{1}{\pi^{n/2}} L_{M-1}^{(n/2)}(|\mathbf{x}|^2)\, e^{-|\mathbf{x}|^2}$$

with the generalized Laguerre polynomial $L_{M-1}^{(n/2)}$.

3.3. Symmetric generating functions

Here we list some other formulas for basis functions η with $\mathcal{F}\eta(0) = 1$ which are symmetric with respect to the coordinate planes $x_i = 0$,

$$\eta(x_1,\ldots,x_i,\ldots,x_n) = \eta(x_1,\ldots,-x_i,\ldots,x_n)\,, \quad i = 1,\ldots,n\,.$$

For the resulting generating functions the moment condition (2.5) is valid with $N = 2M$.

(i) $\displaystyle \tilde{\eta}_{\mathbf{D}}(\mathbf{x}) := \sum_{j=1}^{M} \prod_{\substack{l=1 \\ l\neq j}}^{M} \frac{\mathcal{D}_l\,\mathcal{D}_j^{-n/2}}{\mathcal{D}_l - \mathcal{D}_j}\, \eta\!\left(\frac{\mathbf{x}}{\sqrt{\mathcal{D}_j}}\right)$

    for any M-tuple $\mathbf{D} = (\mathcal{D}_1,\ldots,\mathcal{D}_M)$, $\mathcal{D}_j > 0$, $\mathcal{D}_j \neq \mathcal{D}_l$, $j \neq l$.

(ii) $\displaystyle \tilde{\eta}_M(\mathbf{x}) := \frac{(-1)^{M-1}}{(M-1)!} \left(\frac{d}{d\tau}\right)^{M-1} \left(\tau^{-1-n/2}\eta\!\left(\frac{\mathbf{x}}{\sqrt{\tau}}\right)\right)\Bigg|_{\tau=1}$

(iii) $\displaystyle \hat{\eta}(\mathbf{x}) = \int_1^\infty \eta\!\left(\frac{\mathbf{x}}{\sqrt{\tau}}\right) \frac{d\mu(\tau)}{\tau^{n/2}} \quad \text{with} \quad \int_1^\infty \tau^k\, d\mu(\tau) = \delta_{0k}\,, \quad k = 0,\ldots,M-1\,.$

3.4. Linear combinations of translates

Suppose that η is symmetric and $\mathcal{F}\eta(0) \neq 0$. From the solution $\{a_\beta(\mathcal{D})\}$ of the linear system

$$\sum_{|\beta|<M} a_\beta(\mathcal{D})\beta^{2\alpha} = \left(-\frac{\mathcal{D}}{4\pi^2}\right)^{|\alpha|} \partial^{2\alpha}(\mathcal{F}\eta)^{-1}(0)\,, \quad |\alpha| < M\,.$$

one obtains a generating function satisfying condition A.2 with $N = 2M$ by

$$(3.13) \qquad \tilde{\eta}_M(\mathbf{x},\mathcal{D}) = \sum_{[\beta]<M} 2^{-\sigma(\beta)} a_\beta(\mathcal{D}) \sum_{\{|\mathbf{k}|\}=\beta} \eta\!\left(\mathbf{x} - \frac{\mathbf{k}}{\sqrt{\mathcal{D}}}\right)$$

Here $\{|\mathbf{k}|\}$ denotes the vector $(|k_1|, \ldots, |k_n|)$ and $\sigma(\beta)$ is the number of nonzero components of β. Since

$$\partial^\alpha \mathcal{F}\big(\tilde{\eta}_M(\cdot, \mathcal{D})\big)(\sqrt{\mathcal{D}}\nu) = \sum_{\beta \leq \alpha} \frac{\alpha!}{\beta!\,(\alpha - \beta)!}\, \partial^\beta \mathcal{F}\eta(\sqrt{\mathcal{D}}\nu)\, \partial^{\alpha - \beta}(\mathcal{F}\eta)^{-1}(0)\,,$$

the saturation error of the quasi-interpolant generated by $\tilde{\eta}_M(\mathbf{x}, \mathcal{D})$ satisfies assumption A.3. From (3.13) follows immediately, that it can be written in the form

$$\mathcal{D}^{-n/2} \sum_{\mathbf{m} \in \mathbf{Z}^n} \eta\Big(\frac{\mathbf{x} - h\mathbf{m}}{\sqrt{\mathcal{D}}h}\Big)\Big(\sum_{[\beta] < M} 2^{-\sigma(\beta)} a_\beta \sum_{\{|\mathbf{k}|\} = \beta} u(h(\mathbf{m} - \mathbf{k})) \Big),$$

i.e. a linear combination of the translates of the basis function η can provide any given approximation order up to the saturation error.

4. Semi-analytic cubature formulas

The numerical treatment of potentials and other integral operators with singular kernels arises as a computational task in different fields. Since standard cubature methods are very time-consuming there is an ongoing research to develop new effective algorithms like panel clustering, multipole expansions or wavelet compression based on piecewise polynomial approximations of the density. The effective treatment of integral operators is also one of the main applications of approximate approximation. Here the density of a given integral operator

$$(4.14) \qquad\qquad \mathcal{K}u(\mathbf{x}) = \int_\Omega g(\mathbf{x} - \mathbf{y})\, u(\mathbf{y})\, d\mathbf{y}\,.$$

is replaced by a high order quasi-interpolant with a specially chosen generating function for which the action of the integral operator can be effectively determined. For example, if η is chosen such that an analytic expression of the integral $\mathcal{K}\eta$ is available, which is in general impossible for piecewise polynomials or other finite-element functions, then

$$\mathcal{K}_h u(\mathbf{x}) := \mathcal{K}\mathcal{M}_h u(\mathbf{x}) = h^n \sum_{h\mathbf{m} \in \Omega} u(h\mathbf{m}) \int_{\mathbf{R}^n} g\Big(\sqrt{\mathcal{D}}h\Big(\frac{\mathbf{x} - h\mathbf{m}}{h\sqrt{\mathcal{D}}} - \mathbf{y}\Big)\Big)\, \eta(\mathbf{y})\, d\mathbf{y}$$

is a semi-analytic cubature formula for the integral operator.

    Consider for example the Newton potential

$$\mathcal{N}_3 u(\mathbf{x}) := \frac{1}{4\pi} \int_{\mathbf{R}^3} \frac{u(\mathbf{y})}{|\mathbf{x} - \mathbf{y}|}\, d\mathbf{y}\,.$$

If the density u is replaced by the quasi-interpolant with the generating function $\eta_{2M}(\mathbf{x}) = \pi^{-3/2}\, L_{M-1}^{(3/2)}(|\mathbf{x}|^2)\, e^{-|\mathbf{x}|^2}$ one obtains the semi-analytic formula

$$\mathcal{N}_{3,h}u(\mathbf{x}) := \frac{\mathcal{D}h^2}{4(\pi\mathcal{D})^{3/2}} \sum_{\mathbf{m}\in\mathbf{Z}^3} u(h\mathbf{m}) \left(\frac{\sqrt{\pi}}{|\mathbf{r_m}|}\,\mathrm{erf}(|\mathbf{r_m}|) + e^{-|\mathbf{r_m}|^2} \sum_{j=0}^{M-2} \frac{L_j^{(1/2)}(|\mathbf{r_m}|^2)}{j+1} \right),$$

with $\mathbf{r_m} = (\mathbf{x}/h - \mathbf{m})/\sqrt{\mathcal{D}}$ and the error function erf.

Error estimates for these approximations can be obtained by estimating the quasi-interpolation in weak Sobolev norms, for example. Here the saturation error, which is caused by fast oscillating functions, tends to zero. Therefore the cubature formulas for smoothing integral operators, for example pseudodifferential operators of negative order, become converging. For the particular case of the Newton potential the following estimate holds.

Theorem 4.3. *If $u \in W_p^N(\mathbf{R}^3) \cap W_q^N(\mathbf{R}^3)$, $1 < p < 3/2$, $q = 3p/(3 - 2p)$, $2 \le L \le 2M$, then for any $\varepsilon > 0$ there exists $\mathcal{D} > 0$ such that*

$$\|\mathcal{N}_3 u - \mathcal{N}_{3,h}u\|_{L_q(\mathbf{R}^3)} \le c_\eta\,(\sqrt{\mathcal{D}}h)^L \|u\|_{W_p^L(\mathbf{R}^3)} + \varepsilon h^2 \|u\|_{W_p^{L-1}(\mathbf{R}^3)}$$

and

$$\|\nabla\mathcal{N}_3 u - \nabla\mathcal{N}_{3,h}u\|_{L_q(\mathbf{R}^3)} \le c_\eta\,(\sqrt{\mathcal{D}}h)^L \|u\|_{W_p^L(\mathbf{R}^3)} + \varepsilon h \|u\|_{W_p^{L-1}(\mathbf{R}^3)}\ .$$

There exist several other important integral operators of mathematical physics and appropriate basis functions leading to semi-analytic cubature formulas or to approximations which require only simple one-dimensional quadratures of well-behaved functions (see [11], [15]). Here only two simple examples are mentioned.

The action of the diffraction potential

$$(4.15)\qquad \mathcal{K}\eta_{2M}(\mathbf{x}) = \frac{1}{4\pi} \int_{\mathbf{R}^3} \frac{e^{ik|\mathbf{x}-\mathbf{y}|}}{|\mathbf{x} - \mathbf{y}|}\,\eta_{2M}(\mathbf{y})\,d\mathbf{y}\ ,\qquad \mathrm{Im}\,k \ge 0\,,\ \mathbf{x}\in\mathbf{R}^3\,,$$

is given by the analytic formula

$$\mathcal{K}\eta_{2M}(\mathbf{x}) = \frac{e^{-k^2/4}}{4\pi|\mathbf{x}|} \left(i\sin(k|\mathbf{x}|) + \mathrm{Re}\left(e^{ik|\mathbf{x}|}\mathrm{erf}\left(|\mathbf{x}| + \frac{ik}{2}\right)\right)\right) \sum_{j=0}^{M-1} \frac{(k^2/4)^j}{j!}$$

$$+ \frac{e^{-|\mathbf{x}|^2}}{\pi^{3/2}} \sum_{j=0}^{M-2} \frac{(-1)^j\,H_{2j+1}(|\mathbf{x}|)}{2^{2j+3}\,|\mathbf{x}|} \sum_{r=0}^{M-2-j} \frac{(k^2/4)^r}{(j+r+1)!}\ ,$$

where $H_j(x)$ denotes the Hermite polynomials of order j.

Consider the integral operator of the screen problem

$$(4.16)\qquad\qquad \mathcal{V}u(\mathbf{x}) := \frac{1}{\pi} \int_{\mathbf{R}^2} \frac{u(\mathbf{y})}{|\mathbf{x} - \mathbf{y}|}\,d\mathbf{y}\ ,\qquad \mathbf{x} = (x', x_3)\in\mathbf{R}^3\,.$$

The action of this operator on the function $\eta_{2M}(\mathbf{x}) = \pi^{-1} L_{M-1}^{(1)}(|\mathbf{x}|^2)\, e^{-|\mathbf{x}|^2}$ can be computed with the quadrature of the one-dimensional integral

$$\mathcal{V}\eta_{2m}(\mathbf{x}) = \frac{2}{\pi} \int_0^\infty J_0(|x'|\rho)\, \exp(-\rho^2 - |x_3|\rho) \sum_{j=0}^{M-1} \frac{\rho^{2j}}{j!}\, d\rho\,,$$

where J_0 is the Bessel function.

5.   Cubature of integral operators over bounded domains

Often the computation of potentials is required for finding in some domain a particular solution of a partial differential equation with inhomogeneous right-hand side u. If this function can be extended smoothly outside the domain then the cubature method considered in the previous section provides a high order approximation of such a solution. Here we briefly describe how this method can be extended to the computation of integral operators over a given bounded domain Ω. Since in general the quasi-interpolant $\mathcal{M}_h u$ with uniformly distributed nodes does not approximate u near $\partial\Omega$ one has to refine the mesh towards the boundary. In view of the simplicity of the formulas for $\mathcal{M}_h$ it is useful to take a sequence of uniform meshes in certain boundary layers. In the joint paper with T. Ivanov [3] such a multiscale approximation formula was constructed which is numerically stable for Lipschitz domains. The approximant behaves like $\mathcal{M}_h u$ with the same basis function and compactly supported, smooth functions u if the error is measured in L_p, $1 \leq p < \infty$, or in negative Sobolev-norms. This method was modified in [4] for polyhedral domains where anisotropic mesh refinement leads to a considerable reduction of the numerical costs.

5.1.   Approximate refinement equation

The construction of the boundary layer approximants is based on the following

Theorem 5.4. ([16]) *Suppose that the Fourier transform of $\eta \in \mathcal{S}(\mathbf{R}^n)$ does not vanish, $\mathcal{F}\eta \neq 0$, and that for given $\mu \in (0,1)$ the function $\tilde{\eta} = \mathcal{F}^{-1}\big(\mathcal{F}\eta(\cdot)/\mathcal{F}\eta(\mu\cdot)\big)$ belongs also to $\mathcal{S}(\mathbf{R}^n)$. Then for any $\varepsilon > 0$ there exists $\mathcal{D} > 0$ such that*

$$(5.17) \qquad \eta\Big(\frac{\mathbf{x}}{\sqrt{\mathcal{D}}}\Big) = \mathcal{D}^{-n/2} \sum_{\mathbf{m}\in\mathbf{Z}^n} \tilde{\eta}\Big(\frac{\mu\mathbf{m}}{\sqrt{\mathcal{D}}}\Big)\eta\Big(\frac{\mathbf{x}-\mu\mathbf{m}}{\sqrt{\mathcal{D}}\mu}\Big) + R_{\eta,\mu,\mathcal{D}}(\mathbf{x})\,,$$

where $R_{\eta,\mu,\mathcal{D}} \in \mathcal{S}(\mathbf{R}^n)$ fulfills $|R_{\eta,\mu,\mathcal{D}}| < \varepsilon$.

Relation (5.17) is the approximate counterpart of the two-scale or refinement equation which is well known from wavelet theory. In the next section the approximate refinement equation is used to perform an approximate multiresolution decomposition of linear spaces generated by the translates of η. It leads also to an approximate factorization of quasi-interpolation operators which is the basis of multiscale approximation schemes.

Corollary 5.5. ([3]) *If $\mu^{-1} \in \mathbf{N}$ then (5.17) implies the factorization of the quasi-interpolation operator (2.2)*

$$(5.18) \qquad \mathcal{M}_h = \mathcal{M}_{\mu h} \widetilde{\mathcal{M}}_h + \mathcal{R}_h \,, \quad \|\mathcal{R}_{h,\mu}\| < \varepsilon \,,$$

with the quasi-interpolant

$$\widetilde{\mathcal{M}}_h u(\mathbf{x}) := \mathcal{D}^{-n/2} \sum_{\mathbf{m} \in \mathbf{Z}^n} u(h\mathbf{m}) \, \tilde{\eta}\Big(\frac{\mathbf{x} - h\mathbf{m}}{\sqrt{\mathcal{D}}h} \Big) \,.$$

Since $\tilde{\eta}$ satisfies together with η the assumptions A.1–3 the quasi-interpolation $\widetilde{\mathcal{M}}_h$ provides the same approximation properties as $\mathcal{M}_h$.

5.2. Isotropic boundary layer approximate approximation

The isotropic boundary layer approximation is derived by the iterative application of the factorization (5.18) in boundary layers $\mathcal{Q}_k \subset \Omega$ where the quasi-interpolation on the given uniform mesh of size $h_{k-1} = \mu^{k-1}h$, $\mu^{-1} \in \mathbf{N}$, becomes worse. Here we switch to the finer uniform mesh of size $h_k = h_{k-1}/\mu$ and approximate $u - \mathcal{M}_{h_{k-1}}u$ by a linear combination of the functions $\eta((\cdot - h_k\mathbf{m})/h_k\sqrt{\mathcal{D}})$, where the nodes $h_k\mathbf{m} \in \mathcal{Q}_k$. Since in view of (5.18) $\mathcal{M}_{h_{k-1}}u$ can be expressed in terms of these functions with the accuracy ε the approximation error is determined by the difference of $u - \mathcal{M}_{h_k}u$, which is large only at some sublayer $\mathcal{Q}_{k+1}$ near the boundary. Thus one obtains a boundary layer approximant of the form

$$(5.19) \qquad \mathcal{B}_L u(\mathbf{x}) = \mathcal{D}^{-n/2} \sum_{k=0}^{L} \sum_{h_k\mathbf{m} \in \mathcal{Q}_k} c_{k,\mathbf{m}} \, \eta\left(\frac{\mathbf{x} - h_k\mathbf{m}}{h_k\sqrt{\mathcal{D}}} \right) ,$$

where the coefficients $c_{k,\mathbf{m}}$ are given by

$$(5.20) \qquad c_{k,\mathbf{m}} = \begin{cases} u(h_0\mathbf{m}), & k = 0, \\ u(h_k\mathbf{m}) - (\widetilde{\mathcal{M}}_{h_{k-1}}u)(h_k\mathbf{m}), & k \geq 1. \end{cases}$$

Note that $c_{k,\mathbf{m}}$ is a linear combination of values of u near the point $h_k\mathbf{m}$ with coefficients not depending on h and the number of the boundary layer. Indeed,

$$(\widetilde{\mathcal{M}}_{h_{k-1}}u)(h_k\mathbf{m}) = \mathcal{D}^{-n/2} \sum_{|\mu\mathbf{m} - \mathbf{j}| \leq \widetilde{N}_s} u(h_{k-1}\mathbf{j}) \, \tilde{\eta}\Big(\frac{\mu\mathbf{m} - \mathbf{j}}{\sqrt{\mathcal{D}}} \Big),$$

therefore the coefficients $\tilde{\eta}((\mu\mathbf{m} - \mathbf{j})/\sqrt{\mathcal{D}})$ can be precomputed and used for any initial mesh size h and $k \geq 1$. Furthermore, the width of the boundary layers $\mathcal{Q}_k$ is proportional to h_k with some factor depending only on the basis function η.

A detailed analysis of this multiscale approximation together with numerical results is contained in [3], where the following theorem has been proven.

Theorem 5.6. *Suppose that Ω is a bounded domain with Lipschitz boundary and let $u \in W_p^N(\Omega)$ with $N > n/p$. If $\eta \in \mathcal{S}(\mathbf{R}^n)$ satisfies the assumptions of Theorem 5.4 and the moment condition A.2 then the boundary layer approximant $\mathcal{B}_L$ approximates u with the error*

$$\|u - \mathcal{B}_L u\|_{L_p(\mathbf{R}^n)} \le c_1 (\mathcal{D}h)^N \|\nabla_N u\|_{L_p(\Omega)} + c_2 (\mu^L h)^{1/p} \|u\|_{L_\infty(\Omega)} + \varepsilon \|u\|_{W_p^N(\Omega)} .$$

Thus, by choosing the number of boundary layers L so that $\mu^L = O(h^{Np-1})$ the error for the boundary layer approximation $\mathcal{B}_L u$ is of the form $\mathcal{O}((\mathcal{D}h)^N)) + \varepsilon$.

The cubature formula $\mathcal{K}_h u$ for an integral operator over the bounded domain Ω

$$\mathcal{K}u(\mathbf{x}) = \int_\Omega k(\mathbf{x} - \mathbf{y}) u(\mathbf{y}) d\mathbf{y} ,$$

is obtained from the boundary layer approximation of the density u by

$$\mathcal{K}_h u(\mathbf{x}) := \mathcal{K}\mathcal{B}_L u(\mathbf{x}) = \mathcal{D}^{-n/2} \sum_{k=0}^{L} \sum_{h_k \mathbf{m} \in \mathcal{Q}_k} c_{k,\mathbf{m}} \int_{\mathbf{R}^n} k(\mathbf{x} - \mathbf{y}) \eta \left(\frac{\mathbf{y} - h_k \mathbf{m}}{h_k \sqrt{D}} \right) d\mathbf{y} .$$

Theorem 5.7. *([3]) Let $u \in W_p^N(\Omega)$ with $N > n/p$ and the integral operator $\mathcal{K}$ maps $L_p(\Omega)$ boundedly into $W_p^m(\Omega_1)$. Under the assumptions made above for any $\varepsilon > 0$ there exists $D > 0$ such that*

$$\|\mathcal{K}u - \mathcal{K}_h u\|_{W_p^m(\Omega_1)} \le c_1 (Dh)^N \|\nabla_N u\|_{L_p(\Omega)} + c_2 (\mu^L h)^{1/p} \|u\|_{L_\infty(\Omega)} + \varepsilon \|u\|_{W_p^{N-1}(\Omega)} .$$

If additionally $\mathcal{K} \in \mathcal{L}((W_{p/(p-1)}^m(\Omega))', L_p(\Omega_1))$ then

$$\|\mathcal{K}u - \mathcal{K}_h u\|_{L_p(\Omega_1)} \le (c_1 (Dh)^N + c_2 (\mu^L h)^{1/p+r}) \|u\|_{W_p^N(\Omega)} + \varepsilon \, h^m \, \|u\|_{W_p^{N-1}(\Omega)} ,$$

where $0 < r < m/n$, $r \le (p-1)/p$.

5.3.   Anisotropic boundary layer approximate approximation

The cubature method based on isotropic boundary layer approximations has the advantage, that one gets immediately effective formulas for integrals over Lipschitz domains if only $\mathcal{K}\eta$ is known. On the other hand, the refinement in all directions may lead to a large number of summands in these formulas. But the approximation error increases in the direction towards the boundary, thus one actually needs refinement only in this direction. Therefore a considerable reduction of data points in the boundary layer approximation (5.19) can be derived by anisotropic mesh refinement. This is discussed in [4] for the example of a polyhedral domain Ω in $\mathbf{R}^3$.

Using a partition of unity the function u is decomposed into the sum of functions with support in the interior, near corners, near edges and near faces of Ω, respectively. The approximation of the function with compact support in the interior of Ω can be performed by the usual quasi-interpolant (2.2) with a suitable generating function, whereas for the approximation of the functions supported near the corners of Ω the approximation operator (5.19) can be used. The reduction of data points is reached by constructing the boundary layer approximants for the functions near edges and faces using mesh refinement only in the direction normal to the boundary. Consider for example in $\mathbf{R}_+^3 = \{\mathbf{x} = (x', x_3) \in \mathbf{R}^3 : x_3 \geq 0\}$ a sufficiently smooth function u with bounded support and $u(x', 0) \not\equiv 0$. The construction of high order approximants to u with respect to the sequence of anisotropically distributed mesh points $\{(hm', \mu^k hm_3) : (m', m_3) \in \mathbf{Z}_+^3\}$ is based on a generating function of the tensor product form

$$(5.21) \qquad \eta_3(x', x_3) = \eta_2(x')\, \eta_1(x_3)$$

where η_2 is a suitable generating function and η_1 satisfies the conditions of Theorem 5.4. The approximate refinement equation for this function results in the factorization

$$\mathcal{M}_{h,\mu} = \mathcal{M}_{h,\mu^2}\, \widetilde{\mathcal{M}}_{h,\mu} + \mathcal{R}_{h,\mu}\,, \quad \|\mathcal{R}_{h,\mu}\| < \varepsilon\,,$$

where now

$$\mathcal{M}_{h,\mu}u(\mathbf{x}) := \mathcal{D}^{-3/2} \sum_{\mathbf{m} \in \mathbf{Z}^3} u(h\mathbf{m})\, \eta_3\left(\frac{x' - hm'}{h\sqrt{\mathcal{D}}}, \frac{x_3 - \mu hm_3}{\mu h\sqrt{\mathcal{D}}}\right)$$

$$\widetilde{\mathcal{M}}_{h,\mu}u(\mathbf{x}) := \mathcal{D}^{-3/2} \sum_{\mathbf{m} \in \mathbf{Z}^3} u(h\mathbf{m})\, \eta_2\left(\frac{x' - hm'}{\sqrt{\mathcal{D}}h'}\right) \tilde{\eta}_1\left(\frac{x_3 - \mu hm_3}{\mu h\sqrt{\mathcal{D}}}\right)$$

with the dual function $\tilde{\eta}_1$ from the refinement equation for η_1. Analogously to the isotropic case the approximant is defined as

$$(5.22) \qquad \mathcal{D}^{-3/2} \sum_{k=0}^{L} \sum_{(hm', \mu^k hm_3) \in \mathcal{Q}_k} c_{k,\mathbf{m}}\, \eta_3\left(\frac{x' - hm'}{h\sqrt{\mathcal{D}}}, \frac{x_3 - \mu^k hm_3}{\mu^k h\sqrt{\mathcal{D}}}\right)$$

with the coefficients

$$c_{k,\mathbf{m}} = \begin{cases} u(hm', hm_3), & k = 0 \\ u(hm', \mu^k hm_3) - (\widetilde{\mathcal{M}}_{h,\mu^{k-1}}u)(hm', \mu^k hm_3), & k \geq 1 \end{cases}$$

where $\mathcal{Q}_k$ are layers in $\mathbf{R}^3$ parallel to the plane $\{x_3 = 0\}$. In [4] it is shown that the assertion of Theorem 5.6 remains true for the approximation (5.22) if η satisfies the moment condition (2.5) for given N.

The quasi-interpolation for edge supported functions with the order $O((\mathcal{D}h)^N)$ can be performed using generating functions of the form (5.21) with η_2 satisfying the conditions of Theorem 5.4. The wedge under consideration is affinely transformed to a wedge with right inner angle, and here the construction of the boundary layer approximant is a straightforward extension of the above construction.

Let us note that it is possible to find tensor product generating functions (5.21) where the action of potentials can be computed by simple one-dimensional quadrature (see [4]).

6. Approximate wavelets

The application of wavelet methods to the representation of integral and differential operators is one of the actual research topics in the numerical analysis of solution methods for corresponding operator equations. These methods are based on the multiresolution analysis for fine scale spaces. In numerical procedures one starts with a finite sequence of nested closed subspaces

$$(6.23) \qquad V_0 \subset V_1 \subset \ldots \subset V_d \subset L_2(\mathbf{R^n})$$

with the properties

(i) $f(\mathbf{x}) \in V_0$ if and only if $f(\mathbf{x} - \mathbf{m}) \in V_0$ for any $\mathbf{m} \in \mathbf{Z}^n$;

(ii) $f(\mathbf{x}) \in V_j$ if and only if $f(2\mathbf{x}) \in V_{j+1}$ for any $j = 0, \ldots, d-1$;

(iii) there exists ϕ such that $\{\phi(\cdot - \mathbf{m})\}_{\mathbf{m} \in \mathbf{Z^n}}$ is an L_2-stable basis in V_0.

Then the spaces V_j are spanned by the dilated shifts $\phi(2^j \cdot -\mathbf{m})$, $\mathbf{m} \in \mathbf{Z}^n$, of the scaling function ϕ. The main goal of the multiresolution is to determine a new basis of the space V_d corresponding to the finest grid. To this end V_d is decomposed into the orthogonal sum

$$(6.24) \qquad V_n = V_0 \bigoplus_{j=1}^{d} W_j \, ,$$

where the wavelet space W_j is the orthogonal complement $W_j = V_j \ominus V_{j-1}$. There exist $2^n - 1$ functions $\psi_v \in W_1$, called prewavelets, with the property that the shifts $\{\psi_v(2^j \cdot -\mathbf{x}), \mathbf{m} \in \mathbf{Z}^n, v \in \mathcal{V}'\}$ form an L_2-stable basis in the space W_{j+1}. Here the prewavelets ψ_v are indexed by the set $\mathcal{V}' = \mathcal{V}\backslash\{\mathbf{0}\}$ with $\mathcal{V}$ denoting the set of vertices of the cube $[0, 1]^n$. Thus one obtains a basis of V_d consisting of

$$\{\phi(\cdot - \mathbf{m}), \mathbf{m} \in \mathbf{Z}^n\} \quad \text{and} \quad \{\psi_v(2^j \cdot -\mathbf{m}), \mathbf{m} \in \mathbf{Z}^d, v \in \mathcal{V}', j = 0, \ldots, d-1\} \, .$$

The elements of V_d and operators acting on them are now expanded into the new basis and the computations take place in this system of coordinates, where one hopes to achieve that the computation is faster than in the original system. Since in view of (ii) the scaling function ϕ has to satisfy an exact refinement equation as a rule the approximating functions for the above mentioned applications of wavelet based numerical methods are piecewise polynomials.

The generating functions considered in approximate approximations do not fulfil an exact refinement equation. But in view of Theorem 5.4 a large class of these functions satisfy approximate refinement equations and the error can be controlled by the parameter $\mathcal{D}$. Therefore it is possible to perform an approximate multiresolution analysis of fine scale spaces spanned by the translates of theses functions, i.e. to decompose a fine scale space within some given tolerance into a direct sum of a coarse scale space and wavelet spaces. Then elements of the fine scale space and operators acting on them can be represented within the given tolerance in this new base. The advantage of this approximate wavelet approach lies in the possibility to get efficiently computable sparse approximations for important integral operators, as shown in [16].

6.1. Approximate multiresolution analysis

Consider the closed linear subspaces of $L_2(\mathbf{R}^n)$

$$\mathbf{V}_j := \Big\{ \sum_{\mathbf{m} \in \mathbf{Z}^n} a_{\mathbf{m}}\, \eta_{\mathcal{D}}(2^j \cdot -\mathbf{m}),\ \{a_{\mathbf{m}}\} \in \ell_2(\mathbf{Z}^n) \Big\},$$

where $\eta_{\mathcal{D}} := \eta(\cdot/\sqrt{\mathcal{D}})$ satisfies the assumptions of Theorem 5.4. Since $\mathcal{F}\eta_{\mathcal{D}} \neq 0$ the family of translates $\{\eta_{\mathcal{D}}(2^j \cdot -\mathbf{m})\}_{\mathbf{m}\in\mathbf{Z}^n}$ is an L_2-stable basis in $\mathbf{V}_j$. From the approximate refinement equation (5.17) it is clear that for any $l < j$ the space $\mathbf{V}_l$ is almost included in $\mathbf{V}_j$. In particular, if $\mu = 1/2$ then any element

$$\varphi_j = \sum_{\mathbf{m} \in \mathbf{Z}^n} a_{\mathbf{m}}\eta_{\mathcal{D}}(2^j \mathbf{x} - \mathbf{m}) \in \mathbf{V}_j$$

can be perturbed such that

$$\varphi_j - \sum_{\mathbf{m} \in \mathbf{Z}^n} a_{\mathbf{m}} R_{\eta,2,\mathcal{D}}(2^j \cdot -\mathbf{m}) \in \mathbf{V}_{j+1} .$$

The norm of the perturbation is small for $\mathcal{D}$ large enough, hence for any $\varepsilon > 0$ there exists $\mathcal{D}$ such that

$$(6.25) \qquad \|\varphi_j - P_{j+1}\varphi_j\|_2 \leq \varepsilon \|\varphi_j\|_2 ,$$

where P_{j+1} denotes the L_2-orthogonal projection onto $\mathbf{V}_{j+1}$.

Furthermore, we allow also a small perturbation of the wavelet basis which spans the closed subspace of all functions in $\mathbf{V}_{j+1}$ which are orthogonal to $\mathbf{V}_j$. We assume the existence of functions $\psi_{\mathcal{D},v}$, $v \in \mathcal{V}'$, such that the space $\mathbf{W}_1$ spanned by their integer translates has the property that if $\varphi_1 \in \mathbf{V}_1$ and $\varphi_1 \perp \mathbf{V}_0$ then

$$(6.26) \qquad \|\varphi_1 - Q_1\varphi_1\|_2 \leq \varepsilon \|\varphi_1\|_2 ,$$

where Q_1 is the L_2-orthoprojection onto $\mathbf{W}_1$.

Then it can be easily seen that for fixed d any $\varphi_d \in \mathbf{V}_d$ can be represented within some prescribed tolerance as an element of the multiresolution structure

$$(6.27) \qquad \widetilde{\mathbf{V}}_d := \mathbf{V}_0 \dot{+} \mathbf{W}_1 \dot{+} \ldots \dot{+} \mathbf{W}_d \, .$$

More precisely, under the assumptions (6.25) and (6.26) the following estimate holds:

$$(6.28) \qquad \left\| \varphi_d - P_0 \varphi_d - \sum_{j=1}^{d} Q_j \varphi_d \right\|_2 \leq d \, \frac{3\varepsilon - \varepsilon^2}{1 - \varepsilon} \, \|\varphi_d\|_2 \, .$$

Hence the approximate multiresolution analysis allows to combine the flexible choice of basis functions used in approximate approximations with the interesting features of wavelets, as the localization in both space and frequency domains and vanishing moment properties.

6.2. Example

To illustrate the above approach we consider the example of the one-dimensional Gaussian function $\eta_{\mathcal{D}}(x) = \exp(-x^2/\mathcal{D})$ providing the approximate refinement equation

$$\eta_{\mathcal{D}}(x) = \frac{2}{\sqrt{3\pi\mathcal{D}}} \sum_{m=-\infty}^{\infty} e^{-m^2/3\mathcal{D}} \eta_{\mathcal{D}}(2x - m) - \eta_{\mathcal{D}}(x) \sum_{\nu \neq 0} e^{-3\pi^2 \mathcal{D}\nu^2/4} e^{3\pi i x \nu} \, .$$

To find an approximate wavelet we start with the element of V_1

$$(6.29) \quad \sum_{m=-\infty}^{\infty} (-1)^{m-1} \mu_{m-1} \eta_{\mathcal{D}}(2x - m) \quad \text{with} \quad \mu_m = \int_{\mathbf{R}} \eta_{\mathcal{D}}(x)\, \eta_{\mathcal{D}}(2x + m)\, dx \, ,$$

which is orthogonal to all integer shifts of the scaling function $\eta_{\mathcal{D}}$ and has the Fourier transform

$$c_{\mathcal{D}}\, e^{-\pi i \lambda} e^{-\pi^2 \mathcal{D}\lambda^2/4} \sum_{k=-\infty}^{\infty} \exp(-5\pi^2 \mathcal{D}(\lambda + 2k + 1)^2/4)$$

with some factor $c_{\mathcal{D}}$. It can be easily seen that for $\mathcal{D}$ sufficiently large the function with the Fourier transform

$$c_{\mathcal{D}}\, e^{-\pi i \lambda} e^{-\pi^2 \mathcal{D}\lambda^2/4} \left(\exp(-5\pi^2 \mathcal{D}(\lambda + 1)^2/4) + \exp(-5\pi^2 \mathcal{D}(\lambda - 1)^2/4) \right)$$

provides a very accurate smooth approximation of (6.29). Therefore the approximate wavelet can be defined as

$$(6.30) \qquad \psi_{\mathcal{D}}(x) := e^{-(2x-1)^2/6\mathcal{D}} \cos \frac{5\pi}{6} (2x - 1) \, ,$$

which perturbs the corresponding function in V_1 orthogonal to V_0 by

$$R_{\mathcal{D}}(x) = e^{-(2x-1)^2/6\mathcal{D}} \sum_{k=1}^{\infty} \cos\frac{5\pi}{6}(2k+1)(2x-1)\, e^{-5\pi^2\mathcal{D}(k^2+k)/6} = O\!\left(e^{-5\pi^2\mathcal{D}/3}\right).$$

Hence, any element $\varphi_d \in V_d$, which approximates some function up to a prescribed accuracy with the order $\mathcal{O}(2^{-dN})$, can be expanded within the same accuracy into the new basis

$$\{\eta_{\mathcal{D}}(\cdot - m), m \in \mathbf{Z}\} \quad \text{and} \quad \{\psi_{\mathcal{D}}(2^j \cdot -m), m \in \mathbf{Z}, j = 0,\ldots,d-1\}.$$

Analytic representations of the projection operators P_0 and Q_1 are given in [16].

Besides the good localization the approximate wavelets have an interesting feature. In general its power moments do not vanish, but they are very small and decrease as $\mathcal{D}$ increases. This implies in particular fast decay of integral operators $\mathcal{K}$ applied to them if the kernel $k(\mathbf{x}, \mathbf{y})$ satisfies

$$|\partial_{\mathbf{y}}^{\alpha} k(\mathbf{x},\mathbf{y})| \le c_\alpha\, |\mathbf{x}-\mathbf{y}|^{-(\gamma+|\alpha|)} \quad \text{for some } \gamma > 0 .$$

This effect is shown in Figure 1 giving the graphs of the approximate wavelet ψ_3 and of the Hilbert transform of ψ_3

$$\mathcal{H}\psi_{\mathcal{D}}(x) = \frac{1}{\pi} \int_{-\infty}^{\infty} \frac{\psi_{\mathcal{D}}(y)}{y - x}\, dy$$

$$= \frac{ie^{\,25\pi^2\mathcal{D}/24}}{4}\left(W\!\left(\frac{5\pi\mathcal{D} - 2i(2x-1)}{2\sqrt{6\mathcal{D}}}\right) - W\!\left(\frac{5\pi\mathcal{D} + 2i(2x-1)}{2\sqrt{6\mathcal{D}}}\right)\right),$$

where the function W is defined by

$$W(z) := e^{z^2}\left(\operatorname{erfc}(z) - \operatorname{erfc}(-z)\right) .$$

6.3. The multivariate case

As mentioned above in d dimensions the wavelet space W_1 is spanned by the integer shifts of $2^d - 1$ prewavelets ψ_v. If the scaling function $\eta_{\mathcal{D}}$ used in the approximate resolution analysis is a tensor product of one-dimensional functions

$$\eta_{\mathcal{D}}(\mathbf{x}) = \prod_{j=1}^{d} \phi_{\mathcal{D}}(x_j) ,$$

as for example the Gaussian, then the approximate wavelets are obtained as the tensor products

$$\psi_{\mathcal{D},v}(\mathbf{x}) = w_{v_1}(x_1) \cdots w_{v_d}(x_d) ,\quad v = (v_1, \cdots, v_d) \in \mathcal{V}' ,$$

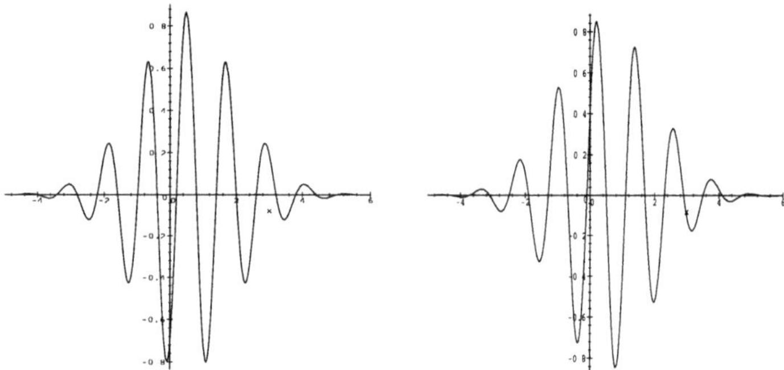

Figure 1: Graph of the approximate wavelet ψ_3 (left) and its Hilbert transform (right)

where $w_0(x) = \phi_\mathcal{D}(x)$ and $w_1(x) = \psi_\mathcal{D}(x)$ denotes the one-dimensional approximate wavelet function associated with $\phi_\mathcal{D}$. Note that for different $v \in \mathcal{V}$ the principal shift invariant spaces $X_\mathbf{v} := \{\psi_{\mathcal{D},v}(\cdot - \mathbf{m}), \mathbf{m} \in \mathbf{Z}^d\}$ are nearly orthogonal. Thus the space of approximate wavelets is the direct sum of the spaces $X_\mathbf{v}$, $v \in \mathcal{V}'$. For more general basis functions the approximate wavelets can be determined by using the extension of formula (6.29)

$$\sum_{\mathbf{m} \in \mathbf{Z}^d} (-1)^{(\mathbf{m},v)} \mu_\mathbf{m}\, \eta_\mathcal{D}(2\mathbf{x} - \mathbf{m} - v)\,, \quad v \in \mathcal{V}'\,,$$

$$\text{where} \quad \mu_\mathbf{m} = \int_{\mathbf{R}^d} \eta_\mathcal{D}(\mathbf{x})\, \eta_\mathcal{D}(2\mathbf{x} + \mathbf{m})\, d\mathbf{x}\,,$$

which gives elements from V_1 orthogonal to V_0. Again, appropriate truncation of the infinite series representations of the Fourier transform of these functions leads to small perturbations. The resulting functions, the multivariate approximate wavelets, have compact analytic representations.

7. Numerical algorithms based upon approximate approximations

The real power of approximate approximations is in the capability to treat multi-dimensional integral operators very efficiently. Therefore it is natural to use it as underlying approximation method in numerical algorithms for solving problems with integro-differential equations. Another very important application of approximate approximations is in the large field of integral equations methods for solving initial and boundary value problems for partial differential equations. V. Maz'ya developed numerical algorithms for some typical examples, which can be found in

[11], [10], [12], [5], [6]. Here I briefly describe the approach of Maz'ya and Karlin to solve evolution equations with nonlocal operators based upon approximate approximations

7.1. Nonlocal evolution equations

In [6] V. Maz'ya and V. Karlin considered the Cauchy-Problem for equations of the general form

$$u_t - P_1(D_{\mathbf{x}})u = P_2(D_{\mathbf{x}})F(\mathbf{x}, t, u, P_3(D_{\mathbf{x}})u) \, , \ t > 0 \, , \ \mathbf{x} \in \mathbf{R}^d \, , \quad u(\mathbf{x}, 0) = \varphi(\mathbf{x}) \, ,$$

where $D_{\mathbf{x}} = (-i\partial/\partial x_1, \ldots, -i\partial/\partial x_n)$. The operators $P_j(D_{\mathbf{x}})$ are convolutions with the symbol $P_j(2\pi\xi)$ and F is supposed to be a smooth function. The equation is discretized in time by a two-parameter finite-difference approximation with time step τ. Then $u(\mathbf{x}, t)$ is approximated by a sequence of functions $u_n(\mathbf{x}) = u(\mathbf{x}, n\tau)$, $n = 0, 1, 2, \ldots$, satisfying

$$\tau^{-1}(u_n - u_{n-1}) - \theta_1 P_1(D_{\mathbf{x}})u_n - (1 - \theta_1)P_1(D_{\mathbf{x}})u_{n-1}$$

$$= P_2(D_{\mathbf{x}})[((1 + \theta_2)F_{n-1}(\mathbf{x}) - \theta_2 F_{n-2}(\mathbf{x})] \, ,$$

where $F_n(\mathbf{x}) = F(\mathbf{x}, n\tau, u_n, P_3(D_{\mathbf{x}})u_n)$ and $0 < \theta_i \le 1$. For $\theta_1 = \theta_2 = 1/2$ the scheme is of second order accuracy. With the notations

$$\mu = \tau\theta_1 \, , \ y = u_n + (\theta_1^{-1} - 1)u_{n-1} \, ,$$

$$f = \theta_1^{-1}u_{n-1} \, , \ g = \tau[(1 + \theta_2)F_{n-1} - \theta_2 F_{n-2}]$$

one derives the following linear problem

$$-\mu P_1(D_{\mathbf{x}})y + y = f + P_2(D_{\mathbf{x}})g \, , \ \mathbf{x} \in \mathbf{R}^d \, ,$$

providing the solution

$$y = f + (R - I)f + P_2 Rg \quad \text{with} \quad R = (I - \mu P_1)^{-1} \, .$$

Replacing f and g by its quasi-interpolants (2.2) one obtains the approximate solution

$$y_h = f + (R - I)\mathcal{M}_h f + P_2 R \mathcal{M}_h g$$
$$= f + \mathcal{D}^{-d/2} \sum_{\mathbf{m} \in \mathbf{Z}^d} f(h\mathbf{m})[(R\eta)\big(\frac{\mathbf{x} - h\mathbf{m}}{\sqrt{\mathcal{D}}h}\big) - \eta\big(\frac{\mathbf{x} - h\mathbf{m}}{\sqrt{\mathcal{D}}h}\big)]$$
$$+ \mathcal{D}^{-d/2} \sum_{\mathbf{m} \in \mathbf{Z}^d} g(h\mathbf{m})(P_2 R\eta)\big(\frac{\mathbf{x} - h\mathbf{m}}{\sqrt{\mathcal{D}}h}\big) \, .$$

Thus the values of y_h at the grid points $x_{\mathbf{k}} = h\mathbf{k}$ are linear combinations of

$$(R\eta)\big(\frac{\mathbf{k} - \mathbf{m}}{\sqrt{\mathcal{D}}}\big) \quad \text{and} \quad (P_2 R\eta)\big(\frac{\mathbf{k} - \mathbf{m}}{\sqrt{\mathcal{D}}}\big) \, ,$$

which may be effectively computed if the generating function η is suitably chosen. By this way Maz'ya and Karlin derive the following explicit scheme to compute the approximate solution $u_{n,h}$ of the Cauchy problem:

$$u_{n,h} = u_{n-1,h} + \theta_1^{-1}(R - I)\mathcal{M}_h u_{n-1,h} + \tau P_2 R \mathcal{M}_h[(1 + \theta_2)F_{n-1} - \theta_2 F_{n-2}] \, .$$

Compared with other explicit schemes for solving time-dependent problems the proposed method is very robust with respect to variations of the ratio between time and spatial discretization. Numerous numerical tests for different equations have shown, that of course the non-linearity in the original equation imposes restrictions to the time step τ, but there exists no strict connection between τ and the mesh size h. In these tests the present method provides an accuracy of $\mathcal{O}(\tau^2 + \tau h^N)$ at each time step, where N is the approximation order of the approximate quasi-interpolation with the generating function η. For $\theta_1 = \theta_2 = 1/2$ the numerical accuracy increases to $\mathcal{O}(\tau^3 + \tau h^N)$. However, the rigorous error analysis of this quite general method remains open. The estimation of the saturation errors, which occur at each time step, is rather involved and understood at present only for some special equations. Here again for $\mathcal{D}$ sufficiently large the saturations errors can be kept below a given error level, as expected also from the numerical experiments.

7.2. Examples

In [6] the method is applied to different nonlinear nonlocal evolution equations, here two examples will be mentioned. The first concerns the Joseph equation

$$(7.31) \qquad u_t + \delta^{-1}u_x + (2\delta)^{-1}\int_{-\infty}^{\infty} u_{yy} \coth\frac{\pi(y - x)}{2\delta}\, dy = -(u^2)_x \, ,$$

which describes the unidirectional propagation of small-amplitude, nonlinear, dispersive, long waves in stratified fluids. Note that the shallow water approximation $(\delta \ll 1)$ of (7.31) is the Korteweg-de-Vries equation, whereas the deep water approximation $(\delta \gg 1)$ is the Benjamin-Ono equation

$$u_t + \mathcal{H}u_{xx} = -(u^2)_x$$

with the Hilbert transform $\mathcal{H}$. In Figure 2 the computational results for Joseph's equation with the initial data $\varphi(x) = \delta\exp(-x^2)$ for $\delta = 0.1, 0.333, 1$, and 10, corresponding to shallow, intermediate, and deep water, are shown. In these computations η was the Gaussian function, $\mathcal{D} = 3$, $h = 0.1$ and $\tau = 0.001$.

Another interesting example which is difficult to solve by using finite-difference or finite-element methods represents the two-dimensional equation of flame front propagation

$$(7.32) u_t + a\Delta^2 u + \epsilon\Delta u + b\Delta\int_{\mathbf{R}^2} \frac{u(\mathbf{y}, t)}{|\mathbf{x} - \mathbf{y}|}\, d\mathbf{y} + cu = -\frac{1}{2}(\nabla u)^2 \, , \ t > 0, \ \mathbf{x} \in \mathbf{R}^2.$$

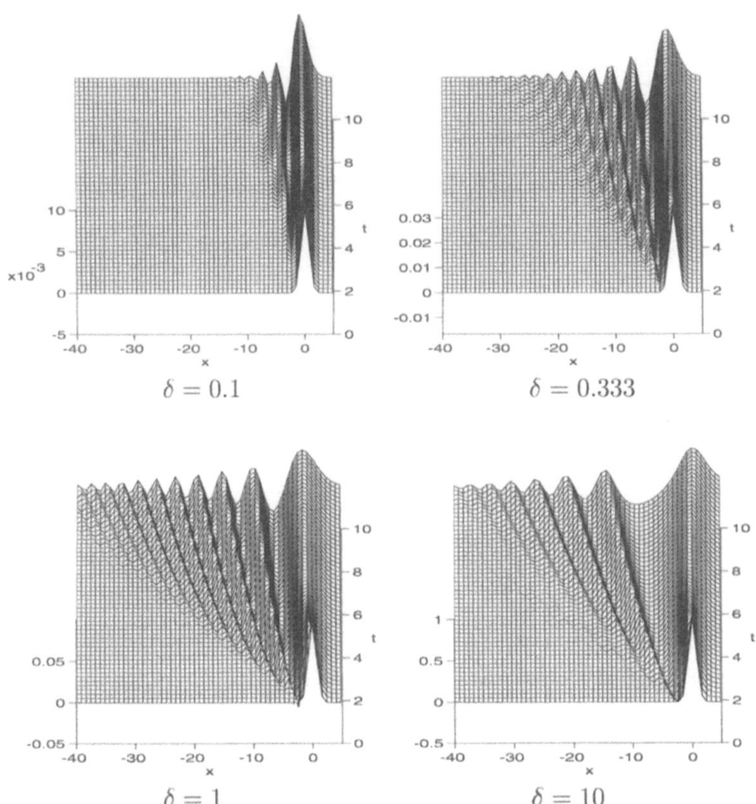

$$\delta = 0.1 \qquad\qquad \delta = 0.333$$

$$\delta = 1 \qquad\qquad \delta = 10$$

Figure 2: Solution of the Joseph equation for different δ

Here

$$P_1(\xi) = -a|\xi|^4 + \epsilon|\xi|^2 + b|\xi| - c,\ P_2(\xi) = 1,\ P_3(\xi) = i\xi,\ F(\mathbf{x},t,u,\mathbf{v}) = -|\mathbf{v}|^2/2\,.$$

With the two-dimensional generating functions (3.12) the occuring integrals are transformed to the zero-order Hankel transform with smooth and rapidly decaying integrands and may be done by using a standard quadrature procedure. In Figure 3 the results of the numerical solution of (7.32) with the initial data

$$\varphi(\mathbf{x}) = 1 + \frac{1}{2}\sin 2x_1\,\sin 2x_2$$

and the parameters $a = 10^{-4}$, $\epsilon = 0.05$, $b = 0.005$ and $c = 1/6$ are given for different time values. The 2π-periodic solution was computed for the discretization parameters $h = \pi/32$ and $\tau = 10^{-3}$.

The given parameters correspond to a linear flame instability and the computations result in a corrugated flame front.

$t = 0$

$t = 1$

$t = 5$

$t = 10$

Figure 3: Level lines of the surface of the flame front for different t

7.3. Further applications

Finally I mention some other problems for which algorithms based on approximate approximations have been applied successfully. The algorithm described above was used to solve non-linear parabolic differential equations. The papers [11], [12], [5] contain also numerical results for different model equations, among them the two-dimensional Navier-Stokes problem.

Approximate approximations were successfully used in [7] to solve hypersingular integral equations of the Peierls type

$$\int_{-\infty}^{\infty} K(x-y)\,u(y)\,dy = F(u(x)), \quad K(x) = -x^{-2} + \kappa(x)\,,$$

with κ being smooth and the integral defined in the finite part Hadamard sense. Integral equations of this type occur in dislocation theory. Due to the efficient cubature formulas critical Peierls stresses were calculated at very high accuracy for a variety of dislocations.

A further application of approximate approximations is an extension of the Boundary Element Method. While usually the numerical solution of boundary integral equations is sought in form of piecewise polynomial functions the so-called Boundary Point Method uses suitable basis functions for which the boundary integrals can be effectively computed. Representing the numerical solution as linear combination of translates of this basis function similar to the quasi-interpolant (2.8) the discretization of the boundary integral equations requires only the computation of boundary potentials whose densities are the basis functions. In [10], [11] the Boundary Point Method is applied to two- and three-dimensional potential problems with smooth boundary, where the surface integration is replaces by the integration over the tangent plane. The obtained second order approximation rate corresponds to the theoretically expected results.

References

[1] C. de Boor and A. Ron, Fourier Analysis of the Approximation Power of Principal Shift-Invariant Spaces, Constr. Approx. **8** (1992), 427–462.

[2] R. K. Beatson and W. A. Light, Quasi-interpolation in the Absence of Polynomial Reproduction, in: *Numerical Methods of Approximation Theory, Vol. 9*, D. Braess and L. L. Schumaker, (eds.), 21–39, Basel: Birkhäuser 1992.

[3] T. Ivanov, V. Maz'ya, and G. Schmidt, Boundary Layer Approximate Approximations and Cubature of Potentials in Domains. WIAS Preprint No. 402, 1998, to appear in Adv. Comp. Math.

[4] ______ Boundary layer approximate approximations for the cubature of potentials to appear in *Mathematical Aspects of Boundary Element Methods*, Pitman Research Notes in Mathematics Series, London: Longman.

[5] V. Karlin and V. Maz'ya, Time-marching algorithms for initial-boundary value problems for semi-linear evolution equations based upon approximate approximations, BIT **35** (1995) no. 4, 548–560.

[6] ______ Time-marching algorithms for non-local evolution equations based upon "approximate approximations", SIAM J. Sci. Comput. **18** (1997), 736–752.

[7] V. Karlin, V. Maz'ya, A. B. Movchan, J. R. Willis, and R. Bullogh, Numerical solution of nonlinear hypersingular integral equations of the Peierls type in dislocation theory, to appear in SIAM J. Appl. Math.

[8] V. Kozlov, A. Levin, and V. Maz'ya, Mathematical algorithm for the reconstruction of the optical distribution of the power density under sondage of a laser beam, Leningrad Research Engineering Institute of the USSR Academy of Sciences, Preprint 35, Leningrad 1990.

[9] V. Maz'ya, A New Approximation Method and its Applications to the Calculation of Volume Potentials. Boundary Point Method, in: 3. DFG-Kolloqium des DFG-Forschungsschwerpunktes "Randelementmethoden", 30 Sep–5 Oct 1991.

[10] V. Maz'ya, Boundary point method, Preprint LiTH-MAT-R-91-44, Linköping University 1991.

[11] ――――Approximate Approximations, in: *The Mathematics of Finite Elements and Applications. Highlights 1993*, J. R. Whiteman (ed.), 77–104, Wiley & Sons, Chichester 1994.

[12] V.Maz'ya and V., Karlin, *Semi-analytic time-marching algorithms for semi-linear parabolic equations*, BIT 34, (1994), 129–147.

[13] V. Maz'ya and G. Schmidt, On Approximate Approximations, Preprint LiTH-MAT-R-94-12, Linköping University 1994.

[14] ――――― On Approximate Approximations using Gaussian Kernels, IMA J. Num. Anal. **16** (1996), 13–29.

[15] ――――― "Approximate Approximations" and the cubature of potentials, Rend. Mat. Acc. Lincei, s. 9, **6** (1995), 161–184.

[16] ――――― Approximate Wavelets and the Approximation of Pseudodifferential Operators, to appear in Appl. Comp. Harm. Anal.

[17] ――――― On Quasi-interpolation with non-uniformly distributed centers on Domains and Manifolds, WIAS Preprint No. 422, Berlin 1998.

[18] E. Stein and G. Weiss, *Introduction to Fourier Analysis on Euclidian Spaces*, University Press, Princeton 1971.

[19] F. Stenger, *Numerical Methods Based on Sinc and Analytic Functions*, Berlin: Springer 1993.

[20] E.N. Vil'chevskaya and S.K. Kanaun, Calculation of elastic fields near thin inclusions and cracks, Leningrad Research Engineering Institute of the USSR Academy of Sciences, Preprint 57, Leningrad 1991.

[21] ―――――Integral equations for a thin inclusion in a homogeneous elastic medium, J. Appl. Math. Mech. 56, No.2, 235–243 (1992), translation from Prikl. Mat. Mekh. 56, No.2 (1992), 275–285.

Weierstrass Institute
for Applied Analysis and Stochastics,
Mohrenstrasse 39,
D – 10117 Berlin,
schmidt@wias-berlin.de

1991 Mathematics Subject Classification: Primary 65–02, 41A30, 41A63, 45K05, 65D32

Submitted: 01.02.1999

Operator Theory:
Advances and Applications, Vol. 109

Maz'ya's work on the biography of Hadamard

ROGER COOKE

I knew Maz'ya's work long before I met him. Whenever I looked in works on functional analysis, potential theory, and elsewhere, it seemed I found some important contribution by Vladimir Maz'ya. Although the thought of meeting him was far from my mind when I was invited to spend part of the winter of 1991 at the Institut Mittag-Leffler, my feeling when I was introduced to him was one of seeing for the first time up close a person you felt you knew from his writing.

As others in this volume have discussed Maz'ya's pure and applied mathematical work better than I can do, I will devote just a few lines to a discussion of his joint work with T. Shaposhnikova on the biography of Hadamard. Shaposhikova's book on Hadamard, co-authored by the late E.M. Polishchuk, had just appeared in the excellent series of biographies published by Nauka press, Moscow. Like all the books in this series, it was carefully researched and solidly written. Its particular strength, I thought when reading it at the time, was the seemingly effortless exposition of the intricacies of Hadamard's work. I wondered as I read the book who had taken the trouble to study so much of Hadamard's enormous output, who was capable of understanding so many diverse areas of mathematics, and summarize it so succinctly. Had I paid more attention at the beginning, I might have guessed. At the end of the preface, Shaposhnikova wrote (Polishchuk had died in 1987):

"Last but not least I would like to express my gratitude (also shared by E.M. Polishchuk) to my husband Vladimir Gilelevich Maz'ya, who has been our constant consultant. His advice has greatly helped in shaping this book".

I did not know at the time that Shaposhnikova and Maz'ya were now permanent residents of Sweden. (There were many Russian citizens at the Institut that year, some of whom continued to live in Russia.) The opportunity to travel in the West meant for them the opportunity to find out more about Hadamard than could be gleaned from published documents. The two lost no time in doing so. Although much of the relevant material unfortunately dissapeared during World War II, they were able to find a great deal of new material nevertheless. They located Hadamard's grandson, Francis Picard and were able to look at the materials collected by Hadamard's daughter Jacqueline, including her autobiography. They interviewed several mathematicians who had known Hadamard personally, and they submitted their drafts of the new and revised material to experts in a number of areas of mathematics and the history of mathematics, and the history of French mathematics in particular. The result, published in English this time, was a triumph of mathematical biography that one can only hope will be repeated for many other great twentieth-century mathematicians. Their book "Jacques

Hadamard, a Universal Mathematician" was published jointly by the American Mathematical Society and the London Mathematical Society as Volume 14 in the History of Mathematics series.

Besides a coherent narrative and clear exposition of the mathematical details, the book contains hundreds of photographs and pictures of documents concerning Hadamard. It will probably turn out to be the definitive biography of Hadamard. I keep it on my shelf and look at it from time to time, reminding myself of what I would like to see in relation to all the great mathematicians of the twentieth century. How wonderful it would be if a similar biograpy could be assembled about Lebesgue, Borel, F. and M. Riesz, Weyl, Kolmogorov, and others. Most of these would surpass the power of any historian of mathematics to write without some consultation; I can't think of a better consultant to recommend than V.G. Maz'ya.

Department of Mathematics,
University of Vermont,
Burlington VT 05401-1455, USA,
cooke@emba.uvm.edu

1991 Mathematics Subject Classification: Primary 01A70, 01A60

Submitted: 19.01.1999

Operator Theory:
Advances and Applications, Vol. 109

Isoperimetric inequalities and capacities on Riemannian manifolds

ALEXANDER GRIGOR'YAN

We discuss extensions of some results of V.G.Maz'ya to Riemannian manifolds. His proofs of the relationships between capacities, isoperimetric inequalities and Sobolev inequalities did not use specific properties of the Euclidean space. His method, transplanted to manifolds, gives a unified approach to such results as parabolicity criteria, eigenvalues estimates, heat kernel estimates, etc.

1. Introduction

One of the important contributions of Maz'ya is the general understanding of the relationship between isoperimetric inequalities and functional inequalities of the Sobolev type in regions of $\mathbb{R}^n$. However, his method is not linked to specific properties of the Euclidean spaces and can be carried over to Riemannian manifolds. This was fully understood by Maz'ya in the 1960s when he wrote in [24, p.153]

> "Let us mention that, despite here we restrict ourself by considering sets in $\mathbb{R}^n$, the method of proof of criteria of [such] type does not use specific properties of the Euclidean space. The proofs can be carried over to the case of singular Riemannian manifolds; the exponents in the embedding theorems will depend not only on measures μ, ν and on the degree of irregularity of the boundary of [region] D but also on the singularities of the metric of the manifold."

Nowadays it is commonly accepted that embedding theorems and Sobolev type inequalities play the crucial role in Analysis on manifolds. A series of deep investigations undertaken during the last two decades, have revealed the fact that practically all important properties of solutions to second order elliptic and parabolic PDE have certain geometric background. For example, such properties as the mean-value inequality and the Harnack inequality are equivalent to some geometric conditions stated in terms of Sobolev and Poincaré inequalities as well as the volume of geodesic balls. On the other hand, estimates of the eigenvalues, Green kernel and heat kernel are closely related to the isoperimetric inequalities.

Maz'ya was one of the first mathematicians who started looking at the properties of solutions through the geometric properties of the domain. His method was based on the systematic use of the level sets of the functions and the co-area formula. The latter was independently discovered by Kronrod [18] (the case of two variables) and Federer [5] (the general case).

In this note, I would like to sketch some results of Maz'ya as they should look for Riemannian manifolds, and their relations with recent developments. Let M denote a smooth connected Riemannian manifold of dimension n. Let $B(x, r)$ denote the open geodesic ball of radius r centered at $x \in M$. Denote by μ the Riemannian volume and by σ the $(n-1)$-dimensional volume on hypersurfaces (that is, the $(n-1)$-Hausdorff measure). Differential operators such as the gradient ∇, the divergence div and the Laplace operator Δ, are those associated with the Riemannian metric.

Given a compact set $F \subset M$ and an open set $G \subset M$ containing F, we call the couple (F, G) a *capacitor*. Each capacitor has p-capacity (where $p \geq 1$) defined by

$$(1.1) \qquad \operatorname{cap}_p(F, G) := \inf_u \int_{G \backslash F} |\nabla u|^p \, d\mu,$$

where the inf is taken over all Lipschitz functions u with compact support in G such that $u = 1$ on F. Clearly, $\operatorname{cap}_p(F, G)$ increases on enlargement of F and on shrinking of Ω. We will write $\operatorname{cap}_p(F)$ for $\operatorname{cap}_p(F, M)$. If F is precompact then we set $\operatorname{cap}_p(F, G) = \operatorname{cap}_p(\overline{F}, G)$.

The following are the key ingredients of Maz'ya's method.

(i) **The estimate of the Dirichlet integral via the capacities of level sets.** Let u be a Lipschitz function on M with compact support. Denote

$$U_t := \{ x \in M : |u(x)| \geq t \}.$$

Then

$$(1.2) \qquad \int_M |\nabla u|^p \, d\mu \geq a_p \int_0^\infty \operatorname{cap}_p(U_t) d(t^p),$$

where

$$(1.3) \qquad a_p = \begin{cases} (p-1)^{p-1} p^{-p}, & \text{if } \quad p > 1, \\ 1, & \text{if } \quad p = 1. \end{cases}$$

See [25, Theorem 3], [26, Section 2.3.1] or [27, p.150-51].

The integral of the function u itself can be estimated via the measure of sets U_t. Indeed, for any $\alpha \geq 1$, the following inequality holds

$$(1.4) \qquad \int_M |u|^{\alpha p} \, d\mu \leq \left(\int_0^\infty \mu(U_t)^{\frac{1}{\alpha}} d(t^p) \right)^\alpha.$$

If $\alpha = 1$ then the equality is attained in (1.4), and it is trivial. See [26, Section 2.3.2] for the proof in the general case.

(ii) **Upper bound of capacity via flux.** Let $G \subset M$ be a precompact open set and $F \subset G$ be compact. Suppose that a Lipschitz function u is defined in $\overline{G \setminus F}$ such that $u = a$ on ∂F and $u = b$ on ∂G where $a < b$ are real constants. For any $t \in [a, b]$, denote $E_t := \{x : u(x) = t\}$ and define the *p-flux* of u through E_t by

$$\mathrm{flux}_p(u, E_t) := \int_{E_t} |\nabla u|^{p-1}\, d\sigma$$

(if E_t is empty then set $\mathrm{flux}_p(u, E_t) = 0$). Then, for all $p > 1$,

$$(1.5) \qquad \mathrm{cap}_p(F, G) \leq \left(\int_a^b \frac{dt}{\mathrm{flux}_p(u, E_t)^{\frac{1}{p-1}}} \right)^{1-p}$$

(this is true also if $a > b$ in which case one should switch the limits in the integral (1.5)). Furthermore, the equality in (1.5) is attained for some function u satisfying the above conditions. See [21], [23, Lemma 1], [26, Lemma 2.2.2/1] or [27, p.149]. In fact, the idea of (1.5) goes back to Pólya and Szegö [29].

(iii) **Lower bound for capacity via the isoperimetric function.** We say that M admits *an isoperimetric function $I(v)$* if, for any precompact open set $\Omega \subset M$ with smooth boundary,

$$(1.6) \qquad \sigma(\partial\Omega) \geq I(\mu(\Omega)).$$

If M admits a positive continuous isoperimetric function $I(v)$ then, for any capacitor (F, G) and $p > 1$,

$$(1.7) \qquad \mathrm{cap}_p(F, G) \geq \left(\int_{\mu(F)}^{\mu(G)} \frac{dv}{I(v)^{\frac{p}{p-1}}} \right)^{1-p}.$$

If $p = 1$ then, instead of (1.7), we have

$$(1.8) \qquad \mathrm{cap}_1(F, G) \geq \inf_{\mu(F) \leq v \leq \mu(G)} I(v).$$

See [21], [26, Corollary 2.2.3/2] or [27, p.150-51].

We will use (i)-(iii) as building blocks to show how to obtain a series of important results in a unified way.

2. Capacity of balls

Fix a point $o \in M$ and denote $B_r = B(o, r)$ and $S_r := \sigma(\partial B_r)$. Let us assume that M is geodesically complete so that any ball is precompact. Let us choose in

142   Grigor'yan

(1.5) $u = \mathrm{dist}(\cdot, o)$. Then $E_t = \partial B_t$ and $|\nabla u| \leq 1$ which implies $\mathrm{flux}_p(u, E_t) \leq S_t$. Hence, (1.5) yields, for all $R > r > 0$,

$$(2.1) \qquad \mathrm{cap}_p(B_r, B_R) \leq \left(\int_r^R \frac{dt}{S_t^{\frac{1}{p-1}}} \right)^{1-p}$$

(let us assume for simplicity $p > 1$ whenever it is needed).

In applications, it is frequently more convenient to use the volume function V_r rather than S_r. Let us show that the following estimate holds

$$(2.2) \qquad \mathrm{cap}_p(B_r, B_R) \leq 2^p \left(\int_r^R \left(\frac{t-r}{V_t - V_r} \right)^{\frac{1}{p-1}} dt \right)^{1-p}.$$

Observe first that $V_t' = S_t$. Hence, (2.2) will follow from (2.1) if we prove the following inequality

$$\int_a^b \frac{dt}{(v')^\gamma} \geq 2^{-1-\gamma} \int_a^b \left(\frac{t-a}{v(t)-v(a)} \right)^\gamma dt,$$

for all monotone increasing functions v on $[a, b]$, where $b > a$ and $\gamma > 0$ (it suffices to take $\gamma = \frac{1}{p-1}$). Without loss of generality, we may assume $a = 0$ and $v(a) = 0$. Denote by $s(t)$ the increasing rearrangement of v' on (a, b) and define the function v^* by

$$v^* := \int_0^t s(\tau) d\tau.$$

Then, by the convexity of v^*,

$$s(t) \leq \frac{v^*(2t) - v^*(t)}{t} \leq \frac{v^*(2t)}{t}.$$

By the properties of rearrangements $v^*(t) \leq v(t)$, whence

$$\int_0^b \frac{dt}{(v')^\gamma} = \int_0^b \frac{dt}{s(t)^\gamma} \geq \int_0^b \left(\frac{t}{v^*(2t)} \right)^\gamma dt \geq 2^{-1-\gamma} \int_0^b \left(\frac{t}{v(t)} \right)^\gamma dt,$$

which was to be proved .

In the case when M possesses a rotation symmetry with respect to o, it is possible to show that (2.1) becomes the equality. Such a manifold is called *spherically symmetric* with the pole o and can be described as follows. Topologically, M can be identified with $\mathbb{R}^n$ (assuming M is geodesically complete and non-compact),

---

The upper bound (2.2) for capacity in the case $p = 2$ was proved by Sturm [30] using a different method. His estimate has the better constant 2 rather than $4 = 2^p$.

and the metric of M can be written in the polar coordinates (ρ, θ) (centered at o) as follows

$$ds^2 = d\rho^2 + \psi^2(\rho)d\theta^2.$$

Here $\theta \in \mathbb{S}^{n-1}$ and $d\theta^2$ is the standard metric on $\mathbb{S}^{n-1}$. Let us denote such a manifold by M_ψ. Function ψ is any positive smooth function on $\mathbb{R}_+$ such that $\psi(0) = 0$ and $\psi'(0) = 1$ (see [8]). It is easy to see that

$$S_r = \omega_n \psi(r)^{n-1} \quad \text{and} \quad V_r = \int_0^r S_t dt$$

so that each function S_r, V_r uniquely determines M_ψ.

For example, if $\psi(\rho) = \rho$ then $M_\psi = \mathbb{R}^n$. If $\psi(\rho) = \sinh\rho$ then $M = \mathbb{H}^n$. If $\psi(\rho) = \sin\rho$ then $M = \mathbb{S}^n$ (in this case, ρ varies in $[0, \pi]$ and M is compact).

As was remarked already, capacity of a ball on M_ψ is computed by

$$(2.3) \qquad \mathrm{cap}_p\left(B_r, B_R\right) = \left(\int_r^R \frac{dt}{S_t^{\frac{1}{p-1}}} \right)^{1-p}.$$

Taking $R = \infty$, one obtains $\mathrm{cap}_p(B_r)$. For example if $M_\psi = \mathbb{R}^n$ then $\mathrm{cap}_p(B_r) = \mathrm{const}\, r^{n-p}$ provided $p < n$, and $\mathrm{cap}_p(B_r) = 0$ if $p \geq n$. If $M_\psi = \mathbb{H}^n$ then the exact computation is generally not possible but one easily gets the estimate

$$\mathrm{cap}_p(B_r) \sim \mathrm{const}\, S_r, \quad r \to \infty.$$

For comparison, in $\mathbb{R}^n$ one has $\mathrm{cap}_p(B_r) = \mathrm{const}\, S_r r^{1-p}$, if $p < n$.

3. Parabolicity of manifolds

A Riemannian manifold M is called *parabolic* if any of the following equivalent properties holds:

(a) any bounded subharmonic function on M is constant;

(b) there is no positive fundamental solution of the Laplace operator on M;

(c) $\mathrm{cap}(F) = 0$ for any compact $F \subset M$;

(d) the Brownian motion on M is recurrent .

It is not quite obvious that (a)–(d) should be equivalent – see [13] for a detailed account of parabolicity and related topics.

Let us say that the manifold M is *p-parabolic* if $\mathrm{cap}_p(F) = 0$ for any compact $F \subset M$. This notion is connected to the properties of the p-Laplace operator defined by

$$\Delta_p u = \mathrm{div}(|\nabla u|^{p-2}\, \nabla u).$$

Indeed, it is known that the p-parabolicity is equivalent to the fact that any bounded p-subharmonic function is constant and to the non-existence of a positive fundamental solution of Δ_p (see [14]).

Finding convenient geometric conditions for parabolicity and non-parabolicity is an important task of potential theory on manifolds. Clearly, the estimates of capacity can give certain conditions of parabolicity.

Assume for simplicity that M is geodesically complete and non-compact (observe that a compact manifold is always p-parabolic). The upper bound of capacity (2.1) implies the following parabolicity test: if

$$(3.1) \qquad \int^{\infty} \frac{dt}{S_t^{\frac{1}{p-1}}} = \infty$$

then $\mathrm{cap}_p(B_r) = 0$ and, therefore, M is p-parabolic. For example, if $S_r = cr^{n-1}$ then (3.1) holds provided $n \leq p$. Therefore, if $n \leq p$ then $\mathbb{R}^n$ is p-parabolic.

The capacity upper bound (2.2) gives the following parabolicity test in terms of the volume growth function: a geodesically complete manifold M is p-parabolic provided

$$(3.2) \qquad \int^{\infty} \left(\frac{t}{V_t}\right)^{\frac{1}{p-1}} dt = \infty.$$

For example, (3.2) holds if $V_r \leq Cr^p$, for large r, or even if $V_r \leq Cr^p \log^{p-1} r$.

The parabolicity condition (3.1) for the case when $p = 2$ and $n = 2$ goes back to Ahlfors [1] and Nevanlinna [28] who proved it for simply connected Riemann surfaces (in this case, the parabolicity of M is equivalent to the fact that M is conformally equivalent to $\mathbb{R}^2$). For general manifolds, (3.1) was noticed by many authors, see for example [9], [19].

The sufficient condition $V_r \leq Cr^2$ for 2-parabolicity was first proved by Cheng and Yau [3]. The 2-parabolicity under the more general hypothesis (3.2) was proved by Karp [16], Varopoulos [31] and the author [10]. The case $p = n$ was studied by Keselman and Zorich [17]. The proof for arbitrary $p > 1$ was obtained by Holopainen [15].

As we have seen, the capacity approach allows to obtain all mentioned parabolicity criteria in a few lines.

Let us turn to non-parabolicity tests. Assume that M admits the isoperimetric function $I(v)$ and let for simplicity $\mu(M) = \infty$. We claim that if

$$(3.3) \qquad \int^{\infty} \frac{dv}{I^{\frac{p}{p-1}}(v)} < \infty$$

then M is p-non-parabolic. Indeed, given (3.3), the capacity estimate (1.7) implies $\text{cap}_p(F) > 0$ for large enough compact F. For example, let $I(v) = cv^{1-1/n}$. Then (3.3) holds if $n > p$. In particular, $\mathbb{R}^n$ is p-non-parabolic provided $n > p$.

The non-parabolicity test (3.3) was obtained by Fernández [7] and by the author [10], for the case $p = 2$. The general case seems to have not been published.

4.   Isoperimetric inequality and Sobolev inequality

Assume that M admits the isoperimetric function $I(v) = cv^{1-1/n}$ like $\mathbb{R}^n$. Then (1.7) and (1.8) imply the following "isoperimetric inequality" for capacity

$$(4.1) \qquad \text{cap}_p(F) \geq c'\mu(F)^{1-p/n},$$

whence by (1.2)

$$(4.2) \qquad \int |\nabla u|^p \, d\mu \geq c' a_p \int_0^\infty \mu(U_t)^{1-p/n} d(t^p).$$

Assume $p < n$ and apply (1.4) with $\alpha = \frac{n}{n-p}$, which yields

$$\left(\int |u|^{\frac{np}{n-p}} d\mu \right)^{\frac{n-p}{n}} \leq \int_0^\infty \mu(U_t)^{1-p/n} d(t^p).$$

Comparing with (4.2), we conclude

$$(4.3) \qquad \int |\nabla u|^p \, d\mu \geq c'' \left(\int |u|^{\frac{np}{n-p}} d\mu \right)^{\frac{n-p}{n}}.$$

This inequality in $\mathbb{R}^n$ was proved by S.L.Sobolev for $p > 1$ and by E.Gagliardo for $p = 1$. Maz'ya [20] as well as Federer and Fleming [6], were first to realize that the Sobolev inequality (4.3) is a consequence of the isoperimetric inequality (1.6). This has made it possible to prove Sobolev inequalities on manifolds where the method of singular integrals originally applied by Sobolev, would not work.

5.   Capacity and the principal frequency

For any precompact region $\Omega \subset M$, let us denote by $\lambda(\Omega)$ the first eigenvalue of the Dirichlet problem in Ω for the Laplace operator. By the variational property,

$$\lambda(\Omega) = \inf_u \frac{\int |\nabla u|^2 \, d\mu}{\int |u|^2 \, d\mu}$$

where the inf is taken over all Lipschitz functions $u \not\equiv 0$ compactly supported in Ω. Analogously, let us define the *principal p-frequency* $\lambda_p(\Omega)$ by

$$(5.1) \qquad \lambda_p(\Omega) := \inf_u \frac{\int |\nabla u|^p \, d\mu}{\int |u|^p \, d\mu}$$

where Ω is any open subset of M. Clearly, $\lambda_p(\Omega)$ is decreasing on enlargement of Ω.

For any open set Ω, let us define *the Maz'ya constant* of Ω as follows:

$$(5.2) \qquad m_p(\Omega) := \inf_{F \subset\subset \Omega} \frac{\mathrm{cap}_p(F, \Omega)}{\mu(F)}.$$

We claim that

$$(5.3) \qquad m_p(\Omega) \geq \lambda_p(\Omega) \geq a_p m_p(\Omega)$$

where a_p is defined by (1.3). The left-hand side inequality in (5.3) easily follows if we take the test function u in definition (5.1) of $\lambda_p(\Omega)$ so that u is equal to 1 on a compact set $F \subset \Omega$, and apply definition (1.1) of capacity. To prove the lower bound of $\lambda_p(\Omega)$ in (5.3), let us apply (1.4) with $\alpha = 1$ and compare it with (1.2), rewritten for Ω instead of M. Then we have

$$\int_\Omega |\nabla u|^p \, d\mu \geq a_p \int_0^\infty \mathrm{cap}_p(U_t, \Omega) d(t^p),$$

and

$$\int_\Omega |u|^p d\mu \leq \int_0^\infty \mu(U_t) d(t^p),$$

whence

$$\frac{\int_\Omega |\nabla u|^p \, d\mu}{\int_\Omega |u|^p d\mu} \geq a_p m_p(\Omega)$$

which was to be proved.

In fact, both the Sobolev inequality (4.3) and the eigenvalue estimate (5.3) are particular cases of a more general theorem of Maz'ya [26, Theorem 2.3.2/1], which covers also the norms of u in Orlicz spaces.

Let us show an example of application of (5.3). If (4.1) is valid then

$$\frac{\mathrm{cap}_p(F, \Omega)}{\mu(F)} \geq c' \mu(F)^{-p/n} \geq c' \mu(\Omega)^{-p/n},$$

whence

$$m_p(\Omega) \geq c' \mu(\Omega)^{-p/n}$$

and, by (5.3),

$$\lambda_p(\Omega) \geq c' a_p \mu(\Omega)^{-p/n}.$$

6. Cheeger's inequality

Inequality (5.3) provides a powerful tool for estimating principal frequencies. Let us set in the next discussion $p = 2$ and omit p from all notation (say, cap means cap_2 etc.). Then (5.3) takes the form

$$(6.1) \qquad m(\Omega) \geq \lambda(\Omega) \geq \frac{1}{4} m(\Omega).$$

The Cheeger constant $h(\Omega)$ of the region Ω is defined by

$$h(\Omega) := \inf_{F \subset \Omega} \frac{\sigma(\partial F)}{\mu(F)}$$

where F runs over all compact subsets of Ω with smooth boundary . The Cheeger inequality [2] says that

$$(6.2) \qquad \lambda(\Omega) \geq \frac{1}{4} h^2(\Omega).$$

Let us deduce (6.2) from (6.1). Indeed, given the Cheeger constant $h(\Omega)$, we may say that Ω admits the isoperimetric function $I(v) = h(\Omega)v$. Hence, by (1.7),

$$\mathrm{cap}(F, \Omega) \geq \left(\int_{\mu(F)}^{\mu(\Omega)} \frac{dv}{I(v)^2} \right)^{-1} \geq h^2(\Omega)\mu(F),$$

whence $m(\Omega) \geq h^2(\Omega)$, which together with (6.1) finishes the proof of (6.2).

The ratio of $\lambda(\Omega)/h^2(\Omega)$ can be made arbitrarily large, by choosing "bad" regions Ω. The Maz'ya inequality (5.3) shows that in such situations it is better to use capacity to estimate $\lambda(\Omega)$. We will show an application of that in Section 7.. Note that the estimates (6.1) were proved by Maz'ya as early as in 1962 (see [22, Theorem 1]), long before the question of estimating eigenvalues on manifolds was even raised.

For arbitrary $p > 1$, one gets similarly

$$m_p(\Omega) \geq (p-1)^{1-p} h^p(\Omega),$$

whence, by (5.3),

$$\lambda_p(\Omega) \geq p^{-p} h^p(\Omega).$$

The Cheeger constant $h(\Omega)$ can be regarded as the limiting case of the Maz'ya constant $m_p(\Omega)$ for $p = 1$.

7. Eigenvalues of balls on spherically symmetric manifolds

We assume in this section that $M = M_\psi$ is a spherically symmetric manifold and follow the notation introduced in Section 2.. The question to be discussed here is how to estimate $\lambda(B_R)$ as a function of the radius R?

Let us modify the definition of the Maz'ya constant $m_p(\Omega)$ in the case when $\Omega = B_R$. Namely, in the definition (5.2), let us assume in addition that F runs only over balls B_r with $r < R$. In the definition (5.1) of $\lambda_p(B_R)$, the minimum is attained for the radially symmetric function u. Therefore, in the proof of the Maz'ya inequalities (5.3), if $\Omega = B_R$ then it suffices to use only the radially symmetric test functions. Hence, we obtain

$$(7.1) \qquad m_p(B_R) \geq \lambda_p(B_R) \geq a_p m_p(B_R)$$

where m_p now refers to the modified Maz'ya constant.

Let us assume in the sequel $p = 2$ and suppress p from the notation. Since the capacity $\mathrm{cap}(B_r, B_R)$ is explicitly given by (2.3), that is,

$$\mathrm{cap}\,(B_r, B_R) = \left(\int_r^R \frac{dt}{S_t} \right)^{-1},$$

the modified Maz'ya constant can be computed as follows:

$$(7.2) \qquad m(B_R) = \frac{1}{\Phi(R)}$$

where

$$(7.3) \qquad \Phi(R) := \sup_{r \leq R} \left[V_r \int_r^R \frac{dt}{S_t} \right].$$

Respectively, we obtain the eigenvalue estimates

$$(7.4) \qquad \frac{1}{\Phi(R)} \geq \lambda(B_R) \geq \frac{1}{4\Phi(R)}.$$

For comparison, the Cheeger inequality implies another estimate

$$(7.5) \qquad \lambda(B_R) \geq \frac{1}{4} \left(\inf_{r \leq R} \frac{S_r}{V_r} \right)^2$$

(see [4, Lemma 8.2]), which is generally weaker than (7.5), but in some situation is substantially weaker.

Let us show some examples.

Examples. 1. Let $S_r \asymp r^{N-1}$ for large r, where $N > 0$. Respectively, the volume function is $V_r \asymp r^N$. Simple computation shows that $\Phi(R) \asymp R^2$ for large R whence $\lambda(B_R) \asymp R^{-2}$.

2. Let us take $S_r \asymp r^{-1}$ so that $V_r \asymp \log r$. Then the optimal r in (7.3) is of the order $R/\sqrt{\log R}$, and we find

$$\Phi(R) \asymp R^2 \log R \quad \text{and} \quad \lambda(B_R) \asymp \frac{1}{R^2 \log R}.$$

Note that (7.5) gives the weaker estimate

$$\lambda(B_R) \geq \frac{c}{R^2 \log^2 R}.$$

3. If $S_r \asymp (r \log r)^{-1}$ and, therefore, $V_r \asymp \log \log r$ then

$$\Phi(R) \asymp R^2 \log R \log \log R \quad \text{and} \quad \lambda(B_R) \asymp \frac{1}{R^2 \log R \log \log R},$$

which is again better than (7.5).

8. Heat kernel on spherically symmetric manifolds

By definition, the heat kernel on a Riemannian manifold M is the smallest positive fundamental solution to the heat equation $u_t = \Delta u$ on $\mathbb{R}_+ \times M$. We denote the heat kernel by $p(t, x, y)$ where $t > 0$ and $x, y \in M$. One of the most interesting questions about the heat kernel is obtaining its long time estimates. This question has been attracted much attention in the past decade, and we refer the reader to the surveys [11] and [12] for general overview.

Here, we assume that $M = M_\psi$ is spherically symmetric manifold as above and try to estimate $p(t, o, o)$ as a function of t, as $t \to \infty$. Suppose that we are given a positive non-increasing function $\Lambda(v)$ such that, for each ball B_r,

$$\lambda(B_r) \geq \Lambda(V_r).$$

In other word, Λ provides the lower bound for $\lambda(B_r)$ via the volume of B_r. Then we have, for all $t > 0$,

$$(8.1) \qquad\qquad p(t, o, o) \leq \frac{4}{P(t/2)}$$

where the function P is defined by

$$(8.2) \qquad\qquad t = \int_0^{P(t)} \frac{dv}{v \Lambda(v)}.$$

(see [4, Eq. (8.19)]). Hence, the question of estimating $p(t, o, o)$ amounts to obtaining sharp lower bounds for $\lambda(B_r)$. For example, if $M_\psi = \mathbb{R}^n$ then $\Lambda(v) = cv^{-2/n}$ whence $P(t) = c't^{n/2}$ and $p(t, o, o) \leq Ct^{-n/2}$.

For general M_ψ, the function Λ can be obtained from (7.4). Namely, let us define Λ by

$$\text{(8.3)} \qquad \Lambda(V_R) = \frac{1}{4\Phi(R)}$$

where Φ is defined by (7.3). Then, the formulas (7.3), (8.3), (8.2) and (8.1) give *implicitly* an upper bound for $p(t, o, o)$ via the volume function V_r. This approach is similar to one adopted in [4], but the novelty here is the usage the Maz'ya estimate (7.4) of the eigenvalues rather than (7.5) that was used in [4].

Substituting (8.3) into (8.2), we obtain

$$t = 4 \int_0^{V^{-1}(P(t))} \Phi(r) \frac{S_r \, dr}{V_r}$$

where V^{-1} is the inverse function to $V_{(.)}$. Denote $R = R(t) = V^{-1}(P(t))$. Alternatively, $R(t)$ is defined by

$$\text{(8.4)} \qquad t = 4 \int_0^{R(t)} \Phi(r) \frac{S_r \, dr}{V_r}.$$

Then (8.1) implies

$$\text{(8.5)} \qquad p(t, o, o) \le \frac{4}{V_{R(t/2)}}$$

Let us try and estimate $R(t)$ more explicitly. Since the function $\Phi(r)$ is monotone increasing, (8.4) implies, for all $t > t_0 > 0$,

$$\text{(8.6)} \qquad t - t_0 \le 4\Phi(R) \log \frac{V_R}{V_{R_0}}$$

where $R_0 = R(t_0)$. This inequality can be turned into a lower bound for $R(t)$ which can be then used in (8.5).

For example, let $V_r = \exp(r^\alpha)$ for large r, where $\alpha \in (0, 1)$. Then (7.3) yields, for large R,

$$\Phi(R) \asymp R^{2(1-\alpha)}$$

and (8.6) implies, for large t,

$$R(t) \ge ct^{\frac{1}{2-\alpha}}.$$

Respectively, the heat kernel admits the following upper bound

$$p(t, o, o) \le C \exp\left(-ct^{\frac{\alpha}{2-\alpha}}\right).$$

On the other hand, one can show that there is a matching lower bound here – see [4, p. 42].

If V_r grows polynomially in r then the estimate (8.6) for $R(t)$ is too rough. To get a better estimate, let us assume that

(a) V_r satisfies the doubling property: for all $r > 0$ and some constant C,

$$(8.7) \qquad\qquad V_{2r} \leq C V_r$$

(b) and, for all $r > 0$ and some $c > 0$,

$$(8.8) \qquad\qquad \frac{S_r}{V_r} \geq \frac{c}{r}.$$

We claim that, under (a) and (b), the following estimate holds

$$(8.9) \qquad\qquad p(t, o, o) \asymp \frac{1}{V_{\sqrt{t}}}.$$

Previously, this estimate was known provided the doubling property (a) holds and S_r/V_r is non-increasing function – see [4, Corollary 8.5]. On the contrary, the hypothesis (8.8) does not restrict S_r from above. For example, the graph of S_r against r may have arbitrarily high peeks on small intervals as long as this does not violate the doubling property (8.7).

To prove (8.9), let first observe that (7.3) and (8.8) imply, for any $R > 0$ and some $r \in (0, R)$,

$$(8.10) \qquad \Phi(R) = V_r \int_r^R \frac{dt}{S_t} \leq \int_r^R \frac{V_t dt}{S_t} \leq \frac{1}{2c}\left(R^2 - r^2\right) < \frac{1}{2c}R^2.$$

Then the integral in (8.4) can be estimated as follows:

$$\int_0^R \Phi(r)\frac{S_r dr}{V_r} = \sum_{k=0}^{\infty} \int_{R2^{-(k+1)}}^{R2^{-k}} \Phi(r)\frac{S_r dr}{V_r} \leq \sum_{k=0}^{\infty} \Phi(R2^{-k}) \int_{R2^{-(k+1)}}^{R2^{-k}} \frac{S_r dr}{V_r}$$

$$\leq \sum_{k=0}^{\infty} \Phi(R2^{-k}) \log \frac{V(R2^{-k})}{V(R2^{-(k+1)})} \leq C \sum_{k=0}^{\infty} \Phi(R2^{-k}) \leq C' R^2$$

where we have used (8.7) and (8.10).

Hence, (8.4) implies

$$R(t) \geq c' \sqrt{t},$$

which together with (8.5) and (8.7) yields

$$(8.11) \qquad\qquad p(t, o, o) \leq \frac{C''}{V_{\sqrt{t}}}.$$

We are left to note that, by [4, Theorem 7.2], the upper bound (8.11) and the doubling property (8.7) imply the lower bound in (8.9).

References

[1] AHLFORS L.V., *Sur le type d'une surface de Riemann*, C.R. Acad. Sci. Paris, **201** (1935) 30–32.

[2] CHEEGER J., *A lower bound for the smallest eigenvalue of the Laplacian*, in: Problems in Analysis: A Symposium in honor of Salomon Bochner, Princeton University Press. Princeton, 1970. 195–199.

[3] CHENG S.Y., YAU S.-T., *Differential equations on Riemannian manifolds and their geometric applications*, Comm. Pure Appl. Math., **28** (1975) 333–354.

[4] COULHON T., GRIGOR'YAN A., *On-diagonal lower bounds for heat kernels on non-compact manifolds and Markov chains*, Duke Math. J., **89** no.1, (1997) 133–199.

[5] FEDERER H., *Curvature measures*, Trans. Amer. Math. Soc., **93** (1959) no.3, 418–491.

[6] FEDERER H., FLEMING W.H., *Normal and integral currents*, Ann. Math., **72** (1960) 458–520.

[7] FERNÁNDEZ, J.L., *On the existence of Green's function on Riemannian manifolds*, Proc. AMS, **96** (1986) 284–286.

[8] GREENE R., WU W., *Function theory of manifolds which possess a pole*, Lecture Notes Math 699, Springer, 1979.

[9] GRIGOR'YAN A., *On the existence of a Green function on a manifold*, (in Russian) Uspekhi Matem. Nauk, **38** (1983) no.1, 161–162. Engl. transl. Russian Math. Surveys, **38** (1983) no.1, 190–191.

[10] GRIGOR'YAN A., *On the existence of positive fundamental solution of the Laplace equation on Riemannian manifolds*, (in Russian) Matem. Sbornik, **128** (1985) no.3, 354–363. Engl. transl. Math. USSR Sb., **56** (1987) 349–358.

[11] GRIGOR'YAN A., *Heat kernel on a non-compact Riemannian manifold*, in: 1993 Summer Research Institute on Stochastic Analysis, ed. M.Pinsky et al., Proceedings of Symposia in Pure Mathematics, **57** (1994) 239–263.

[12] GRIGOR'YAN A., *Estimates of heat kernels on Riemannian manifolds*, to appear in Proceedings of the Instructional Conference on Spectral Theory and Geometry, Edinburgh, April 1998

[13] GRIGOR'YAN A., *Analytic and geometric background of recurrence and non-explosion of the Brownian motion on Riemannian manifolds*, to appear in Bull. Amer. Math. Soc.

[14] HOLOPAINEN I., *Positive solutions of quasilinear elliptic equations on Riemannian manifolds*, Proc. London Math. Soc. (3), **65** (1992) 651–672.

[15] HOLOPAINEN I., *Volume growth, Green's functions and parabolicity of ends*, to appear in Duke Math. J.

[16] KARP L., *Subharmonic functions, harmonic mappings and isometric immersions*, in: Seminar on Differential Geometry, ed. S.T.Yau, Ann. Math. Stud. 102, Pinceton, 1982.

[17] KESELMAN V.M., ZORICH V.A., *On conformal type of a Riemannian manifold*, Funct. Anal. Appl., **30** (1996) 106–117.

[18] KRONROD A.S., *On functions of two variables*, (in Russian) Uspechi Matem. Nauk, **5** (1950) no.1, 24–134.

[19] LYONS T., SULLIVAN D., *Function theory, random paths and covering spaces*, J. Diff. Geom., **19** (1984) 299–323.

[20] MAZ'YA V.G., *Classes of domains and embedding theorems for functional spaces*, (in Russian) Dokl. Acad. Nauk SSSR, **133** no.3, (1960) 527–530. Engl. transl. Soviet Math. Dokl., **1** (1961) 882–885.

[21] MAZ'YA V.G., *The p-conductivity and theorems on imbedding certain function spaces into C-space*, (in Russian) Dokl. Acad. Nauk SSSR, **140** (1961) 299–302. Engl. transl. Soviet Math. Dokl., **2** (1961) 1200–1203.

[22] MAZ'YA V.G., *The negative spectrum of the n-dimensional Schrödinger operator*, (in Russian) Dokl. Acad. Nauk SSSR, **144** no.4, (1962) 721–722. Engl. transl. Soviet Math. Dokl., **3** (1962) 808–810.

[23] MAZ'YA V.G., *On the theory of the n-dimensional Schrödinger operator*, (in Russian) Izv. Acad. Nauk SSSR, Ser. Mat., **28** (1964) 1145–1172.

[24] MAZ'YA V.G., *Embedding theorems and their applications*, (in Russian) in: Trudy Simposiuma po teoremam vlozheniya, Baku 1966 god, ed. L.D.Kudryavzev, Nauka, Moskow, 1970.

[25] MAZ'YA V.G., *On certain intergal inequalities for functions of many variables*, (in Russian) Problemy Matematicheskogo Analiza, Leningrad. Univer., **3** (1972) 33–68. Engl. transl. J. Soviet Math., **1** (1973) 205–234.

[26] MAZ'YA V.G., *Sobolev spaces*, (in Russian) Izdat. Leningrad Gos. Univ. Leningrad, 1985. Engl. transl. Springer-Verlag, 1985.

[27] MAZ'YA V.G., *Classes of domains, measures and capacities in the theory of differentiable functions*, Encyclopaedia Math. Sci. 26, Springer, Berlin, 1991.

[28] NEVANLINNA R., *Ein Satz über offene Riemannsche Flächen*, Ann. Acad. Sci. Fenn. Part A., **54** (1940) 1–18.

[29] PÓLYA G., SZEGÖ, *Isoperimetric inequalities in mathematical physics*, Princeton University Press, Princeton, 1951.

[30] STURM K-TH., *Sharp estimates for capacities and applications to symmetrical diffusions*, Probability theory and related fields, **103** (1995) no.1, 73–89.

[31] VAROPOULOS N.TH., *Potential theory and diffusion of Riemannian manifolds*, in: Conference on Harmonic Analysis in honor of Antoni Zygmund. Vol I, II, Wadsworth Math. Ser., Wadsworth, Belmont, Calif., 1983. 821–837.

Department of Mathematics
Imperial College
London SW7 2BZ
United Kingdom
a.grigoryan@ic.ac.uk

1991 Mathematics Subject Classification: Primary 58G32

Submitted: 28.01.1999

Operator Theory:
Advances and Applications, Vol. 109

Multipliers of differentiable functions and their traces

TATYANA SHAPOSHNIKOVA

Abstract: The paper starts with a short survey of sufficient and/or necessary conditions for a function to be a multiplier between spaces of differentiable functions which stem from Maz'ya's well-known isoperimetric criteria for Sobolev embeddings. Another subject touched here is traces and extensions of multipliers. It is shown, in particular, that the multipliers in the space of Bessel potentials $H_p^l(\mathbf{R}^n)$, $\{l\} > 0$, $1 < p < \infty$, are traces of multipliers in a certain class of differentiable functions in $\mathbf{R}^{n+s}$ with a weighted mixed norm.

1. Introduction

By a multiplier acting from one Banach function space S_1, into another, S_2, one means a function which defines a bounded linear mapping of S_1 into S_2 by pointwise multiplication. Thus, any pair S_1, S_2 generates the space of multipliers $M(S_1 \to S_2)$. (I shall write MS instead of $M(S \to S)$.)

A theory of multipliers between different spaces of Sobolev type has been developed by Vladimir Maz'ya and myself since 1979. I recall that in the late seventies the word "multiplier" was usually associated with that of Fourier. A deep theory of L_p-Fourier multipliers created by Marcinkiewicz, Mikhlin, Hörmander *et al.* was especially popular. As for the multipliers acting in pairs of differentiable functions, only a few isolated results were known.

By that time V. Maz'ya had understood that his necessary and sufficient conditions for Sobolev embeddings, stated in terms of isoperimetric inequalities between measures and capacities, opened broad perspectives in the study of such multipliers [7]. He introduced me to this area when I asked him about a possible direction of research, after I had felt a weakening of my interest in the numerical analysis I was working in. Soon, advancing in the problem of traces of Sobolev multipliers, I became involved, together with Vladimir, in the absorbing work on new function spaces which is still going on with sporadic intermissions. During one of them we married and the theory of multipliers became our family business.

By 1983 we were able to summarize our results in a book. It was first written in Russian and sent to Nauka, the publishing house of the Soviet Academy of Science. When after a year the book was rejected, I translated it into English and our friend Gaetano Fichera smuggled it to the West. He recommended it for publication in Pitman's "Monographs and Studies in Mathematics", where it appeared in 1985. A bit later, we managed to publish an extended version with the

Leningrad University Publishing House. In the sequel I refer only to the English edition [10].

2. Description and properties of multipliers

The norm in $M(S_1 \to S_2)$ is defined as

$$(2.1) \qquad \|\gamma\|_{M(S_1 \to S_2)} = \sup\{\|\gamma u\|_{S_2} : \ \|u\|_{S_1} \le 1\}.$$

Let, as usual, $W_p^k(\mathbf{R}^n)$ be the Sobolev space of functions in $L_p(\mathbf{R}^n)$ whose partial derivatives of order k also belong to $L_p(\mathbf{R}^n)$, and let $C_0^\infty(\mathbf{R}^n)$ be the space of infinitely differentiable functions with compact support.

I start with describing some norms in $M(W_p^m(\mathbf{R}^n) \to W_p^l(\mathbf{R}^n))$, equivalent to (2.1), which have explicit analytic forms (see [10, Ch. 1]). The simplest cases are $mp > n$, $p > 1$ and $m \ge n$, $p = 1$, when

$$(2.2) \qquad \|\gamma; \mathbf{R}^n\|_{M(W_p^m \to W_p^l)} \sim \sup_{x \in \mathbf{R}^n} \|\gamma; B_1(x)\|_{W_p^l}.$$

Here and elsewhere $B_\rho^{(n)}(x)$ is the ball with radius ρ centered at $x \in \mathbf{R}^n$; if x is the origin, I write simply $B_\rho^{(n)}$. The equivalence $a \sim b$ means that a/b is bounded and separated from zero by positive constants depending on n, p, m, and l.

A rather simple norm can be introduced in $M(W_1^m(\mathbf{R}^n) \to W_1^l(\mathbf{R}^n))$ for $m < n$. Namely, for $m < l$,

$$
\begin{aligned}
\|\gamma; & \mathbf{R}^n\|_{M(W_1^m \to W_1^l)} \\
(2.3) \qquad & \sim \sup_{\substack{x \in \mathbf{R}^n \\ \rho \in (0,1)}} \rho^{m-n} \|\nabla_l \gamma; B_\rho^{(n)}(x)\|_{L_1} + \sup_{x \in \mathbf{R}^n} \|\gamma; B_1^{(n)}(x)\|_{L_1},
\end{aligned}
$$

and for $m = l$ the last term should be replaced by $\|\gamma; \mathbf{R}^n\|_{L_\infty}$.

For the case $mp \le n$, $p > 1$, I give three expressions for the norm in the space $M(W_p^m(\mathbf{R}^n) \to W_p^l(\mathbf{R}^n))$. They look so different that one cannot see their equivalence directly. These norms seem to be equally useful, each having its own range of applications.

One of the equivalent norms in $M(W_p^m(\mathbf{R}^n) \to W_p^l(\mathbf{R}^n))$ for the just mentioned case can be described in terms of the capacity of a compact subset e of $\mathbf{R}^n$:

$$\mathrm{cap}(e, W_p^m) = \inf\{\|u; \mathbf{R}^n\|_{W_p^m}^p : \ u \in C_0^\infty(\mathbf{R}^n), \ u \ge 1 \text{ on } e\}.$$

In fact, let $mp \le n$, $p > 1$. If $m > l$ we have

$$
\begin{aligned}
\|\gamma; & \mathbf{R}^n\|_{M(W_p^m \to W_p^l)} \\
(2.4) \qquad & \sim \sup_{\substack{e \subset \mathbf{R}^n \\ \mathrm{diam}(e) \le 1}} \frac{\|\nabla_l \gamma; e\|_{L_p}}{\left(\mathrm{cap}\left(e, W_p^m\right)\right)^{1/p}} + \sup_{\substack{e \subset \mathbf{R}^n \\ \mathrm{diam}(e) \le 1}} \frac{\|\gamma; e\|_{L_p}}{\left(\mathrm{cap}\left(e, W_p^{m-l}\right)\right)^{1/p}},
\end{aligned}
$$

and for $m = l$ the second supremum should be replaced by $||\gamma; \mathbf{R}^n||_{L_\infty}$. As Verbitsky showed [10, Sec. 2.6], the second term in the right-hand side can be replaced by

$$\sup_{x \in \mathbf{R}^n} ||\gamma; B_1^{(n)}(x)||_{L_1}.$$

The proof of relation (2.2) is straightforward, whereas that of (2.3) and (2.4) rests upon results by Maz'ya [6], [8], [9] and Adams [1] on the trace inequality

$$(2.5) \qquad \int_{\mathbf{R}^n} |u(x)|^p d\mu \leq c||u; \mathbf{R}^n||_{W_p^m}^p, \quad u \in C_0^\infty(\mathbf{R}^n),$$

where μ is an arbitrary measure in $\mathbf{R}^n$.

From results of Kerman and Saywer [5] and Maz'ya and Verbitsky [13] concerning incquality (2.5) it follows that the boundedness of the first supremum in (2.4) is equivalent to each of the conditions

$$(2.6) \qquad \sup_{Q \subset \mathbf{R}^n} \left[\frac{\int_Q (J_m \chi_Q |\nabla_l \gamma|^p(x))^{p/(p-1)} dx}{\int_Q |\nabla_l \gamma|^p dx}\right]^{(p-1)/p} < \infty,$$

and

$$(2.7) \qquad \sup_{x \in \mathbf{R}^n} \frac{J_m(J_m |\nabla_l \gamma|^p)^{p/(p-1)}(x)}{J_m |\nabla_l \gamma|^p(x)} < \infty.$$

Here Q is an arbitrary cube of edge length ≤ 1, J_m is the Bessel potential of order m, i.e. the operator $(-\Delta + 1)^{-m/2}$, and χ_Q is the characteristic function of Q.

The above mentioned equivalent representations for the norm in $M(W_p^m(\mathbf{R}^n) \to W_p^l(\mathbf{R}^n))$ with $mp \leq n$, $p > 1$, $m > l$, lead to either necessary or sufficient conditions which are easier to use. One sufficient condition follows directly from (2.4) and the lower estimates for the capacity of a compact set e with $\operatorname{diam}(e) < 1$

$$\operatorname{cap}(e, W_p^m) \geq c \begin{cases} (\operatorname{mes}_n e)^{1-pm/n} & \text{for } mp < n, \\ |\log(2^{-n}\operatorname{mes}_n e)|^{1-p} & \text{for } mp = n, \end{cases}$$

where mes_n is the Lebesgue measure.

By Fefferman-Phong [3], the first supremum in (2.4) is dominated by

$$(2.8) \qquad \sup_{\substack{x \in \mathbf{R}^n \\ \rho \in (0,1)}} \rho^{m-n/q} ||\nabla_l \gamma; B_\rho^{(n)}(x)||_{L_q},$$

where $q > p$. On the other hand, the value (2.8) with $q = p$ is not less than the first supremum in (2.4). In this way we arrive at noncoinciding sufficient and necessary conditions for $\gamma \in M(W_p^m(\mathbf{R}^n) \to W_p^l(\mathbf{R}^n))$ with $mp < n$, $p > 1$.

An obvious sufficient condition for (2.6) and (2.7) is the boundedness of the nonlinear Bessel potential

$$J_m(J_m |\nabla_l \gamma|^p)^{1/(p-1)}$$

158 Shaposhnikova

(for the theory of nonlinear potentials see [4], [2]). When $p = 2$ the last requirement is

$$\sup_{x \in \mathbf{R}^n} J_{2m} |\nabla_l \gamma|^2 (x) < \infty$$

which is similar to the Kato condition frequently used in the theory of Schrödinger operator (see, for example, [23]).

In the recent paper [12] V. Maz'ya and myself proved that the maximal algebra in $M\left(W_p^m\left(\mathbf{R}^n\right) \to W_p^l\left(\mathbf{R}^n\right)\right)$, $m \geq l$, $p > 1$, consists of $\gamma \in W_{p,\mathrm{loc}}^l\left(\mathbf{R}^n\right)$ with the finite norm

$$\sup_{\substack{e \subset R^n \\ \mathrm{diam}(e) \leq 1}} \frac{\|\nabla_l \gamma; e\|_{L_p}}{\left(\mathrm{cap}\left(e, W_p^m\right)\right)^{1/p}} + \|\gamma; \mathbf{R}^n\|_{L_\infty} .$$

Here and elsewhere the notation S_{loc} stands for the space $\{u : \eta u \in S(\mathbf{R}^n) \text{ for all } \eta \in C_0^\infty(\mathbf{R}^n)\}$. Similarly, the norm in the maximal algebra in $M(W_1^m(\mathbf{R}^n) \to W_1^l(\mathbf{R}^n))$ is

$$\sup_{\substack{x \in \mathbf{R}^n \\ \rho \in (0,1)}} \rho^{m-n} \|\nabla_l \gamma; B_\rho^{(n)}(x)\|_{L_1} + \|\gamma; \mathbf{R}^n\|_{L_\infty}.$$

In the cases $mp > n$, $p > 1$ and $m \geq n$, $p = 1$, these norms can be simplified as

$$\sup_{x \in \mathbf{R}^n} \|\nabla_l \gamma; B_1^{(n)}(x)\|_{L_p} + \|\gamma; \mathbf{R}^n\|_{L_\infty}.$$

The proof in [12] is based on the following new pointwise interpolation inequality

$$|\nabla_k u(x)| \leq c \left(\mathcal{M} u(x)\right)^{1-k/m} \left(\mathcal{M} \nabla_m u(x)\right)^{k/m}, \quad k < m,$$

where $\mathcal{M}$ is the Hardy-Littlewood maximal operator.

Now I briefly comment on equivalent representations for the norm of a multiplier between other spaces of differentiable functions. For a non-integer $l > 0$ the space $W_p^l(\mathbf{R}^n)$, $p \geq 1$, is introduced as the completion of $C_0^\infty(\mathbf{R}^n)$ in the norm

$$\|D_{p,l} u; \mathbf{R}^n\|_{L_p} + \|u; \mathbf{R}^n\|_{L_p},$$

where

$$D_{p,l} u(x) = \left(\int_{\mathbf{R}^n} |\nabla_{[l]} u(x+h) - \nabla_{[l]} u(x)|^p |h|^{-n-p\{l\}} dh\right)^{1/p}.$$

Replacing here $D_{p,l} u(x)$ by

$$S_l u(x) = \left(\int_0^\infty \left[\int_{B_1^{(n)}} |\nabla_{[l]} u(x+t\theta) - \nabla_{[l]} u(x)| d\theta\right]^2 t^{-1-2\{l\}} dt\right)^{1/2}$$

one arrives at an equivalent normalization of the Bessel potential space $H_p^l(\mathbf{R}^n)$, $p > 1$ [24]. To obtain an analytic representation for the norm in $M(W_p^m(\mathbf{R}^n) \to$

$W_p^l(\mathbf{R}^n))$, $p \geq 1$, with non-integer l, one should replace $\nabla_l \gamma$ by $D_{p,l}\gamma$ in the right-hand sides of (2.3) and (2.4) (cf. [10, Ch. 3]). Using $S_l\gamma$ instead of $\nabla_l\gamma$ in (2.4) and replacing $\nabla_l\gamma$ by $J_l\gamma$ in (2.6) and (2.7), one arrives at norms in $M(H_p^m(\mathbf{R}^n) \to H_p^l(\mathbf{R}^n))$, $p > 1$ (cf. [10, Ch.2], [5], and [13]). Whenever the capacity is involved, it should be replaced by that generated by the corresponding space of fractional order.

In [10, Sec. 3.9] one can find a similar characterization of the space $MB_{p,p}^l(\mathbf{R}^n)$, where $B_{p,q}^l(\mathbf{R}^n)$ is the Besov space with the norm

$$\left(\int_{\mathbf{R}^n} ||\nabla_{[l]}u(\cdot + h) - \nabla_{[l]}u(\cdot); \mathbf{R}^n||_{L_p}^q |h|^{-n-q\{l\}}\,dh\right)^{1/q} + ||u; \mathbf{R}^n||_{W_p^{[l]}}.$$

Recently, Sickel [22] described multipliers in the Triebel-Lizorkin space $F_{p,q}^s(\mathbf{R}^n)$.

The following condition from [10, Sec. 3.3.1] is sufficient for $\gamma \in M(W_p^m(\mathbf{R}^n) \to W_p^l(\mathbf{R}^n))$ with $mp < n$, $p > 1$, $m > l$:

$$(2.9) \qquad\qquad \gamma \in B_{\sigma,\infty,\mathrm{unif}}^\lambda(\mathbf{R}^n),$$

where $\sigma \geq p$, $\lambda = n/\sigma - m + l$, $\lambda > l$, $\{\lambda\} > 0$. By S_{unif} I denote the space endowed with the norm

$$||v; \mathbf{R}^n||_{S_{\mathrm{unif}}} = \sup_{z \in \mathbf{R}^n} ||v\eta_z; \mathbf{R}^n||_S,$$

where $\eta_z(x) = \eta(z - x)$, η is an arbitrary function in $C_0^\infty(\mathbf{R}^n)$, $\eta = 1$ on $B_1^{(n)}$, and S is a Banach function space. If $m = l$ one should add $\gamma \in L_\infty(\mathbf{R}^n)$ to the condition (2.9).

In (2.9) the "number of derivatives" λ exceeds l. One can obtain sufficient conditions for a function γ to belong to the class $M(W_p^m(\mathbf{R}^n) \to W_p^l(\mathbf{R}^n))$, which are stated in terms of the space $B_{p,q,\mathrm{unif}}^l(\mathbf{R}^n)$:

$$(2.10) \qquad\qquad \gamma \in B_{\sigma,p,\mathrm{unif}}^l(\mathbf{R}^n) \quad \text{for } m > l,$$

where $\sigma \in [n/m, \infty)$, $mp < n$, or $\sigma \in (p, \infty)$, $mp = n$;

$$(2.11) \qquad\qquad \gamma \in (B_{\sigma,p,\mathrm{unif}}^l \cap L_\infty)(\mathbf{R}^n) \quad \text{for } m = l,$$

where $\sigma \in [n/l, \infty)$, $lp < n$, or $\sigma \in (p, \infty]$, $lp = n$ (see [10, Sec. 3.3.2]). Since the diminishing of exponent q leads to narrowing of $B_{\sigma,q}^\lambda(\mathbf{R}^n)$ and the diminishing of λ leads to expansion of this space, (2.11) is not comparable with (2.9).

Putting $\sigma = \infty$ in (2.11) one arrives at the following simple sufficient condition for $\gamma \in MW_p^l(\mathbf{R}^n)$ (and hence for $\gamma \in M(W_p^m(\mathbf{R}^n) \to W_p^l(\mathbf{R}^n))$) stated in terms of the modulus of continuity ω of the vector-valued function $\nabla_{[l]}\gamma$:

$$\int_0^1 \left(\frac{\omega(t)}{t^{\{l\}+1/p}}\right)^p dt < \infty.$$

One shows by using a gap Fourier series that even this, rather rough, condition is precise in a certain sense [10, Sec. 3.3].

Sufficient conditions of different type can be obtained in terms of the pointwise decrease of the Fourier transform $F\gamma$. For example, if

$$(F\gamma)(\xi) = O(1 + |\xi|^{m-l-n})$$

with $m > l$, then $\gamma \in M(W_p^m(\mathbf{R}^n) \to W_p^l(\mathbf{R}^n))$, $1 < p \leq 2$, $n > 2m$ [10, Sec. 3.3.1].

I shall not dwell upon various topics in the theory of multipliers developed in [10] such as description of multipliers in domains, the essential norm and compactness of multipliers, the spectrum of multipliers in $H_p^l(\mathbf{R}^n)$ and its dual $H_{p'}^{-l}(\mathbf{R}^n)$, embeddings of spaces of multipliers, the composition $\varphi(\gamma)$ of a Hölder continuous function φ of one variable and of a multiplier $\gamma \in M(W_p^m(\mathbf{R}^n) \to W_p^l(\mathbf{R}^n))$, (p, l)-diffeomorphisms preserving Sobolev spaces and (p, l)-manifolds on which such spaces can be correctly defined, an implicit function theorem for multipliers, etc. I also only mention that multipliers proved to be useful in numerous applications to differential and integral operators (see [10], [15], [16], [20]), especially in L_p-theory of elliptic boundary value problems with non-smooth coefficients and for domains with non-smooth boundary (see [10], [19], [21]). Instead, I shall concentrate on trace and extension theorems for multipliers.

Let $W_{p,\beta}^k(\mathbf{R}^{n+s})$ be the completion of $C_0^\infty(\mathbf{R}^{n+s})$ in the norm

$$||U; \mathbf{R}^{n+s}||_{W_{p,\beta}^k} = \left(\int_{\mathbf{R}^{n+s}} |y|^{p\beta} |\nabla_{k,z} U(z)|^p dz\right)^{1/p} + ||U; \mathbf{R}^{n+s}||_{L_p},$$

where $\mathbf{R}^{n+s} = \{z = (x, y) : x \in \mathbf{R}^n, y \in \mathbf{R}^s\}$. It is known that for non-integer $l > 0$ the space $W_p^l(\mathbf{R}^n)$ is the space of traces on $\mathbf{R}^n$ of functions in $W_{p,\beta}^k(\mathbf{R}^{n+s})$, where k is an integer, $k > l$, and $\beta = k - l - s/p$. The next theorem, proved in [10, Ch. 5], establishes an analogous assertion for the corresponding spaces of multipliers.

Let T be the extension operator defined by

$$(2.12) \qquad (T\gamma)(x, y) = \int_{\mathbf{R}^n} \zeta(t)\gamma(x + |y|\xi)d\xi,$$

where ζ is a function in the space $(W_{1,\mathrm{loc}}^k \cap L_1)(\mathbf{R}^n)$ subject to the normalization condition

$$(2.13) \qquad \int_{\mathbf{R}^n} \zeta(\xi)d\xi = 1.$$

Further, let

$$(2.14) \qquad \mathcal{C}_{\mu,\nu} := \int_{\mathbf{R}^n} (1 + |x|)^\nu \sum_{j=0}^{\mu} \sup_{\partial B_{|x|}^{(n)}} |\nabla_j \zeta(x)|(1 + |x|)^j dx,$$

where μ is a positive integer and $\nu > 0$. We shall also use the spherical coordinates (ρ, ω) in $\mathbf{R}^s$: $\rho = |y|$ and $\omega = y/|y|$.

Theorem 1. (i) *Let* $\Gamma \in MW_{p,\beta}^k(\mathbf{R}^{n+s})$, *where k is an integer, $1 \le p < \infty$, and $\beta p > -s$. Then, for almost all $x \in \mathbf{R}^n$, there exists the $L_p(\partial B_1^{(s)})$-limit*

$$\gamma(x) = \lim_{\rho \to 0} \Gamma(x, \rho, \cdot).$$

Moreover, the function γ belongs to the space $MW_p^l(\mathbf{R}^n)$ with $l = k - \beta - s/p$ and the estimate

$$\|\gamma; \mathbf{R}^n\|_{MW_p^l} \le c \|\Gamma; \mathbf{R}^{n+s}\|_{MW_{p,\beta}^k}$$

is valid.

(ii) *Let $\gamma \in MW_p^l(\mathbf{R}^n)$, where l is positive non-integer and $1 \le p < \infty$. Further, let $T\gamma$ be the extension of γ to $\mathbf{R}^{n+s}$ defined by (2.12) with the kernel ζ subject to $\mathcal{C}_{k+r,l+r} < \infty$ and*

$$(2.15) \qquad \int_{\mathbf{R}^n} x^\alpha \zeta(x) dx = 0, \quad 0 < |\alpha| \le [l] + r,$$

where r is a non-negative integer. Then $\nabla_{r,z}(T\gamma) \in MW_{p,\beta}^k(\mathbf{R}^{n+s})$, where $\beta = k - l - s/p$, k is an integer, $k > l$. Moreover, the inequality

$$\|\nabla_{r,z}(T\gamma); \mathbf{R}^{n+s}\|_{MW_{p,\beta}^k} \le c\mathcal{C}_{k+r,l+r} \|\nabla_{r,x}\gamma; \mathbf{R}^n\|_{MW_p^l}.$$

is valid.

Remark 1. Analogous assertion holds if $\mathbf{R}^{n+s}$ is replaced by $\mathbf{R}_+^{n+1} = \{z = (x, y) : x \in \mathbf{R}^n, y > 0\}$. The power moment condition (2.15) should be omitted in this case.

The following theorem contains a characterization of traces of functions in the multiplier space $MH_p^l(\mathbf{R}^{n+s})$.

Theorem 2. [18] (i) *Let $\Gamma \in MH_p^l(\mathbf{R}^{n+s})$, where $l - s/p$ is a positive non-integer number and $1 < p < \infty$. Then the function $\gamma(x) = \Gamma(x, 0)$ belongs to the space $MW_p^{l-s/p}(\mathbf{R}^n)$ and the estimate*

$$\|\gamma; \mathbf{R}^n\|_{MW_p^{l-s/p}} \le c \|\Gamma; \mathbf{R}^{n+s}\|_{MH_p^l}$$

holds.

(ii) *Let $\gamma \in MW_p^{l-s/p}(\mathbf{R}^n)$, where $l - s/p$ is a positive non-integer, and $1 < p < \infty$. Further, let $T\gamma$ be the extension of γ to $\mathbf{R}^{n+s}$ defined by (2.12), subject to conditions (2.13), and (2.15) with $r = 0$. Then $T\gamma \in MH_p^l(\mathbf{R}^{n+s})$ and the estimate*

$$\|T\gamma; \mathbf{R}^{n+s}\|_{MH_p^l} \le c\mathcal{C}_{[l],l} \|\gamma; \mathbf{R}^n\|_{MW_p^{l-s/p}}$$

is valid.

Now I formulate a trace and extension theorem for multipliers between two Sobolev spaces.

Theorem 3. [11] *Let m and l be integers, $m \geq l \geq 1$, $1 < p < \infty$, and $pm \leq n$.*
(i) If

$$\Gamma \in M(W_p^m(\mathbf{R}_+^n) \to W_p^l(\mathbf{R}_+^n))$$

and γ is the trace of Γ on $\mathbf{R}^{n-1}$, then

$$\gamma \in M(W_p^{m-1/p}(\mathbf{R}^{n-1}) \to W_p^{l-1/p}(\mathbf{R}^{n-1}))$$

and the estimate

$$\|\gamma; \mathbf{R}^{n-1}\|_{M(W_p^{m-1/p} \to W_p^{l-1/p})} \leq c\|\Gamma; \mathbf{R}_+^n\|_{M(W_p^m \to W_p^l)}$$

holds.

(ii) Let T be the extension operator defined by (2.12) and subject to (2.13). Suppose that

$$\int_{\mathbf{R}^{n-1}} (|\xi|^{l-1/p} + |\xi|^{l-m}) \sum_{i=0}^{l+1} \sup_{\partial B_{|\xi|}^{(n-1)}} |\nabla_i \zeta(\xi)|(1 + |\xi|)^i d\xi < \infty.$$

Then the operator T :

$$M(W_p^{m-1/p}(\mathbf{R}^{n-1}) \to W_p^{l-1/p}(\mathbf{R}^{n-1})) \to M(W_p^m(\mathbf{R}_+^n) \to W_p^l(\mathbf{R}_+^n))$$

is bounded.

Remark 2. Note that the same result is also valid for $mp > n$, $p > 1$ and $m \geq n$, $p = 1$, because $M(W_p^m(\mathbf{R}_+^n) \to W_p^l(\mathbf{R}_+^n)) = W_{p,\mathrm{unif}}^l(\mathbf{R}_+^n)$ in these cases (see (2.2)).

Clearly, for $m = l$ the last theorem coincides with the special case $\beta = 0$ in Theorem 1.

3. Multipliers in the space of Bessel potentials as traces of multipliers

3.1. Statement of the result

The goal of this subsection is to show that multipliers in the space $H_p^l(\mathbf{R}^n)$ are traces of multipliers in a weighted Sobolev space $L_{p,\beta}^k(\mathbf{R}^{n+s})$ defined as the completion of $C_0^\infty(\mathbf{R}^{n+s})$ in the norm

$$\left(\int_{\mathbf{R}^n} \left(\int_{\mathbf{R}^s} |y|^{2\beta} |\nabla_{k,z} U|^2 dy\right)^{p/2} dx\right)^{1/p} + \left(\int_{\mathbf{R}^n} \left(\int_{\mathbf{R}^s} |y|^{2\beta} |U|^2 dy\right)^{p/2} dx\right)^{1/p}.$$

Let the first term be denoted by $< U >_{p,\beta,k}$. For $k > r$, $\beta > r - s/p$, by Hardy's inequality one has

$$(3.1) \qquad\qquad < U >_{p,\beta-r,k-r} \leq c < U >_{p,\beta,k} .$$

The following auxiliary assertion shows that elements of $H_p^l(\mathbf{R}^n)$ are traces on $\mathbf{R}^n$ of functions in $L_{p,\beta}^k(\mathbf{R}^{n+s})$ (see [14] for $0 < l < 1$, the general case is treated in a similar way).

Lemma 1. (i) *Let* $U \in L_{p,\beta}^k(\mathbf{R}^{n+s})$, *where* k *is an integer*, $2\beta > -s$, *and* $1 < p < \infty$. *Then, for almost all* $x \in \mathbf{R}^n$, *there exists the limit*

$$U(x,0) = \lim_{\rho \to +0} \int_{\partial B_1^{(s)}} U(x;\rho,\omega)d\omega.$$

Moreover, $U(\cdot,0) \in H_p^l(\mathbf{R}^n)$ *with* $l = k - \beta - s/2$, $\{l\} > 0$ *and*

$$(3.2) \qquad\qquad \|U(\cdot,0); \mathbf{R}^n\|_{H_p^l} \leq c\|U; \mathbf{R}^{n+s}\|_{L_{p,\beta}^k}.$$

(ii) *Let* $u \in H_p^l(\mathbf{R}^n)$, $l > 0$, $1 < p < \infty$. *There exists a linear continuous extension operator:*

$$H_p^l(\mathbf{R}^n) \ni u \to U \in L_{p,\beta}^k(\mathbf{R}^{n+s}),$$

where k *is an integer,* $k > l$, *and* $\beta = k - l - s/2$.

The main result of this section runs as follows.

Theorem 4. (i) *Let* $\Gamma \in ML_{p,\beta}^k(\mathbf{R}^{n+s})$, *where* k *is an integer,* $1 < p < \infty$, $2\beta > -s$, *and let* $k - \beta - s/2$ *be a positive non-integer number. Then the function* $\gamma = \Gamma(x,0)$ *belongs to the space* $MH_p^l(\mathbf{R}^n)$ *with* $l = k - \beta - s/2$ *and the estimate*

$$\|\gamma; \mathbf{R}^n\|_{MH_p^l} \leq c\|\Gamma; \mathbf{R}^{n+s}\|_{ML_{p,\beta}^k}$$

is valid.

(ii) *Let* $\gamma \in MH_p^l(\mathbf{R}^n)$, *and let* $T\gamma$ *be the extension of* γ *to* $\mathbf{R}^{n+s}$ *defined by* (2.12), *subject to conditions* (2.13) *and* (2.15) *with* $r = 0$. *Then* $T\gamma \in ML_{p,\beta}^k(\mathbf{R}^{n+s})$ *with an integer* k, $k > l$, *and* $\beta = k - l - s/2$. *Moreover,*

$$\|T\gamma; \mathbf{R}^{n+s}\|_{ML_{p,\beta}^k} \leq c\mathcal{C}_{k,l}\|\gamma; \mathbf{R}^n\|_{MH_p^l},$$

where $\mathcal{C}_{k,l}$ *is defined by* (2.14).

3.2. An auxiliary estimate for the extension operator

Let T be the extension operator defined by (2.12). Suppose its kernel ζ is subject to (2.13) and (2.15) with $r = 0$.

The next lemma will be used in Section 3.3.

Lemma 2. *For any positive non-integer* $\delta > 0$ *and any integer* $q > \delta$, *the estimate*

$$(3.3) \qquad \left(\int_{\mathbf{R}^{n+s}} |y|^{2(q-\delta)-s} |\nabla_{q,z}(Tu)|^2 dy \right)^{1/2} \leq c\, C_{q,\delta} S_\delta u(x)$$

is valid, where $C_{q,\delta}$ is defined by (2.14).

Proof. Let τ, ρ, and ω be n-dimensional multi-indices, and let σ be an s-dimensional multi-index such that $|\tau| + |\sigma| = q$, $\rho = 0$, $\omega = \tau$, if $|\tau| \leq \delta$, and $\rho = \tau - \omega$, $|\omega| = [\delta]$, $\omega < \tau$, if $|\tau| > \delta$. We introduce the notation

$$\mathcal{R}_\omega(h,x) = D^\omega u(x+h) - \sum_{|\nu| < [\delta] - |\omega|} D^{\omega+\nu} u(x) \frac{h^\nu}{\nu!}.$$

Using the identity

$$D_y^\sigma \left(|y|^{-n-|\rho|} \int_{\mathbf{R}^n} (D^\rho \zeta) \left(\frac{\xi - x}{|y|} \right) (\xi - x)^\nu d\xi \right)$$

$$= D_y^\sigma \left(|y|^{|\nu|-|\rho|} \int_{\mathbf{R}^n} D^\rho \zeta(\xi) \xi^\nu d\xi \right) = 0,$$

we obtain

$$D_x^\tau D_y^\sigma \int_{\mathbf{R}^n} \zeta(\eta) u(x+|y|\eta) d\eta = D_x^\rho D_y^\sigma \int_{\mathbf{R}^n} \zeta(\eta) D_x^\omega u(x+|y|\eta) d\eta$$

$$= D_y^\sigma \left(|y|^{-n-|\rho|} \int_{\mathbf{R}^n} (D^\rho \zeta) \left(\frac{\xi - x}{|y|} \right) D^\omega u(\xi) d\xi \right)$$

$$= D_y^\sigma \left(|y|^{-n-|\rho|} \int_{\mathbf{R}^n} (D^\rho \zeta) \left(\frac{\xi - x}{|y|} \right) \mathcal{R}_\omega(\xi - x, x) d\xi \right).$$

Clearly,

$$\left| D_y^\sigma \left(|y|^{-n-|\rho|} (D^\rho \zeta) \left(\frac{\xi - x}{|y|} \right) \mathcal{R}_\omega(\xi - x, x) \right) \right|$$

$$\leq |y|^{\delta - |\rho| - |\sigma| - |\omega|} \varphi \left(\frac{\xi - x}{|y|} \right) \frac{|\mathcal{R}_\omega(\xi - x, x)|}{|\xi - x|^{\delta - |\omega| + n}},$$

where φ is a function admitting the estimate

$$(3.4) \qquad \varphi(\xi) \leq c |\xi|^{\delta - |\omega| + n} \sum_{i=0}^{|\sigma|} |\nabla_{i+|\rho|} \zeta(\xi)| (|\xi|^i + 1).$$

Since $|\rho| + |\omega| = |\tau|$, $|\tau| + |\sigma| = q$, we arrive at the inequality

$$\int_{\mathbf{R}^s} |y|^{2(q-\delta)-s} |D_x^\tau D_y^\sigma (Tu)|^2 dy$$

$$\leq c \int_{\mathbf{R}^s} \left(\int_{\mathbf{R}^n} \varphi \left(\frac{\xi - x}{|y|} \right) \frac{|\mathcal{R}_\omega(\xi - x, x)|}{|\xi - x|^{\delta - |\omega| + n}} d\xi \right)^2 \frac{dy}{|y|^s}.$$

Passing to the spherical coordinates $t = |\xi - x|$ and $\theta = (\xi - x)t^{-1}$, one can write the right-hand side as

$$c \int_0^\infty \left(\int_0^\infty \int_{\partial B_1^{(n)}} \varphi\left(\frac{t\theta}{\lambda}\right) \frac{|\mathcal{R}_\omega(t\theta, x)|}{t^{\delta - |\omega|}} \frac{dt}{t} d\theta \right)^2 \frac{d\lambda}{\lambda}.$$

This expression does not exceed

$$(3.5) \qquad c \int_0^\infty \left(\int_0^\infty Q\left(\frac{t}{\lambda}\right) g(t) \frac{dt}{t} \right)^2 \frac{d\lambda}{\lambda},$$

where

$$Q(t) = \sup_{\theta \in \partial B_1^{(n)}} \varphi(t\theta)$$

and

$$g(t) = t^{|\omega| - \delta} \int_{\partial B_1^{(n)}} |\mathcal{R}_\omega(t\theta, x)| d\theta.$$

By Minkowski's inequality (3.5) is not greater than

$$c \left(\int_0^\infty Q(t) \frac{dt}{t} \right)^2 \int_0^\infty g(t)^2 \frac{dt}{t}.$$

This and (3.4) imply

$$\left(\int_{\mathbf{R}^s} |y|^{2(q-\delta)-s} |D_x^\tau D_y^\sigma (Tu)|^2 dy \right)^{1/2}$$
$$(3.6) \qquad \leq c\mathcal{C}_{q,\delta} \left(\int_0^\infty t^{2(|\omega|-\delta)-1} \left(\int_{\partial B_1^{(n)}} |\mathcal{R}_\omega(t\theta, x)| d\theta \right)^2 dt \right)^{1/2}.$$

For $0 < \delta < 1$ we have $\omega = 0$ and $\mathcal{R}_0(t\theta, x) = u(x + t\theta) - u(x)$. Therefore, for such δ, the right-hand side of (3.6) is $c\mathcal{C}_{q,\delta}S_\delta u(x)$. Hence we need to consider only $\delta > 1$. Since

$$\mathcal{R}_\omega(t\theta, x) =$$
$$([\delta] - |\omega|) \int_0^1 \sum_{|\nu| = [\delta] - |\omega|} \frac{(t\theta)^\nu}{\nu!} \left((D^{\nu+\omega}u)(x + ht\theta) - D^{\nu+\omega}u(x) \right) (1-h)^{[\delta] - |\omega| - 1} dh,$$

we obtain

$$\left(\int_{\mathbf{R}^s} |y|^{2(q-\delta)-s} |\nabla_{q,z}(Tu)|^2 dy \right)^{1/2}$$
$$\leq c\mathcal{C}_{q,\delta} \left(\int_0^\infty t^{-2\{\delta\}-1} \left(\int_{\partial B_1^{(n)}} \int_0^1 \sum_{|\alpha|=[\delta]} |(D^\alpha u)(x + ht\theta) - D^\alpha u(x)| dh \, d\theta \right)^2 dt \right)^{1/2}.$$

By Minkowski's inequality the right-hand side is dominated by

$$c\mathcal{C}_{q,\delta}\int_0^1\left(\int_0^\infty t^{-2\{\delta\}-1}\left(\int_{\partial B_1^{(n)}}\sum_{|\alpha|=[\delta]}|(D^\alpha u)(x+ht\theta)-D^\alpha u(x)|d\theta\right)^2 dt\right)^{1/2}dh.$$

Making the change of variable $t\to h^{-1}\tau$ for any $h\in(0,1)$, we find that the last value is equal to

$$c\mathcal{C}_{q,\delta}\int_0^1 h^{\{\delta\}}\left(\int_0^\infty \tau^{-2\{\delta\}-1}\left(\int_{\partial B_1^{(n)}}\sum_{|\alpha|=[\delta]}|(D^\alpha u)(x+\tau\theta)-D^\alpha u(x)|d\theta\right)^2 d\tau\right)^{1/2}dh$$

and is not greater than $\mathcal{C}_{q,\delta}S_\delta u(x)$. The proof is complete.

3.3.   Proof of Theorem 4

Now we are in a position to prove the main result of Section 3.

(i) The existence of the trace γ of the function $\Gamma\in ML_{p,\beta}^k(\mathbf{R}^{n+s})$ follows from the inclusion $\Gamma\in L_{p,\beta,\mathrm{loc}}^k(\mathbf{R}^{n+s})$ and Lemma 1. Let $U\in L_{p,\beta}^k(\mathbf{R}^{n+s})$ and let $u(x)=U(x,0)$. We have

$$\|\gamma u;\mathbf{R}^n\|_{H_p^l}\le c\|\Gamma U;\mathbf{R}^{n+s}\|_{L_{p,\beta}^k}\le c\|\Gamma;\mathbf{R}^{n+s}\|_{ML_{p,\beta}^k}\|U;\mathbf{R}^{n+s}\|_{L_{p,\beta}^k}.$$

The result follows from part (ii) of Lemma 1.

(ii) Let μ, ϵ, and ν, be arbitrary multi-indices of dimensions n, s, and $n+s$. Clearly,

$$\begin{aligned}
&< UT\gamma;\mathbf{R}^{n+s} >_{p,\beta,k}\\
(3.7)\qquad &\le c\sum_{|\nu|+|\mu|+|\epsilon|=k}< |D_z^\nu U||D_x^\mu D_y^\epsilon(T\gamma)|;\mathbf{R}^{n+s} >_{p,\beta,0}.
\end{aligned}$$

By $\sum^{(1)}$ and $\sum^{(2)}$ we denote the sums of those terms in (3.7) for which $|\nu|<l$ and $|\nu|>l$, respectively. Since

$$(3.8)\qquad |D_x^\mu D_y^\epsilon(T\gamma)(z)|\le c\mathcal{C}_{k,l}\|\gamma;\mathbf{R}^n\|_{L_\infty}|y|^{-|\mu|-|\epsilon|}$$

(see [10, Sec. 5.1.3]), one has

$$\sum^{(2)}\le c\mathcal{C}_{k,l}\|\gamma;\mathbf{R}^n\|_{L_\infty}< U;\mathbf{R}^{n+s} >_{p,\beta-k+|\nu|,|\nu|}.$$

By (3.1) the right-hand side does not exceed

$$c\mathcal{C}_{k,l}\|\gamma;\mathbf{R}^n\|_{L_\infty}< U;\mathbf{R}^{n+s} >_{p,\beta,k}.$$

Now let $|\nu| < l$. Put

$$\mathcal{Q}_\nu U(z) = D_z^\nu U(z) - \sum_{|\tau| \le [l]-|\nu|} (D_y^\tau D_z^\nu U)(x,0)\frac{y^\tau}{\tau!}.$$

Then

$$\sum{}^{(1)} \le \left(\int_{\mathbf{R}^n} \left(\int_{\mathbf{R}^s} |y|^{2(k-l)-s} \sum_{|\nu|+|\mu|+|\epsilon|=k} |\mathcal{Q}_\nu U|^2 |D_x^\mu D_y^\epsilon(T\gamma)|^2 dy \right)^{p/2} dx \right)^{1/p}$$

$$+ \left(\int_{\mathbf{R}^n} \left(\int_{\mathbf{R}^s} |y|^{2(k-l)-s} \sum_{|\nu|+|\mu|+|\epsilon|=k} \sum_{i=0}^{[l]-|\nu|} |y|^{2i} |D_x^\mu D_y^\epsilon(T\gamma)|^2 \times \right.\right.$$

$$\left.\left. \times |\nabla_{i,y} D_z^\nu U(x,0)|^2 dy \right)^{p/2} dx \right)^{1/p}.$$

Let us denote the first and the second terms on the right by A and B, respectively. Since

$$\mathcal{Q}_\nu U(z) = ([l]-|\nu|+1) \sum_{|\tau|=[l]-|\nu|+1} \frac{y^\tau}{\tau!} \int_0^1 (D_y^\tau D_z^\nu) U(x,ty)(1-t)^{[l]-|\nu|} dt,$$

we have the estimate

$$|\mathcal{Q}_\nu U(z)| \le c\, |y|^{[l]-|\nu|+1} \int_0^1 |(\nabla_{[l]+1,z} U)(x,ty)| dt.$$

Hence, by using (3.8) and Minkowski's inequality we find that A is majorized by

$$c\mathcal{C}_{k,l} \|\gamma; \mathbf{R}^n\|_{L_\infty} \int_0^1 \left(\int_{\mathbf{R}^n} \left(\int_{\mathbf{R}^s} |y|^{2(1-\{l\})-s} |(\nabla_{[l]+1,z} U)(x,ty)|^2 dy \right)^{p/2} dx \right)^{1/p} dt.$$

Therefore,

$$A \le c\mathcal{C}_{k,l} \|\gamma; \mathbf{R}^n\|_{L_\infty} < U >_{p,\beta,k}.$$

By Lemma 2 with $q = k - j$ and $r = l - |\nu| - i$ we find that B is not greater than

$$c\mathcal{C}_{k,l} \sum_{i=0}^{[l]-|\nu|} \left(\int_{\mathbf{R}^n} |(\nabla_i D^\nu) U(x,0)|^p |S_{l-i-|\nu|}\gamma(x)|^p dx \right)^{1/p}$$

$$\le c\mathcal{C}_{k,l} \sum_{i=0}^{[l]-|\nu|} \|S_{l-i-|\nu|}\gamma; \mathbf{R}^n\|_{M(H_p^{l-i-|\nu|} \to L_p)} \|\nabla_{i+|\nu|} U(\cdot,0); \mathbf{R}^n\|_{H_p^{l-i-|\nu|}},$$

which by theorem 2.2.7/1 [10] does not exceed

$$(3.9) \qquad c\,\mathcal{C}_{k,l}\|\gamma;\mathbf{R}^n\|_{MH_p^l}\|U;\mathbf{R}^{n+s}\|_{L_{p,\beta}^k}.$$

Using the estimates obtained for the values A and B we find that (3.9) is the majorant for the norm $<UT\gamma>_{p,\beta,k}$. It remains to note that (3.8) with $\mu = \epsilon = 0$ implies

$$\left(\int_{\mathbf{R}^n}\left(\int_{\mathbf{R}^s}|y|^{2(k-l)-s}|UT\gamma|^2 dy\right)^{p/2}dx\right)^{1/p}$$

$$\leq c\,\mathcal{C}_{k,l}\|\gamma;\mathbf{R}^n\|_{L_\infty}\left(\int_{\mathbf{R}^n}\left(\int_{\mathbf{R}^s}|y|^{2(k-l)-s}|U|^2 dy\right)^{p/2}dx\right)^{1/p}.$$

The proof is complete.

References

[1] Adams, D. R., *On the existence of capacitary strong type estimates in $\mathbf{R}^n$*, Ark. Mat. **14** (1976) 125–140.

[2] Adams, D. R., Hedberg, L. I., *Function spaces and potential theory*, Springer, 1996.

[3] Fefferman, C., *The uncertainty principle*, Bull. Amer. Math. Soc. **9** (1983) 129–206.

[4] Havin, V. P., Maz'ya, V. G., *Nonlinear potential theory*, Russian Math. Surveys **27** (1972), No.6, 71–148.

[5] Kerman, R, Saywer, E. T., *The trace inequality and eigenvalue estimates for Schrödinger operators*, Ann. Inst. Fourier (Grenoble), **36** (1986), 207–228.

[6] Maz'ya, V. G., *On some integral inequalities for functions of several variables*, Problemi Mat. Anal., Leningrad Univ., 1972, No. 3, 33–69.

[7] Maz'ya, V. G., *Multipliers in Sobolev spaces*, In the book: Application of function theory and functional analysis methods to problems of mathematical physics. Pjatoe Sovetsko-Čehoslovackoe Soveščanie, 1976, 181–189. Novosibirsk, 1978.

[8] Maz'ya, V. G., *Summability, with respect to an arbitrary measure, of functions from Sobolev-Slobodeckii spaces*, Zap. Nauchn. Sem. Leningrad. Otdel. Mat. Inst. Steklov. (LOMI) **92** (1979), 192–202.

[9] Maz'ya, V. G., *Sobolev spaces*, Springer-Verlag, 1985.

[10] Maz'ya, V. G., Shaposhnikova, T. O., *Theory of multipliers in spaces of differentiable functions*, Monographs and Studies in Mathematics **23**, Pitman, 1985.

[11] Maz'ya, V. G., Shaposhnikova, T. O., *Traces and extensions of multipliers in pairs of Sobolev spaces*. To appear in the book: Complex Analysis, Operator Theory, and Related Topics: S.A.Vinogradov – In Memoriam, Birkhäuser, 1999.

[12] Maz'ya, V. G., Shaposhnikova, T. O., *On pointwise interpolation inequalities for derivatives*. To appear in Mathematica Bohemica, 1999.

[13] Maz'ya, V. G., Verbitsky, I. E.*Capacitary inequalities for fractional integrals, with applications to partial differential equations and Sobolev multipliers*, Ark. Mat. **33** (1995), No. 1, 81–115.

[14] Shaposhnikova, T. O., *Equivalent norms in spaces with fractional or functional smoothness,* Sibir. Mat. J. **21** (1980) 184–196.

[15] Shaposhnikova, T. O., *Bounded solutions of elliptic equations as multipliers in spaces of differentiable functions,* Zapiski Nauchn. Seminar. LOMI, **194** (1986) 165–176.

[16] Shaposhnikova, T. O., *On solvability of quasilinear elliptic equations in spaces of multipliers,* Izvestia Vyssh. Uch. Zav. Math. **8** (1987) 74–81.

[17] Shaposhnikova, T. O., *Multipliers in the space of Bessel potentials as traces of multipliers in weighted classes,* (In Russian) Trudy Tbilis. Mat. Inst. **88** (1988) 60–63.

[18] Shaposhnikova, T. O., *Traces of multipliers in the space of Bessel potentials,* Matem. Zametki, **46** (1989) 100–109.

[19] Shaposhnikova, T. O., *Applications of multipliers in Sobolev spaces to L_p-coercivity of the Neumann problem,* Dokl. Acad. Nauk USSR **305** (1989) 786–789.

[20] Shaposhnikova, T. O., *On continuity of singular integral operators in Sobolev spaces,* Mathematica Scandinavica **76** (1995) 85–97.

[21] Shaposhnikova, T. O., *Sobolev multipliers in a non-smooth L_p-elliptic theory,* Analysis, numerics and applications of differential and integral equations, Pitman Research Notes in Mathematical Series **379**, 1998, 205–208.

[22] Sickel, W., *On pointwise multipliers for $F_{p,q}^s(\mathbf{R}^n)$ in case $\sigma_{p,q} < s < n/p$.* (To appear)

[23] Simon, B., *Schrödinger semigroups,* Bull. Amer. Math. Soc. **7** (1982) 447–526.

[24] Strichartz, R. S., *Multipliers on fractional Sobolev spaces,* J. Math. and Mech. **16** (1967) No. 9, 1031–1060.

Department of Mathematics,
Linköping University,
S-58183 Linköping, Sweden,
tasha@mai.liu.se

1991 Mathematics Subject Classification: Primary 46E35, 46E25

Submitted: 11.01.1999

Operator Theory:
Advances and Applications, Vol. 109

An asymptotic theory of nonlinear abstract higher order ordinary differential equations

Vladimir Kozlov

This paper is a short summary of results obtained recently together with Vladimir Maz'ya. From 1990 we worked on an asymptotic theory of arbitrary order *linear* differential equations with unbounded operator coefficients in a Hilbert or Banach space. The impetus was given by our previous and simultaneous studies of singularities of solutions to elliptic boundary value problems (see [1]). The first real publication summarizing our linear abstract theory has just appeared as the Springer book [2] (previously we wrote seven preprints concerning this topic). In Introduction to the book we mentioned that reach possibilities for generalizations to *nonlinear* operator equations were left aside. Some

of these possibilities are presented here.

We study solutions to the equation

$$(1) \qquad \mathcal{A}(D_t)u(t) = \sum_{j=0}^{q} D_t^{q-j} \mathcal{N}_j\big(t; u(t), \dots, D_t^{\ell-q}u(t)\big)$$

on a semiaxis $t > t_0$, where $\mathcal{A}(D_t)$ is an ordinary differential operator of order ℓ with unbounded operator coefficients in a Hilbert space and $\mathcal{N}_0, \dots, \mathcal{N}_q$ are nonlinear operators. We write the right-hand side in this form in order to cover more general nonlinearities: the operators $\mathcal{N}_j$ are not assumed to be differentiable. Our conditions on $\mathcal{A}(D_t)$ are dictated by an analogy with linear elliptic operators written in the variational form and are the same as in [2].

Our main concern is with the asymptotic behaviour of solutions as $t \to +\infty$. We show that for a certain class of equations (1) the question of asymptotics can be reduced to that for a finite dimensional dynamical system perturbed by a "weak" non-local nonlinear operator. This is a far-reaching generalization of a similar result for linear differential equations with operator coefficients obtained in [2, Ch.13]. However, since the right-hand side is now written in the generic divergence form, our present result is new also for the linear case and, apparently, even for linear ordinary differential equations with scalar coefficients.

In order to avoid technicalities I give a very approximate description of a general result concerning a solution $u(t)$ to (1). Let u be subject to a growth condition of the type

$$(2) \qquad \|u\|_{W^{\ell-q}(t,t+1)} \leq M(t) \qquad \text{for large positive } t,$$

where $W^{\ell-q}$ is a Sobolev space and the majorant $M(t)$ behaves like $\exp(-k_0 t)$ in a certain rough sense. Estimates of this type are usually obtained in applications by using specific features of the equation: one relies upon monotonicity properties of differential operators, the maximum modulus principle and more refined tricks. Dealing with an abstract theory we take (2) for granted.

Under some natural assumptions on the nonlinearity we obtain the representation

$$(3) \qquad (u(t), \dots, D_t^{\ell-q-1}u(t)) = \sum_{s=1}^{\kappa} h_s(t)(U_s(t), \dots, D_t^{\ell-q-1}U_s(t)) + \mathbf{v}(t).$$

The vector functions U_s are solutions to $\mathcal{A}(D_t)U(t) = 0$ of the form $\exp(i\lambda_\nu t)P(t)$ with λ_ν being eigenvalues of the pencil $\mathcal{A}(\lambda)$ on the line $\Im\lambda = k_0$ and with P being polynomials with constant vector coefficients. The vector $\vec{h} = (h_1, \dots, h_\kappa)$ satisfies the finite dimensional perturbed dynamical system

$$(4) \qquad\qquad\qquad D_t\vec{h}(t) = \mathbb{N}(t; \vec{h}(t)) + \mathbb{K}[\vec{h}](t) .$$

For the components of $\mathbb{N}$ we have the representation

$$\mathbb{N}_k[\vec{h}](t) = \sum_{j=0}^{q} \langle \mathcal{N}_j(t; \sum_{s=1}^{\kappa} h_s(t)U_s(t), \dots, \sum_{s=1}^{\kappa} h_s(t)D_t^{\ell-q}U_s(t)) \,|\, D_t^{q-j}V_k(t) \rangle$$

where V_k are solutions of the adjoint equation $\mathcal{A}^*(D_t)V(t) = 0$ which are connected with U_s by a biorthogonality condition. Here, $\langle \cdot \,|\, \cdot \rangle$ is the scalar product. By $\mathbb{K}$ we denote a non-local nonlinear operator.

The vector function $\mathbf{v}$ can be regarded as a remainder term. We give estimates which show that $\mathbb{K}$ and $\mathbf{v}$ are weak in a certain sense.

Our main technical tool is a comparison principle which shows that solutions to a certain ordinary differential equation majorize solutions of equation (1) (compare with [2], Sect. 5 and [3]). It is this principle that we use to obtain estimates for the vector function $\mathbf{v}$ and the operator $\mathbb{K}$.

System (4) is the corner stone of our asymptotic theory. On one hand, it can be applied to construct solutions of (1) with the vector $\vec{h}(t)$ asymptotically close to a solution of the dynamical system

$$(5) \qquad\qquad\qquad D_t\vec{\chi}(t) = \mathbb{N}(t; \vec{\chi}(t)) .$$

On the other hand, one can try to show that solutions to (1) subject to the growth restriction (2) have the asymptotic representation (3), where the vector $\vec{h}$ satisfies (5).

The dynamical system (5) is, in general, a very complicated object containing an arbitrary nonlinear term, so that even the appearance of a chaotic attractor is not excluded. Therefore, in order to deduce explicit asymptotical formulae for solutions of (1) we must restrict classes of nonlinearities.

For example, let $\mathcal{N}_j(t; 0; \dots, 0) = 0$, $j = 0, \dots, q$, and let m_0 be the maximum length of the Jordan chains corresponding to the eigenvalues of $\mathcal{A}(\lambda)$ on the line

$\Im\lambda = k_0$. Denoting by $\rho(t)$ a certain Lipschitz seminorm of the vector operator $(\mathcal{N}_0,\dots,\mathcal{N}_q)$ on the class of functions subject to

$$(6) \qquad \|u\|_{W^{\ell-q}(t,t+1)} \le c\, t^{m_0-1}\exp(-k_0 t) \qquad \text{for } t > t_0,$$

and assuming that

$$(7) \qquad \int_{t_0}^{+\infty} t^{2m_0-2}\rho(t)dt < \infty,$$

we can show that every solution to (1) subject to (6) has the same polynomial exponential asymptotics

$$u(t) \sim \sum_{s=1}^{\kappa} c_s U_s(t)$$

as in the linear constant coefficients case. Condition (7) is precise (see [2, Sec. 14.4]).

We plan to consider more applications of the above asymptotic approach in our forthcoming paper.

References

[1] Kozlov, V., Maz'ya, V. and Rossmann J.: *Elliptic boundary value problems in domains with point singularities*, Mathematical Surveys and Monographs **52**, Amer. Math. Soc., 1997.

[2] Kozlov, V., Maz'ya, V.: *Differential Equations with Operator Coefficients* (with Applications to Boundary Value Problems for Partial Differential Equations), Monographs in Mathematics, Springer-Verlag, 1999.

[3] Kozlov, V., Maz'ya, V.: Comparison principles for nonlinear operator differential equations in Banach spaces, Differential Operators and Spectral Theory (M. Sh. Birman's 70th Anniversary Collection), Amer. Math. Soc. Transl., Ser. 2, **189**, 1999.

Department of Mathematics
University of Linköping
SE-581 83 Linköping, Sweden,
vlkoz@clapton.mai.liu.se

1991 Mathematics Subject Classification: Primary 34G20, 34G05

Submitted: 28.02.1999

Operator Theory:
Advances and Applications, Vol. 109
© 1999 Birkhäuser Verlag Basel/Switzerland

Sobolev spaces for domains with cusps

Sergei Poborchi

Abstract: This paper presents extension, embedding and trace theorems for Sobolev spaces for domains with cusps, which have been established in joint works by Maz'ya and the author during the last years.

1. Introduction

In this article I review extension, embedding and trace theorems obtained jointly by Maz'ya and myself for Sobolev spaces of functions on domains with cusps, which are the simplest non-Lipschitz domains frequently used in applications. The main object is the Sobolev space $W_p^l(\Omega)$, where Ω is a domain in $\mathbb{R}^n$ with the vertex of an outer (or inner) peak on the boundary.

As early as in 1960, Maz'ya [6] was the first to extend the Sobolev-Gagliardo embedding theorem to the power cusp

$$\Omega = \{x = (y, z) \in \mathbb{R}^n : z \in (0, 1), |y| < z^\lambda\}$$

with $\lambda > 1$ (I shall call it the λ-cusp). He showed that the Sobolev space $W_p^1(\Omega)$ for the λ-cusp is continuously embedded into $L_q(\Omega)$ with sharp exponent

$$q = \big(1 + \lambda(n - 1)\big)p/\big(1 + \lambda(n - 1) - p\big) \quad \text{if} \quad 1 \le p < 1 + \lambda(n - 1).$$

In [6] this result was just an example showing that Maz'ya's general isoperimetric criteria for Sobolev's embeddings could be effectively verified for particular domains. Many other properties of Sobolev's spaces for cuspidal domains can be found in the book [7].

We began our joint work 15 years ago being interested in Sobolev spaces for "good" domains which depend on small or large parameters in such a way that the limit domain is "bad". We were mainly concerned with the speed of degeneration of embedding, extension and trace operators when the parameters tend to their limits. We were motivated by applications to the asymptotic theory of elliptic boundary value problems for domains with singularly perturbed boundaries, and were able to obtain two-sided parameter dependent estimates for the norms of extension, embedding and boundary trace operators. Precise results of the same kind for cuspidal domains appeared as a by-product in that work.

Proofs of these results, which are only formulated in this article, can be found in our recent book [13].

2. Extension theorems

Let Ω be a domain in $\mathbb{R}^n$, i.e., an open connected set. For $1 \leq p \leq \infty$ and positive integer l, the Sobolev space $W_p^l(\Omega)$ is defined as the space of those functions in $L_p(\Omega)$ whose gradients of order l also are in $L_p(\Omega)$. This space is endowed with the norm

$$\|u\|_{W_p^l(\Omega)} = \|u\|_{L_p(\Omega)} + \|\nabla_l u\|_{L_p(\Omega)}.$$

Here $\nabla_l u$ is the gradient of u of order l with length $|\nabla_l u| = (\sum_{|\alpha|=l} |D^\alpha u|^2)^{1/2}$.

We say that Ω belongs to the class EW_p^l if there exists a linear continuous extension operator

$$E : W_p^l(\Omega) \to W_p^l(\mathbb{R}^n),$$

i.e., $Eu\big|_\Omega = u$ for all $u \in W_p^l(\Omega)$.

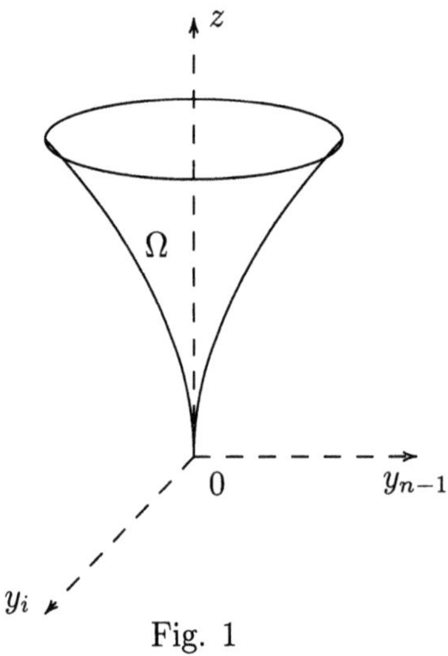

Fig. 1

We are concerned with extendability of Sobolev functions defined on domains with isolated peaks on the boundary. A typical domain with the vertex of an outer peak on the boundary is

(2.1) $$\Omega = \{x = (y, z) \in \mathbb{R}^n : z \in (0, 1), |y| < \varphi(z)\},$$

where φ is an increasing Lipschitz continuous function on $[0, 1]$ such that $\varphi(0) = \lim_{z \to 0} \varphi'(z) = 0$ (see Fig. 1).

One can easily construct examples showing that $\Omega \notin EW_p^l$ if $p < \infty$, $n \geq 2$, and that $\mathbb{R}^2 \setminus \overline{\Omega} \notin EW_p^l$ in the planar case for $p > 1$. On the other hand, Jones' extension theorem [5] says that $\mathbb{R}^n \setminus \overline{\Omega} \in EW_p^l$ if $n > 2$, $1 \leq p \leq \infty$ and $l = 1, 2, \ldots$.

Let σ be a bounded nonnegative measurable function on $\mathbb{R}^n$, which is separated away from zero in the exterior of any ball centered at the origin. By $W_{p,\sigma}^l(\mathbb{R}^n)$ we mean the weighted Sobolev space with norm

$$\|u\|_{W_{p,\sigma}^l(\mathbb{R}^n)} = \|\sigma u\|_{L_p(\mathbb{R}^n)} + \|\sigma \nabla_l u\|_{L_p(\mathbb{R}^n)}.$$

Clearly this weighted space is wider than $W_p^l(\mathbb{R}^n)$. It turns out that there is a linear continuous extension operator: $W_p^l(\Omega) \to W_{p,\sigma}^l(\mathbb{R}^n)$ for an appropriate choice of σ. The following assertion gives precise conditions on the weight σ.

Theorem 2.1. *Let $n \geq 2$ and let Ω be defined by (2.1) with φ satisfying the additional assumption $\varphi(2z) \leq \mathrm{const}\, \varphi(z)$, $z \in (0, 1/2]$. In order that there exist a linear continuous extension operator:*

$$W_p^l(\Omega) \to W_{p,\sigma}^l(\mathbb{R}^n),$$

it is sufficient, and if $\sigma(x)$ depends only on $|x|$ and is nondecreasing in the vicinity of the origin, then it is also necessary that

$$\sigma(x) \leq \begin{cases} c\,(\varphi(|x|)/|x|)^{\min\{l,(n-1)/p\}} & \text{if } lp \neq n-1, \\[2ex] c\left(\dfrac{\varphi(|x|)}{|x|}\right)^l \left|\log\left(\dfrac{\varphi(|x|)}{|x|}\right)\right|^{(1-p)/p} & \text{if } lp = n-1, \end{cases}$$

in a neighborhood of the origin, where c is a positive constant independent of x.

As a simple consequence of the theorem we obtain that $\Omega \in EW_\infty^l$.

Let $p \in (1, \infty)$ and $l = 1, 2, \ldots$. We now turn to the question whether a linear continuous extension operator:

$$(2.2) \qquad\qquad W_p^l(\Omega) \to W_q^l(\mathbb{R}^n)$$

exists for some $q \in [1, p)$. The following theorem gives sharp conditions for the affirmative answer.

Theorem 2.2. *Let Ω be the same as in the preceding theorem. The existence of a linear bounded extension operator, acting as in (2.2) with $q < p$, is equivalent to the inequality*

$$(2.3) \qquad\qquad \int_0^1 \left(\frac{t^\beta}{\varphi(t)}\right)^{n/(\beta-1)} \frac{dt}{t} < \infty,$$

where

$$1/q - 1/p = l(\beta - 1)/(\beta(n-1) + 1) \quad \text{if} \quad lq < n - 1$$

and

$$1/q - 1/p = (n-1)(\beta - 1)/np \quad \text{if} \quad lq > n - 1.$$

In case $lq = n - 1$ the factor $|\log(\varphi(t)/t)|^\gamma$ should be included into the integrand in (2.3) with

$$\gamma = (1 - 1/q)/(1/p - 1/q), \qquad \beta = (np - q)/(q(n - 1)).$$

Example 2.3. Power cusp. Let $\varphi(z) = c\,z^\lambda$, $\lambda > 1$. A linear continuous extension operator, acting as in (2.2) for $q < p$, exists in the following cases.

1° $lq < n - 1$ and

$$(2.4) \qquad\qquad q^{-1} > p^{-1} + l(\lambda - 1)/\big(1 + \lambda(n - 1)\big).$$

2° $lq = n - 1$ and either (2.4) holds or

$$q^{-1} = p^{-1} + l(\lambda - 1)/(1 + \lambda(n - 1)) \quad \text{with} \quad 2q^{-1} < 1 + p^{-1}.$$

3° $lq > n - 1$ and $q^{-1} > p^{-1}(1 + (\lambda - 1)(n - 1)/n)$.

Let $n = 2$ and let Ω be given by (2.1). We now state some results concerning extendability of Sobolev functions defined on $\mathbb{R}^2 \setminus \overline{\Omega}$.

If σ is a weight such that $\sigma|_{\mathbb{R}^2 \setminus \Omega} = 1$,

$$(2.5) \qquad\qquad \sigma(x) \le \mathrm{const}\,\big(\varphi(z)/z\big)^{l - 1/p}$$

for $x = (y, z) \in \Omega$ and small z, then there exists a linear continuous extension operator:

$$W_p^l(\mathbb{R}^2 \setminus \overline{\Omega}) \to W_{p,\sigma}^l(\mathbb{R}^2).$$

Inequality (2.5) is necessary for the existence of the extension operator if $\varphi(2z) \le \mathrm{const}\,\varphi(z)$, $z \in (0, 1/2]$, and $\sigma(x)$ depends only on z for $x \in \Omega$ and is nondecreasing. In particular, we have $\mathbb{R}^2 \setminus \overline{\Omega} \in EW_1^1$.

Consider the planar domain $D = \{x \in \mathbb{R}^2 \setminus \overline{\Omega} : |x| < 2\}$. It turns out that the existence of a bounded linear extension operator:

$$W_p^l(D) \to W_q^l(\mathbb{R}^2), \;\; 1 \le q < p \le \infty,$$

is equivalent to

$$\int_0^1 (t/\varphi(t))^{p(lq-1)/(p-q)}\, t\,dt < \infty.$$

Proofs of the results presented in this section can be found in the papers by Maz'ya and Poborchi [9, 10], and in the book by the same authors [13, Chap. 5]. The fact that $\Omega \in EW_\infty^l$ for Ω of the form (2.1) is due to Whitney [14, 15].

Extension of Sobolev functions with worsening of the class (if a space-preserving extension is impossible) was studied by various authors. Here we mention the papers by Burenkov [2] and Fain [3] where optimal extension operators from wide classes of non-Lipschitz domains were constructed.

3. Embedding theorems

Here we consider summability and continuity properties of Sobolev functions on domains with outer peaks. Below we state assertions using the domain of the form (2.1) as a typical domain of that kind.

The following theorem gives precise conditions on φ and on p, l, q, n for $W_p^l(\Omega)$ to be continuously embedded into $L_q(\Omega)$, $p \leq q$.

Theorem 3.1. *Let Ω be given by (2.1) and let $l \geq 1$ be an integer. Suppose φ satisfies the additional condition $\varphi(2z) \leq \text{const}\, \varphi(z)$, $z \in (0, 1/2]$, in case $l > 1$. When $1 < p \leq q < \infty$, the space $W_p^l(\Omega)$ is continuously embedded into $L_q(\Omega)$ if and only if the expression*

$$\sup_{r \in (0,1)} \left(\int_0^r \varphi(z)^{n-1} dz \right)^{1/q} \left(\int_r^1 \frac{z^{(l-1)p/(p-1)}}{\varphi(z)^{(n-1)/(p-1)}} dz \right)^{1-1/p}$$

is finite. The boundedness of the embedding operator: $W_1^l(\Omega) \to L_q(\Omega)$, $1 \leq q < \infty$, is equivalent to

$$\sup_{r \in (0,1)} r^{l-1} \varphi(r)^{1-n} \left(\int_0^r \varphi(z)^{n-1} dz \right)^{1/q} < \infty.$$

In case $p > 1$ the continuous embedding operator: $W_p^l(\Omega) \to C(\overline{\Omega})$ exists if and only if

$$(3.1) \qquad \int_0^1 z^{(l-1)p/(p-1)} \varphi(z)^{(1-n)/(p-1)} dz < \infty.$$

Finally, the boundedness of the embedding operator: $W_1^l(\Omega) \to C(\overline{\Omega})$ is equivalent to

$$\sup \left\{ r^{l-1} \varphi(r)^{1-n} : r \in (0,1) \right\} < \infty.$$

A proof of the theorem can be found in the book by Maz'ya and Poborchi [13, Sec. 8.2]. See also Maz'ya [7, Sec. 3.3.3, 4.3.5], where the case $l = 1$ was considered.

In the above theorem, the exponent q, providing the continuous embedding $W_p^l(\Omega) \subset L_q(\Omega)$, is generally less than Sobolev's exponent $np/(n - lp)$ if $lp < n$. It turns out that q can be improved by using weighted L_q-spaces with weight functions equal to zero at the vertex of the peak. In particular, q may be equal to Sobolev's exponent.

Let Ω be given by (2.1). Suppose that σ is a bounded positive function on Ω which depends only on z and is separated away from zero in the exterior of any neighborhood of the origin. By $L_{q,\sigma}(\Omega)$, $q \geq 1$, we mean the weighted space with norm $\|u\|_{L_{q,\sigma}(\Omega)} = \|\sigma u\|_{L_q(\Omega)}$.

The following assertion gives necessary and sufficient conditions for the continuity of the embedding $W_p^l(\Omega) \subset L_{q,\sigma}(\Omega)$ (see Maz'ya and Poborchi [13, Sec. 8.2.3]).

180   Poborchi

Theorem 3.2. *Under the hypotheses of Theorem 3.1, assume also that $1 \le p \le q < \infty$ and that $q \le np/(n - lp)$ if $lp < n$. The space $W_p^l(\Omega)$ is continuously embedded into $L_{q,\sigma}(\Omega)$ if and only if the expression*

$$
\sup_{0 < r < 1} \left(\int_0^r \varphi(z)^{n-1} \sigma(z)^q dz \right)^{1/q} \left(\int_r^1 \frac{z^{(l-1)p/(p-1)}}{\varphi(z)^{(n-1)/(p-1)}} dz \right)^{1-1/p}
$$

is finite (when $p = 1$, the second factor is replaced by $r^{l-1}\varphi(r)^{1-n}$).

Example 3.3. Power cusp. Let Ω be defined by (2.1) with $\varphi(z) = \text{const } z^\lambda$, $\lambda > 1$. Put $\sigma(z) = z^\nu$. Then the space $W_p^l(\Omega)$, $lp < n$, is continuously embedded into $L_{q,\sigma}(\Omega)$ with $q = np/(n - lp)$ if and only if $\nu \ge l(\lambda - 1)(n - 1)/n$.

We now state a theorem giving criteria for the compactness of the embeddings $W_p^l(\Omega) \subset L_q(\Omega)$ and $W_p^l(\Omega) \subset C(\overline{\Omega})$ (see Maz'ya and Poborchi [13, Sec. 8.3]).

Theorem 3.4. *Let the assumptions of Theorem 3.1 hold. If $1 < p \le q < \infty$, then a necessary and sufficient condition for the embedding operator: $W_p^l(\Omega) \to L_q(\Omega)$ to be compact is that*

$$
\lim_{r \to +0} \left(\int_0^r \varphi(z)^{n-1} dz \right)^{1/q} \left(\int_r^1 \frac{z^{p(l-1)/(p-1)} dz}{\varphi(z)^{(n-1)/(p-1)}} \right)^{1-1/p} = 0.
$$

The compactness of the embedding operator: $W_1^l(\Omega) \to L_q(\Omega)$, $1 \le q < \infty$, is equivalent to

$$
\lim_{r \to +0} r^{l-1} \varphi(r)^{1-n} \left(\int_0^r \varphi(z)^{n-1} dz \right)^{1/q} = 0.
$$

In case $p > 1$ the compactness of the embedding $W_p^l(\Omega) \subset C(\overline{\Omega})$ is equivalent to its boundedness and to (3.1). Furthermore, the equality

$$
\lim_{r \to +0} r^{l-1} \varphi(r)^{1-n} = 0
$$

is necessary and sufficient for the space $W_1^l(\Omega)$ to be compactly embedded into $C(\overline{\Omega})$.

For λ-cusps, the conclusions of Theorems 3.1 and 3.4 can be written explicitly, and we arrive at the well-known results (cf. Maz'ya [6], [7, Sec. 4.9], Adams [1, Sec. 5.35, 5.36]).

Example 3.5. Let Ω be the same as in Example 3.3. The space $W_p^l(\Omega)$ $(p > 1)$ is continuously embedded into $C(\overline{\Omega})$ if and only if $lp > 1 + \lambda(n - 1)$ and this embedding is compact. The space $W_1^l(\Omega)$ is compactly embedded into $C(\overline{\Omega})$ if and only if $l > 1 + \lambda(n - 1)$, and in case $l = 1 + \lambda(n - 1)$ the noncompact embedding

$W_1^l(\Omega) \subset C(\overline{\Omega})$ holds. If $lp = 1 + \lambda(n-1)$, the compact embedding $W_p^l(\Omega) \subset L_q(\Omega)$ is valid for any $q < \infty$. In case $lp < 1 + \lambda(n-1)$ the embedding $W_p^l(\Omega) \subset L_q(\Omega)$ is continuous if and only if

$$q \leq \left(1 + \lambda(n-1)\right)p / \left(1 + \lambda(n-1) - lp\right),$$

and this embedding is compact if and only if the last inequality is strict.

4. Boundary values of Sobolev functions

Let Ω be a domain in $\mathbb{R}^n$ and $p \in [1, \infty)$. By $TW_p^1(\Omega)$ we mean the space of the traces $u\big|_{\partial\Omega}$ of the functions $u \in W_p^1(\Omega)$. The norm $\| \cdot \|_{TW_p^1(\Omega)}$ is defined as the norm in the factor-space $W_p^1(\Omega)/\mathring{W}_p^1(\Omega)$, $\mathring{W}_p^1(\Omega)$ being the closure of $C_0^\infty(\Omega)$ in $W_p^1(\Omega)$. In other words

$$\|f\|_{TW_p^1(\Omega)} = \inf\{\|u\|_{W_p^1(\Omega)} : u \in W_p^1(\Omega),\ u\big|_{\partial\Omega} = f\}.$$

According to Gagliardo's theorem [4], $TW_p^1(\Omega) = W_p^{1-1/p}(\partial\Omega)$ for $p \in (1, \infty)$ and $TW_1^1(\Omega) = L_1(\partial\Omega)$ if Ω is a bounded Lipschitz domain. When a domain has cusps on the boundary, Gagliardo's theorem generally fails. In case Ω is a planar bounded domain with an isolated cusp and $p > 1$, the description of the space $TW_p^1(\Omega)$ was given by Yakovlev [16, 17].

Here we again consider the domain of the form (2.1) as a typical domain with the vertex of an outer peak on the boundary. It turns out (see Maz'ya and Poborchi [11, 12], [13, Chap. 7]) that the boundary values of functions in $W_p^1(\Omega)$ and in $W_p^1(\mathbb{R}^n \setminus \overline{\Omega})$ can be characterized in terms of the finiteness of the norm

$$\langle f \rangle_{p,\partial\Omega} = \left(\int\limits_{\partial\Omega} |f(x)|^p q(x) ds_x + \iint\limits_{\partial\Omega \times \partial\Omega} |f(x) - f(\xi)|^p Q(x,\xi) ds_x ds_\xi \right)^{1/p},$$

where q and Q are nonnegative weight functions and ds_x, ds_ξ the area elements on $\partial\Omega$.

Let Ω be given by (2.1) for $n > 2$ and let $p \in (1, \infty)$. Suppose that f is a function on $\partial\Omega$ vanishing outside a small neighborhood of the origin.

This function f belongs to $TW_p^1(\Omega)$ if and only if $\langle f \rangle_{p,\partial\Omega} < \infty$, where $x = (y, z)$, $\xi = (\eta, \zeta)$, $0 \leq q(x) \leq \text{const}\, \varphi'(z)$,

$$Q(x,\xi) = \begin{cases} |x - \xi|^{2-n-p} & \text{if } |z - \zeta| < \varphi(z) + \varphi(\zeta),\ z, \zeta \in (0,1), \\ 0 & \text{otherwise.} \end{cases}$$

In addition, the norm $\langle f \rangle_{p,\partial\Omega}$ is equivalent to $\|f\|_{TW_p^1(\Omega)}$.

A necessary and sufficient condition for f to belong to $TW_p^1(\mathbb{R}^n \setminus \overline{\Omega})$ is that $\langle f \rangle_{p,\partial\Omega} < \infty$ with

$$q(x) = \begin{cases} \varphi(z)^{1-p} & \text{for } 1 < p < n-1, \\ \left(\varphi(z)|\log(\varphi(z)/z)|\right)^{1-p} & \text{for } p = n-1, \\ \varphi(z)^{2-n} & \text{for } p > n-1, \end{cases}$$

and $Q(x,\xi) \neq 0$ only if $z, \zeta \in (0,1)$. For these pairs $x, \xi \in \partial\Omega$, Q is defined as follows. If $p < n-1$, then

$$Q(x,\xi) = |x - \xi|^{2-n-p}.$$

In the case $p \geq n-1$ and $|x - \xi| < \varphi(z) + \varphi(\zeta)$ the weight Q is determined by the same formula. Finally, if $|x - \xi| \geq \varphi(z) + \varphi(\zeta)$, then

$$Q(x,\xi) = \frac{\left(\varphi(z) + \varphi(\zeta)\right)^{2(1-p)}}{|x - \xi|} \left(\log\left(1 + \frac{|x - \xi|}{\varphi(z) + \varphi(\zeta)}\right)\right)^{-p}, \quad p = n-1,$$

and

$$Q(x,\xi) = |x - \xi|^{n-p-2}\left(\varphi(z)\varphi(\zeta)\right)^{2-n}, \quad p > n-1.$$

The norm $\langle f \rangle_{p,\partial\Omega}$ with these weights q, Q is equivalent to $\|f\|_{TW_p^1(\mathbb{R}^n \setminus \overline{\Omega})}$. If $p = n-1$, some additional restrictions are imposed on φ (not excluding power cusps). A function f defined on $\partial\Omega$ and vanishing outside a small neighborhood of the origin is in $TW_1^1(\Omega)$ if and only if $\langle f \rangle_{1,\partial\Omega} < \infty$ with

$$0 \leq q(x) \leq \text{const } \varphi'(z),$$

$$Q(x,\xi) = \begin{cases} \left(\varphi(z) + \varphi(\zeta)\right)^{1-n} & \text{for } z, \zeta \in (0,1), \ |z - \zeta| < \varphi(z) + \varphi(\zeta), \\ 0 & \text{otherwise.} \end{cases}$$

Furthermore, the norms $\|f\|_{TW_1^1(\mathbb{R}^n \setminus \overline{\Omega})}$ and $\langle f \rangle_{1,\partial\Omega}$ are equivalent.

The characterization of the space $TW_1^1(\mathbb{R}^n \setminus \overline{\Omega})$ is the same as for Lipschitz domains:

$$TW_1^1(\mathbb{R}^n \setminus \overline{\Omega}) = L_1(\partial\Omega).$$

In all cases mentioned above there exists a bounded extension operator:

$$TW_p^1(\Omega) \to W_p^1(\Omega) \quad \text{and} \quad TW_p^1(\mathbb{R}^n \setminus \overline{\Omega}) \to W_p^1(\mathbb{R}^n \setminus \overline{\Omega}).$$

This operator is linear for $p > 1$ and nonlinear for $p = 1$. One can easily obtain from the above results that the space $TW_p^1(\mathbb{R}^n \setminus \overline{\Omega})$ is continuously embedded into $TW_p^1(\Omega)$ for $p \in [1, \infty)$, $n > 2$, and Ω defined by (2.1). Hence the space of the traces on $\partial\Omega$ of the functions in $W_p^1(\mathbb{R}^n)$ coincides with $TW_p^1(\mathbb{R}^n \setminus \overline{\Omega})$. In particular, this implies that there is a bounded extension operator: $W_p^1(\mathbb{R}^n \setminus \overline{\Omega}) \to W_p^1(\mathbb{R}^n)$.

We now turn to the study of the trace space $TW_p^1(\Omega)$ for another class of Non-Lipschitz domains that may have cusps as shown in Fig. 2, Fig. 3.

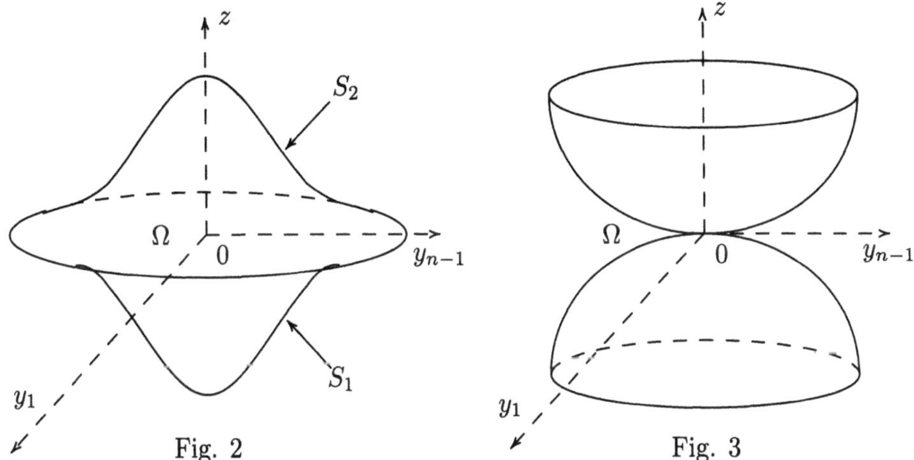

Fig. 2   Fig. 3

Below we give a description of the class of domains. Let G be a domain in $\mathbb{R}^{n-1}$ and let φ_1, φ_2 be Lipschitz continuous functions on $\overline{G}$ such that $\varphi_1 < \varphi_2$ on G and $\varphi_1 = \varphi_2$ on ∂G. We put

$$(4.1) \qquad \Omega = \left\{ x = (y, z) \in \mathbb{R}^n : y \in G, \ z \in \big(\varphi_1(y), \varphi_2(y)\big) \right\}, \quad n \geq 2.$$

In particular, G may coincide with $\mathbb{R}^{n-1}$. In this case Ω is a layer between two Lipschitz graphs.

The lower and upper parts of $\partial\Omega$ are denoted by S_1 and S_2, respectively, i.e.,

$$S_i = \{(y, \varphi_i(y)) : \ y \in G\}, \quad i = 1, 2.$$

Clearly, every function $u \in W_p^1(\Omega)$, $p \in [1, \infty)$, has the traces $u\big|_{S_1}, u\big|_{S_2}$. Moreover, $u\big|_{S_i}$ can be described by Gagliardo's theorem in a neighborhood of any point in S_i, $i = 1, 2$.

Let $p \in (1, \infty)$ and let f be a function on $\partial\Omega$ such that $f\big|_{S_i} \in L_{p,loc}(S_i)$, $i = 1, 2$. Then $f \in TW_p^1(\Omega)$ if and only if the norm on the right part of the following relation is finite, and, moreover,

$$\|f\|_{TW_p^1(\Omega)} \sim \left\{ \sum_{i=1}^{2} \int_{S_i} |f(x)|^p \varphi(y) ds_x \right.$$

$$+ \int_G |f(y, \varphi_2(y)) - f(y, \varphi_1(y))|^p \, \varphi(y)^{1-p} dy$$

$$\left. + \sum_{i=1}^{2} \iint_{\{x, \xi \in S_i : |x-\xi| < a(\varphi(y) + \varphi(\eta))\}} |f(x) - f(\xi)|^p \frac{ds_x ds_\xi}{|x - \xi|^{n+p-2}} \right\}^{1/p},$$

184   Poborchi

where $x = (y, z), \xi = (\eta, \zeta)$, ds_x, ds_ξ are the area elements on $\partial\Omega$, $\varphi = \varphi_2 - \varphi_1$, and a is a positive constant (which can be written explicitly) depending only on n and the Lipschitz constant for φ.

An analogous result holds for $p = 1$:

$$\|f\|_{TW_1^1(\Omega)} \sim \sum_{i=1}^{2} \int_{S_i} |f(x)| \varphi(y) ds_x$$

$$+ \int_G |f(y, \varphi_2(y)) - f(y, \varphi_1(y))| dy$$

$$+ \sum_{i=1}^{2} \iint_{\{x, \xi \in S_i : |x-\xi| < a(\varphi(y)+\varphi(\eta))\}} |f(x) - f(\xi)| \frac{ds_x ds_\xi}{(\varphi(y) + \varphi(\eta))^{n-1}}.$$

Suppose that G in Definition (4.1) is bounded and Lipschitz. Then the space $TW_p^1(\mathbb{R}^n \setminus \overline{\Omega})$ for $p \in (1, \infty)$ is described by the relation

$$\|f\|_{TW_p^1(\mathbb{R}^n \setminus \overline{\Omega})} \sim \left\{ \sum_{i=1,2} \int_{S_i} |f(x)|^p ds_x \right.$$

$$+ \sum_{i=1,2} \iint_{S_i \times S_i} |f(x) - f(\xi)|^p \frac{ds_x ds_\xi}{|x - \xi|^{n+p-2}}$$

$$\left. + \int_G |f(y, \varphi_2(y)) - f(y, \varphi_1(y))|^p \frac{dy}{\operatorname{dist}(y, \partial G)^{p-1}} \right\}^{1/p},$$

and the space $TW_1^1(\mathbb{R}^n \setminus \overline{\Omega})$ coincides with $L_1(\partial\Omega)$.

For the proofs of the above relations for Ω given by (4.1) and for the complementary domain, see Maz'ya, Netrusov, and Poborchi [8], and Maz'ya and Poborchi [13, Chap. 6].

References

[1] ADAMS, R. A.: *Sobolev Spaces*, Academic Press, New York, 1975

[2] BURENKOV, V. I.: On a method of extension of differentiable functions, *Trudy Mat. Inst. Steklov.* **140** (1976), 27–67 (Russian)

[3] FAIN, B. L.: On smooth preserving extension of functions in Sobolev spaces for irregular domains, *Dokl. Akad. Nauk SSSR* **285** (1985), 296–301 (Russian)

[4] GAGLIARDO, E.: Caratterizzazioni delle tracce sulla frontiera relative ad alcune classi di funzioni in più variabili, *Rend. Sem. Mat. Univ. Padova* **27** (1957), 284–305

[5] JONES, P. W.: Quasiconformal mappings and extendability of functions in Sobolev spaces, *Acta Math.* **147** (1981), 71–88

[6] MAZ'YA, V. G.: Classes of regions and embedding theorems for function spaces, *Dokl. Akad. Nauk SSSR* **133** (1960), 527–530 (Russian). English translation: *Soviet Math.* **1** (1960), 882–885.

[7] ______: *Sobolev Spaces*, Springer-Verlag, Berlin New York, 1985

[8] MAZ'YA, V. G., NETRUSOV, YU. V. and POBORCHI, S. V.: Boundary values of functions in Sobolev spaces for some non-Lipschitz domains, *Preprint* LiTH-MAT-R-97-02, Linköping University, 1997. To appear in *Algebra and Analysis*

[9] MAZ'YA, V. G. and POBORCHI, S. V.: Extension of functions in Sobolev classes to the exterior of a domain with the vertex of a peak on the boundary I, *Czech. Math. Journ.* **36**:111 (1986), 634–661 (Russian)

[10] ______: Extension of functions in Sobolev classes to the exterior of a domain with the vertex of a peak on the boundary II, *Czech. Math. Journ.* **37**:112 (1987), 128–150 (Russian)

[11] ______: On traces of functions with summable gradient in a domain with the vertex of a peak on the boundary, *Matem. Zametki* **45**:1 (1989), 57–65 (Russian)

[12] ______: Traces of functions in Sobolev spaces on the boundary of a domain with a peak, in *Sovremennye problemy geometrii i analiza, Trudy Instituta Matematiki* **14** 182–208, Nauka, Novosibirsk, 1989 (Russian). English translation: *Siberian Adv. Math.* **1**:3 (1991), 75–101

[13] ______: *Differentiable Functions on Bad Domains*, World Scientific, Singapore, 1997

[14] WHITNEY, H.: Analytic extensions of differentiable functions defined in closed sets, *Trans. Amer. Math. Soc.* **36** (1934), 63–89

[15] ______: Functions differentiable on the boundaries of regions, *Ann. of Math.* **35**:3 (1934), 482–485

[16] YAKOVLEV, G. N.: *Dokl. Akad. Nauk SSSR* **140**:1 (1961), 73–76 (Russian)

[17] ______: The Dirichlet problem for a domain with non-Lipschitz boundary, *Differ. Uravn.* **1**:8 (1965), 1085–1098 (Russian)

Research Institute for Mathematics and Mechanics
St.Petersburg State University
Bibliotechnaya pl.2
St.Petersburg, Peterhoff,
198904 Russia,
Sergei.Poborchi@paloma.spbu.ru

1991 Mathematics Subject Classification: Primary 31-02, 35-02

Submitted: 01.02.1999

Operator Theory:
Advances and Applications, Vol. 109
© 1999 Birkhäuser Verlag Basel/Switzerland

Extension theorems for Sobolev spaces

Victor Burenkov

Abstract: This paper contains a short survey of extension theorems for Sobolev spaces (leaving aside various variants and generalizations of Sobolev spaces) with emphasis on the estimates for the minimal norm of an extension operator and on extensions with deterioration of properties for degenerate domains.

1. Introduction

I have known Professor V.G. Maz'ya for more than 30 years. Unfortunately we never had the opportunity to work together. However, both of us were well aware of the activities and the research of each of us, and supported each other throughout the years. I highly appreciate his wide basic approach to analysis and differential equations which leads to obtaining excellent results in many directions. Below some results of V.G. Maz'ya and his co-authors related to the extension theory will be discussed in the framework of a survey dedicated to extension theorems for Sobolev spaces.

Let Ω be an open set, $l \in \mathbb{N}$, $1 \leq p \leq \infty$. We shall discuss the problem of extension for the Sobolev spaces $W_p^l(\Omega)$ of all functions $f \in L_p(\Omega)$ for which the weak gradient $\nabla_l f = \{D^\alpha f\}_{|\alpha|=l}$ exists on Ω and

$$\|f\|_{W_p^l(\Omega)} = \|f\|_{L_p(\Omega)} + \|\nabla_l f\|_{L_p(\Omega)} < \infty.$$

Here $\|\nabla_l f\|_{L_p(\Omega)} = \left(\int_\Omega |\nabla_l f|^p \, dx\right)^{1/p}$ and $|\nabla_l f| = \left(\sum_{|\alpha|=l} |D^\alpha f|^2\right)^{1/2}$.

We say that Ω belongs to the class EW_p^l if there exists a bounded linear extension operator

$$E : W_p^l(\Omega) \to W_p^l(\mathbb{R}^n),$$

i.e., $Ef|_\Omega = f$ for all $f \in W_p^l(\Omega)$.

For all the open sets $\Omega \subset R^n$ which will be considered below, the norm $\|f\|_{W_p^l(\Omega)}$ is equivalent to the norm

$$\|f\|_{V_p^l(\Omega)} = \sum_{k=0}^{m} \|\nabla_k f\|_{L_p(\Omega)}.$$

Hence $EW_p^l = EV_p^l$. However results related to the estimates for the norms of extension operators depend essentially on the choice of the norm.

We shall also consider the semi-normed Sobolev spaces $w_p^l(\Omega)$ of all functions $f \in L_1^{loc}(\Omega)$, for which

$$\|f\|_{w_p^l(\Omega)} = \|\nabla_l f\|_{L_p(\Omega)} < \infty.$$

2.　Extensions with preservation of class

A domain $\Omega \subset \mathbb{R}^n$ is an ε, δ-domain, where $\varepsilon \in (0, \infty)$, $\delta \in (0, \infty]$, if any two points $x, y \in \Omega$ such that $|x - y| < \delta$, can be joined by a rectifiable arc $\gamma \subset \Omega$ satisfying the inequalities:

$$l(\gamma) \leq \frac{|x - y|}{\varepsilon}, \quad \mathrm{dist}\,(z, \partial\Omega) \geq \varepsilon \frac{|x - z| \cdot |z - y|}{|x - y|},$$

where $l(\gamma)$ is the length of γ and z is an arbitrary point in γ.

We shall write $\Omega \in \mathcal{A}$ if Ω is an ε, δ-domain for some $\varepsilon, \delta \in (0, \infty)$.

One can easily verify that each domain Ω with a Lipschitz ($\equiv Lip\,1$) boundary belongs to $\mathcal{A}$ and that for each $0 < \gamma < 1$ there exist domains with $Lip\,\gamma$-boundaries such that $\Omega \notin \mathcal{A}$.

If $n = 2$, then $\Omega \in \mathcal{A}$ if, and only if, $\partial\Omega$ is a quasicircle , i.e., the image of a circle under a quasiconformal map of the plane onto itself.

Theorem 2.1. *Let $l \in \mathbb{N}$, $1 \leq p \leq \infty$.*
1. $\mathcal{A} \subset EW_p^l$.
2. If $\Omega \in \mathcal{A}$, then there exists a bounded linear extension operator

$$E : W_p^l(\Omega) \to W_p^l(\mathbb{R}^n) \cap C^\infty(\mathbb{R}^n \setminus \overline{\Omega})$$

such that
(2.1) $$\|\,\mathrm{dist}\,(x, \partial\Omega)^{|\alpha|-l} D^\alpha(Ef)\|_{L_p(\mathbb{R}^n \setminus \overline{\Omega})} \leq c\,\|f\|_{W_p^l(\Omega)},$$

where $|\alpha| > l$ and $c > 0$ is independent of f. The exponent $|\alpha| - l$ is sharp.
Next suppose that $n = 2$ and Ω is a simply connected domain.
3. If $p = 2$, then $\mathcal{A} = EW_2^1$.
4. If $p \neq 2$, then $\mathcal{A} \neq EW_p^1$.

Many authors have contributed to this theorem. Statement 1 : independently V.M. Babič [1] and S.M. Nikol'skii [41] ($1 \leq p \leq \infty$, Ω with a C^l-boundary); A.P. Calderón [24] ($1 < p < \infty$, Ω with a Lipschitz boundary); E.M. Stein [46] ($1 \leq p \leq \infty$, Ω with a Lipschitz boundary); P.W. Jones [28] ($1 \leq p \leq \infty$, $\Omega \in \mathcal{A}$). Statement 2 : V.I. Burenkov [5, 6] (Ω with a Lipschitz boundary); V.I. Burenkov and E.M. Popova [22] ($\Omega \in \mathcal{A}$); E.M. Popova [45] proved that in (2.1) the factor $\mathrm{dist}(x, \partial\Omega)^{|\alpha|-l}$ cannot be replaced by $\mathrm{dist}\,(x, \partial\Omega)^{|\alpha|-l}\mu(\mathrm{dist}\,(x, \partial\Omega))$, where μ is a positive non-decreasing function on $(0, \infty)$ such that $\lim_{t \to 0+} \mu(t) = 0$. Statement 3 : S.K. Vodop'yanov, V.M. Gol'dshtein and T.G. Latfullin [47] (see also the survey paper by S.K. Vodop'yanov, V.M. Gol'dshtein and Yu.G. Reshetnyak [48]); V.M. Gol'dshtein [27] proved that if $n = 2$ and both Ω and $\mathbb{R}^2 \setminus \overline{\Omega}$ are in EW_p^1, then $\Omega \in \mathcal{A}$. Statement 4: V.G. Maz'ya [33]. The domain $\Omega \subset \mathbb{R}^2$, which is such that $\Omega, \mathbb{R}^2 \setminus \overline{\Omega} \in \mathcal{A}$ and $\Omega \in EW_p^1$ for $1 \leq p < 2$, $\mathbb{R}^2 \setminus \overline{\Omega} \in EW_p^1$ for $2 < p \leq \infty$, is presented below:

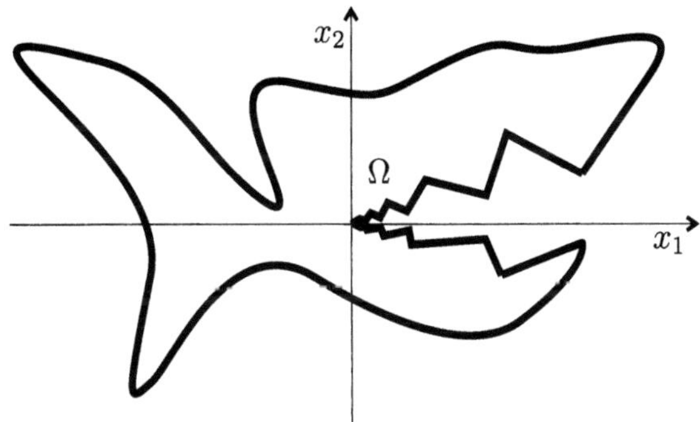

"Jaws". This figure is taken from the cover
of the Russian edition of the book
"Sobolev spaces" by V.G. Maz'ya [33].

The domains $\Omega_+ = \{x \in \Omega, x_2 > 0\}$ and $\Omega_- = \Omega \setminus \overline{\Omega}_+$ are quasiisometric images of the domain G, which is the difference of the rectangle $R = \{x \in \mathbb{R}^2, -1/3 < x_1 < 1, 0 < x_2 < 1/3\}$ and $T = \bigcup_{k=1}^{\infty} t_k$, where t_k are closed isosceles right triangles, whose hypotenuses are the segments $[2^{-k}, 2^{-k+1}]$.

We also note that the description of Ebv and EBV, where $bv(\Omega)$ is the space of functions $f \in L_1^{loc}(\Omega)$, whose distributional derivatives are finite measures on Ω, and $BV(\Omega) = L_1(\Omega) \cap bv(\Omega)$, which, in a certain sense, are close to $w_1^1(\Omega)$, $W_1^1(\Omega)$ respectively, were obtained by Yu.D. Burago and V.G. Maz'ya [3, 33]. (See Theorem 3.1 below.)

From the results of H. Whitney [49] it follows that if $\Omega \subset \mathbb{R}^n$ is a domain satisfying the inequality $d_\Omega(x, y) \leq c|x - y|$, where $c > 0$ is independent of x and y, and $d_\Omega(x, y)$ is the intrinsic metric in Ω, i.e., the infimum of the lengths of arcs in Ω joining x to y, then $\Omega \in EW_\infty^l$, $l \in \mathbb{N}$. For planar bounded finitely connected domains the converse holds. This was proved by V.N. Konovalov [30, 31]. (See also N. Zobin [50].) In [50] it is also shown that for $n > 2$ the converse is not valid.

I would like to emphasize that the description of EW_p^l for all admissible values of n, l and p is still an open (and a difficult) problem, and it is a **challenge** for those who work in the theory of function spaces. I, personally, promise to pay 1,000 DM to the mathematician who will give a complete solution of this problem.

Theorem 2.2. *Let $l \in \mathbb{N}, 1 \le p \le \infty$ and let Ω be an open set with a Lipschitz boundary.*

1. There exists $\delta > 0$ and a bounded linear extension operator

$$(2.2) \qquad\qquad E : w_p^l(\Omega) \to w_p^l(\Omega^\delta).$$

(In general, Ω^δ, the δ-neighbourhood of Ω, cannot be replaced by $\mathbb{R}^n$.)

2. If Ω is bounded, then there exists a linear extension operator

$$(2.3) \qquad\qquad E : w_p^l(\Omega) \to w_p^l(\mathbb{R}^n)$$

(in general, unbounded).

3. If Ω is a bounded domain, then there exists a bounded linear extension operator (2.3).

Theorem 2.2 is proved by the author [7, 8, 10].

A detailed exposition of the extension theory for Sobolev spaces on open sets with Lipschitz boundaries, containing the proofs of all statements related to the extension procedure, is given in [13, 14].

3. Estimates for the minimal norm of an extension operator

The problem is to estimate, in the simplest cases to find,

$$(3.1) \quad \inf_E \|E\|_{W_p^l(\Omega) \to W_p^l(\mathbb{R}^n)}, \quad \inf_E \|E\|_{V_p^l(\Omega) \to V_p^l(\mathbb{R}^n)} \quad \text{or} \quad \inf_E \|E\|_{w_p^l(\Omega) \to w_p^l(\mathbb{R}^n)},$$

where infinum is taken with respect to all extension operators $E : W_p^l(\Omega) \to W_p^l(\mathbb{R}^n)$, $E : V_p^l(\Omega) \to V_p^l(\mathbb{R}^n)$, $E : w_p^l(\Omega) \to w_p^l(\mathbb{R}^n)$ respectively.

The value of $\inf_E \|E\|_{\widetilde{W}_2^1(B_1) \to \widetilde{W}_2^1(\mathbb{R}^n)}$, where $\|f\|_{\widetilde{W}_2^1(\Omega)} = \left(\int_\Omega (|f|^2 + |\nabla f|^2)\, dx\right)^{1/2}$ and B_1 is the unit ball in $\mathbb{R}^n$, was evaluated by S.G. Mikhlin [40].

Theorem 3.1.

$$\inf_E \|E\|_{bv(\Omega) \to bv(\mathbb{R}^n)} = 1 + \sup_{G \subset \Omega} \frac{\tau_\Omega(G)}{P_\Omega(G)},$$

where $P_\Omega(G)$ is the perimeter of G with respect to Ω, i.e., $P_\Omega(G) = \|\chi_{G \cap \Omega}\|_{bv(\Omega)}$, where $\chi_{G \cap \Omega}$ is the characteristic function of $G \cap \Omega$, and

$$\tau_\Omega(G) = \inf_{H : H \cap \Omega = G} P_{\mathbb{R}^n \setminus \Omega}(H).$$

Corollary 3.2.

$$\inf_E \|E\|_{bv(B_1) \to bv(\mathbb{R}^n)} = \frac{n v_n}{2 v_{n-1}},$$

where $v_n = \operatorname{meas} B_1$.

Theorem 3.1 was proved by Yu.D. Burago and V.G. Maz'ya [3].

We note that, in general, for $l > 1$, the quantities (3.1) have different behaviour. Thus in the one-dimensional case

$$\inf_{E} \|E\|_{V_p^l(a,b)\to V_p^l(-\infty,\infty)} \asymp 1 + (b-a)^{-1/p}$$

(V.G. Maz'ya and S.V. Poborchi [38]), i.e., for some $A, B > 0$, which are independent of a and b,

$$A(1 + (b-a)^{-1/p}) \leq \inf \|E\|_{V_p^l(a,b)\to V_p^l(-\infty,\infty)} \leq B(1 + (b-a)^{-1/p}),$$

whilst

$$\inf_{E} \|E\|_{W_p^l(a,b)\to W_p^l(-\infty,\infty)} \asymp 1 + (b-a)^{-l+1-1/p}$$

(V.I. Burenkov and A.L. Gorbunov [20]).

Theorem 3.3. 1. *There exist $c_1, c_2 > 1$ such that for all $l \in \mathbb{N}$, $1 \leq p \leq \infty$*

$$c_1^l(1 + l^l(b-a)^{-l+1-1/p}) \leq \inf_{E} \|E\|_{W_p^l(a,b)\to W_p^l(-\infty,\infty)} \leq c_2^l(1 + l^l(b-a)^{-l+1-1/p}).$$

2. *If $(a,b) = (0,\infty)$, then for all $l \in \mathbb{N}$, $1 \leq p \leq \infty$*

$$(3.2) \qquad 0.3 \cdot 2^l \, l^{-\frac{1}{2p}} \leq \inf_{E} \|E\|_{W_p^l(0,\infty)\to W_p^l(-\infty,\infty)} \leq 8^l.$$

(If $l - 1$, then $\inf_E \|E\|_{W_p^1(0,\infty)\to W_p^1(-\infty,\infty)} = 2^{1/p}$.)

For $b - a \geq 1$ statement 1 was established by the author [9], it's proof for the general case is given in [20]. The second inequality (3.2) is proved in [20], the first inequality (3.2) was obtained by V.I. Burenkov and G.A. Kalyabin [21, 13].

The multidimensional generalization of Theorem 3.3 has the following form [19, 20].

Theorem 3.4. 1. *Let $\Omega \subset \mathbb{R}^n$ be a bounded open set with a Lipschitz boundary. Then there exist $c_3(\Omega), c_4(\Omega) > 0$ such that for all $l \in \mathbb{N}$, $1 \leq p \leq \infty$*

$$c_3(\Omega)^l l^l \leq \inf_{E} \|E\|_{W_p^l(\Omega)\to W_p^l(\mathbb{R}^n)} \leq c_4(\Omega)^l l^l.$$

2. *Let $\Omega = \{x \in \mathbb{R}^n : x_n < \varphi(x_1,\ldots,x_{n-1})\}$, where the function φ satisfies a Lipschitz condition with the constant M. Then there exist $c_5(M), c_6(M) > 1$ such that*

$$c_5(M)^l \leq \inf_{E} \|E\|_{W_p^l(\Omega)\to W_p^l(\mathbb{R}^n)} \leq c_6(M)^l.$$

In the multidimensional case there are only few results in which two-sides estimates of the minimal norm of an extension operator sharp with respect to the geometry of Ω are known. One of them was formulated in Theorem 3.1, another one, formulated below, was obtained by G.A. Kalyabin [29].

Theorem 3.5. *Let $1 < p < \infty$ and let $\Omega \subset \mathbb{R}^2$ be a convex domain such that* $\operatorname{diam} \Omega \leq 1$. *Then*

$$
\inf_E \|E\|_{W_p^1(\Omega) \to W_p^1(\mathbb{R}^n)} \asymp
\begin{cases}
(\operatorname{meas} \Omega)^{-1/p} (\operatorname{diam} \Omega)^{\frac{2-p}{p}}, & 1 < p < 2, \\
(\operatorname{meas} \Omega)^{-1/2} \left(\log \frac{2}{\operatorname{diam} \Omega} \right)^{-1/2}, & p = 2, \\
(\operatorname{meas} \Omega)^{-1/p}, & 2 < p < \infty.
\end{cases}
$$

Next we pass to the case of parameter dependent domains.

M.V. Paukshto [43] has proved that $\inf_E \|E\|_{W_2^1(V_\alpha) \to W_2^1(B_1)} \asymp \alpha^{-1/2}$, where $V_\alpha = \{(r, \varphi) \in \mathbb{R}^2 : 0 < r < 1, \ 0 < \varphi < \alpha\}$, $0 < \alpha < \frac{\pi}{2}$.

V.G. Maz'ya and S.V. Poborchi have carried out a series of investigations of the behaviour of the $\inf_E \|E\|_{V_p^l(\varepsilon\Omega) \to V_p^l(\mathbb{R}^n)}$, where $\Omega \in EV_p^l$ is a bounded domain and, for $\varepsilon > 0$, $\varepsilon\Omega = \{\varepsilon x : x \in \Omega\}$ [38, 39]. Below we formulate some of their results.

We denote

$$
\gamma(\varepsilon) = \inf\{\|f\|_{V_p^l(\mathbb{R}^n)} : f \in V_p^l(\mathbb{R}^n), f = 1 \text{ a.e. on } \varepsilon\Omega\}.
$$

Since $\|Ef\|_{V_p^l(\mathbb{R}^n)} \leq \|E\|_{V_p^l(\varepsilon\Omega) \to V_p^l(\mathbb{R}^n)} \|f\|_{V_p^l(\varepsilon\Omega)}$, it follows that

$$
\|E\|_{V_p^l(\varepsilon\Omega) \to V_p^l(\mathbb{R}^n)} \geq \gamma(\varepsilon)(\operatorname{meas}(\varepsilon\Omega))^{-1/p}.
$$

This simple estimate appeared to be sufficiently sharp and, together with the estimates for $\gamma(\varepsilon)$, it was a base for obtaining estimates below in the cases under consideration. (It was also used in the proof of Theorem 3.5.)

Theorem 3.6. *Let $l \in \mathbb{N}$, $1 \leq p \leq \infty$ and let $\Omega \in EV_p^l$ be a bounded domain. Then*

$$
\inf_E \|E\|_{V_p^l(\varepsilon\Omega) \to V_p^l(\mathbb{R}^n)} \asymp
\begin{cases}
\varepsilon^{-l}, & l < \frac{n}{p}, \\
\varepsilon^{-l} |\log \varepsilon|^{\frac{1-p}{p}}, & l = \frac{n}{p}, \\
\varepsilon^{-\frac{n}{p}}, & l > \frac{n}{p}.
\end{cases}
$$

Theorem 3.7. *Let $l \in \mathbb{N}$, $1 \leq p \leq \infty$, $l < \frac{n}{p} + 1$ and let $\Omega \in EV_p^l$ be a bounded domain. Then*

$$
\inf_E \|E\|_{V_p^l(\varepsilon\Omega) \to V_p^l(\mathbb{R}^n)} \sim \gamma(\varepsilon)(\operatorname{meas}(\varepsilon\Omega))^{-1/p}
$$

as $\varepsilon \to 0+$.

Theorem 3.8. *Let $\Omega \subset \mathbb{R}^2$ be a simply connected bounded domain with a Lipschitz boundary, containing the origin. Then*

$$
\inf_E \|E\|_{V_2^1(\varepsilon\Omega) \to V_2^1(\mathbb{R}^2)} \sim \left(\frac{2\pi}{\operatorname{meas} \Omega} \right)^{1/2} \varepsilon^{-1} |\log \varepsilon|^{-1/2}
$$

as $\varepsilon \to 0+$.

For $n \geq 3$ a similar statement was obtained, involving the Wiener capacity of Ω.

Finally we list other results of such type, which could be found in the book by V.G. Maz'ya and S.V. Poborchi [39]:

i) estimates for $\inf_E \|E\|_{V_p^l(\mathbb{R}^n \setminus \varepsilon\bar{\Omega}) \to V_p^l(\mathbb{R}^n)}$,

ii) estimates in case of extension with zero boundary conditions,

iii) estimates in case of the interior and of the exterior of a thin cylinder,

iv) estimates in case of some domains, depending on two parameters.

4. Extensions with deterioration of class

Suppose that an open set $\Omega \notin EW_p^l$. Next we discuss what kind of weaker extension property could have the spaces $W_p^l(\Omega)$. There are three main possibilities: existence of bounded extension operators

$$(4.1) \qquad E : W_p^l(\Omega) \to W_p^m(\mathbb{R}^n),$$

where $m < l$,

$$(4.2) \qquad E : W_p^l(\Omega) \to W_q^l(\mathbb{R}^n),$$

where $q < p$, and

$$(4.3) \qquad E : W_p^l(\Omega) \to W_{p,\rho}^l(\mathbb{R}^n),$$

where ρ is a bounded positive measurable function such that $\rho(x) = 1$ for $x \in \Omega$,

$$\|f\|_{W_{p,\varrho}^l(\mathbb{R}^n)} = \|f\|_{L_p(\mathbb{R}^n)} + \|\nabla_l f\|_{L_{p,\varrho}(\mathbb{R}^n)}$$

and $\|g\|_{L_{p,\varrho}(\mathbb{R}^n)} = \|\rho g\|_{L_p(\mathbb{R}^n)}$.

Theorem 4.1. *Let $l \in \mathbb{N}$, $1 \leq p \leq \infty$ and let Ω be an open set with a Lip γ-boundary, where $0 < \gamma < 1$. Then there exists a bounded linear extension operator (4.1), where $m = \gamma l$. The exponent γl is sharp.*

Here $W_p^{\gamma l}(\mathbb{R}^n)$ is the Sobolev-Slobodetskiĭ space, i.e., if $\gamma l \in \mathbb{N}$, then $W_p^{\gamma l}(\mathbb{R}^n)$ is the ordinary Sobolev space, whilst if $\gamma l \notin \mathbb{N}$, then $W_p^{\gamma l}(\mathbb{R}^n)$ is the space of fractional order of smoothness, which we shall define below as a particular case of more general spaces. Theorem 4.1 was proved by the author [5, 6, 11]. For $1 < p < \infty$ it can be also derived from the extension theorems with preservation of smoothness for the anisotropic Sobolev-Slobodetskiĭ spaces $W_p^{l,\gamma l,\dots,\gamma l}(\Omega)$ (for theorems of such type see, for example, the book by O.V. Besov, V.I. Il'in and S.M. Nikol'skiĭ [2]).

Let $1 \leq \theta \leq \infty$, $\sigma \in \mathbb{N}$. We shall say that a function $\lambda \in \Lambda_{\theta,\sigma}$ if it is positive and nondecreasing on $(0,\infty)$ and is such that

i) $\lim_{t \to 0+} \lambda(t) = 0$,

ii) $t^{\sigma - 1/\theta}(\lambda(t))^{-1} \in L_\theta(0,1)$,

iii) $t^{-1/\theta}(\lambda(t))^{-1} \in L_\theta(1,\infty)$.

If $\sigma = 1$, we shall write Λ_θ for $\Lambda_{\theta,1}$.

Furthermore, let $1 \leq p, \theta \leq \infty$, $\sigma \in \mathbb{N}$ and $\lambda \in \Lambda_{\theta,\sigma}$. We shall say that a function $f \in B_{p,\theta}^{\lambda(\cdot),\sigma}(\mathbb{R}^n)$, the Nikol'skiĭ-Besov space with generalized smoothness, if f is measurable on $\mathbb{R}^n$ and

$$\|f\|_{B_{p,\theta}^{\lambda(\cdot),\sigma}(\mathbb{R}^n)} = \|f\|_{L_p(\mathbb{R}^n)} + \left(\int_{\mathbb{R}^n} \left(\frac{\|\Delta_h^\sigma f\|_{L_p(\mathbb{R}^n)}}{\lambda(|h|)} \right)^\theta \frac{dh}{|h|^n} \right)^{1/\theta} < \infty.$$

(If $\theta = \infty$, then the integral should be replaced by the appropriate supremum.) We shall also write $B_{p,\theta}^{\lambda(\cdot)}(\mathbb{R}^n)$ for $B_{p,\theta}^{\lambda(\cdot),1}(\mathbb{R}^n)$ and $H_p^{\lambda(\cdot)}(\mathbb{R}^n)$ for $B_{p,\infty}^{\lambda(\cdot)}(\mathbb{R}^n)$.

If $\lambda(t) = t^l \in \Lambda_{\theta,\sigma}$ ($\Leftrightarrow \sigma > l$), then $B_{p,\theta}^{t^l,\sigma}(\mathbb{R}^n) \equiv B_{p,\theta}^l(\mathbb{R}^n)$ — the standard Nikol'skiĭ-Besov space. (In this case the definition does not depend on $\sigma > l$.) Finally, if $l > 0$, $l \notin \mathbb{N}$, then $W_p^l(\mathbb{R}^n) \equiv B_{p,p}^l(\mathbb{R}^n)$.

The space with generalized smoothness, defined above, could be useful in case of open sets with arbitrarily strong degeneration.

The first question, which arizes in this case is whether the extension with preservation of at least some smoothness exists.

Theorem 4.2. *Let $l \in \mathbb{N}$, $1 \leq q \leq p < \infty$ and let $\Omega \subset \mathbb{R}^n$ be a bounded open set. Then the following are equivalent:*

i) there exist $\nu \in \Lambda_\infty$ and a bounded extension operator $E : W_p^l(\Omega) \to H_q^{\nu(\cdot)}(\mathbb{R}^n)$,

ii) there exists $\nu \in \Lambda_\infty$ such that the operator $E_0 : W_p^l(\Omega) \to H_q^{\nu(\cdot)}(\mathbb{R}^n)$ is bounded, where E_0 is the operator of extending function by 0 outside Ω,

iii) the embedding $W_p^l(\Omega) \subset L_q(\Omega)$ is compact.

Corollary 4.3. *If $q < p$, then for each bounded open set $\Omega \subset \mathbb{R}^n$ there exists $\nu \in \Lambda_\infty$ such that the operator $E_0 : W_p^l(\Omega) \to H_q^{\nu(\cdot)}(\mathbb{R}^n)$ is bounded.*

Corollary 4.4. *For each bounded open set with a continuous boundary there exists $\nu \in \Lambda_\infty$ such that the operator $E_0 : W_p^l(\Omega) \to H_p^{\nu(\cdot)}(\mathbb{R}^n)$ is bounded.*

Theorem 4.2 was established by the author [12]. Corollary 4.3 follows since for all bounded open sets $\Omega \subset \mathbb{R}^n$ the embedding $W_p^l(\Omega) \subset L_q(\Omega)$ is compact. Corollary 4.4 follows since for all bounded open sets $\Omega \subset \mathbb{R}^n$ with a continuous boundary the embedding $W_p^l(\Omega) \subset L_p(\Omega)$ is compact. More delicate conditions ensuring the existence of an extension with preservation of some smoothness follow from the results on the compactness of the embedding $W_p^l(\Omega) \subset L_p(\Omega)$, which were studied in detail by V.G. Maz'ya. (See, for example, [32, 33, 34]).

Theorem 4.2 only states, in appropriate cases, the existence of a function ν, which characterizes smoothness. Certainly, the next question is to find ν explicitly. A result in this direction was obtained by V.I. Burenkov and T.V. Verdiev [23] for the spaces $W_p^{\lambda(\cdot)}(\Omega) \equiv B_{p,p}^{\lambda(\cdot)}(\Omega) \supset W_p^1(\Omega)$.

Let $1 \le p < \infty$, $\lambda \in \Lambda_p$, $\omega \in \Lambda_\infty$, and $\Omega \subset \mathbb{R}^n$ be an open set with a H^ω-boundary, i.e., locally Ω is a subgraph of a function φ, satisfying $|\varphi(u) - \varphi(v)| \le \omega(|u - v|)$, where u, v are in appropriate parallelepiped $V \subset \mathbb{R}^{n-1}$. (We note that, for $0 < \gamma \le 1$, $\bigcup_{M \ge 0} H^{Mt^\gamma} = Lip\,\gamma$.) Moreover, let ω have a continuous derivative and

$$\lim_{h \to +0} \lambda(\omega(h))^p \frac{\omega(h)}{h\,\omega'(h)} = 0.$$

Define a function ν by

$$\nu(x) := \max\left\{ \lambda(\omega(x)) \left[\frac{\omega(x)}{x\omega'(x)} \right]^{1/p}, \ \lambda(x) \right\}, \quad x > 0.$$

If the functions

$$\lambda^*(x) = \left(\int_x^\infty \lambda(h)^{-p} \frac{dh}{h} \right)^{-1/p}, \quad \nu^*(x) = \left(\int_x^\infty \nu(h)^{-p} \frac{dh}{h} \right)^{-1/p}$$

satisfy the conditions

$$\limsup_{x \to +0} \frac{\int_0^x \lambda^*(h)^{-p}\,dh}{x\lambda^*(x)^{-p}} < \infty, \quad \limsup_{x \to +0} \frac{\int_0^x \nu^*(h)^{-p}\,dh}{x\nu^*(x)^{-p}} < \infty,$$

then it is proved in [23] that $\nu \in \Lambda_p$ and $E_0 : W_p^{\lambda(\cdot)}(\Omega) \to W_p^{\nu(\cdot)}(\mathbb{R}^n)$. In particular, if Ω has a $Lip\,\gamma$-boundary, where $0 < \gamma \le 1$, then $E_0 : W_p^l(\Omega) \to W_p^{\gamma l}(\mathbb{R}^n)$ if $l < 1/p$.

For $n = 2$ and $\Omega = \{(x_1, x_2) \in \mathbb{R}^2 : |x_1| < 1, \ \omega(|x_1|) < x_2 < 1\}$ a similar result was obtained earlier by V.I. Burenkov and W.D. Evans [15, 16] as an example of application of some Hardy-type inequalities for differences.

Theorem 4.5. *Let $l \in \mathbb{N}$, $1 \le p \le \infty$ and let $\Omega \subset \mathbb{R}^n$ be an open set with a $Lip\,\gamma$-boundary, where $0 < \gamma < 1$. Then, for each $q \in [1, q^0)$, where $q^0 = p(1 + pl(1/\gamma - 1))^{-1}$, there exists a bounded linear extension operator (4.2) The exponent q^0 is sharp.*

Theorem 4.6. *Let $l \in \mathbb{N}$, $1 \le p \le \infty$ and let $\Omega \subset \mathbb{R}^n$ be an open set with a $Lip\,\gamma$-boundary, where $0 < \gamma < 1$. Then there exists a bounded linear extension operator (4.3), where $\rho(x) = \mathrm{dist}\,(x, \partial\Omega)^{l(1/\gamma - 1)}$ if $x \notin \Omega$ and $\mathrm{dist}\,(x, \partial\Omega) \le 1$ and $\rho(x) = 1$ if $x \in \Omega$ or $x \notin \Omega$ but $\mathrm{dist}\,(x, \partial\Omega) > 1$. The exponent $l(1/\gamma - 1)$ is sharp.*

Theorems 4.5 and 4.6 are proved by B.L. Fain [26]. Moreover, they are proved for more general open sets $\Omega \subset \mathbb{R}^n$ satisfying a degenerate condition of $\varepsilon - \delta$ type.

Next we consider domains $\Omega \subset \mathbb{R}^n$ with isolated peaks on the boundary. A typical domain with the vertex of an outer peak on the boundary is

$$\Omega = \{(\bar{x}, x_n) \in \mathbb{R}^n : 0 < x_n < 1, \ |\bar{x}| < \varphi(x_n)\}.$$

Suppose that φ is an increasing Lipschitz continuous function on $[0,1]$ such that $\varphi(0) = \lim_{t\to 0+}\varphi'(t) = 0$ and $\varphi(2t) \le c\varphi(t)$, $t \in (0,1/2]$, where c is independent of t. V.G. Maz'ya and S.V. Poborchi [35, 36, 37, 39] have obtained necessary and sufficient conditions on φ ensuring the existence of a bounded linear extension operator (4.2), where $q < p$. In the case, in which $\rho(x)$ depends only on $|x|$ and is non-decreasing in a neighbourhood of the origin, in [35, 36, 37, 39] necessary and sufficient conditions on φ have been established ensuring the existence of a bounded linear extension operator (4.3). Similar results have been obtained for the planar domain with an inner peak $D = \{x \in \mathbb{R}^2 \setminus \overline{\Omega} : |x| < 2\}$.

These results are discussed in detail in the survey by S.V. Poborchi in the same Proceedings, dedicated to Sobolev spaces on domains with cusps.

If $\Omega \subset \mathbb{R}^n$ is a bounded domain with a Lipschitz boundary, then, by Theorem 2.2, there exists a bounded extension operator $E : w_p^l(\Omega) \to w_p^l(\mathbb{R}^n)$, but it is easy to verify that there does not exist an extension operator E, which is bounded both as

$$E : L_p(\Omega) \to L_p(\mathbb{R}^n), \quad E : w_p^l(\Omega) \to w_p^l(\mathbb{R}^n).$$

However, it is possible to construct an extension operator with a certain deterioration of these properties:

$$(4.4) \qquad E : L_p(\Omega) \to L_{p,\rho}(\mathbb{R}^n), \quad E : w_p^l(\Omega) \to w_p^l(\mathbb{R}^n).$$

Theorem 4.7. *Let $l \in \mathbb{N}$, $1 \le p \le \infty$, and let Ω be a bounded domain with a Lipschitz boundary. Moreover, let ρ be a bounded positive measurable function on $\mathbb{R}^n$ such that $\rho(x) = 1$ if $x \in \Omega$.*

An extension operator E, bounded in the sense (4.4), exists if, and only if,

$$\left\| \rho(x_1,\ldots,x_n) \sum_{i=1}^{n}(1 + |x_i|^{l-1}) \right\|_{L_p(\mathbb{R}^n)} < \infty.$$

Theorem 4.7 was proved by author [8, 10].

If $\Omega \subset \mathbb{R}^n$ is unbounded it could happen that there does not exist a bounded extension operator $E : w_p^l(\Omega) \to w_p^l(\mathbb{R}^n)$. This was found out by Yu.S. Nikol'skii [42]. In this connection a question arose about the existence of a bounded extension operator $E : w_p^l(\Omega) \to w_{p,\rho}^l(\mathbb{R}^n)$, where

$$\|f\|_{w_{p,\rho}^l(\mathbb{R}^n)} = \|\nabla_l f\|_{L_{p,\rho}^l(\mathbb{R}^n)}$$

and $\rho(x) = 1$ if $x \in \Omega$.

Theorem 4.8. *Let $l \in \mathbb{N}$, $1 \le p \le \infty$, $m < n$ and let $\Omega = \Omega_m \times \mathbb{R}^{n-m}$, where $\Omega_m \subset \mathbb{R}^m$ is a bounded domain with a Lipschitz boundary. Moreover, let $\rho(x) = \mu(x_1,\ldots,x_m)$, where μ is a bounded positive measurable function on $\mathbb{R}^m$ such that $\mu(x_1,\ldots,x_m) = 1$ if $(x_1,\ldots,x_m) \in \Omega_m$.*

A bounded extension operator

$$E : w_p^l(\Omega) \to w_{p,\rho}^l(\mathbb{R}^n)$$

exists if, and only if,

$$\left\| \mu(x_1, \ldots, x_m) \sum_{i=1}^{m} (1 + |x_i|^{l-1}) \right\|_{L_p(\mathbb{R}^m)} < \infty.$$

Theorem 4.8 was proved by the author and B.L.Fain ([4], $l = 1$, Ω — a strip; [17], $l \in \mathbb{N}$, Ω — a strip; [26], general case).

References

[1] BABIČ, V. M.: On the extension of functions, *Uspekhi Mat. Nauk* **8** (1953), 111–113 (Russian)

[2] BESOV, O. V., IL'IN, V.P. and NIKOL'SKIĬ, S. M.: *Integral representation of functions and embedding theorems*, 1-st ed. – "Nauka", Moscow, 1975 (Russian); 2-nd ed. –"Nauka", Moscow, 1996 (Russian); English transl. of 1-st ed., Vols. 1, 2, Wiley, 1979

[3] BURAGO, YU. D. and MAZ'YA, V. G.: Some problems of the potential theory and function theory for domains with irregular boundaries, *Zap. Nauchn. Semin. Leningr. Otd. Mat. Inst. Steklov* **3** (1967), 1–152 (Russian); English transl. in *Seminars in Math., Steklov Math. Inst., Leningrad* **3** (1969)

[4] BURENKOV, V. I.: *Some properties of classes of differentiable functions in connection with embedding and extension theorems*, Ph.D. thesis, Moscow, Steklov Math. Inst. (1966), 145 pp. (Russian)

[5] ______: On the extension of functions with preservation and with deterioration of the differential properties, *Dokl. Akad. Nauk SSSR* **224** (1975), 269–272 (Russian); English transl. in *Soviet Math. Dokl.* **16** (1975)

[6] ______: On a certain method for extending differentiable functions, *Trudy Mat. Inst. Steklov* **140** (1976), 27–67 (Russian); English transl. in *Proc. Steklov Inst. Math.* **140** (1976)

[7] ______: On the extension of functions with preservation of semi-norm, *Dokl. Akad. Nauk SSSR* **228** (1976), 779–782 (Russian); English transl. in *Soviet Math. Dokl.* **17** (1976)

[8] ______: *Investigation of spaces of differentiable functions with irregular domain*, D.Sc. thesis, Moscow, Steklov Math. Inst. (1982), 312 pp. (Russian)

[9] ______: On estimates of the norms of extension operators, *9-th All-Union school on the theory of operators in function spaces*. Ternopol. Abstracts (1984), 19–20 (Russian)

[10] ______: Extension of functions with preservation of Sobolev semi-norm, *Trudy Mat. Inst. Steklov* **172** (1985), 81–95 (Russian); English transl. in *Proc. Steklov Inst. Math.* **172** (1985)

[11] ———: *Sobolev spaces on domains,* B.G. Teubner, Stuttgart-Leipzig, Teubner-Texte zur Mathematik, Band 137, 1998

[12] ———: Compactness of embeddings for Sobolev and more general spaces and extensions with preservation of some smoothness (to appear)

[13] ———: Extension theory for Sobolev spaces on open sets with Lipschitz boundaries. I (to appear in the Proceedings of the spring school on function spaces, held in Prague in June 1998)

[14] ———: Extension theory for Sobolev spaces on open sets with Lipschitz boundaries. II (to appear)

[15] BURENKOV, V.I. and EVANS, W.D.: Weighted Hardy inequality for differences and the compactness of the embedding for domains with arbitrarily strong degeneration, *Dokl. Akad. Nauk SSSR* **355** (1997), 583–585 (Russian); English transl. in *Dokl. Akad. Nauk SSSR* **55** (1997)

[16] ———: Weighted Hardy-type inequalities for differences and the extension problem for spaces with generalised smoothness, *J. London Math. Soc.* **(2) 57** (1998), 209–230

[17] BURENKOV, V. I. and FAIN, B. L.: On the extension of functions in Sobolev spaces from a strip with deteriorations of class, *Deposited in VINITY Ac.Sci. USSR*, No 2511-74 (1975), 12 pp. (Russian)

[18] BURENKOV, V. I. and GOL'DMAN M. L.: On extension of functions in L_p, *Trudy Mat. Inst. Steklov* **150** (1979), 31–51 (Russian); English transl. in *Proc. Steklov Inst. Math.* **150** (1979)

[19] BURENKOV, V. I. and GORBUNOV A. L.: Sharp estimates for the minimal norm of an extension operator for Sobolev spaces, *Dokl. Akad. Nauk SSSR* **330** (1993), 680–682 (Russian); English transl. in *Soviet Math. Dokl.* **47** (1993)

[20] ———: Sharp estimates for the minimal norm of an extension operator for Sobolev spaces, *Izvestiya Ross. Akad. Nauk* **61** (1997), 1–44 (Russian); English transl. in *Izvestiya: Mathematics* **61**

[21] BURENKOV, V. I. and KALYABIN, G. A.: Lower estimates of the norms of extension operators for Sobolev spaces on the halfline (to appear in Math. Nachr. in 1998)

[22] BURENKOV, V. I. and POPOVA, E. M.: On improving extension operators with the help of the operators of approximation with preservation of the boundary values, *Trudy Mat. Inst. Steklov* **173** (1986), 50–54 (Russian); English transl. in *Proc. Steklov Inst. Math.* **173** (1986)

[23] BURENKOV, V.I. and VERDIEV, T.V. Extension by zero of functions in spaces with generalized smoothness for degenerate domains (to appear in *Trudy Mat. Inst. Steklov*)

[24] CALDERÓN, A. P.: Lebesgue spaces of differentiable functions and distributions, *Proc. Symp. Pure Math.* **IV** (1961), 33–49

[25] FAIN, B. L.: The extension of functions from an infinite cylinder, *Trudy Mat. Inst. Steklov* **140** (1976), 277–284 (Russian); English transl. in *Proc. Steklov Inst. Math.* **140** (1976)

[26] ______: On extension of functions in Sobolev spaces for irregular domains with preservation of the smoothness exponent, *Dokl. Akad. Nauk SSSR* **285** (1985), 296–301 (Russian); English transl. in *Soviet Math. Dokl.* **30** (1985)

[27] GOL'DSHTEIN, V. M.: Extension of functions with first generalized derivatives from planar domains, *Dokl. Akad. Nauk SSSR* **257** (1981), 268–271 (Russian); English transl. in *Soviet Math. Dokl.* **23** (1981)

[28] JONES, P. W.: Quasiconformal mappings and extendability of functions in Sobolev spaces, *Acta Math.* **147** (1981), 71–88

[29] KALYABIN, G. A. The least norm estimates for certain extension operators from convex planar domains, *Conference in Mathematical Analysis and Applications in Honour of Lars Inge Hedberg's Sixtieth Birthday. Abstracts.* Linkoping, Sweden (1996), 55–56.

[30] KONOVALOV, V. N.: Description of traces for some classes of functions of several variables, Preprint 84.21. *Math Inst. Ac. Sci. Ukraine.* Kiev, 1984 (Russian)

[31] ______ A criterion for extension of Sobolev spaces $W_\infty^{(r)}$ on bounded planar domains, *Dokl. Akad. Nauk SSSR* **289** (1986), 36–39 (Russian); English transl. in *Soviet Math. Dokl.* **33** (1986)

[32] MAZ'YA, V. G.: Classes of regions and embedding theorems for function spaces, *Dokl. Akad. Nauk SSSR* **133** (1960), 527–530 (Russian). English transl. in *Soviet Math. Dokl.* **1** (1960)

[33] ______: *Sobolev spaces*, LGU, Leningrad, 1985 (Russian); English transl. , *Springer-Verlag, Springer Series in Soviet Mathematics*, 1985

[34] ______: Classes of domains, measures and capacities in the theory of spaces of differentiable functions, *Contemprorary problems in mathematics. Fundamental directions,* **26.** *Analysis - 3.* VINITI, Moscow (1988), 158–228 (Russian); English transl. in *Encyclopedia Math. Sci.* **26.** Analysis 3. Springer-Verlag (1991), 141–211

[35] MAZ'YA V. G. and POBORCHI S. V.: On extension of functions in Sobolev spaces to the exterior of a domain with a peak vertex on the boundary, *Dokl. Akad. Nauk SSSR* **275** (1984), 1066–1069 (Russian); English transl. in *Soviet Math. Dokl.* **29** (1984)

[36] ______: Extension of functions in Sobolev spaces to the exterior of a domain with a peak vertex on the boundary. I, Czech. Math. J. **36**: 111 (1986), 634–661 (Russian)

[37] ______: Extension of functions in Sobolev spaces to the exterior of a domain with a peak vertex on the boundary. II, Czech. Math. J. **37**: 112 (1987), 128–150 (Russian)

[38] ______: Extension of functions in Sobolev spaces on parameter dependent domains, *Math. Nachr.* **178** (1996), 5–41.

[39] ______: *Differentiable functions on bad domains*, World Scientific Publishing, 1997

[40] MIKHLIN, S. G.: *Konstanten in einigen Ungleichungen der Analysis*, B.G. Teubner, Leipzig, Teubner-Texte zur Mathematik, Band 35, 1981

[41] NIKOL'SKIĬ, S. M.: On the solutions of the polyharmonic equation by a variational method, *Dokl. Akad. Nauk SSSR* **88** (1953), 409–411 (Russian)

[42] NIKOL'SKIĬ, YU. S.: Boundary values of functions from weight classes, *Dokl. Akad. Nauk SSSR* **164** (1965), 503–506 (Russian); English transl. in *Soviet Math. Dokl.* **6** (1965)

[43] PAUKSHTO, M. V.: On the asymptotic behaviour of the best constant of the extension of functions from the angle, *Bull. Acad. Polon. Sci. Sér. Sci. Mat.* **30** : 1–2 (1982), 79–83

[44] POPOVA, E. M.: Supplements to the theorem of V.I. Burenkov on approximation of functions in Sobolev spaces with preservation of boundary values, In: *"Differential equations and functional analysis"*, Peoples' Friendship University of Russia, Moscow (1985), 86–98 (Russian)

[45] ______: On improving extension operators with the help of the operators of approximation with preservation of the boundary values, In: *"Function spaces and applications to differential equations"*, Peoples' Friendship University of Russia, Moscow (1992), 154–165 (Russian)

[46] STEIN, E. M.: *Singular integrals and differentiability properties of functions*, Princeton Univ. Press, 1970.

[47] VODOP'YANOV S. K., GOL'DSHTEIN V. M. and LATFULLIN, T. G.: A criterion for extension of functions in class L_2^1 from unbounded planar domains, *Sibirsk. Mat. Zh.* **34** (1979), 416–419 (Russian); English transl. in *Siberian Math. J.* **34** (1979)

[48] VODOP'YANOV S. K., GOL'DSHTEIN and RESHETNYAK YU. G.: On geometric properties of functions with first generalized derivatives, *Uspekhi Mat. Nauk* **34** (1979), no. **1 (205)**, 3–74 (Russian); English transl. in *Russian Math. Surveys* **20** (1985).

[49] WHITNEY, H.: Analyticic extension of differentiable functions defined in closed sets, *Trans. Amer. Math. Soc.* **36** (1934), 63–89

[50] ZOBIN, N. Whitney's problem: extendability of functions and intrinsic metric, *C. R. Ac. Sci. Paris* **320**: 1 (1995), 781–786

School of Mathematics
Cardiff University
23 Senghennydd Road
Cardiff, CF2 4YH, United Kingdom,
burenkov@cardiff.ac.uk

1991 Mathematics Subject Classification: Primary 31-02, 35-02

Submitted: 10.02.1999

Operator Theory:
Advances and Applications, Vol. 109
© 1999 Birkhäuser Verlag Basel/Switzerland

Contributions of V.G. Maz'ya to analysis of singularly perturbed boundary value problems

A.B. Movchan

1.   Introduction

It is a great pleasure and honour for me to contribute this article to the volume published on the occasion of the 60th birthday of Prof. V.G. Maz'ya. The objective of this paper is to give a review of contributions of Maz'ya and his colleagues to the development and justification of the method of compound asymptotic expansions. This work created a broad and powerful asymptotic theory, which opened new perspectives in the study of fields in singularly perturbed domains.

Formulations of singularly perturbed problems involve a small parameter, say ε, in such a way that the problem degenerates in some sense as $\varepsilon \to 0$: for example, the equation reduces its order or a part of the boundary reduces its dimension. This article deals with boundary value problems posed in singularly perturbed domains, whose boundaries may include blunted angles, cones and edges, small holes, narrow slits, thin bridges etc. Singular perturbations associated with differential equations containing a small parameter near the high-order derivatives are not considered here. The readers who are interested in analysis of the latter type of problems are referred to the classical work of Vishik and Lyusternik [28] and to the book of Nayfeh [26].

The singularly perturbed problems are difficult to treat numerically, and the standard numerical methods fail to produce satisfactory accuracy. On the other hand, mathematical statements connected with perturbations of domains occur naturally in electromagnetism, hydrodynamics and mechanics of solids. Therefore, singularly perturbed problems attract attention of pure and applied mathematicians as well as engineers. Examples of formal algorithms and applications can be found in the books on asymptotic analysis by Van Dyke [27] and Cole [4]. However, no mathematical justification did exist at that time for asymptotic approximations of fields in singularly perturbed domains. The asymptotic solutions, which existed in the literature, were related to particular examples of singular perturbations, and they could be considered as pieces of art rather than applications of a unified technique.

Such a technique had been developed by Maz'ya, Nazarov and Plamenevskii starting from 1979 (see [16]). Their results are summarized in the two-volume monograph [23]. Algorithms presented in that book develop multi-scale compound

asymptotic expansions and give a rigorous approach to asymptotic approximations of solutions to boundary value problems posed in domains with singularly
perturbed boundaries.

In what follows I shall emphasize main ideas of compound asymptotic expansions
but shall mention only a few of Maz'ya and his colleagues' numerous applications of
this technique. The interested reader can consult the book [23][1], the forthcoming
monograph [11] and the original articles [8], [9], [13]–[22], [12]–[25].

2.   Domain with a small hole

To give an idea of the method of compound asymptotic expansions developed by
Maz'ya, Nazarov and Plamenevskii, I consider a simple example of a singularly
perturbed domain shown in Fig. 1.

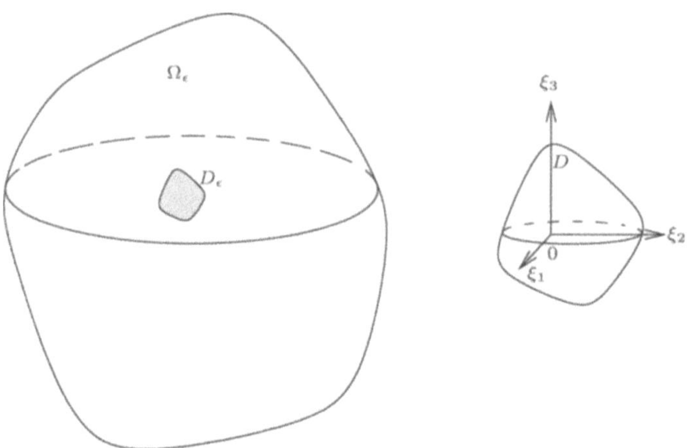

Figure 1: Domain with a small hole, and the scaled region $\mathbb{R}^3 \setminus \overline{D}$.

Let Ω and D be two bounded domains with sufficiently smooth boundaries $\partial\Omega$
and ∂D. Assume that the origin O is the interior point for both Ω and D. We
introduce a small positive parameter ε and define the sets D_ε and Ω_ε by

$$D_\varepsilon = \{\mathbf{x} : \varepsilon^{-1}\mathbf{x} \in D\}, \quad \Omega_\varepsilon = \Omega \setminus \overline{D}_\varepsilon,$$

where $\overline{D}_\varepsilon$ denotes the closure of D_ε. As $\varepsilon \to 0$, the domain Ω_ε degenerates into a
set Ω with the origin being removed.

I begin with the Dirichlet problem

$$(1) \qquad\qquad \Delta u(\mathbf{x}, \varepsilon) = 0, \quad \mathbf{x} \in \Omega_\varepsilon;$$

____

[1] The two volumes of the original edition have been published in German in 1991 and 1992,
and the English edition is to appear in Birkhäuser Verlag.

$$(2) \qquad u(\mathbf{x}, \varepsilon) = \Phi(\mathbf{x}), \quad \mathbf{x} \in \partial\Omega;$$

$$(3) \qquad u(\mathbf{x}, \varepsilon) = \varphi(\varepsilon^{-1}\mathbf{x}), \quad \mathbf{x} \in \partial D_\varepsilon,$$

where Φ and φ are smooth functions defined on $\partial\Omega$ and ∂D respectively. It is straightforward to verify that one cannot neglect the boundary condition (3) and uniformly approximate the field u by the solution of the model problem independent of ε

$$\Delta v(\mathbf{x}) = 0, \quad \mathbf{x} \in \Omega;$$
$$v(\mathbf{x}) = \Phi(\mathbf{x}), \quad \mathbf{x} \in \partial\Omega.$$

Namely, the solution of the boundary layer type, say $w(\varepsilon^{-1}\mathbf{x})$, is required to compensate for an error produced by the field v in the boundary condition (3).

First, let us describe the model problems for a particular case of three dimensions. The leading-order part of the error mentioned is $v(0) - \varphi(\varepsilon^{-1}\mathbf{x})$, and hence the field $w(\boldsymbol{\xi})$ is defined as a solution of the following boundary value problem posed in the scaled domain $\omega = \mathbb{R}^3 \setminus \overline{D}$

$$(4) \qquad \Delta w(\boldsymbol{\xi}) = 0, \ \boldsymbol{\xi} \in \omega; \quad w(\boldsymbol{\xi}) = \varphi(\boldsymbol{\xi}) - v(0), \ \boldsymbol{\xi} \in \partial D,$$

where $\boldsymbol{\xi} = \varepsilon^{-1}\mathbf{x}$. The boundary layer has the asymptotics

$$(5) \qquad w(\boldsymbol{\xi}) \sim \sum_{k=1}^{\infty} \rho^{-k} w^{(k)}(\theta) \ \text{ as } \rho \to \infty;$$

here $\rho = |\boldsymbol{\xi}|$, $\theta = \boldsymbol{\xi}|\boldsymbol{\xi}|^{-1}$, and $w^{(k)}$ are smooth functions on the unit sphere. Hence, being incorporated into the asymptotic approximation, the quantity $v(\mathbf{x}) + w(\varepsilon^{-1}\mathbf{x})$ leaves an error of order $O(\varepsilon)$ in the boundary condition (2) on $\partial\Omega$. Taking into account that the difference $v(\mathbf{x}) - v(0)$ evaluated on ∂D_ε has the order $O(\varepsilon)$, one can use the maximum principle to conclude that $u(\mathbf{x}, \varepsilon) - v(\mathbf{x}) - w(\varepsilon^{-1}\mathbf{x}) = O(\varepsilon)$ in the closure $\overline{\Omega}_\varepsilon$. Proceeding in the same way, one constructs the total asymptotic expansion

$$u(\mathbf{x}, \varepsilon) \sim \sum_{j=0}^{\infty} \varepsilon^j \left(v_j(\mathbf{x}) + w_j(\varepsilon^{-1}\mathbf{x}) \right),$$

where the quantities v_j and w_j satisfy certain model problems independent of the small parameter ε, and they admit the following representations

$$v_j(\mathbf{x}) = \sum_{k=0}^{N-1} r^k v_j^{(k)}(\theta) + O(r^N), \quad \text{as } r \to 0,$$

and

$$w_j(\boldsymbol{\xi}) = \sum_{k=1}^{N-1} \rho^{-k} w_j^{(k)}(\theta) + O(\rho^{-N}), \text{. as } \rho \to \infty;$$

here $r = |\mathbf{x}|$, and the coefficients $v_j^{(k)}$, $w_j^{(k)}$ are smooth functions defined on the unit sphere.

The model problems mentioned have the form

$$(6) \qquad \Delta v_k(\mathbf{x}) = 0, \quad \mathbf{x} \in \Omega; \quad v_k(\mathbf{x}) = -\sum_{j=1}^{k} r^{-j} w_{k-j}^{(j)}(\theta), \quad \mathbf{x} \in \partial\Omega,$$

$$(7) \qquad \Delta w_k(\boldsymbol{\xi}) = 0, \quad \boldsymbol{\xi} \in \omega; \quad w_k(\boldsymbol{\xi}) = -\sum_{j=0}^{k} \rho^{j} v_{k-j}^{(j)}(\theta), \quad \boldsymbol{\xi} \in \partial\omega,$$

and the remainder of the asymptotic approximation is small

$$\mathbf{u}(\mathbf{x}, \varepsilon) - \sum_{j=0}^{N} \varepsilon^{j} \{v_j(\mathbf{x}) + w_j(\varepsilon^{-1}\mathbf{x})\} = O(\varepsilon^{N+1}).$$

The approximation above is called *the compound asymptotic expansion.*

There is an alternative approach of *matched asymptotic expansions* implemented by Il'in [5]–[7]. This method is efficient, as much as the method of compound asymptotic expansions, for the study of the leading-order asymptotics; however, it requires more effort when the total asymptotic expansion of the solution is constructed.

The above example illustrates the method of compound asymptotic expansions in the simplest case when the discrepancy terms occur in the boundary conditions only, and all the model problems are uniquely solvable. One faces a slight complication in analysis of problems involving non-homogeneous operators. To see that, consider equations (1)–(3) where the Laplace equation (1) is replaced by

$$(8) \qquad \Delta u(\mathbf{x}, \varepsilon) - u(\mathbf{x}, \varepsilon) = 0, \quad \mathbf{x} \in \Omega_\varepsilon.$$

The presence of the discrepancy terms in the equation will require some modification of the asymptotic scheme. It will involve *the discrepancy rearrangement* between different model problems. The asymptotic expansion for u includes solutions of Dirichlet boundary value problems for the Poisson equation in $\omega = \mathbb{R}^3 \setminus \overline{D}$ (one needs to construct a boundary layer which decays at infinity) and Dirichlet boundary value problems for the operator $\Delta - 1$ in the domain Ω. After the first iteration one can see that $v_0(\mathbf{x}) + w_0(\varepsilon^{-1}\mathbf{x})$ leaves a discrepancy $-w_0(\varepsilon^{-1}\mathbf{x})$ in equation (8) that cannot be compensated by solutions v_j of model problems in Ω. It is natural to use the functions defined in ω to remove these error terms. The coefficients v_j and w_j satisfy the following model problems

$$\Delta v_j(\mathbf{x}) - v_j(\mathbf{x}) = r^{-1} w_{j-1}^{(1)}(\theta) + r^{-2} w_{j-2}^{(2)}(\theta), \quad \mathbf{x} \in \Omega;$$

$$v_j(\mathbf{x}) = -\sum_{p=1}^{j} r^{-p} w_{j-p}^{(p)}(\theta), \quad \mathbf{x} \in \partial\Omega,$$

and

$$\Delta w_j(\boldsymbol{\xi}) = W_{j-2}(\boldsymbol{\xi}), \quad \boldsymbol{\xi} \in \omega;$$

$$w_j(\boldsymbol{\xi}) = -\sum_{p=0}^{j} \rho^p v_{j-p}^{(p)}(\theta), \quad \boldsymbol{\xi} \in \partial\omega.$$

Here, the functions $v_j^{(p)}$ are similar to the ones in the expansion (7), and $w_j^{(1)}, w_j^{(2)}, W_j$ are the coefficients in the representation of w_j

$$w_j(\boldsymbol{\xi}) = \rho^{-1} w_j^{(1)}(\theta) + \rho^{-2} w_j^{(2)}(\theta) + W_j(\boldsymbol{\xi}),$$

where $W_j(\boldsymbol{\xi}) = O(\rho^{-3})$ as $\rho \to \infty$.

Another complication may occur when model problems, involved in the asymptotic algorithm, have non-trivial kernel and co-kernel. For example, let the Dirichlet condition (2) of the problem (1)–(3) be replaced by

$$(9) \qquad \frac{\partial u}{\partial n}(\mathbf{x}, \varepsilon) = \Psi(\mathbf{x}), \quad \mathbf{x} \in \partial\Omega,$$

where $\partial/\partial n$ denotes the differentiation with respect to the unit outward normal. Then the model problem in Ω

$$(10) \qquad \Delta v(\mathbf{x}) = 0, \quad \mathbf{x} \in \Omega; \quad \frac{\partial v}{\partial n}(\mathbf{x}) = \Psi(\mathbf{x}), \quad \mathbf{x} \in \partial\Omega,$$

has a bounded solution if and only if the right-hand side Ψ satisfies the orthogonality condition

$$(11) \qquad \int_{\partial\Omega} \Psi(\mathbf{x}) ds_{\mathbf{x}} = 0,$$

and then the solution v is defined up to an arbitrary additive constant. The model problem of the boundary layer type has the form (4), and it is uniquely solvable in the class of functions vanishing at infinity (see (5)). On the other hand, the mixed boundary value problem (1), (9), (3) has a solution with finite energy even for the situation when the right-hand side does not satisfy (11). In this case, the model problem (10) may have to be modified to the form

$$(12) \qquad \Delta v(\mathbf{x}) + \beta\delta(\mathbf{x}) = 0, \quad \mathbf{x} \in \Omega; \quad \frac{\partial v}{\partial n}(\mathbf{x}) = \Psi(\mathbf{x}), \quad \mathbf{x} \in \partial\Omega,$$

where

$$\beta = -\int_{\partial\Omega} \Psi(\mathbf{x}) ds_{\mathbf{x}},$$

δ is the Dirac delta function, and the field v is singular (of order $O(|\mathbf{x}|^{-1})$) at the origin. The boundary layer $w(\boldsymbol{\xi})$ is then constructed in such a way that it decays like $O(|\boldsymbol{\xi}|^{-2})$ as $|\boldsymbol{\xi}| \to \infty$, and the additive constant in the representation of the solution of (4) is evaluated from the condition that the first term in the asymptotic expansion (5) vanishes.

It is straightforward to extend the algorithm to the case when the space dimension is greater than 3 – the asymptotic approximation of the solution will have the form similar to the one described above. However, the structure of the asymptotic expansion will have to be modified if a boundary value problem with Dirichlet boundary condition on D_ε is posed in a bounded two-dimensional domain Ω_ε. In particular, for a mixed boundary value problem the outer expansion (in domain Ω) involves solutions of the model problems of the same type as (12), but in contrast with the case of three dimensions the function v has a logarithmic singularity at the origin. As a result, the asymptotic approximation of the field u involves rational functions of $\log \varepsilon$. Namely,

$$u(\mathbf{x}, \varepsilon) = \sum_{j=0}^{N} \varepsilon^j \sum_{k=0}^{j+1} \alpha(\varepsilon)^k (v_{j,k}(\mathbf{x}) + w_{j,k}(\varepsilon^{-1}\mathbf{x})) + O(\varepsilon^{N+1}),$$

where N is a non-negative integer, and the quantity $\alpha(\varepsilon)$ has the form

$$\alpha(\varepsilon) = ((2\pi)^{-1} \log \varepsilon + c)^{-1}, \quad c = \text{const};$$

the coefficients $v_{j,k}$ and $w_{j,k}$ represent solutions of model problems in Ω and ω, respectively.

3. General asymptotic theory by Maz'ya, Nazarov and Plamenevskii

The above examples illustrate the general asymptotic scheme developed by Maz'ya, Nazarov and Plamenevskii [23]. They considered general boundary value problems, elliptic in the Douglis-Nirenberg sense, in domains singularly perturbed at a finite number of interior or boundary points. Coefficients of differential operators admit multi-scaled asymptotic expansions. The main result is an extensive asymptotic theory: complete asymptotic expansions are constructed with coefficients obtained from solving model problems in finite or unbounded (defined in scaled coordinates) domains independent of the small parameter. It is worth noting that, to a large extent, the progress became possible because in 1960s–1970s a theory of elliptic boundary value problems in unbounded domains and domains with point singularities was developed by Kondrat'ev, Maz'ya, Plamenevskii et al.

The construction of the asymptotic algorithms uses *the index theorem*: the function spaces for solutions and right-hand sides of model problems can be chosen in such a way that

$$(13) \qquad\qquad \text{ind}\{\mathcal{L}\} = \sum_{k} \text{ind}\{\mathcal{L}^{(k)}\},$$

where $\mathcal{L}$ and $\mathcal{L}^{(k)}$ denote the operators of the singularly pertubed problem and the operators of model problems, respectively; $\operatorname{ind}\{\mathcal{L}\} = \dim\ker\{\mathcal{L}\} - \dim\operatorname{coker}\{\mathcal{L}\}$. If, for example, the operator $\mathcal{L}$ is invertible then, in accordance with (13), the number of unknown constants in solutions of model problems is compensated by the same number of solvability conditions.

The justification of the asymptotic algorithm is highly non-trivial, and the success is achieved by reducing the original singularly perturbed boundary value problem to a matrix operator form, so that the corresponding matrix is written as a sum of the diagonal part $\operatorname{diag}\{\mathcal{L}^{(1)},\dots,\mathcal{L}^{(k)}\}$ and a small (in some sense) matrix. The important part of the algorithm is the procedure of the discrepancy rearrangement between different model problems explained in the previous section.

In the second volume of the book [23], singular perturbations near multi-dimensional manifolds are investigated. Here, by the way, a new effect occurs: model problems arise with infinite dimensional co-kernel. Hence, the solutions of the model problems are not available. The authors overcome ingeniously this difficulty by using approximate solutions of those problems.

A significant part of the work of Maz'ya, Nazarov and Plamenevskii is related to asymptotic analysis of functionals defined on solutions of singularly perturbed problems. They found various asymptotics of the energy integrals, stress intensity factors, capacities and eigenvalues. I do not discuss this area of their research here referring to the book [23].

4. Asymptotics of solutions of boundary integral equations under a small perturbation of a corner

In this section, I want to say a few words about an interesting application of the method of compound asymptotic expansions to the theory of boundary integral equations on singularly perturbed contours. This application can be found in the paper [13] by Maz'ya and R. Mahnke, his former post-graduate student from Rostock.

Let $\Omega \in \mathbb{R}^2$ be a bounded domain with an angular point on its contour $\partial\Omega$. A domain Ω_ε with smooth boundary is obtained by a small (of size ε) perturbation of $\partial\Omega$ near the vertex of the angle. Let ψ be a smooth function in $\mathbb{R}^2$. It is a classical approach to solve the Dirichlet problem

$$\Delta u_\varepsilon = 0 \;\; \text{in} \;\; \Omega_\varepsilon, \qquad u_\varepsilon = \psi \;\; \text{on} \;\; \partial\Omega_\varepsilon,$$

by expressing the function u_ε in the form of a double-layer potential

$$(14) \qquad u_\varepsilon(\mathbf{x}) = \frac{1}{2\pi}\int_{\partial\Omega_\varepsilon}\mu_\varepsilon(\mathbf{y})\frac{\partial}{\partial n_\mathbf{y}}\log|\mathbf{x}-\mathbf{y}|ds_\mathbf{y}.$$

This leads to the well-known boundary integral equation

$$\frac{1}{2}\mu_\varepsilon + T\mu_\varepsilon = \psi.$$

The operator T is the direct value of the double-layer potential (14). The density function μ_ε is expressed in terms of solutions of auxiliary boundary value problems which enables one to obtain its asymptotic representation as $\varepsilon \to 0$. A similar approach was extensively used by Maz'ya in his earlier work on boundary integral equations for piece-wise smooth domains. The following two scale asymptotic representations are the main results of Maz'ya and Mahnke [13]:

$$\mu_\varepsilon(\mathbf{x}) \sim \begin{cases} \big(\psi(\mathbf{x}) - v(\mathbf{x}) - \varepsilon^{\pi/(2\pi-\alpha)} w_+(\mathbf{x}/\varepsilon)\big)\,|_{\partial\Omega_\varepsilon} & \text{if } 0 < \alpha < \pi \\ \big(\psi(\mathbf{x}) - v(\mathbf{x}) - \varepsilon^{\pi/\alpha} w_-(\mathbf{x}/\varepsilon)\big)\,|_{\partial\Omega_\varepsilon} & \text{if } \pi \leq \alpha < 2\pi, \end{cases}$$

where α denotes the size of the angle. The functions v, w_+ and w_- are solutions of certain exterior Neumann problems for model domains. The remainder function can be estimated uniformly in the norm of $C(\partial\Omega_\varepsilon)$ by $O(\varepsilon^\gamma)$ with $\gamma > 1$.

Clearly, this is only an example of a general method for studying the asymptotic behaviour of solutions to boundary integral equations applicable to the multidimensional case and other boundary value problems.

5.   Compound asymptotics for homogenization problems

The technique of compound asymptotic expansions also proved to be useful in the study of homogenization problems. For example, Maz'ya and Nazarov [14] applied this approach to the analysis of the limit passage for solutions of boundary value problems where smooth domains are approximated by polygons. They uncovered the asymptotic character of the famous polygon-circle paradox in the theory of thin elastic plates: when a thin circular plate is approximated by polygons with freely supported edges, the limit solution does not satisfy the free support conditions on the circle. They also showed new instability effects. For details, the reader is referred to Chapter 18 of the book [23].

Another problem of the same nature was considered in a more recent work of Maz'ya and a young mathematician from Rostock M. Hänler [12]. The authors study a smooth bounded convex domain $\Omega \subset \mathbb{R}^2$ and introduce a *porous layer* bounded by two smooth curves Γ_i and Γ_e located at a constant distance, say l, apart from each other (see Fig. 2). The interior of Γ_i is denoted by Ω^i, and the notation Ω^e stands for the domain bounded by $\partial\Omega$ and Γ_e. The porous layer is modelled by a set of N thin strips connecting the domains Ω^i and Ω^e. Let

$$\Omega_N = \Omega^i \cup T_1 \cup \cdots \cup T_N \cup \Omega^e,$$

and let u solve the following boundary value problem

$$\Delta u(\mathbf{x}) = f(\mathbf{x}), \quad \mathbf{x} \in \Omega_N, \quad \text{supp}(f) \subset \Omega^i \cup \Omega^e,$$

$$\frac{\partial u}{\partial n} = 0, \quad \mathbf{x} \in \partial\Omega_N,$$

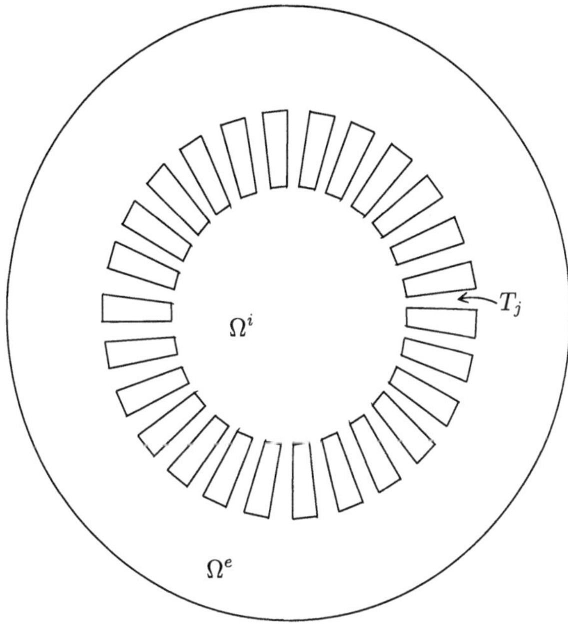

Figure 2: Disintegrating domain.

where the right-hand side f is smooth, and it is assumed to satisfy the orthogonality condition

$$\int_{\Omega_N} f(\mathbf{x})d\mathbf{x} = 0.$$

In terms of physical applications, the quantity u can be interpreted as a temperature, and the above model describes a steady-state heat flow in the heat exchanger. Let U denote the homogenized solution of the problem. Maz'ya and Hänler showed that it satisfies a certain boundary value problem with non-local contact conditions across the interface layer. Namely,

$$\Delta U(\mathbf{x}) = f(\mathbf{x}), \quad \mathbf{x} \in \Omega^i \cup \Omega^e,$$

$$\frac{\partial U}{\partial n}(\mathbf{x}) = \alpha_i(\mathbf{x})(U(\mathbf{x}) - U(\mathbf{x} + l\mathbf{n})), \quad \mathbf{x} \in \Gamma_i,$$

$$\frac{\partial U}{\partial n}(\mathbf{x}) = \alpha_e(\mathbf{x})(U(\mathbf{x} - l\mathbf{n}) - U(\mathbf{x})), \quad \mathbf{x} \in \Gamma_e,$$

$$\frac{\partial U}{\partial n}(\mathbf{x}) = 0, \quad \mathbf{x} \in \partial\Omega.$$

Here, $\mathbf{n}$ is the unit outward normal with respect to Ω^i, l is the constant thickness of the porous layer, and the transmission coefficients α_i and α_e are defined from the analysis of the asymptotic expansions of u in the thin strips and expansions of the field inside Ω^i and Ω^e.

I want to conclude this section by mentioning the work of Maz'ya and Slutskii [24] which is also included in Chapter 19 of the second volume of the book [23]. The paper [24] presents analysis of ordinary differential equations on an ε-periodic curvilinear mesh, which leads to a solution of an elliptic partial differential equation as the main (homogenized) term in the complete asymptotic expansion. The difference equations in a lattice of springs are studied in the work [25], where total asymptotic expansions of solutions are constructed, and the equations of elasticity are derived as a result of homogenization.

6.　Boundary value problems in 3D-1D multi-structures

I finish this survey article with the topic of our joint work with V.A. Kozlov and V.G. Maz'ya (see the monograph [11]). This topic is the asymptotic theory of fields in multi-structures, i.e. domains dependent on a small parameter ε in such a way that the limit region, as $\varepsilon \to 0$, consists of subsets of different space dimensions.

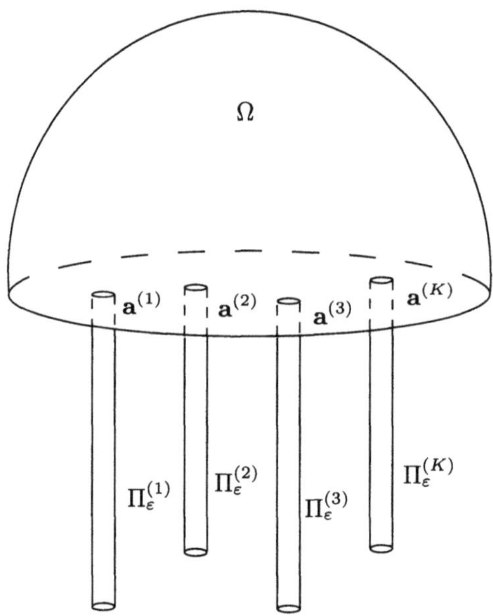

Figure 3: The 3D-1D multi-structure investigated in [8], [9].

The mathematical study of fields in multi-structures was initiated by Ciarlet and Destuynder [3], and a wide range of results obtained by a variational approach are presented in the comprehensive monograph by Ciarlet [2]. Kozlov, Maz'ya and Movchan [8]-[11] used another approach based upon the method of compound asymptotic expansions and developed an asymptotic theory of boundary value problems in multi-structures with junctions of one-dimensional and three-dimensional (3D-1D) elements (see Fig. 3); until recently, the question of modelling of 3D-1D junctions was open. We also present a spectral analysis of boundary value problems in multi-structures obtaining explicit asymptotic formulae for the first eigenvalues and corresponding eigenfunctions.

References

[1] CIARLET, P.G.: *Plates and junctions in elastic multi-structures*, Masson, Paris, 1990.

[2] CIARLET, P.G.: *Mathematical elasticity. Volume II: theory of plates*, Elsevier, 1997.

[3] CIARLET, P.G. and DESTUYNDER, P.: A justification of the two-dimensional plate model, *J. Mécanique* **18** (1979), 315–344.

[4] COLE, J.D.: *Perturbation methods in applied mathematics*, Blaisdell, Waltham, 1968.

[5] IL'IN, A.M.: Boundary value problem for a second order elliptic differential equation in a domain with a slit, I. Two-dimensional case, *Matem. Sbornik* **99** (1976), No 4, 514–537 (in Russian).

[6] IL'IN, A.M.: Boundary value problem for a second order elliptic differential equation in a domain with a thin slit, II. Domain with a small hole, *Matem. Sbornik* **103** (1977), No 2, 265–284 (in Russian).

[7] IL'IN, A.M.: *Matching of asymptotic expansions of solutions of boundary value problems*, Translations of Mathematical Monographs, 102, American Mathematical Society, Providence, 1992.

[8] KOZLOV, V.A., MOVCHAN, A.B. and MAZ'YA, V.G.: Asymptotic analysis of a mixed boundary value problem in a multi-structure, *Asymptotic analysis* **8** (1994), 105–143.

[9] KOZLOV, V.A., MOVCHAN, A.B. and MAZ'YA, V.G.: Asymptotic representation of elastic fields in a multi-structure, *Asymptotic analysis* **11** (1995), 343–415.

[10] KOZLOV, V.A., MOVCHAN, A.B. and MAZ'YA, V.G.: Fields in non-degenerate 1D-3D elastic multi-structures, *LiTH-MAT-R-96-14, 1996, Preprint, Linköping University*.

[11] KOZLOV, V.A., MAZ'YA, V.G. and MOVCHAN, A.B.: *Asymptotic analysis of fields in multi-structures*, Oxford University Press (to appear).

[12] MAZ'YA, V.G. and HÄNLER, M.: Approximation of solutions to the Neumann problem in disintegrating domains. *Math. Nachr.* **162** (1993), 261–278.

[13] MAZ'YA, V.G. and MAHNKE, R.: Asymptotics of the solution of a boundary integral equation under a small perturbation of a corner, *Zeitschrift für Analysis und ihre Anwendungen* **11** (1992), No 2, 173–182.

[14] MAZ'YA, V.G. and NAZAROV, S.A.: The limit passage paradox in solutions of boundary value problems in approximation of smooth domains by polygons, *Izv. Acad. Nauk SSSR* **50** (1986), No 6, 1156–1177 (in Russian).

[15] MAZ'YA, V.G. and NAZAROV, S.A.: On the singularities of solutions of the Neumann problem at a conical point. *Siberian Mathem. Journal* **30** (1989), No 3, 52–63 (in Russian).

[16] MAZ'YA, V.G., NAZAROV, S.A. and PLAMENEWSKII, B.A.: On asymptotics of
solutions to boundary value problems with variation of a domain near conical points.
Doklady Acad. Nauk SSSR **249** (1979), No 1, 94–96 (in Russian).

[17] MAZ'YA, V.G., NAZAROV, S.A. and PLAMENEWSKII, B.A.: On asymptotics of solutions of elliptic boundary value problems in singularly perturbed domains. *Problemy
Matem. Analiza* **8** (1981), 72–153 (in Russian).

[18] MAZ'YA, V.G., NAZAROV, S.A. and PLAMENEWSKII, B.A.: On asymptotics of
solutions of the Dirichlet problem in a three-dimensional domain with a thin body
excluded. *Doklady Acad. Nauk SSSR* **256** (1981), No 1, 37–39 (in Russian).

[19] MAZ'YA, V.G., NAZAROV, S.A. and PLAMENEWSKII, B.A.: The absence of the
De Giorgi theorem for elliptic equations with complex coefficients. *Zap. Nauchn.
Seminar. LOMI* **115** (1982), 156–168 (in Russian).

[20] MAZ'YA, V.G., NAZAROV, S.A. and PLAMENEWSKII, B.A.: On the singularities of
solutions of the Dirichlet problem in the exterior of a slender cone. *Matem. Sbornik*
122 (1983), No 4, 435–456 (in Russian).

[21] MAZ'YA, V.G., NAZAROV, S.A. and PLAMENEWSKII, B.A.: Evaluation of the
asymptotic form of the intensity coefficients on approaching corner or conical points.
U.S.S.R. Comput.Maths.Math.Phys. **23** (1983), No 2, 50–58.

[22] MAZ'YA, V.G., NAZAROV, S.A. and PLAMENEWSKII, B.A.: Asymptotic expansions
of eigenvalues of boundary value problems for the Laplace operator in domains with
small holes. *Mathematics Izvestija* **48** (1984), No 2, 347–371 (in Russian).

[23] MAZJA, W.G., NAZAROW, S.A. and PLAMENEWSKII, B.A.: *Asymptotische Theorie elliptischer Randwertaufgaben in singulär gestörten Gebieten.* Akademie-Verlag,
Berlin, B.1, 1991; B.2, 1992.

[24] MAZ'YA, V.G. and SLUTSKII, A.S.: Homogenization of a differential operator on a
fine periodic curvilinear mesh, *Math. Nachr.* **133** (1986), 107–133.

[25] MAZ'YA, V.G. and SLUTSKII, A.S.: Homogenization of difference equations with
rapidly oscillating coefficients, *Seminar Analysis 1986/87, Akad. Wiss. DDR. Inst.
Math.* 63–92, Berlin, 1987.

[26] NAYFEH, A.H.: *Perturbation methods,* New York, Chichester: Wiley, 1973.

[27] VAN DYKE, M.D.: *Perturbation methods in fluid mechanics,* Academic Press, New
York, 1964.

[28] VISHIK, M.I. and LYUSTERNIK, L.A.: Regular perturbation and boundary layer for
differential equations with a small parameter. *Uspekhi Matem. Nauk* **15** (1957), No
3, 3–80 (in Russian).

*Department of Math. Sciences, University of Liverpool,
Liverpool, L69 3BX, United Kingdom,
abm@liv.ac.uk*

Submitted: 10.03.1999

Operator Theory:
Advances and Applications, Vol. 109

Asymptotic analysis of a mixed boundary value problem in a singularly degenerating domain

JAN ÅSLUND

A mixed boundary value problem for the Laplacian in a multi-structure is considered. The multi-structure consists of a thin-wall tube connected to a three-dimensional body, and the thickness of the wall is of order ε, where ε is a small parameter. The principal term in a compound multi-scaled asymptotic expansion is constructed.

1.   Introduction

In the present paper we consider a mixed boundary value problem in a 2D-3D multi-structure consisting of a thin-wall tube connected to a three-dimensional domain. The results are taken from the licentiate thesis Åslund [1], written under supervision of Professor Maz'ya, where a complete compound asymptotic expansion for the solution was constructed. Here we will construct the principal term in the expansion. The main object is to illustrate the basic steps in the asymptotic analysis.

We use a modified version of the approach in Kozlov, Maz'ya and Movchan [3], where 1D-3D multi-structures were considered. A comprehensive survey of asymptotic analysis of elastic multi-structures and further references can be found in the monograph Ciarlet [2]. See also the monographs [4] and [5], by Maz'ya, Nazarov and Plamenevsky, where the mathematical foundation of the method of compound asymptotic expansions was laid down.

2.   Formulation of the problem

The multi-structure Ω_ε is the union of a three-dimensional domain Ω and a thin-wall tube D_ε; see Figure 1. The thickness of the wall in D_ε is of order ε, where ε is a small parameter. The multi-structure is singularly degenerating as ε tends to zero, and the limit region is the union of the three-dimensional domain Ω and the two-dimensional limit surface D_0.

We will consider the following mixed boundary value problem:

$$(2.1) \qquad -\Delta_x v_\varepsilon(x) \;=\; \mathcal{F}(x,\varepsilon), \quad x \in \Omega_\varepsilon,$$

$$(2.2) \qquad \frac{\partial v_\varepsilon}{\partial n_x}(x) \;=\; \mathcal{G}(x,\varepsilon), \quad x \in \partial\Omega_\varepsilon \setminus \overline{S}_\varepsilon,$$

$$(2.3) \qquad v_\varepsilon(x) \;=\; \mathcal{H}(x,\varepsilon), \quad x \in S_\varepsilon,$$

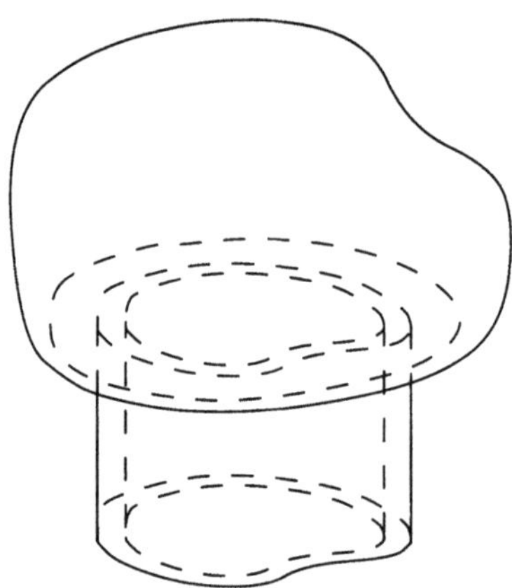

Figure 1: The domain Ω_ε

where S_ε denotes the base surface in the bottom region, and construct the principal term in the formal compound multi-scaled asymptotic expansion

$$(2.4) \qquad v_\varepsilon \sim \sum_{k=0}^{\infty} \varepsilon^k v_k(x,\varepsilon).$$

This expansion can be justified by showing that for any N and $\delta > 0$ the estimate

$$(2.5) \qquad \left\| v_\varepsilon(x) - \sum_{k=0}^{N} \varepsilon^k v_k(x,\varepsilon) \right\|_{H^1(\Omega_\varepsilon)} \leq C_N \varepsilon^{N-\delta}$$

holds. Moreover, by using this expansion one can obtain the asymptotic relation

$$(2.6) \qquad \lambda_1 = \varepsilon |\Gamma| l^{-1} |\Omega|^{-1} + O(\varepsilon^2 |\log \varepsilon|)$$

for the first eigenvalue of the spectral problem

$$
\begin{aligned}
-\Delta_x v_\varepsilon(x) &= \lambda v_\varepsilon(x), & x \in \Omega_\varepsilon, \\
\frac{\partial v_\varepsilon}{\partial n_x}(x) &= 0, & x \in \partial\Omega \setminus \overline{S}_\varepsilon, \\
v_\varepsilon(x) &= 0, & x \in S_\varepsilon.
\end{aligned}
$$

Here $|\Gamma|$ denotes the length of the curve, $|\Omega|$ the measure of the domain and l the length of the tube. Now we give a more precise description of the multi-structure. We let Ω denote a three-dimensional domain located in the half-space $\mathbf{R}^3_- = \{x \in \mathbf{R}^3 : x_3 < 0\}$, with smooth boundary $\partial\Omega$. Furthermore, we assume that some part of the boundary $\partial\Omega$ is located in the plane $x_3 = 0$, and that this

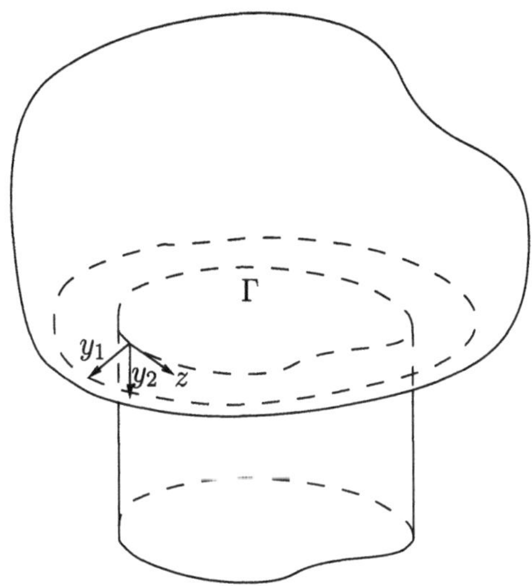

Figure 2: The limit region

part contains a smooth and closed curve denoted by Γ. Let z denote the arc length on Γ and introduce Cartesian coordinates (y_1, y_2) orthogonal to Γ, so that y_2 coincides with the coordinate x_3; see Figure 2. In a neighbourhood of Γ we introduce polar coordinates (r, ϑ), such that the relation $(y_1, y_2) = r(\cos\vartheta, \sin\vartheta)$ holds.

Let D_ε denote the thin-wall tube

$$\{x : \varepsilon^{-1}y_1 \in (-1/2, 1/2),\ y_2 \in [0, l)\}.$$

Now, we define the multi-structure by

$$\Omega_\varepsilon = \Omega \cup D_\varepsilon.$$

We let $S^\pm$ denote the lateral surfaces

$$\{x : \varepsilon^{-1}y_1 = \pm 1/2,\ y_2 \in (0, l)\}$$

of the thin-wall tube D_ε.

3.   The leading order approximation

The terms in the compound asymptotic expansion have the form

$$(3.1) \qquad \begin{aligned} v_k \;=\;& \varepsilon^{-1}(1 - \kappa(\xi_2))C_k(\log\varepsilon) + \mu(\xi)u_k(x, \log\varepsilon) \\ &+ \kappa(\xi_2)(\varepsilon^{-1}V_k(y_2, z, \log\varepsilon) + \varepsilon U_k(\zeta, y_2, z, \log\varepsilon)) \\ &+ \varepsilon^{-1}\Xi(x_3)W_k(\eta, z, \log\varepsilon) + \chi(y)w_k(\xi, z, \log\varepsilon). \end{aligned}$$

and we will construct the principal term v_0. Here μ, κ, χ and Ξ denote smooth cut-off functions, and we have introduced scaled coordinates $(\xi_1, \xi_2) = \varepsilon^{-1}(y_1, y_2)$, $\zeta = \varepsilon^{-1}y_1$ and $(\eta_1, \eta_2) = \varepsilon^{-1}(y_1, l - y_2)$. We assume that the coordinates (y_1, y_2) are well defined on the support of χ, and that $\chi(y) = 1$ in a neighbourhood of the junction region. We assume further that χ vanish near $\partial\Omega \setminus \{x : x_3 = 0\}$ and S_ε, and that $\partial\chi/\partial n_x = 0$ on $\partial\Omega_\varepsilon$.

The function Ξ is chosen so that $\Xi(x_3) = 1$ for $x_3 \geq 2l/3$, and $\Xi(x_3) = 0$ for $x_3 \leq l/3$. In a neighbourhood of the junction region the function μ is defined so that $\mu(\xi) = 1$ in $\{\xi : \xi_2 \leq 0, \ |\xi| \geq 1\}$, and $\mu(\xi) = 0$ in $\{\xi : \xi_2 \geq 0\}$ and $\{\xi : \xi_2 \leq 0, \ |\xi| \leq 1/2\}$. In the remaining parts of Ω and D_ε, the function μ is extended by 1 and 0 respectively. We assume further that $\partial\mu/\partial n_x = 0$ on $\partial\Omega_\varepsilon$. The function κ is chosen so that $\kappa(\xi_2) = 1$ for $\xi_2 \geq 1$, and $\kappa(\xi_2) = 0$ for $\xi_2 \leq 1/2$.

The functions in (3.1) depend polynomially on $\log\varepsilon$, and are solutions of certain boundary value problems posed in ε-independent model domains. The domains that we consider are the three-dimensional domain Ω, the limit surface D_0, the scaled cross-section $I = (-1/2, 1/2)$ of the thin-wall tube, the scaled cross-section of the bottom region

$$\Pi^+ = \{\eta \in \mathbf{R}^2 : \eta_1 \in (-1/2, 1/2), \ \eta_2 > 0\}$$

and the scaled cross-section of the junction region, denoted by ω, which is the union of the semi-strip

$$\{\xi \in \mathbf{R}^2 : \xi_1 \in (-1/2, 1/2), \ \xi_2 \geq 0\}$$

and the half plane

$$\mathbf{R}^2_- = \{\xi \in \mathbf{R}^2 : \xi_2 < 0\}.$$

We assume that the right-hand sides admit similar multi-scaled asymptotic representations:

$$(3.2) \quad \mathcal{F}(x, \varepsilon) \sim \sum_{k=0}^{\infty} \varepsilon^k \Big\{ \mu(\xi)f_k(x, \log\varepsilon) + \varepsilon^{-1}\kappa(\xi_2)F_k(\zeta, y_2, z, \log\varepsilon)$$
$$+ \varepsilon^{-2}\chi(y)\varphi_k(\xi, z, \log\varepsilon) + \varepsilon^{-3}\Xi(x_3)\Phi_k(\eta, z, \log\varepsilon) \Big\},$$

$$(3.3) \quad \mathcal{G}(x, \varepsilon) \sim \sum_{k=0}^{\infty} \varepsilon^k \Big\{ \mu(\xi)g_k(x, \log\varepsilon) + \kappa(\xi_2)G_k(\zeta, y_2, z, \log\varepsilon)$$
$$+ \varepsilon^{-1}\chi(y)\psi_k(\xi, z, \log\varepsilon) + \varepsilon^{-2}\Xi(x_3)\Psi_k(\eta, z, \log\varepsilon) \Big\},$$

and

$$(3.4) \quad \mathcal{H}(x, \varepsilon) \sim \sum_{k=0}^{\infty} \varepsilon^{k-1}\Lambda_k(\eta, z, \log\varepsilon).$$

Here the remainders in the first two expansion can be estimated in the corresponding L_2 norms, and in the last expansion we use the $H^{1/2}(S_\varepsilon)$ norm. In order to construct the leading term v_0 in expansion (2.4), we have to take into account the leading terms in the expansions for the right-hand sides. In Ω we have the leading approximation

$$(3.5) \qquad \varepsilon^{-1}C_0 + u_0(x)$$

for the solution. The function u_0 is defined as the solution of the boundary value problem

$$
\begin{aligned}
-\Delta_x u_0(x) &= f_0(x), && \text{in } \Omega, \\
\frac{\partial u_0}{\partial n_x}(x) &= g_0(x), && \text{on } \partial\Omega \setminus \Gamma,
\end{aligned}
$$

with the orthogonality condition

$$\int_\Omega u_0(x)\, dx = 0,$$

and the prescribed singular behaviour

$$u_0(x) = c_0(z)\pi^{-1}\log(1/r) + O(1), \quad r \to 0,$$

along Γ. Here c_0 is a smooth function that will be specified later on, and the boundary value problem has a unique solution if and only if the condition

$$(3.6) \qquad \int_\Omega f_0(x)\, dx + \int_{\partial\Omega\setminus\Gamma} g_0(x)\, ds_x + \int_\Gamma c_0(z)\, dl_z = 0,$$

is fulfilled. We substitute the term μu_0 into the original boundary value problem and obtain

$$-\Delta_x(\mu u_0) = \mu f_0 - [\Delta_x, \mu]u_0, \quad \text{in } \Omega_\varepsilon,$$

and

$$\frac{\partial}{\partial n_x}(\mu u_0) = \mu g_0 \quad \text{on } \partial\Omega_\varepsilon \setminus \overline{S}_\varepsilon.$$

After the change of variables $\xi = \varepsilon^{-1}y$, the discrepancy term has the asymptotics

$$[\Delta_x, \mu]u_0(x) \sim \varepsilon^{-2}[\Delta_\xi, \mu]\{c_0(z)\pi^{-1}\log(1/\varrho) + c_0(z)\pi^{-1}\log(1/\varepsilon) + d_0(z)\},$$

as $\varepsilon \to 0$, where d_0 is a smooth function, and $\varrho = |\xi|$. Moreover, the term $\varepsilon^{-1}(1 - \kappa)C_0$ causes the discrepancy

$$(3.7) \qquad \varepsilon^{-3}\kappa''(\xi_2)C_0$$

in equation (2.1).

218   Åslund

The leading order approximation in the thin-wall tube D_ε has the form

$$(3.8) \qquad \varepsilon^{-1} V_0(y_2, z) + \varepsilon U_0(\zeta, y_2, z)$$

where the function U_0 is the solution of the following boundary value problem posed on the scaled-cross section $I = (-1/2, 1/2)$:

$$
\begin{aligned}
-\partial_\zeta^2 U_0(\zeta, y_2, z) &= F_0(\zeta, y_2, z) - \int_{-1/2}^{1/2} F_0(\zeta, y_2, z)\, d\zeta \\
&\quad - G_0^+(y_2, z) - G_0^-(y_2, z), \quad \zeta \in I, \\
\pm\partial_\zeta U_0(\zeta, y_2, z) &= G_0^\pm(y_2, z), \quad \zeta = \pm 1/2,
\end{aligned}
$$

where $G_0^\pm = G_0$ on $S^\pm$, with the orthogonality condition

$$\int_{-1/2}^{1/2} U_0\, d\zeta = 0,$$

for uniqueness, and with y_2 and z as parameters. The second function V_0 is the solution of the boundary value problem

$$
\begin{aligned}
(3.9) \qquad -(\partial_{y_2}^2 + \partial_z^2) V_0(y_2, z) &= \int_{-1/2}^{1/2} F_0(\zeta, y_2, z)\, d\zeta \\
&\quad + G_0^+(y_2, z) + G_0^-(y_2, z), \quad \text{on } D_0 \\
V_0(0, z) &= h_0^J(z) \\
V_0(l, z) &= h_0^B(z)
\end{aligned}
$$

on the limit surface. Here the Dirichlet data h_0^J and h_0^B are yet undetermined, smooth functions.

The principal terms of the discrepancies in equation (2.1) caused by (3.8) are

$$(3.10) \qquad [\partial_{\xi_2}^2, \kappa]\left(\varepsilon^{-3} V_0(0, z) + \varepsilon^{-2} \xi_2 \partial_{y_2} V_0(0, z) \right).$$

Moreover, we have to take into account the discrepancy

$$(3.11) \qquad \varepsilon^{-1} V_0(l, z) = \varepsilon^{-1} h_0^B(z)$$

in the Dirichlet boundary condition (2.3) on the bottom surface S_ε.

In order to obtain relations between the undetermined functions c_0, h_0^B, h_0^J and the constant C_0 we analyse the bottom and junction regions, and we begin with the bottom region.

We have to compensate for the discrepancy term (3.11) and the leading terms related to the bottom region in the expansions of the right-hand sides, and we introduce the so called bottom layer

$$(3.12) \qquad \varepsilon^{-1} W_0(\eta, z)$$

that is the solution of the boundary value problem

$$
\begin{aligned}
-\Delta_\eta W_0(\eta, z) &= \Phi_0(\eta, z), \quad \text{in } \Pi^+ \\
\frac{\partial W_0}{\partial n_\eta}(\eta, z) &= \Psi_0(\eta, z), \quad \text{on } \partial\Pi^+ \setminus \overline{S}, \\
W_0(\eta, z) &= \Lambda_0(\eta, z) - h_0^B(z), \quad \text{on } S,
\end{aligned}
$$

with z as a parameter, and where S denotes the base of the semi-strip Π^+. We seek a solution that decays exponentially at infinity and such a solution exists if and only if the relation

$$(3.13) \quad h_0^B(z) = \int_{\Pi^+} \eta_2 \Phi_0(\eta, z) \, d\eta + \int_{\partial\Pi^+ \setminus \overline{S}} \eta_2 \Psi_0(\eta, z) \, ds_\eta + \int_S \Lambda_0(\eta, z) \, ds_\eta$$

holds, and this gives the Dirichlet data for the function V_0 at the bottom region.

Finally we consider the junction region. The discrepancies of order $O(\varepsilon^{-3})$ in (3.7) and (3.10) are cancelled out by letting

$$(3.14) \qquad\qquad h_0^J(z) = C_0$$

where $h_0^J(z)$ is the Dirichlet data for V_0 at the junction region.

As a consequence we have the representation

$$V_0(y_2, z) = (1 - l^{-1} y_2) C_0 + \tilde{V}_0(y_2, z),$$

and

$$\partial_{y_2} V_0(0, z) = -C_0 l^{-1} + \partial_{y_2} \tilde{V}_0(0, z),$$

where $\tilde{V}_0$ is the solution of the boundary value problem (3.9), with $h_0^J = 0$ and h_0^B given by (3.13).

The discrepancies of order $O(\varepsilon^{-2})$ and the leading terms related to the junction region in the expansion of the right-hand sided are now compensated by the junction layer

$$w_0(\eta, z, \log \varepsilon),$$

that is the solution of the boundary value problem

$$
\begin{aligned}
-\Delta_\xi w_0(\xi, z, \log \varepsilon) &= [\Delta_\xi, \mu]\{c_0(z)\pi^{-1}(\log(1/\varrho) + \log(1/\varepsilon)) + d_0(z)\} \\
&\quad + [\partial_{\xi_2}^2, \kappa]\{\xi_2 \partial_{y_2} V_0(0, z)\} \\
&\quad + \varphi_0(\xi, z) + a_0(z, \log \varepsilon)\kappa''(\xi_2) \quad \text{in } \omega, \\
\frac{\partial}{\partial n_\xi} w_0(\xi, z, \log \varepsilon) &= \psi_0(\xi, z), \quad \text{on } \partial\omega,
\end{aligned}
$$

with z as a parameter, where we have introduced the additional term

$$a_0(z, \log \varepsilon)\kappa''(\xi_2) = (a_0^{(0)}(z) + a_0^{(1)}(z)\log \varepsilon)\kappa''(\xi_2)$$

which can be compensated for when the next term in the asymptotic expansion is constructed. We seek a solution that vanishes at infinity, and such a solution exists if and only if the right-hand sides in the boundary value problem, denoted by φ and ψ respectively, satisfy the conditions

$$\int_\omega \varphi\, d\xi + \int_{\partial\omega} \psi\, ds_\xi = 0$$

and

$$\int_\omega \varphi w_h\, d\xi + \int_{\partial\omega} \psi w_h\, ds = 0,$$

where w_h is any solution of the homogeneous boundary value problem with the asymptotic behaviour

$$w_h = \pi^{-1}\log(1/\varrho) + O(\varrho^{-1}), \quad \varrho \to \infty$$

in the half-plane and

$$w_h = \xi_2 + O(1), \quad \xi_2 \to \infty$$

in the semi-strip.

These conditions lead to the relations

$$(3.15) \quad c_0(z) = -C_0 l^{-1} + \partial_{y_2}\tilde{V}_0(0,z) + \int_\omega \varphi_0(\xi,z)\, d\xi + \int_{\partial\omega} \psi_0(\xi,z)\, ds_\xi$$

and

$$\begin{aligned}
a_0(z,\log\varepsilon) \;=\; & c_0(z)\pi^{-1}\left(\int_\omega w_h(\xi)[\Delta_\xi,\mu]\log(1/\varrho)\, d\xi + \log(1/\varepsilon)\right) + d_0(z) \\
& +(\partial_{y_2}\tilde{V}_0(0,z))\int_\omega (\kappa''(\xi_2)\xi_2 + 2\kappa'(\xi_2))w_h(\xi)\, d\xi \\
& + \int_\omega \varphi_0(\xi,z)w_h(\xi)\, d\xi + \int_{\partial\omega} \psi_0(\xi,z)w_h(\xi)\, ds_\xi.
\end{aligned}$$

By substituting (3.15) into (3.6) we obtain

$$\begin{aligned}
C_0 \;=\; & |\Gamma|^{-1}l\bigg\{\int_\Omega f_0(x)\, dx + \int_{\partial\Omega\backslash\Gamma} g_0(x)\, ds \\
& +\partial_{y_2}\left(\int_\Gamma \tilde{V}_0(y_2,z)\, dl_z\right)\bigg|_{y_2=0} + \int_\Gamma I_0(z)\, dl_z\bigg\},
\end{aligned}$$

Now we have calculated the constant C_0 and then c_0 is given by (3.15) and h_0^J by (3.14), and we have completed the construction of the principal term:

$$\begin{aligned}
v_0 \;=\; & \varepsilon^{-1}(1-\kappa(\xi_2))C_0 + \mu(\xi)u_0(x) + \kappa(\xi_2)\left(\varepsilon^{-1}V_0(y_2,z) + \varepsilon U_0(\zeta,y_2,z)\right) \\
& +\varepsilon^{-1}\Xi(x_3)W_0(\eta,z) + \chi(y)w_0(\xi,z,\log\varepsilon).
\end{aligned}$$

The subsequent terms in the expansion depend polynomially on $\log\varepsilon$. Furthermore, we have to consider more general model problems. However, the steps in the analysis are almost the same.

References

[1] Åslund, J. *Asymptotic Analysis of a Mixed Boundary Value Problem for the Laplacian in a 2D-3D Multi-Structure*, Licentiate Thesis, LIU-TEK-LIC-1998:76, Linköping, 1998.

[2] Ciarlet, P.G. *Mathematical elasticity. Volume II: Theory of Plates*, Elsevier,1997.

[3] Kozlov, V.A., Maz'ya, V.G., Movchan, A.B. *Asymptotic Analysis of Fields in Multi-Structures*, Oxford University Press, 1999.

[4] Maz'ya, V.G., Nazarov, S.A., Plamenevsky, B.A. *Asymptotische Theorie elliptischer Randwertaufgaben in singulär gestörten Gebieten* **1**, Akademie-Verlag, Berlin, 1991.

[5] Maz'ya, V.G., Nazarov, S.A., Plamenevsky, B.A. *Asymptotische Theorie elliptischer Randwertaufgaben in singulär gestörten Gebieten* **2**, Akademie-Verlag, Berlin, 1992.

Department of Mathematics,
Linköpings universitet,
S-581 83 Linköping, Sweden,
jaasl@mai.liu.se

1991 Mathematics Subject Classification: Primary 35C20; Secondary 35J25

Submitted: 22.02.1999

Operator Theory:
Advances and Applications, Vol. 109

A history of the Cosserat spectrum

ALEXANDER KOZHEVNIKOV

The paper is a brief survey of the 100-year history of the Cosserat spectrum in elasto-statics which was first studied by Eugène and François Cosserat between 1898 and 1901 and later by Vladimir Maz'ya and Solomon Mikhlin.

1. Introduction

It is my pleasure to dedicate this paper to Vladimir Maz'ya on the occasion of his 60th birthday. I was prompted to write it by the short note "On the Cosserat spectrum of the equations of the theory of elasticity" by Maz'ya and Mikhlin published in 1967 [25]. That article was Maz'ya's only joint paper with Mikhlin although they were very close for many years. While it is only a drop in the ocean of Maz'ya's forty-year work, it deserves special attention because it was at the beginning of an interesting recent development in spectral analysis of partial differential operators.

The paper deals with the Cosserat eigenvalues, i.e. those values of a parameter ω, which admit non-trivial solutions to the Dirichlet problem:

$$(1.1) \qquad \Delta \mathbf{u} + \omega \operatorname{grad} \operatorname{div} \mathbf{u} = \mathbf{0}, \quad \mathbf{u}|_{\partial\Omega} = \mathbf{0}$$

in a bounded domain $\Omega \subset \mathbb{R}^3$ with a sufficiently smooth boundary $\partial\Omega$. Here $\mathbf{u} = (u_1, u_2, u_2)$ is a vector-function, and Δ is the Laplace operator. Since $\Delta \mathbf{u}$ and $\operatorname{grad} \operatorname{div} \mathbf{u}$ are both second order operators, the spectral problem (1.1) is of a rather unconventional form and therefore the standard methods cannot be directly applied.

Notice that the equation $\Delta \mathbf{u} + \omega \operatorname{grad} \operatorname{div} \mathbf{u} = \mathbf{0}$ may be viewed as the Lamé equation governing the equilibrium of a homogeneous isotropic linear elastic body with the displacement vector $\mathbf{u}$ and with the constant ω related to the Lamé constants λ, μ and to the Poisson ratio σ by

$$\omega := \frac{\lambda + \mu}{\mu} := \frac{1}{1 - 2\sigma}.$$

Since from the physical point of view all admissible values of σ belong to the interval $(-1, 1/2)$, the half-axis $(1/3, \infty)$ is the range of ω. Therefore, the Cosserat spectrum, those values of ω for which the uniqueness is violated, is possible only on the closed ray $[-\infty, 1/3]$.

The eigenvalue problem (1.1) was first investigated by the well-known French mathematicians Eugène and François Cosserat (see [10, v. 3, p. 428]), who published nine papers on this problem [1]–[9] between 1898 and 1901. The paper [25] by V. Maz'ya and S. Mikhlin and two papers by S. Mikhlin [26], [27] initiated a new period in the development of this topic, almost forgotten by the 1960s. The period is still continuing, and new results on the Cosserat spectrum have been published recently: [15]–[20], [22], [33]–[35], [39].

The aim of this paper is to review this field of research.

2. The first boundary-value problem of elastostatics

The eigenvalue problem (1.1) was studied by E. and F. Cosserat in [1]–[9] where they explicitly determined the Cosserat spectrum for some certain types of domain Ω: a ball $\{x \in \mathbb{R}^3 : |x| < R\}$, a spherical shell $\{x \in \mathbb{R}^3 : R_1 < |x| < R_2\}$, an ellipsoid (for these domains, the variables can be separated). The papers by Cosserat mentioned above are short notes practically with few proof. Nevertheless, the finite-multiple eigenvalues were found for the domains indicated above, and it was proved that they converge to the point $\omega = -2$. Moreover, the finite intervals containing all finite-multiple eigenvalues were also indicated: $[-3, -2]$ for the ball, and $[-3, -1]$ for the spherical shell.

The work by Cosserat was continued by V. Maz'ya and S. Mikhlin [25]–[31] who investigated the Cosserat spectrum for arbitrary domains.

In view of the fact that the equation $\Delta \mathbf{u} + \omega \mathrm{grad}\, \mathrm{div}\mathbf{u} = \mathbf{0}$ is the Lamé equation of elastostatics, any uniqueness theorem in elasticity automatically demarcates regions of the real axis in which the Cosserat spectrum cannot lie. For example, the classical result that the first boundary value problem (with displacements given on the boundary) has a unique solution when $\lambda + 2\mu > 0$, $\mu > 0$, immediately implies (see S. Mikhlin [26]) that the Cosserat spectrum for that case is confined to the interval $[-\infty, -1]$ and may have only three limit points $\omega = -\infty, -2, -1$.

In the paper [27] S. Mikhlin indicated that among Cosserat eigenvalues there are two special points $\omega = -1$ and $\omega = \infty$ which are *infinite-multiple eigenvalues*, and the corresponding eigenspaces consist of functions $\mathbf{u}$ satisfying respectively the relations $\mathrm{div}\mathbf{u} = 0$ and $\mathrm{rot}\mathbf{u} = 0$.

The next important step was taken by V. Maz'ya and S. Mikhlin in the paper [25], where they showed that the points $\omega = \infty$ and $\omega = -1$ are *isolated* points of the Cosserat spectrum.

To obtain the result concerning the point $\omega = \infty$ it was shown in [25] that the solution to the boundary value problem

$$(2.1) \qquad \Delta \mathbf{u} + \omega \mathrm{grad}\, \mathrm{div}\mathbf{u} = \mathbf{f}, \quad \mathbf{u}|_{\partial\Omega} = \mathbf{0}$$

can be represented in the form of a power series in $\varepsilon := \omega^{-1}$ which strongly converges in an appropriate Sobolev space for all sufficiently large values of ω.

Substituting $p := \varepsilon\,\mathrm{div}\mathbf{u}$ into the first equation (2.1) and rewriting it as a system $\Delta\mathbf{u}+\mathrm{grad}p = \mathbf{f}$, $\mathrm{div}\mathbf{u} = \varepsilon^{-1}p$, the authors looked for a solution $(\mathbf{u},p)$ in the form of a power series

$$(2.2) \qquad \mathbf{u} := \sum_{n=0}^{\infty} \varepsilon^n \mathbf{u}^{(n)}, \quad p := \sum_{n=0}^{\infty} \varepsilon^n p^{(n)}.$$

This leads to the recurrent relations for the coefficients

$$\Delta\mathbf{u}^{(0)} + \mathrm{grad}p^{(0)} = \mathbf{f}, \quad \mathrm{div}\mathbf{u}^{(0)} = 0,$$

$$n \geq 1, \quad \Delta\mathbf{u}^{(n)} + \mathrm{grad}p^{(n)} = \mathbf{0}, \quad \mathrm{div}\mathbf{u}^{(n)} = p^{(n-1)}.$$

The convergence of the series (2.2) is not trivial. By the way, independently and even later in 1975 M. C. Pelissier [36] and in 1979 R. Temam [37] proved that the power series gives an asymptotic expansion of the solution.

To prove that $\omega = -1$ is an isolated point of the Cosserat spectrum V. Maz'ya and S. Mikhlin [25] showed that the solution of the first boundary value problem for the equation $\Delta\mathbf{u} + \omega\mathrm{grad}\,\mathrm{div}\mathbf{u} = \mathbf{f}$ can be represented in the form of a power series in $\epsilon := (1+\omega)\,\omega^{-1}$, which strongly converges in an appropriate Sobolev space for all sufficiently small values of ϵ, i.e. for all points ω belonging to a sufficiently small neighborhood of the point -1. More precisely, the substitution $\mathbf{w} := (1+\omega)\,\mathbf{u}$, $\mathbf{q} :=\mathrm{rot}\mathbf{w}$ reduces the equation $\Delta\mathbf{u} + \omega\mathrm{grad}\,\mathrm{div}\mathbf{u} = \mathbf{f}$ to a system

$$\Delta\mathbf{w} + \mathrm{rot}\mathbf{q} = \mathbf{f}, \quad \mathrm{rot}\mathbf{w} = \epsilon\mathbf{q}.$$

Then, looking for a solution $(\mathbf{w}, \mathbf{q})$ in the form of power series,

$$\mathbf{w} := \sum_{n=0}^{\infty} \epsilon^n \mathbf{w}^{(n)}, \quad \mathbf{q} := \sum_{n=0}^{\infty} \epsilon^n \mathbf{q}^{(n)}$$

the authors obtained the recurrence relations for the coefficients

$$\Delta\mathbf{w}^{(0)} + \mathrm{rot}\mathbf{q}^{(0)} = \mathbf{f}, \quad \mathrm{rot}\mathbf{w}^{(0)} = \mathbf{0}, \quad \mathrm{div}\mathbf{q}^{(0)} = 0,$$

$$n \geq 1, \quad \Delta\mathbf{w}^{(n)} + \mathrm{rot}\mathbf{q}^{(n)} = \mathbf{0}, \quad \mathrm{rot}\mathbf{w}^{(n)} = \mathbf{q}^{(n-1)}, \quad \mathrm{div}\mathbf{q}^{(n)} = 0.$$

Thus, the function $\mathbf{u}$ is represented in the form of the Laurent series

$$(2.3) \qquad \mathbf{u} = -\epsilon^{-1}\mathbf{w}^{(0)} + \sum_{n=0}^{\infty} \epsilon^n \left(\mathbf{w}^{(n)} - \mathbf{w}^{(n+1)}\right).$$

In addition to the series expansions (2.2) and (2.3), V. Maz'ya and S. Mikhlin in the paper [25] (see also [31, Sect. 3, (3.32), (3.33)]) obtained the following two interesting inequalities in the Sobolev space $\left[H_0^1\left(\Omega\right)\right]^3$:

$$(2.4) \qquad D\left(\mathbf{u}\right) \leq C_1 \int_{\Omega} \left(\mathrm{div}\mathbf{u}\right)^2 dx \quad \text{for } \mathbf{u} \in \ker\mathrm{div}^{\perp},$$

and

$$(2.5) \qquad D\left(\mathbf{u}\right) \le C_2 \int_\Omega \left|\mathrm{rot}\mathbf{u}\right|^2 dx \quad \text{for } \mathbf{u} \in \ker \mathrm{rot}^\perp.$$

Here, $\ker\mathrm{div}$ and $\ker\mathrm{rot}$ denote the subspaces of all vector-functions satisfying $\mathrm{div}\mathbf{u} = 0$ and $\mathrm{rot}\mathbf{u} = 0$, respectively, wheareas $\ker\mathrm{div}^\perp$ and $\ker\mathrm{rot}^\perp$ are their orthogonal complements in $\left[H_0^1\left(\Omega\right)\right]^3$ with respect to the scalar product

$$(2.6) \qquad D\left(\mathbf{u},\mathbf{v}\right) := \sum_{j,k=1}^{3} \int_\Omega \partial_k u_j \partial_k v_j dx.$$

The Maz'ya-Mikhlin inequalities (2.4) and (2.5) were recently reproved by W. Velte [39].

Concerning other Cosserat eigenvalues it was shown by S. Mikhlin [31] that they are isolated finite-multiple eigenvalues with a unique limit point $\omega = -2$. The proof follows from an integral equation to which L. Lichtenstein [23] (see also the monograph by E. Trefftz [38]) reduced the first boundary problem of elastostatics. This equation has the form

$$(2.7) \qquad \theta\left(x\right) + \frac{\varkappa}{4\pi}\int_{\partial\Omega} \rho\frac{\partial^2 G}{\partial\rho\partial\mathbf{n}}\theta\left(x\right)d_y S = \frac{2}{2+\omega}\theta_0\left(x\right), \quad x \in \partial\Omega.$$

Here $\mathbf{u}$ is the unknown displacement vector; $\mathbf{u}_0$ is a vector with harmonic components in Ω and coinciding with $\mathbf{u}$ on $\partial\Omega$; $\theta :=\mathrm{div}\mathbf{u}, \theta_0 :=\mathrm{div}\mathbf{u}_0$; $\varkappa := -\omega/\left(\omega+2\right)$; $G\left(x,y\right)$ is the Green function of the Dirichlet problem for the Laplace equation; ρ is the radius vector going from y to x, and also its length.

It was also proved by S. Mikhlin [31] that the spectral problem (1.1) generates a bounded self-adjoint non-negative operator in the space $\left[H_0^1\left(\Omega\right)\right]^3$ with the inner-product (2.6). As a result, this operator has a countable system of (Cosserat) eigenvectors, orthogonal in the metric of the Dirichlet integral (2.6) and, moreover, this system is complete in the Sobolev space $\left[H_0^1\left(\Omega\right)\right]^3$ as well as in the space $\left[L_2\left(\Omega\right)\right]^3$.

A new result concerning the Cosserat spectrum of the first boundary problem of elastostatics appeared almost thirty years later in the paper [19] by A. Kozhevnikov and T. Skubachevskaya. It deals with the asymptotic distribution of the finite-multiple eigenvalues for the spectral problem (1.1). It was shown that there are

$$(2.8) \qquad C\varepsilon^{-2} + o\left(\varepsilon^{-2}\right) \quad \text{as } \varepsilon \to 0$$

finite-multiple Cosserat eigenvalues outside the ε-neighborhood of the limit point $\omega = -2$, where

$$(2.9) \qquad C = \frac{1}{32\pi}\int_{\partial\Omega} \left(3k_1^2 + 3k_2^2 + 2k_1 k_2\right)dS.$$

Here k_1 and k_2 are the principal curvatures of the surface $\partial\Omega$ and dS is the infinitesimal surface element. The formulae (2.8), (2.9) give a negative answer to the question formulated by S. Mikhlin [31, Sect. 3, p.66] and repeated by W. Velte [39, Sect. 2, p. 851], (i.e. whether the point $\omega = -2$ could be an infinite-multiple isolated eigenvalue).

3. The second and other boundary-value problems

To state the second and other boundary value problems we introduce some notations. Let $\partial_j f := \partial f/\partial x_j$, $\gamma_0 f := f|_{\partial\Omega}$ and $\mathbf{n} = (n_1, n_2, n_3)$ be the unit interior normal vector field defined near $\partial\Omega$.

We consider the following linearized basic boundary value problems of elastostatics [21], [26, Sect. 3]:

$$(3.1) \qquad \begin{aligned} &\text{(i)} \quad \Delta\mathbf{u} + \omega\operatorname{grad}\operatorname{div}\mathbf{u} = \mathbf{f} \quad \text{for} \ x \in \Omega, \\ &\text{(ii)} \quad T_k^\omega \begin{pmatrix} \mathbf{u} \\ \operatorname{div}\mathbf{u} \end{pmatrix} = \mathbf{g} \qquad \text{for} \ x \in \partial\Omega, \end{aligned}$$

Here $\mathbf{u}, \mathbf{f}, \mathbf{g}$ are vector-functions with range in $\mathbb{R}^3$, T_k^ω $(k = 1, 2, 3, 4)$ are boundary operators describing respectively the four basic boundary value problems of elasticity: with given displacements $\mathbf{u}$ $(k = 1)$ and stresses $(k = 2)$ on the boundary or the problem of hard $(k = 3)$ and soft contact $(k = 4)$. For the problem of hard contact $(k = 3)$ the normal component of the displacement and the tangential component of the stress vector are given on the boundary. For the problem of soft contact $(k = 4)$ the normal component of the stress vector and the tangential component of the displacement are given on the boundary. More precisely,

$$(3.2) \qquad T_1^\omega \begin{pmatrix} \mathbf{u} \\ \operatorname{div}\mathbf{u} \end{pmatrix} := \gamma_0\mathbf{u},$$

$$(3.3) \qquad T_2^\omega \begin{pmatrix} \mathbf{u} \\ \operatorname{div}\mathbf{u} \end{pmatrix} := \chi\mathbf{u} + (\omega - 1)\,\gamma_0\,(\operatorname{div}\mathbf{u})\,\mathbf{n},$$

$$(3.4) \qquad T_3^\omega \begin{pmatrix} \mathbf{u} \\ \operatorname{div}\mathbf{u} \end{pmatrix} := (\chi\mathbf{u})_\tau + \gamma_0\mathbf{u_n},$$

$$(3.5) \qquad T_4^\omega \begin{pmatrix} \mathbf{u} \\ \operatorname{div}\mathbf{u} \end{pmatrix} := (\gamma_0\mathbf{u})_\tau - (\chi\mathbf{u})_\mathbf{n} - (\omega - 1)\,\gamma_0\,(\operatorname{div}\mathbf{u})\,\mathbf{n},$$

where $\mathbf{u_n} := \left(\sum_{j=1}^3 u_j\,n_j\right)\mathbf{n}$, $\mathbf{u}_\tau$ is the tangential component of $\mathbf{u}$, i.e. $\mathbf{u}_\tau := \mathbf{u} - \mathbf{u_n}$, χ is a special first order boundary operator defined via the strain vector $\chi\mathbf{u}$ with components

$$(\chi\mathbf{u})_i = \gamma_0 \left(\sum_{j=1}^3 (\partial_i u_j + \partial_j u_i)\,n_j \right) \qquad (i = 1, 2, 3).$$

We use the unified notation T_k^ω, although the operators T_1^ω and T_3^ω do not depend on ω.

The Cosserat spectrum consists of the set of numbers ω for which the homogeneous ($\mathbf{f} = \mathbf{0}, \mathbf{g} = \mathbf{0}$) problem (3.1) admits non-trivial solutions $\mathbf{u}$.

In view of the fact that the equation (i) of (3.1) is the Navier equation of elastostatics, any uniqueness theorem in elasticity automatically demarcates regions of the real axis in which the Cosserat spectrum cannot lie. For example, the classical result that the second boundary value problem (with stresses given on the boundary) has a unique solution when $3\lambda + 2\mu > 0$, $\mu > 0$, immediately implies (see S. Mikhlin [31]) that the Cosserat spectrum for that case must be confined to the the the interval $\left[-\infty, \frac{1}{3}\right]$. The same result holds for the third and fourth problems. The reader may find complete surveys of the theory of uniqueness in the monograph by R. J. Knops and L. E. Payne [14], by V. Kupradze et al [21] and in the comprehensive article by M. E. Gurtin [13].

The second eigenvalue problem (3.1) ($k = 2$) was first investigated by E. and F. Cosserat in [1]–[9] when Ω is a ball and a spherical shell. For these domains the finite-multiple eigenvalues were calculated and it was proved that they converge to $\omega = 0$. The finite intervals containing all finite-multiple eigenvalues were also indicated: $[0, 1/3]$ for the ball and $[-1, 1/3]$ for the spherical shell. The results of Cosserat as well as further investigations by S. Mikhlin [26]–[31] concerning the Cosserat spectrum were summarized in the review paper by S. Mikhlin [31] (see also the monograph by S. Mikhlin, N. Morozov, M. Paukshto [32, Appendix I]).

It was shown by S. Mikhlin [31] that all points of the Cosserat spectrum of the second boundary-value problem are isolated eigenvalues except for $\omega = 0$ which may be a unique limit point of finite-multiple eigenvalues (the uniqueness in the case of the second boundary problem was noted later in the paper [15] by A. Kozhevnikov). Moreover, S. Mikhlin proved that the points $\omega = -1$, $\omega = -\infty$ are isolated infinite-multiple eigenvalues and that all other Cosserat eigenvalues belong to the ray $(-\infty, 1/3]$.

The last result was refined just recently by A. Kozhevnikov [18]. It was found that the point $\omega = -1$ is actually a lower bound of the finite-multiple Cosserat eigenvalues for any convex domain Ω. This means that all finite-multiple Cosserat eigenvalues of the second boundary value problem are contained in the interval $(-1, 1/3]$.

It was indicated in the previous section that many facts about Cosserat spectrum may be obtained from the Lichtenstein integral equation (2.7) which holds for the first boundary value problem. A reduction of all other boundary value problems of elastostatics to an integral equation on $\partial\Omega$ similar to the Lichtenstein equation was found by A. Kozhevnikov [15], [16]. As a result, it was proved that the Cosserat spectrum of the third and the fourth boundary value problems consists of isolated eigenvalues except for $\omega = -1$ which may be a unique limit point of the finite-multiple eigenvalues.

The above-mentioned facts concerning Cosserat spectrum may be summarized in the form of the following table:

Number of Boundary-Value Problem	Interval Containing Finite-Multiple Eigenvalues	Limit Point of Eigenvalues	Isolated Infinite-Multiple Eigenvalues
1	$(-\infty, -1)$	$\omega = -2$	$\omega = \infty, -1$
2	$\begin{cases} (-\infty, 1/3], \\ (-1, 1/3], \quad \text{if } \Omega \text{ is convex} \end{cases}$	$\omega = 0$	$\omega = \infty, -1$
3, 4	$(-\infty, 1/3]$	$\omega = -1$	$\omega = \infty$

A related and natural question on the asymptotic distribution of the finite-multiple Cosserat eigenvalues tending to their unique limit point was investigated by A. Kozhevnikov and T. Skubachevskaya [19] for all four basic boundary value problems of elastostatics. By a reduction of the basic boundary value problems of elastostatics to an integral equation, the principal symbol of the corresponding integral (pseudo-differential) operator acting on $\partial\Omega$ was calculated. This principal symbol was presented in terms of the principal curvatures and the coefficients of the first quadratic form of the surface $\partial\Omega$. Further, the principal term of the eigenvalue asymptotics of the integral operator was found. As a result, the asymptotic distribution (similar to (2.8), (2.9)) of the finite-multiple Cosserat eigenvalues tending to their unique limit point was obtained in view of a relation between the Cosserat eigenvalues and the eigenvalues of the integral operator.

The Maz'ya-Mikhlin power series expansion (2.2) for the solution of the first boundary value problem of elastostatics in the neighborhood of the point $\omega = \infty$ was recently further elaborated by A. Kozhevnikov and O. Lepsky [20]. It was proved that the series converges for all four boundary value problems of elasto-statics and for any values of ω lying outside the minimal interval with the center at the origin and of radius $r \geq 1$, which contains all of the Cosserat eigenvalues. In particular, the values of r for some domains are known, due to the classical results by E. and F. Cosserat [1]-[4] (see also S. Mikhlin [26]). For the ball and for the spherical shell we have $r = 3$ for the cases of the first $(k = 1)$ problem and $r = 1$ for the case of the second $(k = 2)$ problem (3.1), respectively. For the ellipsoid $\{x \in \mathbb{R}^3 : x_1^2/a_1^2 + x_2^2/a_2^2 + x_3^2/a_3^2 < 1\}$ $(a_1 \geq a_2 \geq a_3)$ we have by [2] $r \geq 1 + (a_1/a_2)^2 + (a_1/a_3)^2$ in the case of the first $(k = 1)$ boundary value problem (3.1). Possible applications of the Maz'ya-Mikhlin power series expansion will be discussed in the next section.

Some of the above results are also valid for external domains (see S. Mikhlin [31] and A. Kozhevnikov [17]).

4.   Application and other related results

The Maz'ya-Mikhlin power series (2.2) for the first boundary value problem as well as their analogs for other problems may by considered as a theoretical foundation for the well-known "optical-polarization" method or the "photo-elasticity" method for determining the static stress of a three-dimensional body. As is known, this method allows one, with the help of experiments, to find stresses for "models", i.e. bodies prepared from special plastics. These stresses are considered as approximations of stresses for the bodies of natural materials. The Poisson ratios for materials used nowadays as model materials are approximately $1/2$ $(\omega \simeq \infty)$. Using the first term of the power series, considered in this paper, we can calculate (instead of finding experimentally) static stresses for such "models". Moreover, using additional terms of the series we can calculate more precise approximations of static stresses for the bodies of natural materials, whenever $r < \omega$.

Since, by [18], the Cosserat eigenvalues except $\omega = -\infty$ for the second boundary value problem belong to the interval $[-1, 1/3]$ for any convex domain Ω, we see that the power series expansions (2.2) converge for all $|\omega| > 1$, and in particular for $\omega > 1$. In view of the relation $\omega = 1/(1 - 2\sigma)$, the series converge for all values of the Poisson ratio $\sigma \in (0, 1/2]$. The conservation energy law leads to the well-known inequality for the Poisson ratio, $-1 < \sigma < 1/2$, which holds for any real elastic medium. Since elastic media with negative σ are not known, the above power series converge for all really possible elastic media.

The "coefficients" of the power series (2.2) are solutions of the stationary linearized completely non-homogeneous Stokes problem. According to [37, Ch. I, Sect. 6], since many efficient algorithms are known for solving the Stokes problem, and since the discretization of (3.1) leads to a very ill-conditioned matrix for large ω (materials with the Poisson ratio σ near to $1/2$), one can try to compute $\mathbf{u}$ for large ω by using Stokes equations.

A new application of the Cosserat eigenvalues was obtained by X. Markenscoff and M. Paukshto [33],[34] and by M. Paukshto [35]. It was found that any Cosserat eigenvector $\mathbf{u}$ may be visualized as displacement in an equilibrium thermoelastic problem with temperature distributions $T \sim \mathrm{div}\mathbf{u}$. Using the fact that the collection of the divergences of the Cosserat eigenvectors is complete in the space $L_2(\Omega)$, the solution of the equilibrium equations of thermoelasticity under arbitrary temperature distribution was expanded in a rapidly convergent series of eigenvectors. It was proved that the Cosserat eigenvectors can be characterized as stationary points of the Helmholtz free energy (and elastic energy) under the constraint that the thermal part of free energy $\int_\Omega T^2 dx$ is constant.

A new unconventional spectral problem related to the Cosserat spectral problem was studied by M. Levitin [22]. He investigated a generalized Cosserat spectrum, i.e. the spectrum of the Dirichlet problem for the equation in Ω

$$\Delta \mathbf{u} + a(z)\,\mathrm{grad}\,\mathrm{div}\mathbf{u} + b(z)\,\mathbf{u} = \mathbf{0},$$

where z is a complex spectral parameter and $a(z)$, $b(z)$ are given rational meromorphic functions. A description of the generalized Cosserat spectrum was announced in terms of the sets of zeros and poles of some functions depending on $a(z)$ and $b(z)$. By [22], the generalized Cosserat spectrum (as well as the classical one) consists of either isolated infinite-multiple eigenvalues or of isolated finite-multiple eigenvalues or of the limit points of the latter.

Recently W. Velte [39] obtained some inequalities related to the Cosserat spectrum. It was proved that if p and $\mathbf{q} = (q_1, q_2, q_3)$ satisfy the equations

$$\mathrm{rot}\,\mathbf{q} = -p \quad \text{and} \quad \mathrm{div}\,\mathbf{q} = 0,$$

then the following inequalities hold:

$$\int_\Omega p^2\, dx \le (C_1 - 1) \int_\Omega |\mathbf{q}|^2\, dx, \qquad \text{if} \quad \int_\Omega p\, dx = 0,$$

and

$$\int_\Omega |\mathbf{q}|^2\, dx \le (C_2 - 1) \int_\Omega p^2\, dx, \qquad \text{if} \quad \mathbf{q} \cdot \mathbf{n} = 0 \quad \text{on} \quad \partial\Omega.$$

Here C_1 and C_2 are the same constants as in the Maz'ya-Mikhlin inequalities (2.4) and (2.5). It is interesting that the constants are related to the Cosserat finite-multiple eigenvalues for the first boundary value problem. Namely, for any Cosserat finite-multiple eigenvalue ω for the optimal constants C_1 and C_2 the following inequality holds:

$$(C_2 - 1)^{-1} \le -(\omega + 1) \le C_1 - 1.$$

There is a close relation between notions such as ellipticity, Fredholm property and Cosserat and essential spectrum. It is clear that the points of accumulation of the Cosserat eigenvalues are those values of $\omega \in \mathbb{C}$ where the operator of the boundary problem (3.1) is not a Fredholm operator. It is interesting that the points $\omega = -1, \infty$, where the ellipticity condition of the Lamé equation $\Delta \mathbf{u} + \omega \mathrm{grad}\,\mathrm{div}\,\mathbf{u} = \mathbf{f}$ fails, are isolated infinite-multiple Cosserat eigenvalues. The points where the complementary or Shapiro-Lopatinskii condition fails are limit points for the finite-multiple Cosserat eigenvalues. This fact was proved by S. Mikhlin [31] for the first and partly for the second boundary problems and by A. Kozhevnikov for the other problems [16].

A related question for the operators generated by the mixed-order systems which are elliptic in the sense of Douglis-Nirenberg was investigated by G. Grubb and G. Geymonat [11], and by G. Geymonat and E. Sanchez-Palencia [12]. It was proved that those values of the spectral parameter $\lambda \in \mathbb{C}$ where the Fredholm property fails (the essential spectrum) are exactly the same values of λ where the ellipticity condition is not satisfied.

At the conclusion of this short review the author hopes that the history of the Cosserat spectrum will be continued and we will be able to see many new and interesting papers on the topic.

References

[1] Eugène et François Cosserat, Sur les équations de la théorie de l'élasticité, C.R. Acad. Sci. (Paris) 126 (1898), 1089–1091.

[2] Eugène et François Cosserat, Sur les fonctions potentielles de la théorie de l'élasticité, C.R. Acad. Sci. (Paris) 126 (1898), 1129–1132.

[3] Eugène et François Cosserat, Sur la déformation infiniment petite d'un ellipsoide élastique, C.R. Acad. Sci. (Paris) 127 (1898), 315–318.

[4] Eugène et François Cosserat, Sur la solution des équations de l'élasticité dans le cas où les valeurs des inconnues à la frontière sont données, C.R. Acad. Sci. (Paris) 133 (1901), 145–147.

[5] Eugène et François Cosserat, Sur une application des fonctions potentielles de la théorie de l'élasticité, C.R. Acad. Sci. (Paris) 133 (1901), 210–213.

[6] Eugène et François Cosserat, Sur la déformation infiniment petite d'un corps élastique soumis à des forces données, C.R. Acad. Sci. (Paris) 133 (1901), 271–273.

[7] Eugène et François Cosserat, Sur la déformation infiniment petite d'une enveloppe sphérique élastique, C.R. Acad. Sci. (Paris) 133 (1901), 326–329.

[8] Eugène et François Cosserat, Sur la déformation infiniment petite d'un ellipsoide élastique soumis à des efforts données sur la frontière, C.R. Acad. Sci. (Paris) 133 (1901), 361–364.

[9] Eugène et François Cosserat, Sur un point critique particulier de la solution des équations de l'élasticité dans le cas où les efforts sur la frontière sont données, C.R. Acad. Sci. (Paris) 133 (1901), 382–384.

[10] Dictionary of Scientific Biography, Charles Scribner's Sons, New York, 1971.

[11] G. Grubb, G. Geymonat, The essential spectrum of elliptic systems of mixed order, Math. Ann. 227 (1977), 247–276.

[12] G. Geymonat, E. Sanchez-Palencia, On the vanishing viscosity limit for acoustic phenomena in a bounded region, Archiv for Rat. Mech. and Anal., 75 (1981), 257–268.

[13] M. E. Gurtin, The linear theory of elasticity. In: Flügge S., Truesdell C. (ed.) Handbuch der Physik, vol. YIa/2, pp. 347–424, Springer, Berlin, 1972.

[14] R. J. Knops, L. E. Payne, Uniqueness theorems in linear elasticity, Springer, New York, 1971.

[15] A. N. Kozhevnikov, On the second and third boundary value problems of the static elasticity theory, Sov. Math. Dokl., 38 (1989), 427–430.

[16] A. N. Kozhevnikov, The basic boundary value problems of the static elasticity theory and their Cosserat spectrum, Mathematische Zeitschrift, 213 (1993), 241–274.

[17] A. Kozhevnikov, On the first stationary boundary-value problem of elasticity in weighted Sobolev spaces in exterior domains of $\mathbb{R}^3$, Applied Mathematics and Optimization, 34 (1996), 183–190.

[18] A. Kozhevnikov, On a lower bound for the Cosserat eigenvalues, Applicable Analysis (to appear).

[19] A. Kozhevnikov, T. Skubachevskaya, Some applications of pseudo-differential operators to elasticity, Hokkaido Math. J., 26, no. 2 (1997), 297–322.

[20] A. Kozhevnikov, O. Lepsky, Power Series Solutions to Basic Stationary Boundary Value Problems of Elasticity, Integral Equations and Operator Theory, 31, (1998), 449–469

[21] V. D. Kupradze, T. G. Gegelia, T. V. Burchuladze, H. O. Basheleishvili, Three-dimensional problems of elasticity and thermoelasticity, North-Holland, Amsterdam, New York, 1979.

[22] Levitin, M.R., On the spectrum of a generalized Cosserat problem, C. R. Acad. Sci. Paris, 315, Série I (1992), 925–930

[23] L. Lichtenstein, Über die erste Randwertaufgabe der Elastizitätstheorie, Math. Zeit. 20 (1924), 21–28.

[24] A. E. H. Love, A treatise on the mathematical theory of elasticity, Cambridge University Press, London, 1934.

[25] V. G. Mazya, S. G. Mikhlin, On the Cosserat spectrum of the equations of the theory of elasticity, Vestnik Leningrad Univ. Math. No. 3 (1967), 58–63.

[26] S. G. Mikhlin, On Cosserat functions. In: Problemy matem. analiza. Kraev. zadachi i integr. uravneniya (Problems of mathematical analysis. Boundary value problems and integral equations) Izdat. Leningrad Univ. 1966, 56–59.

[27] S. G. Mikhlin, Further investigation of Cosserat functions, Vestnik Leningrad. Univ., no. 2 (1967), 98–102.

[28] S. G. Mikhlin, Some properties of the Cosserat spectrum of spatial and plane problems of the theory of elasticity, Vestnik Leningrad. Univ., no. 7 (1970), 31–45.

[29] S. G. Mikhlin, The Cosserat spectrum of static problems of the theory of elasticity and its application. In: Problemy mekhaniki tverdogo deformiruemogo tela (Problems of the mechanics of a solid deformable body), Sudostroenie, Leningrad 1970, 265–271.

[30] S. G. Mikhlin, The Cosserat spectrum of problems of the theory of elasticity for infinite domains. In: Issledovaniya po uprugosti i plastichnosti (Investigations on elasticity and plasticity), 1973, no. 9.

[31] S. G. Mikhlin, The spectrum of a family of operators in the theory of elasticity, Russian Math. Surveys., 28, no. 3 (1973), 45–88.

[32] Mikhlin S.G., Morozov, N.F., Paukshto, M.V., The Integral Equations of the Theory of Elasticity, Teubner, Stuttgart, 1995.

[33] X. Markenscoff, M. Paukshto, On the Cavities and Rigid Inclusions Correspondence and the Cosserat Spectrum, Math. Nachr. 177 (1996), 183–188.

[34] X. Markenscoff, M. Paukshto, The Cosserat spectrum in the theory of elasticity and applications, Proc. R. Soc. Lond. A (1998) 454, 631–643.

[35] M. Paukshto, On Some Applications of Integral Equations in Elasticity, Proc.of 4th Int. Conference in Integral Methods in Science and Engineering, IMSE 96, Oulu, Finland.

[36] M. C. Pelissier, Résolution numérique de quelques problèmes raides en mécanique des milieux faiblement compressibles, Calcolo, 12 (1975), 275–314.

[37] R. Temam, Navier-Stokes equations. Theory and numerical analysis, North-Holland, Amsterdam, New York, Oxford (1979).

[38] E. Trefftz, Die mathematische Theorie der Elastizität, Teubner, Leipzig 1931.

[39] W. Velte, On inequalities of Friedrichs and Babuška-Aziz in Dimension Three, Zeitschrift für Analysis und ihre Anwendungen, 17 (1998), 843–857.

Department of Mathematics
University of Haifa
Haifa 31905, Israel,
kogevn@mathcs2.haifa.ac.il

1991 Mathematics Subject Classification: Primary 35P05, 35P15, 35P20, 73C02, 35J45, 35J55, 35Q30

Submitted: 17.02.1999

Operator Theory:
Advances and Applications, Vol. 109
© 1999 Birkhäuser Verlag Basel/Switzerland

Boundary integral equations on a contour with cusp

ALEXANDER SOLOVIEV

Abstract: Theorems on boundary integral equations in the logarithmic potential theory obtained jointly with V. Maz'ya are presented in this paper.

1.   Introduction

In this paper I review some results obtained together with V.Maz'ya, which concern boundary integral equations of the Dirichlet and Neumann problems for the Laplace operator on domains with cusps.

   A classical approach to solving these problems is based on a representation of their solutions in the form of the double layer potential $W\sigma$ and the simple layer potential $V\tau$. For the interior Dirichlet problem with a function φ in the boundary condition and for the exterior Neumann problem with boundary data ψ the densities of potentials can be found from the boundary integral equations

$$(1.1) \qquad \pi\sigma + W\sigma = \varphi$$

and

$$(1.2) \qquad \pi\tau + \left(\frac{\partial}{\partial n}V\right)\tau = \psi,$$

where $\partial/\partial n$ is the derivative with respect to the outward normal to the boundary contour. For a plane domain with Lyapunov's boundary Γ the operators W and $\frac{\partial}{\partial n}V$ are compact in the space of continuous functions $C(\Gamma)$. Therefore the Fredholm alternative holds for either equation (1.1) and (1.2). If the domain has corner points on Γ then the double layer potential is not a compact operator in $C(\Gamma)$. However, as shown by J. Radon (see [1]), for contours with bounded rotation without cusps, the Fredholm theory can be applied to equations (1.1) and (1.2) considered in $C(\Gamma)$ and its conjugate space $C(\Gamma)^*$, respectively. J. Radon introduced the Fredholm radius for an operator T acting in a Banach space B as the maximum R such that Fredholm's theorems are valid for $|\lambda| < R$ for each of the equations

$$\sigma - \lambda T\sigma = \varphi \quad \text{and} \quad \tau - \bar{\lambda}T^*\tau = \psi \quad (\varphi \in B, \psi \in B^*).$$

Here the operator T^*, adjoint to T, acts in the dual space B^*. For a boundary curve Γ with bounded rotation and without cusps J. Radon proved that the Fredholm radius of the double layer potential in $C(\Gamma)$ is greater than 1.

A necessary and sufficient condition for the integral operator in (1.1) to be Fredholm in $L^p(\Gamma)$ for Γ in the just mentioned class was obtained by V. Shelepov (see [2]), whose paper was based on the earlier work by Ya. Lopatinskii [3]. Lopatinskii studied (1.1) in a weighted space of summable functions under the assumption that the contour has a finite number of corner points and has no cusps. By using the Gohberg-Krein formula (see [4]) he found the index of $\pi I - W$.

A different and quite general method of studying boundary integral equations on irregular curves and surfaces was suggested by V. Maz'ya (see [5]). His approach does not use theories of the Fredholm and singular integral operators and is based on representations of solutions to integral equations by solutions of certain auxiliary boundary value problems. Therefore a theory of elliptic boundary value problems in domains with irregular boundaries can be used to get solvability theorems for boundary integral equations.

We shortly describe this approach under the assumption that Ω is bounded by a smooth curve Γ with a single singular point O. We take a Cartesian coordinate system in which the point O coincides with the origin. Under certain general conditions on the function φ there exist harmonic extensions $u^{(i)}$ onto Ω and $u^{(e)}$ onto Ω' of φ satisfying

$$(1.3) \quad \varphi(z) = \frac{1}{2\pi} \lim_{\varepsilon \to 0} \int_{\{\Gamma : |\zeta| > \varepsilon\}} \left(\frac{\partial u^{(i)}}{\partial n_\zeta}(\zeta) - \frac{\partial u^{(e)}}{\partial n_\zeta}(\zeta) \right) \log \frac{1}{|z - \zeta|} ds_\zeta + u^{(e)}(\infty)$$

on $\Gamma \setminus \{O\}$. Let $v^{(e)}$ denote a solution of the Neumann problem in Ω', vanishing at infinity, with the function $\partial u^{(i)}/\partial n$ on $\Gamma \setminus \{O\}$ in the boundary condition. We can choose $v^{(e)}$ so that for $w = v^{(e)} - u^{(e)} + u^{(e)}(\infty)$ on $z \in \Gamma \setminus \{O\}$ the equality

$$(1.4) \quad \pi w(z) - \lim_{\varepsilon \to 0} \int_{\{\Gamma : |\zeta| > \varepsilon\}} w(\zeta) \frac{\partial}{\partial n_\zeta} \log \frac{1}{|z - \zeta|} ds_\zeta = -2\pi(\varphi(z) - u^{(e)}(\infty))$$

holds. Solutions of equations (1.1) and (1.2) are constructed by means of (1.3) and (1.4). So, the function

$$\sigma = \frac{1}{2\pi}(v^{(e)} - \varphi)$$

is a solution of (1.1). A solution of (1.2) can be obtained as follows. Let us introduce the solution $v^{(e)}$ of the Neumann problem in Ω' with boundary data ψ, vanishing at infinity, and the harmonic extension $u^{(i)}$ onto Ω of the restriction of $v^{(e)}$ to $\Gamma \setminus \{O\}$. Under sufficiently general assumptions on the function ψ we can select $v^{(e)}$ and $u^{(i)}$ so that the density

$$\tau = \frac{1}{2\pi}\left(\frac{\partial u^{(i)}}{\partial n} - \psi \right)$$

satisfies (1.2).

Using this method we (see [6]–[8]) obtained asymptotic formulae for solutions of integral equations of the logarithmic potential theory near cusps on the boundary contour. Moreover, this approach permitted us to find, for each integral equation under consideration, a pair of function spaces such that the corresponding integral operator maps one space onto another (see [9]–[11]).

I state some of our results in the sequel.

2. Integral equations in weighted Sobolev spaces

1. Let Ω is a plane simply connected domain with compact closure and let $\Gamma \setminus \{O\}$ belong to the class C^2. We say that O is an outward (inward) peak if Ω (the exterior domain Ω') is given near O by the inequalities $\kappa_-(x) < y < \kappa_+(x)$, $0 < x < \delta$, where

$$x^{-\mu-1}\kappa_\pm(x) \in C^2([0,\delta]), \quad \lim_{x\to +0} x^{-\mu-1}\kappa_\pm(x) = \alpha_\pm$$

with $\mu > 0$ and $\alpha_+ > \alpha_-$.

If $|q|^\beta \varphi \in L_p(\Gamma)$ then we say that φ belongs to $\mathcal{L}_{p,\beta}(\Gamma)$. The norm in this space is given by

$$\| \varphi \|_{\mathcal{L}_{p,\beta}(\Gamma)} = \| \, |q|^\beta \varphi \, \|_{L_p(\Gamma)} \, .$$

Let $\mathcal{L}^1_{p,\beta}(\Gamma)$ be the space of absolutely continuous functions on $\Gamma \setminus \{O\}$ with the finite norm

$$\| \varphi \|_{\mathcal{L}^1_{p,\beta}(\Gamma)} = \| \varphi' \|_{\mathcal{L}_{p,\beta}(\Gamma)} + \| \varphi \|_{\mathcal{L}_{p,\beta-1}(\Gamma)},$$

where φ' is the derivative with respect to the arc length. We introduce the space $\mathfrak{N}_{p,\beta}(\Gamma)$ of absolutely continuous functions φ on $\Gamma \setminus \{O\}$ with the finite norm

$$\| \varphi \|_{\mathfrak{N}_{p,\beta}(\Gamma)} = \left(\int_0^\delta |\varphi_+(x) - \varphi_-(x)|^p x^{p(\beta-\mu)} \, dx \right)^{1/p} + \| \varphi \|_{\mathcal{L}^1_{p,\beta+1}(\Gamma)},$$

where $\varphi_\pm = \varphi|_{\Gamma_\pm}$. By $\mathfrak{N}^{-1}_{p,\beta}(\Gamma)$ we denote the space of functions on $\Gamma \setminus \{O\}$ represented in the form $\varphi = \psi'$, where $\psi \in \mathfrak{N}_{p,\beta}(\Gamma)$ and $\psi(z_0) = 0$ with a fixed point $z_0 \in \Gamma \setminus \{O\}$. A norm on $\mathfrak{N}^{-1}_{p,\beta}(\Gamma)$ is defined by

$$\| \varphi \|_{\mathfrak{N}^{-1}_{p,\beta}(\Gamma)} = \| \psi \|_{\mathfrak{N}_{p,\beta}(\Gamma)} \, .$$

Also we introduce the space $\mathfrak{N}^{(+)}_{p,\beta}(\mathfrak{d})$ of absolutely continuous functions φ on $\Gamma \setminus \{O\}$ with the finite norm

$$\| \varphi \|_{\mathfrak{N}^{(+)}_{p,\beta}(\Gamma)} = \left(\int_0^\delta |\varphi_+(x) + \varphi_-(x)|^p x^{p(\beta-\mu)} \, dx \right)^{1/p} + \| \varphi \|_{\mathcal{L}^1_{p,\beta+1}(\Gamma)} \, .$$

The following two statements, concerning auxiliary boundary value problems, are crucial for the implementation of Maz'ya's scheme.

Theorem 2.1. *Let Ω have an outward peak and let φ belong to the space $\mathfrak{N}_{p,\beta}(\Gamma)$, where $0 < \beta + p^{-1} < \min\{\mu, 1\}$. Then there exists a harmonic extension u of φ onto Ω with normal derivative in $\mathfrak{N}_{p,\beta}^{-1}(\Gamma)$ satisfying*

$$\| \partial\varphi/\partial n \|_{\mathfrak{N}_{p,\beta}^{-1}(\Gamma)} \le c \, \| \varphi \|_{\mathfrak{N}_{p,\beta}(\Gamma)},$$

where the constant c does not depend on φ.

A proof of this theorem can be found in [12] (see also [9]). Before stating the following result we introduce the number

$$m = \left[\mu - \beta - p^{-1} + 2^{-1} \right].$$

Theorem 2.2. *Let Ω have an inward peak and let $\varphi \in \mathfrak{N}_{p,\beta}^{(+)}(\Gamma)$, where*

$$0 < \beta + p^{-1} < \min\{\mu, 1\}, \quad \mu - \beta - p^{-1} + 2^{-1} \notin \mathbf{N}.$$

Then there exists a harmonic extension of φ onto Ω' with normal derivative in the space $\mathcal{L}_{p,\beta+1}(\Gamma)$ such that the conjugate function g, $g(z_0) = 0$ with a fixed point $z_0 \in \Gamma \setminus \{O\}$, has the representation

$$\sum_{k=1}^{m} c_k(\varphi) \operatorname{Re} z^{k-1/2} + g^{\#}(z),$$

where $c_k(\varphi)$ are linear continuous functionals in $\mathfrak{N}_{p,\beta}^{(+)}(\Gamma)$ and $g^{\#}$ satisfies

$$\|g^{\#}\|_{\mathfrak{N}_{p,\beta}(\Gamma)} \le c \|\varphi\|_{\mathfrak{N}_{p,\beta}^{(+)}(\Gamma)}$$

with c independent on φ.

For the proof of this theorem see [11].

2. As in a classical approach we represent a solution of problem (1.1) in the form

$$u(z) = W\sigma(z) \quad (z = x + iy \in \Omega),$$

where $W\sigma$ is the double layer potential

$$W\sigma(z) = \int_{\Gamma} \sigma(q) \frac{\partial}{\partial n_q} \log \frac{1}{|z - q|} ds_q.$$

Then the density σ satisfies the equation

$$(\pi I - W)\sigma = -\varphi \quad \text{on } \Gamma \setminus \{O\}.$$

Theorem 2.3. *Let Ω have an outward peak and let*

$$0 < \beta + p^{-1} < \min\{\mu, 1\}, \quad \beta + p^{-1} \neq 1/2.$$

Then the operator

$$\mathcal{L}^1_{p,\beta+1}(\Gamma) \ni \sigma \xrightarrow{\;\;W\;\;} (\pi I - W)\sigma \in \mathfrak{N}_{p,\beta}(\Gamma)$$

is surjective. Moreover,
  a) $\ker W = \{0\}$ *for* $0 < \beta + p^{-1} < 1/2$;
  b) $\dim \ker W = 1$ *for* $1/2 < \beta + p^{-1} < 1$.
In the last case6s

$$\ker W = \left\{ t \operatorname{Re} \frac{1}{\gamma_0} \right\},$$

where $t \in \mathbf{R}$ and γ_0 is the conformal mapping of Ω' onto $\mathbf{R}^2_+$, normalized by the conditions $\gamma_0(0) = 0$, $\gamma_0(\infty) = i$.

This theorem is proved in [10], where also an integral equation of the Neumann problem in a domain with outward peak is considered.

Let $\mathfrak{P}(\Gamma)$ denote the space of restrictions to $\Gamma \setminus \{O\}$ of real-valued functions of the form $p(z) = \sum_{k=0}^m t^{(k)} \operatorname{Re} z^k$. A norm of p is defined by

$$\| p \|_{\mathfrak{P}(\Gamma)} = \sum_{k=0}^m |t^{(k)}|.$$

The space $\mathfrak{M}_{p,\beta}(\Gamma)$ is defined as the direct sum of $\mathfrak{N}^{(+)}_{p,\beta}(\Gamma)$ and $\mathfrak{P}(\Gamma)$.

By $\mathfrak{M}^{-1}_{p,\beta}(\Gamma)$ we denote the space of functions on $\Gamma \setminus \{O\}$ represented in the form $\varphi = \psi'$, where $\psi \in \mathfrak{M}_{p,\beta}(\Gamma)$ and $\psi(z_0) = 0$ at a fixed point $z_0 \in \Gamma \setminus \{O\}$. We supply $\mathfrak{Y}_{p,\beta}(\Gamma)$ with the norm

$$\|\varphi\|_{\mathfrak{M}^{-1}_{p,\beta}(\Gamma)} = \|\psi\|_{\mathfrak{M}_{p,\beta}(\Gamma)}.$$

If Ω has an outward peak we represent the solution of the exterior Neumann problem with boundary data ψ as

$$u(z) = (V\sigma)(z) - \sum_{k=1}^m t^{(k)} \mathcal{R}^{ext}_k(z), \quad z \in \Omega^c.$$

Here

$$(V\sigma)(z) = \int_\Gamma \sigma(q) \log \frac{|z|}{|z-q|} ds_q$$

and

$$\mathcal{R}^{ext}_k(z) = \operatorname{Re}\left(\frac{z\, z_0}{z_0 - z} \right)^{k-1/2}, \quad k = 1, \ldots, m,$$

where z_0 is a fixed point in Ω. The density σ and the vector $t = \left(t^{(1)}, \ldots, t^{(m)}\right)$ are defined by

$$\pi\sigma - \left(\frac{\partial}{\partial n}V\right)\sigma + \sum_{k=1}^{m} t^{(k)}\frac{\partial}{\partial n}\mathcal{R}_k^{ext} = -\psi \ \text{ on } \ \Gamma\setminus\{O\}.$$

The following theorem and some results concerning the integral equation of the Dirichlet problem in a domain with an inward peak are given in [11].

Theorem 2.4. *Let Ω have an outward peak and let*

$$0 < \beta + p^{-1} < \min\{\mu, 1\}, \ \ \mu - \beta - p^{-1} + 2^{-1} \notin \mathbf{N}.$$

Then the operator

$$(2.1) \ \ \mathcal{L}_{p,\beta+1}(\Gamma) \times \mathbf{R}^m \ni (\sigma, t) \longmapsto \pi\sigma - \left(\frac{\partial}{\partial n}V\right)\sigma + \sum_{k=1}^{m} t^{(k)}\frac{\partial}{\partial n}\mathcal{R}_k^{ext} \in \mathfrak{M}_{p,\beta}^{-1}(\Gamma)$$

is surjective. If $\mu - \beta - p^{-1} + 2^{-1} \in \mathbf{N}$, the operator (2.1) is not Fredholm.

A different way of solving the Dirichlet problem in Ω with a boundary function φ consists in the representation of the solution as

$$u(z) = V\sigma(z) + c, \ \ z \in \Omega,$$

where $c \in \mathbf{R}$. The density σ and the constant c satisfy the equation

$$V\sigma + c = \varphi \ \text{ on } \ \Gamma\setminus\{O\}.$$

Theorem 2.5. *Let Ω have either outward or inward peak and let*

$$0 < \beta + p^{-1} < \min\{\mu, 1\}, \ \ \beta + p^{-1} \neq 1/2.$$

Then the operator

$$\mathcal{L}_{p,\beta+1}(\Gamma) \times \mathbf{R} \ni (\sigma, t) \xrightarrow{\ \mathfrak{V}\ } V\sigma + c \in \mathfrak{N}_{p,\beta}(\Gamma)$$

is surjective and
   a) $\ker \mathfrak{V} = \{0\}$ for $0 < \beta + p^{-1} < 1/2$;
   b) $\dim \ker \mathfrak{V} = 1$ for $1/2 < \beta + p^{-1} < 1$.
If Ω has an outward peak and $1/2 < \beta + p^{-1} < 1$ then

$$\ker \mathfrak{V} = \left\{\frac{t}{\pi}\frac{\partial}{\partial n}\operatorname{Im}\frac{1}{\gamma^{(out)}}, \ t\operatorname{Im}\frac{1}{\gamma^{(out)}(\infty)}\right\},$$

where $t \in \mathbf{R}$ and $\gamma^{(out)}$ is the conformal mapping of Ω^c onto $\mathbf{R}_+^2$ normalized by $\gamma^{(out)}(0) = 0$ and $\gamma^{(out)}(\infty) = i$.

If Ω has an inward peak and $1/2 < \beta + p^{-1} < 1$, then

$$\ker \mathfrak{V} = \left\{ \frac{t}{\pi} \frac{\partial}{\partial n} \operatorname{Im} \frac{1}{\gamma^{(in)}}, 0 \right\},$$

where $t \in \mathbf{R}$ and $\gamma^{(in)}$ is the conformal mapping of Ω onto $\mathbf{R}_+^2$ normalized by the conditions $\gamma^{(in)}(0) = 0$ and $\gamma^{(in)}(z_0) = i$ with a fixed point $z_0 \in \Omega$.
In the case $\beta + p^{-1} = 1/2$ the operator $\mathfrak{V}$ is not Fredholm.

This theorem is proved in [9].

3. Now, we consider the so called direct version of the method of boundary integral equations when solutions of the Dirichlet and Neumann problems in Ω are represented directly by solutions of boundary integral equations. These equations are obtained from the integral representation for a harmonic function:

$$u(z) = \frac{1}{2\pi} \left(V \frac{\partial u}{\partial n} \right)(z) - \frac{1}{2\pi} (W u)(z), \quad z \in \Omega.$$

For the Dirichlet problem in Ω with boundary data φ the normal derivative $\partial u / \partial n$ on $\Gamma \setminus \{O\}$ solves the integral equation of the first kind

$$(2.2) \qquad\qquad V\gamma = \pi\varphi + W\varphi$$

on $\Gamma \setminus \{O\}$. The following two theorems and other theorems concerning the direct method can be found in [12].

Theorem 2.6. *Let Ω have an inward peak, and let $0 < \beta + p^{-1} < \min\{\mu, 1\}$ and $\beta + p^{-1} \neq 1/2$. Then equation (2.2) has a solution $\gamma \in \mathcal{L}_{p,\beta+1}(\Gamma)$ for every $\varphi \in \mathcal{L}_{p,\beta+1}^1(\Gamma)$. This solution is unique for $0 < \beta+p^{-1} < 1/2$ and the homogeneous equation (2.2) has a one-dimensional space of solutions for $1/2 < \beta+p^{-1} < 1$ given by*

$$\left\{ t \frac{\partial}{\partial n} \operatorname{Im} \frac{1}{\gamma^{(in)}} \right\},$$

where $t \in \mathbf{R}$ and $\gamma^{(in)}$ is the conformal mapping Ω onto $\mathbf{R}_+^2$ subject to the conditions $\gamma^{(in)}(0) = 0$ and $\gamma^{(in)}(z_0) = i$ with a fixed point $z_0 \in \Omega$.

For the Neumann problem in Ω the normal derivative $\partial u / \partial n$ of the solution u is defined by the boundary data ψ. The trace σ of u on $\Gamma \setminus \{O\}$ satisfies the equation

$$(2.3) \qquad\qquad \pi\sigma + W\sigma = V\psi.$$

Theorem 2.7. *Let Ω have an outward peak, and let $0 < \beta + p^{-1} < \min\{\mu, 1\}$ and $\mu - \beta - p^{-1} + 2^{-1} \notin N$. Then equation (2.3) with $\psi \in \mathfrak{N}_{p,\beta}^{-1}(\Gamma)$ has the unique solution σ in $\mathfrak{N}_{p,\beta}(\Gamma)$ satisfying $\int_\Gamma \sigma\, ds = 0$.*

References

[1] RADON, J.: Über die Randwertaufgaben beim logarithmischen Potential, *S.-B. Akad. Wiss. Wien Math.-Nat. Kl.* Abt. 2a, 128:7 (1919), 1123–1167

[2] SHELEPOV V. YU.: On the index of an integral operator of the potential type in the space L^p. *Dokl. Akad. Nauk SSSR* 186:6 (1969), 1266–1270 (Russian)

[3] LOPATINSKII, YA. B.: On some type of singular integral equations. *Teor. i prikl. matem., L'vovskii un-t* II (1963), 53–57 (Ukrainian)

[4] GOHBERG, I. C. and KREIN, M. G.: Systems of integral equations on a half-line with kernels depending on the difference of arguments. *UMN* 13:2 (1958), 3–72 (Russian)

[5] MAZ'YA, V. G.: The integral equations of potential theory in domains with piecewise smooth boundaries. *UMN* 36:4 (1981), 229–230 (Russian)

[6] MAZ'YA, V. G. and SOLOVIEV, A. A.: On an integral equation for the Dirichlet problem in a plane domain with cusps on the boundary. *Mat. Sbornik* 180:9 (1989), 1211–1233 (Russian)

[7] ______: On the boundary integral equation of the Neumann problem for a domain with a peak. *Trudy Leningrad. Mat. Ob.* 1 (1990), 109–134 (Russian)

[8] MAZ'YA, V. and SOLOVIEV, A.: Boundary integral equations of the logarithmic potential theory for domains with peaks. *Rend. Mat. Acc. Lincei* (Ser. 9) 6 (1995), 211–236

[9] ______: L_p-theory of a boundary integral equation on a cuspidal contour. *Appl. Anal.* 65 (1997), 289–305

[10] ______: L_p-theory of boundary integral equations on a contour with outward peak. *Integral Equations and Operator Theory* 32 (1998), 75–100

[11] ______: L_p-theory of boundary integral equations on a contour with inward peak. *Zeitschrift für Analysis und ihre Anwendungen* 17:3 (1998), 641–673

[12] ______: L_p-theory of direct boundary integral equations on a contour with peak (to appear)

Department of Mathematics
Chelyabinsk State University
Chelyabinsk
454016 Russia
alsol@cgu.chel.su

1991 Mathematics Subject Classification: Primary 31A10, 45A05

Submitted: 22.02.1999

Operator Theory:
Advances and Applications, Vol. 109
© 1999 Birkhäuser Verlag Basel/Switzerland

On Maz'ya type inequalities for convolution operators

STEFAN EILERTSEN

Abstract. A presentation of my results concerning an inequality for convolution operators invented by Vladimir Maz'ya.

1. Introduction

Let $\Gamma(x) = \gamma|x|^{4-n}$ denote a fundamental solution of the biharmonic operator Δ^2 in $\mathbf{R}^n$. In 1977 Maz'ya [3], [4] proved and used the inequality

$$(1.1) \quad \Re \int_{\mathbf{R}^n} \Delta^2 u(x) \cdot \overline{u(x)} \Gamma(x)\, dx \geq \frac{1}{2}|u(0)|^2 + c\sum_{l=1}^{2} \int_{\mathbf{R}^n} |\nabla_l u(x)|^2 |x|^{2l-n}\, dx,$$

for $u \in C_0^\infty(\mathbf{R}^n)$ to obtain a sufficient condition for the Wiener regularity of a boundary point with respect to this operator. The proof only covers the dimensions $4, 5, 6, 7$, due to the fact that the left integral can take on negative values for higher dimensions. These results were extended, in 1983, to the polyharmonic operator $(-\Delta)^m$, $m > 2$ for dimensions $2m$, $2m + 1$ and $2m + 2$ by Maz'ya-Donchev [7].

If $n > 4$ it follows from Parseval's formula that the positivity of the left integral in (1.1) is equivalent to

$$(1.2) \quad \Re \int_{\mathbf{R}^n} \int_{\mathbf{R}^n} \frac{\psi(x)}{\psi(x-y)} f(x)\overline{f(y)}\, dx\, dy > 0, \quad 0 \neq f \in C_0^\infty(\mathbf{R}^n),$$

where $\psi(x) = |x|^4$ (or $\psi(x) = |x|^{2m}$ in case of the polyharmonic operator). Let us say that a function ψ possessing this positivity property is admissible. It is not at all clear which functions are admissible. In 1996, Maz'ya proposed the above inequality (1.2) as a topic for my master thesis in the one-dimensional case with ψ being a positive polynomial. The inequality proved to be interesting and very sensitive with respect to the choice of ψ. Later I considered the inequality in $\mathbf{R}^n$ for the function $\psi(x) = |x|^{2\lambda}$ and real λ. This corresponds to the fractional Laplacian $(-\Delta)^\lambda$ and lead to regularity results for that operator.

Here I shall give a short survey of my results concerning (1.2). The proofs of these and related results can be found in [1] and [2].

2. One-dimensional polynomials

We begin by considering the polynomials $(1+x^2)^m$. They turn out to be admissible if and only if $m = 1, 2, 3$.

For $m = 1, 2$, the positivity can be easily checked by means of Parseval's formula: Clearly

$$
\begin{aligned}
(2.1) \qquad \Re \int_{\mathbf{R}^1} \int_{\mathbf{R}^1} & \frac{1 + x^2}{1 + (x-y)^2} f(x)\overline{f(y)}\, dx\, dy \\
&= \frac{1}{2} \int_{\mathbf{R}^1} \int_{\mathbf{R}^1} \left(1 + \frac{1 + 2xy}{1 + (x-y)^2} \right) f(x)\overline{f(y)}\, dx\, dy \\
&= \frac{1}{2}|\widehat{f}(0)|^2 + \frac{1}{\pi} \int_{\mathbf{R}^1} \left(|\widehat{f}(\xi)|^2 + 2|\widehat{f'}(\xi)|^2 \right) \widehat{K}(\xi)\, d\xi,
\end{aligned}
$$

where $\widehat{K}$ is the Fourier transform of $1/(1+x^2)$, which is positive. For the case $m = 2$, the positivity follows similarly from the decomposition

$$
\frac{(1+x^2)^2 + (1+y^2)^2}{(1+(x-y)^2)^2} = 1 + \frac{4xy}{1+(x-y)^2} + \frac{1 + 2x^2 y^2}{(1+(x-y)^2)^2}.
$$

It appears that the case $m = 3$ is more involved. Proceeding similarly as in (2.1) and calculating the Fourier transform of $1/(1+x^2)^m$, $m = 1, 2, 3$ we arrive at the identity

$$
\begin{aligned}
\Re \int_{\mathbf{R}^1} \int_{\mathbf{R}^1} & \frac{(1+x^2)^3}{(1+(x-y)^2)^3} f(x)\overline{f(y)}\, dx\, dy \\
&= \frac{1}{2}|\widehat{f}(0)|^2 + \frac{1}{32} \int_{\mathbf{R}^1} e^{-|\xi|} \big((\xi^2 + 3|\xi| + 3)|\widehat{f}|^2 + 48|\widehat{f'}|^2 \\
&\quad + 3(-\xi^2 + 9|\xi| + 9)|\widehat{f''}|^2 + 2(\xi^2 + 3|\xi| + 3)|\widehat{f'''}|^2 \big)\, d\xi.
\end{aligned}
$$

The presence of the negative number -3 makes it nontrivial that the last expression is positive. In fact, by using a certain function f, it can be shown that if in the above integral the number -3 is replaced by $-3 - \epsilon$ then it can take on negative values, for any $\epsilon > 0$. However, in the present case the positivity can be proved by using the following identity for a real function $h \in C_0^\infty[0, \infty)$,

$$
\begin{aligned}
4h(0)^2 + 2\int_0^\infty & e^{-x}\left(x^2 h^2 + 6(1-x)(h')^2 + 3x(2-x)(h'')^2 + 2x^2(h''')^2 \right) dx \\
&= \int_0^\infty x^2 e^{-x} \left(3(h'' - h' + h)^2 + (2h''' - 3h'' + 3h' - h)^2 \right) dx.
\end{aligned}
$$

To verify this identity, one can expand the right side and integrate by parts several times.

We shall now see that when $m \geq 4$ the polynomial $\psi(x) = (1 + x^2)^m$ no longer satisfies (1.2). Let $\phi_A \in C_0^\infty((0, A+1))$ be real and ω be a fixed real number. We define the functions f_A by its Fourier transform

$$\widehat{f_A}(\xi) = e^{|\xi|/2}\phi_A(|\xi|)\cos(\omega\xi/2).$$

The functions ϕ_A can be chosen in such a way that the following asymptotic formula is valid when A tends to infinity,

$$\Re \int_{\mathbf{R}^1} \int_{\mathbf{R}^1} \frac{(1+x^2)^m}{(1+(x-y)^2)^m} f_A(x)\overline{f_A(y)}\, dx\, dy$$
$$= 2^{-3m}\Re\left(3 + \omega^2 - 2\omega i\right)^m A^m/m! + O(A^{m-1}).$$

The range of the argument for $3 + \omega^2 - 2\omega i$ is $[-\pi/6, \pi/6]$ so if $m \geq 4$ then $(3 + \omega^2 - 2\omega i)^m$ takes on values with a negative real part and so the integral can be negative.

In contrast to this, it can be shown that if the positive sequence $(a_j)_1^n$ grows sufficiently fast then the polynomial

$$\psi(x) = (a_1 + x^2)(a_2 + x^2)\ldots(a_n + x^2)$$

satisfies (1.2).

It is curious that, while the polynomial $(1 + x^2)^2$ is admissible, the similar polynomial $1 + x^4$ is not: To obtain necessary algebraic conditions for a polynomial to have the positivity property, one can let the test function approximate a distribution with support in a finite number of points. For instance if we let f tend to $\alpha\delta_t - \delta_t'$, where δ_t is the Dirac measure at t and α is a real constant, we have

$$\int_{\mathbf{R}^1} \int_{\mathbf{R}^1} \frac{p(x)}{p(x-y)} f(x)\overline{f(y)}\, dx\, dy \to \left(\alpha + \frac{\partial}{\partial x}\right)\left(\alpha + \frac{\partial}{\partial y}\right) \frac{p(x)}{p(x-y)}\bigg|_{x=y=t}.$$

For a polynomial with $p'(0) = 0$, the last expression equals

$$\alpha^2 p(t) + \alpha p'(t)p(0)^{-1} + p(t)p''(0)p(0)^{-2}.$$

This being nonnegative for all real α is equivalent to

$$(2.2) \qquad\qquad 4p''(0)p(t)^2 \geq p(0)p'(t)^2.$$

Since the polynomial $1 + x^4$ is not subject to condition (2.2), it cannot have the positivity property. This polynomial is also excluded by another necessary condition. Namely, the range of Fourier transform of $1/p$ must be contained in the right half plain.

The last mentioned condition also gives another striking example of how sensitive the inequality (1.2) is. Despite the evident positivity for $1 + x^2$ which is seen in formula (2.1), the polynomial $1 + (x + \epsilon)^2$ is not admissible for any $\epsilon \neq 0$.

3.   The functions $|x|^{2\lambda}$ in $\mathbf{R^n}$

We denote the quadratic form

$$I_\lambda(f) = \int_{\mathbf{R^n}} \int_{\mathbf{R^n}} \frac{|x|^{2\lambda}}{|x-y|^{2\lambda}} f(x)\overline{f(y)}\, dx\, dy.$$

By considering the results of Maz'ya and Maz'ya-Donchev, along with the easily proved fact that $\Re I_\lambda(f) \geq 0$ for $\lambda = 0, 1$, one is lead to assume that the positivity holds for

(3.1) $$\lambda \in (0,1] \cup [n/2 - 1, n/2)$$

but not for λ in some interval $I \subset (1, n/2-1)$. This turns out to be true. However, for non integer λ's, quite different methods from those used in [4] and [5] have to be applied.

By Parseval's formula we have

$$I_\lambda(f) = \int_{\mathbf{R^n}} (-\Delta)^\lambda \widehat{f}(x) \cdot \overline{\widehat{f}(x)}\, \Gamma(x)\, dx,$$

where Γ is a fundamental solution of the operator $(-\Delta)^\lambda$. The following identity, valid for $0 < \lambda < 1$,

(3.2) $$\Re \int_{\mathbf{R^n}} (-\Delta)^\lambda u(x) \cdot \overline{u(x)}\Gamma(x)\, dx$$

$$= \frac{1}{2}|u(0)|^2 + A_\lambda \int_{\mathbf{R^n}} \int_{\mathbf{R^n}} \frac{|u(x) - u(y)|^2}{|x-y|^{n+2\lambda}} |x|^{2\lambda-n}\, dx\, dy,$$

where $A_\lambda > 0$ ensures the positivity for the left interval in (3.1).

A very interesting fact is that the positivity property of I_λ in $\mathbf{R^n}$ is symmetric about the point $\lambda = n/4$. Thus positivity also for the right interval in (3.1) follows from formula (3.2). For example, the positivity of the Maz'ya-Donchev case $(-\Delta)^m$, $n = 2m+1$ follows from the case of the fractional operator $(-\Delta)^{1/2}$ and vice versa. The symmetry originates from the relation

$$I_\lambda(f) = I_{n/2-\lambda}(\Gamma \cdot (-\Delta)^\lambda \widehat{f}), \quad 0 < \lambda < n/2,$$

for a real f in $C_0^\infty(\mathbf{R^n})$.

It is more complicated to obtain the appropriate minorant for the right interval in (3.1). (I refer to a right-hand side corresponding to those in (1.1) and (3.2), needed in the study of boundary regularity.) Let us now describe a method for that purpose which also leads to necessary conditions for positivity. Writing $r = |x|$, $x' = x/r$, a function $f \in C_0^\infty(\mathbf{R^n})$ can be expanded in a sum

$$f(x) = \sum_{j=0}^{\infty} f_j(r)S_j(r,x')$$

converging in the sense of $L^2(\mathbf{R^n})$, such that $f_j \in C_0^\infty(\mathbf{R^1})$ and for a fixed $r \geq 0$, the function $S_j(r, x')$ is a normalized spherical harmonic function of degree j. This decomposition of f linearizes the form I_λ in the sense that

$$\Re I_\lambda(f) = \frac{1}{2\pi} \sum_{j=0}^\infty \int_{-\infty}^\infty |\widehat{g_j}(\xi)|^2 \Phi_j^\lambda(\xi) \, d\xi,$$

where $g_j(t) = e^{nt} f_j(e^t)$ and Φ_j^λ are complicated functions to be described now. Let ω be the area of the n-dimensional unit ball, $(\lambda)_m = \Gamma(\lambda + m)/\Gamma(\lambda)$ and

$$\varphi_t(\xi) = \frac{t}{t^2 + \xi^2}, \quad \text{for } t > 0, \quad \varphi_0 = \pi\delta.$$

Then the functions Φ_j^λ are given by the, for those λ we consider, absolutely convergent series

$$\Phi_j^\lambda = \omega \sum_{m=0}^\infty \frac{(\lambda)_{m+j}(\lambda - n/2 + 1)_m}{(n/2)_{m+j} m!} (\varphi_{2m+j} + \varphi_{2m+j+2\lambda}).$$

It can immediately be seen that $\Phi_j^\lambda(\xi) \geq 0$ for λ in the right interval of (3.1), so we have a new proof for this case. It must be said that the method I actually used in [2] involves linearizing a more general form than I_λ.

We now turn to describe necessary conditions for positivity. The necessity that $\lim_{\xi \to 0} \Phi_0^\lambda(\xi) \geq 0$ provides a particularly strong condition. (I conjecture that actually the nonnegativity of the limit is also sufficient.) However, the expansion of the limit we arrive at from the definition of Φ_0^λ is quite unpleasant but fortunately the following formula is true,

$$\lim_{\xi \to 0} \Phi_0^\lambda(\xi) = \omega \sum_{m=1}^\infty \left(\frac{(\lambda)_m}{m(n/2)_m} - \frac{\lambda - 1}{m(m + \lambda - 1)} \right).$$

From the requirement that this expression be nonnegative, the simple condition

$$2\lambda^2 > (\lambda - 1)n$$

for the positivity of I_λ follows easily. Both these conditions have the above mentioned symmetry about the point $\lambda = n/4$. The last condition shows that positivity cannot hold in an interval $(a_n, n/2 - a_n)$, which is nonempty for $n \geq 8$, where $a_n \to 1$ as $n \to \infty$.

The topic of such Maz'ya type inequalities contains a number of unanswered questions. One interesting problem is the question of the actual connection between an inequality like (1.1) and the Wiener type regularity. For instance, is there a corresponding sufficient Wiener type condition for Δ^2 in dimension 8 to those obtained in [3] for the dimensions $4, 5, 6, 7$?

References

[1] S. Eilertsen, *On weighted positivity of ordinary differential operators*, to appear in Journal of inequalities and applications.

[2] S. Eilertsen, *On weighted positivity and the Wiener regularity of a boundary point for the fractional laplacian*, to appear.

[3] V. G. Maz'ya, *On the behavior near the boundary of solutions to the Dirichlet problem for the biharmonic operator*, Dokl. Akad. Nauk SSSR, **18:4** (1977), 15–19.

[4] V. G. Maz'ya, *Behavior of solutions to the Dirichlet problem for the biharmonic operator at a boundary point*, Equadiff IV, Lecture Notes in Math. **703**, 250–262, Springer-Verlag, Berlin-Heidelberg, (1979).

[5] V. G. Maz'ya, *On the Wiener Type Regularity of a Boundary Point for the Polyharmonic Operator*, to appear in Appl. Anal.

[6] V. G. Maz'ya, *Sobolev Spaces*, Springer-Verlag, Berlin-New York, (1985).

[7] V. G. Maz'ya, T. Donchev, *On the Wiener regularity of a boundary point for the polyharmonic operator*, Dokl. Bolg. Akad. Nauk **36:2** (1983), 177–179; English translation, Amer. Math. Soc. Transl. **(2)137** (1987), 53–55.

[8] V. G. Maz'ya, M. Otelbaev, *Embedding theorems and the spectrum of a pseudodifferential operator*, Sib. Mat. Zh. **18** (1977) 1073–1087 (Russian). English translation: Siberian Math. J. **18** (1977) 758–769.

Department of Mathematics,
Linköping University,
S-581 83 Linköping, Sweden,
steil@mai.liu.se

1991 Mathematics Subject Classification: Primary 35S15, 35B05, 31B30, 26D10, 31C15

Submitted: 16.02.1999

Operator Theory:
Advances and Applications, Vol. 109

Sharp constants and maximum principles for elliptic and parabolic systems with continuous boundary data

GERSHON KRESIN

Abstract: This review is concerned with results obtained in a series of joint papers by V. Maz'ya and the author, and it covers a number of topics. First, explicit formulae for essential norms of the elastic and hydrodynamic double layer potentials are discussed for boundaries, having vertices and edges, and these norms are considered in the space of continuous vector-valued functions. Secondly, the explicit expressions for the best constants in estimates of solutions of some systems and equations of mathematical physics are surveyed. Third, criteria for the validity of the maximum modulus principle are stated for solutions of elliptic and parabolic systems. The fourth topic deals with necessary and sufficient conditions providing the maximum norm principle for parabolic systems. Here the norm is defined as Minkowski's functional of a convex body.

0. Introduction

This paper is a survey of results on the best constants in some inequalities relating the Fredholm property for boundary integral operators and properties of uniform norms of solutions to linear elliptic and parabolic systems with continuous boundary data. These are joint results with Vladimir Maz'ya published in [26–34, 39].

Before passing to mathematics I would like to say a few words of personal character. It was S.G. Mikhlin who introduced me to V.G. Maz'ya in Leningrad in 1975 after I defended my Master Thesis. I was fortunate to find a teacher in Maz'ya, not only because he is "an encyclopedia of mathematics", as Mikhlin once mentioned to me, but in many other respects as well. When meeting (more seldom now than I would like) we discuss mathematics as well as literature, history, politics, and I am often amazed how keen his analysis and intuition can be. At times, cases of judofobia were a subject of our conversations, since he and myself, as many others, felt the pressure of anti-semitism in the former Soviet Union.

In Israel, congratulating on birthdays, it is a custom to say "Until 120!". The best of wishes never fall out of fashion, none the least on memorable days. So, this I wish V.G. Maz'ya, with all my heart.

The first section of this survey presents the results from [31] relating the theory of matrix integral operators of the double layer potential type in the space of continuous vector-functions. These operators include, in particular, regular integral operators arising in the study of the first boundary value problem in linear elastostatics and hydrodynamics of a viscous incompressible fluid.

A classical part of the theory of harmonic, elastic and hydrodynamic potentials is based on the restriction of sufficient smoothness of the boundary of a domain $\mathcal{D} \subset \mathbb{R}^n$. This assumption enables one to apply the Fredholm theory to the regular boundary integral equations due to the compactness of the corresponding boundary integral operators. The problems arising in applications require the study of equations on boundaries with singularities (for instance, angular points if $n = 2$ and edges or conical points if $n = 3$). Turning to non-smooth curves or surfaces, some difficulties arise because the corresponding boundary integral operators are noncompact.

Considering the logarithmic harmonic potentials of the single and double layer on curves of "bounded rotation" J. Radon has introduced [50, 51] the notions of the *essential norm* and the *Fredholm radius* of a bounded operator.

The essential norm $\|L\|$ of a linear bounded operator L acting on a Banach space $\mathfrak{B}$ is defined as

$$\|L\| = \inf_{\{\mathcal{K}\}} \|L - \mathcal{K}\|_{\mathfrak{B} \to \mathfrak{B}},$$

where $\{\mathcal{K}\}$ is the set of all linear compact operators on $\mathfrak{B}$.

The Fredholm radius $\mathcal{R}(L)$ of the operator L is the radius of the largest circle on the complex λ-plane centered at $\lambda = 0$ inside which $I + \lambda L$ is a Fredholm operator.

The essential norm $\|L\|$ is related to $\mathcal{R}(L)$ by the inequality

$$(0.1) \qquad\qquad\qquad \mathcal{R}(L) \geq \|L\|^{-1}.$$

It was shown by I. Gohberg and A. Marcus [18] that the Fredholm radius of the operator L, acting in a Banach space $\mathfrak{B}$ with the norm $\|\cdot\|_0$, satisfies

$$\mathcal{R}(L) = \sup\left\{\|L\|^{-1} : \|\cdot\| \in \mathcal{N}(\|\cdot\|_0)\right\},$$

where $\|L\|$ is calculated with respect to a norm $\|\cdot\|$ in the set $\mathcal{N}(\|\cdot\|_0)$ of all norms in $\mathfrak{B}$, equivalent to $\|\cdot\|_0$.

The formulae for the essential norm of the double layer potential in the space $C(\partial\mathcal{D})$ were obtained in the works of J. Radon [51] for $n = 2$, Ju. Burago and V. Maz'ya [8], J. Kral [23] for $n > 2$.

Below we give representations for the norm and the essential norm of matrix integral operators of the double layer potential type in the space $C(\partial\mathcal{D})$ of continuous m-component vector-functions with the norm $\|\phi\| = \sup\left\{|\phi(x)| : x \in \partial\mathcal{D}\right\}$, where $|\cdot|$ is the Euclidean length of a vector. In particular, we give formulae for the essential norm of the regular integral operators of elastostatics and hydrodynamics under the assumption that $\partial\mathcal{D}$ has angular points if $n = 2$ and has edges or conical points if $n = 3$. It follows from our results that the reciprocal of the essential norm of the integral operators of elastostatics and hydrodynamics is not the best constant in the lower estimate (0.1) for the Fredholm radius of these operators, contrary to the case of the logarithmic double layer potential on curves of bounded rotation (see J. Radon [51]).

The second section of the survey contains formulas for the best constants in inequalities expressing the maximum modulus theorem for some classical equations and systems of mathematical physics in a half-space (see [26, 27, 29, 39]). For example, we shall give a representation for the best constant in the inequality for suprema of the elastic displacement vector inside a half-space and on its boundary. We shall also present formulas for the best constants in similar inequalities for viscoelastic displacements, velocity vector in a viscous incompressible flow, in Miranda's inequality for the modulus of gradient of a biharmonic function in a half-space, and in inequality for the components of stress tensor in a half-plane.

It should be noted that all abovementioned best constants are greater that one. This means, in particular, that the classical maximum modulus principle is not valid for solutions of the Lamé and Stokes systems in a half-space. This fact was noticed for an elastic ball by G. Polya [47] as early as in 1930.

Other sections of the present survey are concerned with criteria for validity of maximum principles for linear elliptic and parabolic systems of the second order. Theorems on extremal properties of solutions of the second order elliptic and parabolic equations, constituting the contents of the classical maximum principles, find numerous and various applications when investigating uniqueness and approximation of solutions to boundary value problems, or estimating eigenvalues, etc.

For the uniformly elliptic equation in a bounded domain $\mathcal{D} \in \mathbb{R}^n$

$$(0.2) \qquad \sum_{j,k=1}^{n} a_{jk}(x) \frac{\partial^2 u}{\partial x_j \partial x_k} - \sum_{j=1}^{n} a_j(x) \frac{\partial u}{\partial x_j} - a_0(x)u = 0$$

with bounded coefficients and with $a_0(x) \geq 0$ in $\mathcal{D}$, the best known are the two classical forms of theorems on extremal properties of solutions: the weak maximum principle and the strong maximum principle. According to a formulation of the weak maximum principle, the global maximum of the modulus of a solution $u \in C^2(\mathcal{D}) \cap C(\overline{\mathcal{D}})$ to equation (0.2) is attained at the boundary of a domain, in other words,

$$(0.3) \qquad \max_{\overline{\mathcal{D}}} |u| = \max_{\partial \mathcal{D}} |u|.$$

By the strong maximum principle a non-constant solution u cannot attain a non-negative maximum and a nonpositive minimum inside $\mathcal{D}$.

The weak and the strong forms of the maximum principle have been also found for the uniformly parabolic equation in a cylinder $Q_T = \mathcal{D} \times (0, T]$

$$(0.4) \qquad \frac{\partial u}{\partial t} - \sum_{j,k=1}^{n} a_{jk}(x,t) \frac{\partial^2 u}{\partial x_j \partial x_k} + \sum_{j=1}^{n} a_j(x,t) \frac{\partial u}{\partial x_j} + a_0(x,t)u = 0$$

with bounded coefficients and with $a_0(x,t) \geq 0$ in Q_T. For instance, according
to the weak maximum principle, if $u \in C^{(2,1)}(Q_T) \cap C(\overline{Q}_T)$ is a solution of (0.4),
then

$$\max_{\overline{Q}_T} \mid u \mid = \max_{\overline{\Gamma}_T} \mid u \mid,$$

where $\Gamma_T = \{(x,t) \in \partial Q_T : 0 \leq t < T\}$.

We remind that all said above is only concerned with scalar equations. However,
during the last years, due to efforts of many mathematicians, various forms of
the maximum principle are being succesfully transferred to second order elliptic
and parabolic systems. From the point of view of structural restrictions most
of the systems considered are *weakly coupled systems* and *systems with scalar* or
diagonal principal part. Let $\mathcal{A}_{jk}, \mathcal{A}_j, \mathcal{A}_0$ be $(m \times m)$-matrix-valued functions, $u =
(u_1, \ldots, u_m)$. The systems

$$(0.5) \qquad \sum_{j,k=1}^{n} \mathcal{A}_{jk}(x) \frac{\partial^2 u}{\partial x_j \partial x_k} - \sum_{j=1}^{n} \mathcal{A}_j(x) \frac{\partial u}{\partial x_j} - \mathcal{A}_0(x)u = 0,$$

$$(0.6) \qquad \frac{\partial u}{\partial t} - \sum_{j,k=1}^{n} \mathcal{A}_{jk}(x,t) \frac{\partial^2 u}{\partial x_j \partial x_k} + \sum_{j=1}^{n} \mathcal{A}_j(x,t) \frac{\partial u}{\partial x_j} + \mathcal{A}_0(x,t)u = 0$$

are said to be weakly coupled, if the matrices $\mathcal{A}_{jk}, \mathcal{A}_j$ are diagonal. The systems
(0.5), (0.6) are called systems with scalar or diagonal principal part if $\mathcal{A}_{jk}$ are
scalar or diagonal matrices, respectively.

A survey devoted to maximum principles for solutions of equations and weakly
coupled systems can be found in the books by M.H. Protter and H.F. Weinberger
[48] and W. Walter [62] which also contain rich bibliographies on this subject.

For systems with a scalar principal part the maximum modulus principle (i.e.,
theorems on extremal properties of the Euclidean length of a vector solution) and
some of their generalizations are established. These results can be found in the
works of P. Szeptycki [60], T. Stys [59], C. Miranda [44], J. Wasowski [63], G.N.
Hille and M.H. Protter [22], and also in the survey by M.H. Protter [49].

It is clear that in the case of systems some new possibilities connected with the
maximum principle arise. These possibilities are, for example due to a variety of
norms in the space of vector-functions $u = (u_1, \ldots, u_m)$. Instead of comparing
norms of a solution u in a domain $\mathcal{D}$ and on its boundary $\partial \mathcal{D}$ (for elliptic systems)
or in the cylinder Q_T and on its parabolic boundary Γ_T (for parabolic systems) it
is possible to compare the maxima of values $\mathcal{M}(u)$ of some homogeneous convex
nonnegative functional $\mathcal{M}$. It is worth noting that a generalization of the maximum
principle for modulus and for components of solutions of systems in the framework
of geometric approach has given rise to developing the theory of invariant sets
for nonlinear elliptic and parabolic systems, in partiular, for reaction-diffusion
equations (see, for instance, [4, 6, 10, 11, 53, 64] and references therein). However,

we shall not characterize this field, since we consider maximum principles and invariant sets only for linear systems and the just mentioned papers on invariant sets do not contain our results as special cases.

Thus a problem arises: for a functional $\mathcal{M}$ given on $\mathbb{R}^m$, find all elliptic systems (0.5), such that the maximum principle

$$(0.7) \qquad \max_{x \in \overline{\mathcal{D}}} \mathcal{M}(u(x)) = \max_{x \in \partial \mathcal{D}} \mathcal{M}(u(x))$$

holds for all solutions $u \in C^2(\mathcal{D}) \cap C(\overline{\mathcal{D}})$ of these systems. An analogous problem can be formulated for parabolic systems (0.6) and the maximum principle

$$(0.8) \qquad \max_{(x,t) \in \overline{Q}_T} \mathcal{M}(u(x,t)) = \max_{(x,t) \in \overline{\Gamma}_T} \mathcal{M}(u(x,t)),$$

where u is an arbitrary solution of (0.6) in $C^{(2,1)}(Q_T) \cap C(\overline{Q}_T)$.

The best possible answer could have been given by necessary and sufficient conditions (on the structure of system), which guarantee the maximum principle (0.7) for elliptic system (0.5) and the maximum principle (0.8) for parabolic system (0.6).

Criteria for validity of the maximum modulus principle for linear elliptic and parabolic systems are given in Sections 3 and 4, respectively. For detailed exposition see [28, 30, 34, 39]. In other words, we state necessary and sufficient conditions for the constant in the inequality between the suprema of the modulus of the solution inside $\mathcal{D}$ and on its boundary $\partial \mathcal{D}$ (for elliptic systems) or inside the cylinder Q_T and on its parabolic boundary Γ_T (for parabolic systems) to be equal unity.

The last Section 5 concerns our latest results (see [32, 33]) on criteria for validity of the maximum norm principles for linear parabolic systems where the norm is understood in a generalized sense, namely as the Minkowski functional of a compact convex body in $\mathbb{R}^m$.

1. The norm and the essential norm of the double layer elastic and hydrodynamic potentials in the space of continuous functions

This section contains some results on the norm and the essential norm of matrix integral operators of the double layer potential type in the space of continuous vector-functions on non-smooth boundaries [31]. Such operators arise in elastostatics and hydrodynamics of a viscous incompressible fluid.

1.1. Introduction

When studying boundary integral equations on irregular curves and surfaces, one encounters difficulties connected with the change of functional-analytic properties of boundary integral operators. For sufficiently smooth boundaries these operators are compact, whereas they lose compactness if the boundary contains a singularity. However, for a rather large class of irregular curves and surfaces boundary integral operators remain bounded.

Let S be a rectifiable curve and let s denote the arc length on $S(0 \leq s \leq l)$. If the angle $\vartheta(s)$ between the positively oriented tangent and the x-axis is a function of bounded variation on $[0, l]$, the curve S is said to be a curve with bounded rotation.

Suppose that S is the boundary of a domain D with compact closure and that u is a harmonic function on D, satisfying the Dirichlet condition

$$u\Big|_{S} = g,$$

with g in the space of continuous functions $C(S)$. When seeking u as a double layer logarithmic potential with unknown density $\varphi \in C(S)$, one arrives at the integral equation

$$\varphi + T\varphi = 2g \quad \text{on} \quad S.$$

Here T is a continuous operator in $C(S)$.

In [51] Radon proved the formula for the essential norm of T in $C(S)$

$$\|T\| = \alpha/\pi,$$

where

$$\alpha = \sup_{0 \leq s \leq \ell} \big|\theta(s+0) - \theta(s-0)\big|.$$

This enabled him to apply the Fredholm theory to the above integral equation provided $\alpha < \pi$, i.e. there are no cusps on S.

Three years earlier Carleman [9] showed that if S is the union of a finite number N of closed arcs $p_j p_{j+1}(p_{N+1} = p_1)$ of the class C^2, then the essential norm of T in the space with the norm

$$\sup_{j} \sup_{x \in S} |x - p_j|^{\kappa_j} |u(x)|, \quad 0 \leq \kappa_j \leq 1,$$

admits the estimate

$$\|T\| \leq \sup_{j} \frac{\sin(|\pi - \alpha_j|\kappa_j)}{\sin(\pi \kappa_j)},$$

where α_j is the angle between the half-tangents to S at p_j. In the case $\kappa_j = 0$, Carleman's estimate coincides with that of Radon.

Carleman [9] also established an analogous result for two-dimensional surfaces S comprised by two surfaces S_1 and S_2 of the class C^2 having a common curved edge

E, which is assumed to be a twice continuously differentiable curve. He proved, in particular, that for the essential norm of T on the space $C(S)$ the estimate

$$\|T\| \leq \sup_{p \in E} |1 - \alpha(p)/\pi|$$

holds, $\alpha(p)$ being the minimum of the two angles made by the tangential planes to S_1 and S_2 at $p \in E$.

In Burago and Maz'ya [8] and Král [23] it was shown that for a multidimensional domain $\mathcal{D}$ with bounded variation of the solid angle $\omega_{\mathcal{D}}(p, \mathbb{B})$ ($p \in \partial\mathcal{D}$, $\mathbb{B} \subset \mathbb{R}^n$) the essential norm of T in $C(\partial\mathcal{D})$ is

$$(1.1) \qquad \frac{2}{\omega_n} \lim_{\delta \to 0} \sup_{p \in \partial\mathcal{D}} \operatorname{var} \omega_{\mathcal{D}}(p, \cdot)\big(B_\delta(p)\big),$$

where $B_\delta(p) = \big\{x \in \mathbb{R}^n : |x - p| < \delta\big\}$ and ω_n is the area of the unit sphere $S^{n-1} = \big\{x \in \mathbb{R}^n : |x| = 1\big\}$.

Many results on the invertibility and the Fredholm property of boundary integral operators on piecewise smooth surfaces are either presented or cited in the papers [5, 12, 14, 15, 19-21, 24, 25, 37, 38, 52, 54, 61].

In the present section we give formulae for the norms and essential norms of boundary integral operators for two- and three-dimensional elastostatics and hydrodynamics in the space of continuous vector-valued functions on the boundary.

We start by introducing the vector-valued double layer potentials of the form

$$(1.2) \qquad W(q) = \int_{\partial\mathcal{D}} K(e_{qx})\phi(x)\omega_{\mathcal{D}}(q, dx)$$

and consider the matrix-valued integral operators appearing in the boundary integral equations generated by these potentials. In (1.2) we have used the notation: $\mathcal{D}$ is an open set in $\mathbb{R}^n$ with the compact closure $\overline{\mathcal{D}}$ and the boundary $\partial\mathcal{D}$; $q \notin \partial\mathcal{D}$; K is a homogeneous of degree 0 even ($m \times m$)-matrix-valued function continuous on S^{n-1}; $e_{qx} = (x - q)|x - q|^{-1}$; ϕ is an element of the space $C(\partial\mathcal{D})$, i.e. ϕ is a continuous m-component vector-valued function on $\partial\mathcal{D}$ with the norm

$$\|\phi\| = \sup \big\{|\phi(x)| : x \in \partial\mathcal{D}\big\}.$$

We normalize the matrix-valued funciton K by

$$\int_{S^{n-1}} K(\sigma)d\sigma = I,$$

where I is the unit matrix of order m.

Note that the elastic and the hydrodynamic potentials are special cases of (1.2).

Similar to [8, 23], we assume that $\mathcal{D}$ has a finite perimeter $P(\mathcal{D})$, $\partial\mathcal{D} = \partial(\mathbb{R}^n \backslash \overline{\mathcal{D}})$ and

$$(A) \qquad \sup_{p \in \partial\mathcal{D}} \operatorname{var} \omega_{\mathcal{D}}(p, \cdot)(\partial\mathcal{D} \backslash p) < \infty.$$

1.2. Matrix integral operators of the double layer potential type

First, we remind some facts connected with the perimeter of a set (see e.g. [16]).

Definition 1.1. *Let $\mathbb{B} \subset \mathbb{R}^n$ be a Borel set and suppose its characteristic function $\chi_{\mathbb{B}}$ belongs to the space $BV(\mathbb{R}^n)$ of locally integrable functions on $\mathbb{R}^n$ whose gradients (in the distributional sense) are finite vector-valued charges on $\mathbb{R}^n$. The total variation of the charge $\nabla\chi_{\mathbb{B}}$ is called the perimeter of the set $\mathbb{B}$ (in the sense of Caccioppoli and De Giorgi) and is denoted by $P(\mathbb{B})$.*

Henceforth we assume that $\mathcal{D}$ is a bounded open set in $\mathbb{R}^n$ with finite perimeter $P(\mathcal{D})$ and that $\partial\mathcal{D} = \partial(\mathbb{R}^n\backslash\overline{\mathcal{D}})$.

Definition 1.2. *A unit vector $\nu(p)$ is referred to as the (outer) normal (in the sense of Federer) to the set $\mathcal{D}$ at the point p if*

$$\lim_{\varepsilon\to 0} \varepsilon^{-n}\mathrm{mes}_n\big\{y:\ y\in\mathcal{D}\cap B_\varepsilon(p),\ (\nu(p),y-p)>0\big\}=0,$$

$$\lim_{\varepsilon\to 0} \varepsilon^{-n}\mathrm{mes}_n\big\{y:\ y\in\mathcal{D}\cap B_\varepsilon(p),\ (\nu(p),y-p)<0\big\}=\frac{v_n}{2},$$

where v_n is the volume of the n-dimensional unit ball.

The set of all points $p\in\partial\mathcal{D}$ at which a normal to $\mathcal{D}$ exists is called the reduced boundary of $\mathcal{D}$ and is denoted by $\partial^*\mathcal{D}$. If the normal at p does not exist we put formally $\nu(p)=0$.

It is known (cf. [16]) that

$$(1.3) \qquad\qquad \nabla\chi_{\mathcal{D}}(\mathbb{B}) = \int_{\mathbb{B}\cap\partial\mathcal{D}} \nu(x)H_{n-1}(dx),$$

where H_{n-1} is the $(n-1)$-dimensional Hausdorff measure in $\mathbb{R}^n$.

Let, for brevity, $r_{px}=|x-p|$.

Definition 1.3. *([8]). The solid angle at which a set $\mathbb{B}\cap\partial\mathcal{D}$ is seen from a point $p\notin\overline{\mathbb{B}}$ is the set function*

$$\omega_{\mathcal{D}}(p,\mathbb{B}) = \begin{cases} -\int_{\mathbb{B}} \nabla\log r_{px}^{-1}\,\nabla\chi_{\mathcal{D}}(dx) & \text{for}\quad n=2, \\[2mm] \frac{1}{2-n}\int_{\mathbb{B}} \nabla r_{px}^{2-n}\nabla\chi_{\mathcal{D}}(dx) & \text{for}\quad n>2. \end{cases}$$

In view of (1.3) one make take the integrals over $\mathbb{B}\cap\partial\mathcal{D}$ and replace $\chi_{\mathcal{D}}(dx)$ by $\nu(x)H_{n-1}(dx)$.

For any $p\in\mathbb{R}^n$ and any Borel sets $\mathbb{B}$ whose closures do not contain p we introduce the matrix-valued set function

$$\Psi_{\mathcal{D}}(p,\mathbb{B}) = \int_{\mathbb{B}\cap\partial\mathcal{D}} K(e_{px})\omega_{\mathcal{D}}(p,dx).$$

Condition (A) enables one to extend $\Psi_{\mathcal{D}}(p,\mathbb{B})$ to all Borel subsets of $\mathbb{R}^n\backslash p$.

Let

$$
(1.4) \qquad \Psi_{\mathcal{D}}(p,p) = \begin{cases} 0 & \text{for} \quad p \notin \partial\mathcal{D}, \\[2mm] \frac{1}{2}I - \Psi_{\mathcal{D}}(p, \partial\mathcal{D}\backslash p) & \text{for} \quad p \in \partial\mathcal{D}. \end{cases}
$$

For $p \in \mathbb{B}$ we put

$$
(1.5) \qquad \Psi_{\mathcal{D}}(p, \mathbb{B}) = \Psi_{\mathcal{D}}(p, \mathbb{B}\backslash p) + \Psi_{\mathcal{D}}(p,p).
$$

Thus the matrix-valued function $\mathbb{B} \to \Psi_{\mathcal{D}}(p, \mathbb{B})$ is defined and is finite for all Borel subsets of $\mathbb{R}^n$.

One can see that $\Psi_{\mathcal{D}}(p, \partial\mathcal{D}) = 0$ for $p \notin \overline{\mathcal{D}}$ and that $\Psi_{\mathcal{D}}(p, \partial\mathcal{D}) = I$ for $p \in \mathcal{D}$. Hence and by (1.4), (1.5) one obtains

$$
\Psi_{\mathcal{D}}(p, \partial\mathcal{D}) = \begin{cases} I & \text{for} \quad p \in \mathcal{D}, \\[2mm] \frac{1}{2}I & \text{for} \quad p \in \partial\mathcal{D}, \\[2mm] 0 & \text{for} \quad p \in \mathbb{R}^n\backslash\overline{\mathcal{D}}. \end{cases}
$$

Another important property of $\Psi_{\mathcal{D}}$ is that $\Psi_{\mathcal{D}}(p,p) = 0$ for $p \in \partial^*\mathcal{D}$.

Let $\phi \in C(\partial\mathcal{D})$. We introduce the vector-valued double layer potential

$$
W(q) = \int_{\partial\mathcal{D}} K(e_{qx})\phi(x)\omega_{\mathcal{D}}(q, dx),
$$

where $q \notin \partial\mathcal{D}$. The function W satisfies the following classical relations.

Theorem 1.1. *The inner and the outer limit values of $W(q)$ are*

$$
W^{(i)}(p) = \frac{1}{2}\phi(p) + \int_{\partial\mathcal{D}} \Psi_{\mathcal{D}}(p, dx)\phi(x),
$$

$$
W^{(e)}(p) = -\frac{1}{2}\phi(p) + \int_{\partial\mathcal{D}} \Psi_{\mathcal{D}}(p, dx)\phi(x).
$$

Theorem 1.1 implies the continuity of the operator T defined by

$$
(T\phi)(p) = 2 \int_{\partial\mathcal{D}} \Psi_{\mathcal{D}}(p, dx)\phi(x)
$$

in the space $C(\partial\mathcal{D})$.

Theorem 1.2. *Let K^* denote the adjoint matrix of K. Then*

$$\|T\| = 2 \sup_{p \in \partial^* \mathcal{D}} \sup_{|z|=1} \int_{\partial \mathcal{D}} |K^*(e_{px})z| |\omega_{\mathcal{D}}|(p, dx).$$

Corollary 1.1. *Let $\mathcal{D}$ be a bounded convex domain in $\mathbb{R}^n$. Then*

$$\|T\| = \sup_{|z|=1} \int_{S^{n-1}} |K^*(\sigma)z| d\sigma.$$

The next assertion extends representation (1.1) for the essential norm of the harmonic double layer potential to the vector case.

Theorem 1.3. *The essential norm of T in $C(\partial \mathcal{D})$ is*

$$2 \lim_{\delta \to 0} \sup_{p \in \partial^* \mathcal{D}} \sup_{|z|=1} \int_{\partial \mathcal{D} \cap B_{\delta}(p)} |K^*(e_{px})z| \, |\omega_{\mathcal{D}}|(p, dx).$$

The next assertion follows from the last theorem.

Theorem 1.4. *Let $\mathcal{D}$ be a bounded convex domain. Then*

$$\|T\| = \sup_{p \in \partial \mathcal{D} \setminus \partial^* \mathcal{D}} \sup_{|z|=1} \left[\int_{S^{n-1}} |K^*(\sigma)z| d\sigma - 2 \int_{\mathcal{E}(p)} |K^*(\sigma)z| d\sigma \right]$$

if $\partial \mathcal{D} \neq \partial^* \mathcal{D}$ and $\|T\| = 0$ if $\partial \mathcal{D} = \partial^* \mathcal{D}$. Here $\mathcal{E}(p)$ is a topological limit of the projections $\mathcal{E}_{\varepsilon}(p)$ of the sets $\mathcal{D} \cap \partial B_{\varepsilon}(p)$ onto the sphere $\partial B_1(p)$ as $\varepsilon \to 0$.

1.3. Boundary integral operators of elastostatics and hydrodynamics

Consider the internal and the external Dirichlet problems for the Lamé system

$$(1.6) \qquad \mu \Delta u + (\lambda + \mu) \mathrm{grad} \, \mathrm{div} \, u = 0 \quad \text{in} \quad \mathcal{D}, \qquad u\big|_{\partial \mathcal{D}} = f,$$

$$(1.7) \qquad \mu \Delta u + (\lambda + \mu) \mathrm{grad} \, \mathrm{div} \, u = 0 \quad \text{in} \quad \mathbb{R}^n \setminus \overline{\mathcal{D}}, \; u\big|_{\partial \mathcal{D}} = f,$$

where λ and μ are the Lamé constants, $u = (u_1, \ldots, u_n)$ is the displacement vector and $f \in C(\partial \mathcal{D})$. Note that in this subsection $m = n$.

Elements of the fundamental solution $\Gamma(x - q)$ of the Lamé system (the Kelvin-Somigliana tensor) are

$$-\frac{\lambda + \mu}{4\pi \mu (\lambda + 2\mu)} \left[\frac{\lambda + 3\mu}{\lambda + \mu} \delta_{ij} \log \frac{1}{r_{qx}} + \frac{(x_i - q_i)(x_j - q_j)}{r_{qx}^2} \right] \quad \text{for} \quad n = 2,$$

$$-\frac{\lambda + \mu}{2\omega_n \mu (\lambda + 2\mu)} \left[\frac{\lambda + 3\mu}{(n-2)(\lambda + \mu)} \delta_{ij} \frac{1}{r_{qx}^{n-2}} + \frac{(x_i - q_i)(x_j - q_j)}{r_{qx}^n} \right] \quad \text{for} \quad n \geq 3,$$

where $1 \leq i, \, j \leq n$, δ_{ij} is the Kronecker delta and ω_n is the area of the unit $(n-1)$-dimensional sphere.

The boundary value problems (1.6), (1.7) can be reduced to the system of integral equations using the elastic double layer potential

$$(1.8) \qquad \mathfrak{W}_\gamma(q) = \int_{\partial D} \left[\mathcal{T}_\gamma(\partial/\partial x, \nu(x)) \Gamma(x - q) \right]^* \phi(x) H_{n-1}(dx), \quad q \notin \partial D,$$

depending on the parameter $\gamma > 0$ (see, for example, [35]). Here

$$\mathcal{T}_\gamma(\partial/\partial x, \nu(x))$$

is an $(n \times n)$-matrix-valued differential operator referred to as the generalized stress operator, * means the transition to the transposed matrix and $\phi \in C(\partial D)$. The value of the operator $\mathcal{T}_\gamma(\partial/\partial x, \nu(x))$ on a vector u is defined by

$$\mathcal{T}_\gamma(\partial/\partial x, \nu(x)) u = (\mu + \gamma)(\nu(x), \nabla) u + (\lambda + \mu - \gamma)\nu(x) \operatorname{div} u + \gamma \nu(x) \times \operatorname{rot} u.$$

For $\gamma = \mu(\lambda + \mu)(\lambda + 3\mu)^{-1}$ we adopt the notation $\mathcal{W}_{el}(q)$ for the potential (1.8). If $\gamma = \mu(\lambda + \mu)(\lambda + 3\mu)^{-1}$, the operator $\mathcal{T}_\gamma(\partial/\partial x, \nu(x))$ is denoted by $\mathcal{N}(\partial/\partial x, \nu(x))$, is called the pseudo-stress operator, and is defined as

$$\mathcal{N}(\partial/\partial x, \nu(x)) u = \frac{2\mu(\lambda + 2\mu)}{\lambda + 3\mu} (\nu(x), \nabla) u + \frac{(\lambda + \mu)(\lambda + 2\mu)}{\lambda + 3\mu} \nu(x) \operatorname{div} u$$

$$+ \frac{\mu(\lambda + \mu)}{\lambda + 3\mu} \nu(x) \times \operatorname{rot} u.$$

For the boundary of the class $C^{1,\alpha}$, $0 < \alpha < 1$, the kernel of the integral operator of the system for the density of the potential $\mathcal{W}_{el}(q)$ has the singularity $O(|q - x|^{\alpha-n})$. Clearly,

$$\mathcal{W}_{el}(q) = \int_{\partial D} \left[\mathcal{N}(\partial/\partial x, \nu(x)) \Gamma(x - q) \right]^* \phi(x) H_{n-1}(dx)$$

or, in vectoral notation,

$$(1.9) \qquad \mathcal{W}_{el}(q) = \frac{2\mu}{\omega_n(\lambda + 3\mu)} \int_{\partial D} \frac{(x - q, \nu(x))\phi(x)}{r_{qx}^n} H_{n-1}(dx)$$

$$+ \frac{n(\lambda + \mu)}{\omega_n(\lambda + 3\mu)} \int_{\partial D} \frac{(x - q, \nu(x))(x - q, \phi(x))(x - q)}{r_{qx}^{n+2}} H_{n-1}(dx),$$

where $q \notin \partial D$, $\phi \in C(\partial D)$.

Consider now the internal and the external Dirichlet problems for the Stokes system

$$(1.10) \qquad \eta \Delta u - \operatorname{grad} p = 0, \ \operatorname{div} u = 0 \ \text{in} \ \mathcal{D}, \qquad u\big|_{\partial D} = f,$$

$$(1.11) \qquad \eta \Delta u - \operatorname{grad} p = 0, \ \operatorname{div} u = 0 \ \text{in} \ \mathbb{R}^n \backslash \overline{\mathcal{D}}, \ u\big|_{\partial D} = f,$$

where η is the coefficient of kinematic viscosity, $u = (u_1, \ldots, u_n)$ is the velocity vector of a fluid and p is the pressure.

The components

$$\left[U^j(x-q),\ P^j(x-q)\right]_{j=1}^{n}$$

of the fundamental solution $U(x-q)$, $P(x-q)$ of the Stokes system are given by

$$U_i^j(x-q) = -\frac{1}{4\pi\eta}\left[\delta_{ij}\log\frac{1}{r_{qx}} + \frac{(x_i-q_i)(x_j-q_j)}{r_{qx}^2}\right] \quad \text{for}\quad n=2,$$

$$U_i^j(x-q) = -\frac{1}{2\omega_n\eta}\left[\frac{\delta_{ij}}{n-2}\delta_{ij}\frac{1}{r_{qx}^{n-2}} + \frac{(x_i-q_i)(x_j-q_j)}{r_{qx}^n}\right] \quad \text{for}\quad n\geq 3,$$

$$P^j(x-q) = -\frac{x_j-q_j}{\omega_n r_{qx}^n}, \ 1\leq i,j\leq n.$$

In the case of the domain $\mathcal{D}$ with boundary of the class $C^{1,\alpha}$ problems (1.10), (1.11) can be reduced to systems of Fredholm integral equations using the hydrodynamical double layer potential

$$\mathcal{W}_{hyd}(q) = \int_{\partial\mathcal{D}}\left[\mathcal{S}(\partial/\partial x,\nu(x))\big(U(x-q),\ P(x-q)\big)\right]\phi(x)H_{n-1}(dx),$$

where $q\notin\partial\mathcal{D}$ and

$$\mathcal{S}(\partial/\partial x,\nu(x))(V,Q) = Q\nu(x) + \eta\left[2(\nu(x),\nabla)V + \nu(x)\times\operatorname{rot}V\right].$$

One can write $\mathcal{W}_{hyd}(q)$ in the form

$$(1.12)\qquad \mathcal{W}_{hyd}(q) = \frac{n}{\omega_n}\int_{\partial\mathcal{D}}\frac{(x-q,\ \nu(x))(x-q,\ \phi(x))(x-q)}{r_{qx}^{n+2}}H_{n-1}(dx),$$

with $q\notin\partial\mathcal{D}$ and $\phi\in C(\partial\mathcal{D})$.

We introduce the vector-valued potential

$$(1.13)\qquad W_\kappa^{(n)}(q) = \frac{(1-\kappa)}{\omega_n}\int_{\partial\mathcal{D}}\frac{(x-q,\ \nu(x))\phi(x)}{r_{qx}^n}H_{n-1}(dx)$$

$$+\frac{n\kappa}{\omega_n}\int_{\partial\mathcal{D}}\frac{(x-q,\ \nu(x))(x-q,\ \phi(x))(x-q)}{r_{qx}^{n+2}}H_{n-1}(dx),$$

where $q\notin\partial\mathcal{D}$, κ is a parameter with values in $[0,1]$, and $\phi\in C(\partial\mathcal{D})$.

We note that in the two and three-dimensional elasticity theory the Lamé constants λ and μ satisfy the conditions $\mu>0$, $3\lambda+2\mu>0$ which lead to the restriction

$$0 < (\lambda+\mu)(\lambda+3\mu)^{-1} < 1.$$

In what follows we assume that the last inequalities are valid.

From (1.9), (1.12) and (1.13) one obtains

$$W_\kappa^{(n)}(q) = \mathcal{W}_{el}(q)\quad\text{for}\quad \kappa = (\lambda+\mu)(\lambda+3\mu)^{-1},$$

$$W_\kappa^{(n)}(q) = \mathcal{W}_{hyd}(q)\quad\text{for}\quad \kappa = 1.$$

Consider the $(n \times n)$-matrix-valued function $K_\kappa^{(n)}(e)$, defined on the sphere S^{n-1}, with the entries

$$\frac{1}{\omega_n}\Big[(1-\kappa)\delta_{ij} + n\kappa(e,e_i)(e,e_j)\Big],$$

e_j being the unit vector directed along the j-th coordinate axis. According to (1.3) and Definition 1.3, the solid angle is

$$\omega_{\mathcal{D}}(q,\mathbb{B}) = \int_{\mathbb{B}\cap\partial\mathcal{D}} \frac{\big(x-q,\nu(x)\big)}{r_{qx}^n} H_{n-1}(dx).$$

Therefore, the potential (1.13) is a particular case of the vector potential (1.2):

$$(1.14) \qquad W_\kappa^{(n)}(y) = \int_{\partial\mathcal{D}} K_\kappa^{(n)}(e_{qx})\phi(x)\omega_{\mathcal{D}}(q,dx).$$

Following (1.4), (1.5), we define the set function $\Psi_{\mathcal{D}}(p,\mathbb{B})$ for $K_\kappa^{(n)}(e)$. Applying Theorem 1.1 to the potential (1.14), we arrive at the equation

$$\phi(p) + \big(T_\kappa^{(n)}\phi\big)(p) = 2f(p)$$

for the internal boundary value problems (1.6), (1.10) and at the equation

$$-\phi(p) + \big(T_\kappa^{(n)}\phi\big)(p) = 2f(p)$$

for the external boundary value problems (1.7), (1.11). The integral operator

$$\big(T_\kappa^{(n)}\phi\big)(p) = 2\int_{\partial\mathcal{D}} \Psi_{\mathcal{D}}(p,dx)\phi(x)$$

acts in $C(\partial\mathcal{D})$ and is continuous.

Theorem 1.5. *Let $\mathcal{D}$ be a convex domain. Then*

$$\|T_\kappa^{(n)}\| = \frac{2\Gamma(\frac{n}{2})}{\sqrt{\pi}\,\Gamma(\frac{n-1}{2})} \int_0^{\pi/2} \big[(1-\kappa)^2 + n\kappa(n\kappa - 2\kappa + 2)\cos^2\vartheta\big]^{1/2} \sin^{n-2}\vartheta\, d\vartheta.$$

In particular,

$$\|T_\kappa^{(2)}\| = \frac{2}{\pi}(1+\kappa)E\Big(\frac{2\sqrt{\kappa}}{1+\kappa}\Big) = 1 + \frac{1}{2^2}\kappa^2 + \frac{1}{2^2 4^2}\kappa^4 + \ldots + \Big[\frac{(2m-3)!!}{2^m m!}\Big]^2\kappa^{2m} + \ldots,$$

where E is the complete elliptic integral of the second kind, and

$$\|T_\kappa^{(3)}\| = \frac{1}{2}\Big[1 + 2\kappa + \frac{(1-\kappa)^2}{\sqrt{3\kappa(\kappa+2)}}\log\frac{1+2\kappa+\sqrt{3\kappa(\kappa+2)}}{1-\kappa}\Big].$$

For the hydrodynamic double layer potential we have

$$||T_1^{(n)}|| = \frac{2}{\sqrt{\pi}} \frac{\Gamma(\frac{n}{2}+1)}{\Gamma(\frac{n+1}{2})},$$

so that

$$||T_1^{(2)}|| = 4/\pi, \quad ||T_1^{(3)}|| = 3/2.$$

The next assertion, generalizing Theorem 1.5, shows that for a convex domain $\mathcal{D}$ the value $||T_\kappa^{(n)}||$ yields the minimum of the norms of the operator $T_\kappa^{(n)}$ in some class of spaces of continuous vector-valued functions on $\partial\mathcal{D}$.

Let $|| \cdot ||_n$ be an arbitrary norm in $\mathbb{R}^n$ and let $|| \cdot ||_n^*$ be the corresponding dual norm. By $C(\partial\mathcal{D}, || \cdot ||_n)$ we denote the space of continuous n-component vector-valued functions on $\partial\mathcal{D}$ with the norm

$$||\phi|| = \sup\left\{||\phi(x)||_n : x \in \partial\mathcal{D}\right\}.$$

Proposition 1.1. *Let $\mathcal{D}$ be a convex domain. Then*

$$||T_\kappa^{(n)}||_{C(\partial\mathcal{D}, ||\cdot||_n)} = \frac{1}{\omega_n} \sup_{||z||_n^*=1} \int_{S^{n-1}} ||(1-\kappa)z + n\kappa e_\sigma(e_\sigma, z)||_n^* d\sigma,$$

where e_σ is the vector connecting the origin with $\sigma \in S^{n-1}$. Moreover,

$$||T_\kappa^{(n)}|| = \min_{||\cdot||_n} ||T_\kappa^{(n)}||_{C(\partial\mathcal{D}, ||\cdot||_n)},$$

where the minimum is taken over the set of all norms in the space $\mathbb{R}^n$.

In the following statement $\mathcal{D}$ is a bounded domain in $\mathbb{R}^2$ with a smooth boundary except for the single angular point. Let β be the opening of the angle calculated in $\mathcal{D}$, $0 \leq \beta \leq 2\pi$, and let $\alpha = \min\{\beta, 2\pi - \beta\}$.

Theorem 1.6. *The equality*

$$(1.15) \qquad [\![T_\kappa^{(2)}]\!] = \frac{2}{\pi}(1 + \kappa)E\left(\frac{\pi - \alpha}{2}, \frac{2\sqrt{\kappa}}{1+\kappa}\right)$$

is valid, where $E(\varphi, k)$ is the elliptic integral of the second kind.

In particular,

$$[\![T_1^{(2)}]\!] = (4/\pi)\cos(\alpha/2).$$

Remark 1.1. V. Shelepov [56] proved that the Fredholm radius of the operator $T_\kappa^{(2)}$ in the space of continuous vector-valued functions is

$$(1.16) \qquad \mathcal{R}(T_\kappa^{(2)}) = \frac{\pi}{\pi - \alpha + \kappa \sin \alpha}.$$

A comparison of (1.15) and (1.16) shows that

$$\mathcal{R}(T_\kappa^{(2)}) > \|T_\kappa^{(2)}\|^{-1}$$

for $\kappa > 0$. We note that for $\kappa = 0$, i.e. for the logarithmic potential T, the essential norm and $\mathcal{R}^{-1}$ are equal, as shown already by Radon [51].

In the following theorem $\mathcal{D}$ is a bounded domain in $\mathbb{R}^3$ and $\partial\mathcal{D}$ is a smooth surface except for the only conic point q. Suppose that the tangent cone $V(q)$ is spherical and denote its opening by β, $0 \le \beta \le 2\pi$. We put $\alpha = \min\{\beta,\, 2\pi - \beta\}$.

Theorem 1.7. *The equality*

$$\|T_\kappa^{(3)}\| = \frac{(1-\kappa)}{\pi}\left[\sqrt{1 + \Gamma^2 \sin^2 \frac{\alpha}{2}}\; E\left(\frac{\Gamma \sin \frac{\alpha}{2}}{\sqrt{1 + \Gamma^2 \sin^2 \frac{\alpha}{2}}}\right) \cos \frac{\alpha}{2}\right.$$

$$\left. + \int_0^{\alpha/2} \frac{1 + \Gamma^2 \cos^2 \vartheta}{\Gamma \cos \vartheta}\, \arcsin \frac{\Gamma \cos \vartheta \cos(\alpha/2)}{\sqrt{1 + \Gamma^2 \cos^2 \vartheta}}\, d\vartheta\right]$$

is valid with

$$\Gamma = (1 - \kappa)^{-1}\sqrt{3\kappa(\kappa + 2)}.$$

In particular,

$$\|T_1^{(3)}\| = \frac{3}{2\pi}(\pi - \alpha + \sin \alpha).$$

In the following assertion $\mathcal{D}$ is a bounded domain in $\mathbb{R}^3$ whose boundary is the union of two smooth surfaces with the common part γ. Suppose the curve γ is smooth. Then at each $\zeta \in \gamma$ two tangent planes are defined. The angle between them, calculated in $\mathcal{D}$, will be denoted by $\beta(\zeta)$, $0 \le \beta(\zeta) \le 2\pi$. Further, we put

$$\alpha = \min_{\zeta \in \gamma}\{\beta(\zeta),\, 2\pi - \beta(\zeta)\}.$$

Theorem 1.8. *The equality*

$$\|T_\kappa^{(3)}\| = \frac{(1-\kappa)}{\pi}\left[\frac{\pi - \alpha}{2} + \int_{\alpha/2}^{\pi/2} \frac{1 + \Gamma^2 \sin^2 \vartheta}{\Gamma \sin \vartheta}\, \arccos\left(1 + \Gamma^2 \sin^2 \vartheta\right)^{-1/2} d\vartheta\right],$$

is valid with the same constant Γ as in Theorem 1.7.

In particular,

$$\|T_1^{(3)}\| = (3/2)\cos(\alpha/2).$$

2. Exact constants in inequalities of maximum principle type for certain systems and equations of mathematical physics

In this section I present some results on the best constants in inequalities for the modulus of solutions to the Lamé and Stokes systems [39], the systems of viscoelasticity theory [26] and the planar deformed state [39], as well as for the modulus of the gradient of a solution of the biharmonic equation [27, 39]. Moreover, I present an assertion according to which the abovementioned constant for the Lamé and Stokes systems is the best possible in some sense [29]. All these systems and the biharmonic equation are considered in a half-space or/and in a half-plane.

2.1. Introduction

When deriving representations for the norm and the essential norm of integral operators of elasticity theory and of hydrodynamics we used the methods which can be also applied to obtain the best constants in inequalities of maximum principle type.

In this subsection we present the values of the best constant $\mathcal{K}(\mathbb{R}^n_+)$ in the inequality

$$(2.1) \qquad |u(x)| \leq \mathcal{K} \sup\{|u(x')| : x' \in \partial\mathbb{R}^n_+\},$$

where $x \in \mathbb{R}^n_+ = \{x = (x_1,\ldots,x_n) : x_n > 0\}$, u is the displacement vector in the Lamé system or the velocity vector of the viscous incompressible fluid in the Stokes system. Besides, we shall give similar constants in inequalities for components of the stress tensor in the plane case, as well as in the inequality for displacements of the viscoelastic medium in a half-space. We conclude the subsection with the theorem on the best constant in the inequality of the Miranda-Agmon type for solutions of higher order elliptic equations, in particular, for the biharmonic equation.

In this section by $C(\overline{\mathbb{R}^n_+})$ we denote the space of bounded and continuous n-component vector-valued functions on $\overline{\mathbb{R}^n_+}$ with the norm $\|u\| = \sup\{|u(x)| : x \in \overline{\mathbb{R}^n_+}\}$. The notation $C(\partial\mathbb{R}^n_+)$ has a similar meaning. By $C^2(\mathbb{R}^n_+)$ we denote the space of n-component vector-valued functions with continuous derivatives up to the second order in $\mathbb{R}^n_+$.

2.2. The Lamé and Stokes systems

In the half-space $\mathbb{R}^n_+$, $n \geq 2$, consider the Lamé system

$$(2.2) \qquad \mu\Delta u + (\lambda + \mu)\operatorname{grad} \operatorname{div} u = 0,$$

and the Stokes system

$$(2.3) \qquad \nu\Delta u - \operatorname{grad} p = 0, \quad \operatorname{div} u = 0,$$

with the boundary condition

$$(2.4) \qquad u\big|_{x_n=0} = f,$$

where λ and μ are the Lamé constants, ν is the kinematic coefficient of viscosity, $f \in C(\partial \mathbb{R}^n_+)$, $u = (u_1, \ldots, u_n)$ is the displacement vector of an elastic medium or the velocity vector of a fluid, and p is the pressure in the fluid.

For the solution $u \in C^2(\mathbb{R}^n_+) \cap C(\overline{\mathbb{R}^n_+})$ of problems (2.2), (2.4) and (2.3), (2.4) we have the representation (see [35, 36])

$$u(x) = \int_{\partial \mathbb{R}^n_+} U_\kappa \left(\frac{y-x}{|y-x|} \right) \frac{x_n}{|y-x|^n} f(y') dy',$$

where $x \in \mathbb{R}^n_+$, $y = (y', 0)$, $y' = (y_1, \ldots, y_{n-1})$. Here $\kappa = 1$ for the Stokes system, $\kappa = (\lambda+\mu)(\lambda+3\mu)^{-1}$ for the Lamé system, and U_κ is the $(n \times n)$-matrix-valued function on $S^{n-1} = \{x \in \mathbb{R}^n : |x| = 1\}$ with the entries

$$\frac{2}{\omega_n} \left[(1-\kappa)\delta_{ij} + n\kappa \frac{(y_i - x_i)(y_j - x_j)}{|y-x|^2} \right],$$

ω_n being the area of the sphere S^{n-1}.

Theorem 2.1. *The exact constant $\mathcal{K}(\mathbb{R}^n_+)$ for the Lamé and the Stokes systems in (2.1) has the form*

$$\mathcal{K}(\mathbb{R}^n_+) = \frac{2\Gamma(\frac{n}{2})}{\sqrt{\pi}\,\Gamma(\frac{n-1}{2})} \int_0^{\pi/2} \left[(1-\kappa)^2 + n\kappa(n\kappa - 2\kappa + 2)\cos^2\vartheta \right]^{1/2} \sin^{n-2}\vartheta\, d\vartheta$$

and the inequality $\mathcal{K}(\mathbb{R}^n_+) > 1$ holds for $\kappa \neq 0$.

In the case $\kappa = 1$, i.e., for an n-dimensional Stokes system,

$$\mathcal{K}(\mathbb{R}^n_+) = \frac{2}{\sqrt{\pi}} \frac{\Gamma(\frac{n}{2}+1)}{\Gamma(\frac{n+1}{2})}.$$

Corollary 2.1. *The equality*

$$\mathcal{K}(\mathbb{R}^3_+) = \frac{1}{2}\left[1 + 2\kappa + \frac{(1-\kappa)^2}{\sqrt{3\kappa(\kappa+2)}} \log \frac{1 + 2\kappa + \sqrt{3\kappa(\kappa+2)}}{1-\kappa} \right]$$

is valid. In particular, $\mathcal{K}(\mathbb{R}^3_+) = 3/2$ for $\kappa = 1$.

Corollary 2.2. *The equality*

$$\mathcal{K}(\mathbb{R}^2_+) = \frac{2}{\pi}(1+\kappa)E\left(\frac{2\sqrt{\kappa}}{1+\kappa} \right) = 1 + \frac{1}{2^2}\kappa^2 + \frac{1}{2^2 4^2}\kappa^4 + \ldots + \left[\frac{(2m-3)!!}{2^m m!} \right]^2 \kappa^{2m} + \ldots$$

is valid, where E is the complete elliptic integral of the second kind. In particular, $\mathcal{K}(\mathbb{R}^2_+) = 4/\pi$ for $\kappa = 1$.

The next assertion generalizes Theorem 2.1 to the case of an arbitrary norm in $\mathbb{R}^n$. Besides, we claim that the value $\mathcal{K}(\mathbb{R}^n_+)$ is minimal in the sense indicated below.

Let $||\cdot||_n$ be an arbitrary norm in $\mathbb{R}^n$ and let $||\cdot||_n^*$ be the corresponding dual norm. By $\mathcal{K}(\mathbb{R}^n_+, ||\cdot||_n)$ we denote the best constant in the inequality

$$||u(x)||_n \leq \mathcal{K}\sup\{||u(x')||_n : x' \in \partial\mathbb{R}^n_+\},$$

where $x \in \mathbb{R}^n_+$ and u is a solution of the Lamé system or of the Stokes system in the class $C^2(\mathbb{R}^n_+) \cap C(\overline{\mathbb{R}^n_+})$.

Proposition 2.1. *The exact constant $\mathcal{K}(\mathbb{R}^n_+, ||\cdot||_n)$ has the form*

$$\mathcal{K}(\mathbb{R}^n_+, ||\cdot||_n) = \frac{1}{\omega_n}\sup_{||z||_n^*=1}\int_{S^{n-1}}||(1-\kappa)z + n\kappa e_\sigma(e_\sigma, z)||_n^* d\sigma,$$

where e_σ is the vector connecting the origin with $\sigma \in S^{n-1}$. Moreover, the equality

$$\mathcal{K}(\mathbb{R}^n_+) = \min_{||\cdot||_n}\mathcal{K}(\mathbb{R}^n_+, ||\cdot||_n),$$

is valid, where the minimum is taken over the set of all norms in the space $\mathbb{R}^n$.

Remark 2.1. It follows from Theorem 2.1 and Proposition 2.1 that for $\kappa \neq 0$ the equality

$$\mathcal{K}(\mathbb{R}^n_+, ||\cdot||_n) = 1,$$

i.e., the maximum norm principle for Lamé and Stokes systems in the half-space $\mathbb{R}^n_+$, is not valid.

2.3. Planar deformed state

Let σ_{11}, σ_{12} and σ_{22} be the components of the stress tensor in the half-plane $\mathbb{R}^2_+$. Consider the system of equations in $\mathbb{R}^2_+$ for the stresses in a planar deformed state (see [45]):

$$\partial\sigma_{11}/\partial x_1 + \partial\sigma_{12}/\partial x_2 = 0,$$
$$\partial\sigma_{12}/\partial x_1 + \partial\sigma_{22}/\partial x_2 = 0,$$
$$\Delta(\sigma_{11} + \sigma_{22}) = 0,$$

with the boundary conditions

$$\sigma_{12}(x_1, 0) = p_1(x_1), \quad \sigma_{22}(x_1, 0) = p_2(x_1),$$

where p_1 and p_2 are continuous and bounded functions on $\partial\mathbb{R}^2_+$.

Theorem 2.2. *The exact constant in the inequality*

$$||(\sigma_{12}^2 + \sigma_{22}^2)^{1/2}||_{C(\overline{\mathbb{R}^2_+})} \leq \mathcal{K}||(\sigma_{12}^2 + \sigma_{22}^2)^{1/2}||_{C(\partial\mathbb{R}^2_+)}$$

is equal to $4/\pi$.

2.4. The system of quasistatic theory of viscoelasticity (the three parametric model)

In this subsection we consider the first boundary value problem for equations of the quasistatic theory of linear viscoelasticity for the medium described by the three parametric model in the half-space $\mathbb{R}^3_+$. We present the best value $\mathcal{K}(t)$ of the coefficient $K(t)$ in the inequality

$$(2.5) \qquad |u(x,t)| \leq K(t)\sup\{|u(y,\tau)| : y \in \partial\mathbb{R}^3_+,\ 0 \leq \tau \leq t\},$$

where $u = (u_1, u_2, u_3)$ is the vector of viscoelastic displacements and $x \in \mathbb{R}^3_+$. In particular, we give formulas for $\mathcal{K}(t)$ in the case of the Maxwell medium and in the case of the medium with purely elastic behaviour under volume compression (the last assumption is often used in applications).

The equilibrium equations for the viscoelastic medium described by the three parametric model under quasistatic assumptions has the form (see [17])

$$(2.6) \quad \int_0^t \left\{ M(t-\tau)\Delta\frac{\partial u(x,\tau)}{\partial\tau} + \left(C(t-\tau) + \frac{M(t-\tau)}{3}\right) \operatorname{grad} \operatorname{div}\frac{\partial u(x,\tau)}{\partial\tau} \right\}d\tau$$

$$+ \left\{ M(t)\Delta u(x,\ +0) + \left(C(t) + \frac{M(t)}{3}\right) \operatorname{grad} \operatorname{div} u(x,\ +0) \right\} = 0,$$

where $M(\tau) = \mu(1 - \alpha + \alpha\exp(-\rho\tau))$, $C(\tau) = \kappa(1 - \beta + \beta\exp(-\theta\tau))$ are relaxation functions under displacement and isotropic compression, respectively, $\mu, \kappa, \alpha, \beta, \rho, \theta$ are parameters of the model ($\mu, \kappa, \rho, \theta > 0$, $0 \leq \alpha, \beta \leq 1$), $u(x,\ +0)$ is the limit of $u(x,t)$ as $t \to +0$. We assume that $\alpha + \beta > 0$.

We supply (2.6) with the boundary condition

$$(2.7) \qquad u\big|_{x_3=0, t>0} = \varphi,$$

where $\varphi = (\varphi_1, \varphi_2, \varphi_3)$ is a continuous and bounded vector-valued function on $\partial\mathbb{R}^3_+ \times \overline{\mathbb{R}^1_+}$.

A bounded in $\mathbb{R}^3_+ \times \mathbb{R}^1_+$ solution of the problem (2.6), (2.7) has the form

$$u(x,t) = \frac{1}{2\pi} \int_{y_3=0} \mathfrak{M}_\gamma(e_{xy})\varphi(y',t)\frac{x_3}{|y-x|^3}dy'$$

$$+ \frac{1}{2\pi} \int_0^t \int_{y_3=0} \mathcal{H}(t-\tau)\mathfrak{N}(e_{xy})\varphi(y',\tau)\frac{x_3}{|y-x|^3}dy'd\tau,$$

where $e_{xy} = (y-x)|y-x|^{-1}$, $y' = (y_1, y_2)$, $\mathfrak{M}_\gamma$ and $\mathfrak{N}$ are (3×3)-matrices with elements $(1-\gamma)\delta_{ij} + 3\gamma(e,e_i)(e,e_j)$ and $-\delta_{ij} + 3(e,e_i)(e,e_j)$, respectively, $\gamma = (\mu + 3\kappa)(7\mu + 3\kappa)^{-1}$, $\mathcal{H}(\tau) = \eta_1\exp(-\xi_1\tau) + \eta_2\exp(-\xi_2\tau)$ with

$$\eta_1 = \frac{c(a + \sqrt{a^2 - b}) - d}{2\sqrt{a^2 - b}}, \ \eta_2 = -\frac{c(a - \sqrt{a^2 - b}) - d}{2\sqrt{a^2 - b}},$$

$$\xi_1 = a + \sqrt{a^2 - b}, \ \xi_2 = a - \sqrt{a^2 - b}.$$

Here

$$a = \frac{7\mu(\theta + \rho - \alpha\rho) + 3\kappa(\theta + \rho - \alpha\theta)}{2(7\mu + 3\kappa)}, \quad b = \frac{\rho\theta(7\mu(1-\alpha) + 3\kappa(1-\beta))}{7\mu + 3\kappa},$$

$$c = \frac{18\kappa\mu(\alpha\rho - \beta\theta)}{(7\mu + 3\kappa)^2}, \quad d = \frac{18\kappa\mu\rho\theta(\alpha - \beta)}{(7\mu + 3\kappa)^2}.$$

Theorem 2.3. *The best constant in (2.5) for the viscoelastic medium in the three parameter model is*

$$\mathcal{K}(t) = \left(1 + \frac{\sqrt{3}}{6}\log(2 + \sqrt{3})\right)\mathcal{F}(t)$$

$$+ \frac{1}{2}\left(1 + 2\gamma + \frac{(1-\gamma)^2}{\sqrt{3\gamma(\gamma + 2)}}\log\frac{1 + 2\gamma + \sqrt{3\gamma(\gamma + 2)}}{1 - \gamma}\right),$$

where

$$\mathcal{F}(t) = \int_0^t |\mathcal{H}(\tau)|d\tau = \int_0^t |\eta_1 \exp(-\xi_1\tau) + \eta_2\exp(-\xi_2\tau)|d\tau.$$

In particular,

$$\mathcal{F}(t) = \eta(1 - \exp(-\xi t))$$

with

$$\eta = \frac{18\kappa\mu|\rho - \theta|}{(7\mu\theta + 3\kappa\rho)(7\mu + 3\kappa)}, \quad \xi = \frac{7\mu\theta + 3\kappa\rho}{7\mu + 3\kappa}$$

for the Maxwell medium ($\alpha = \beta = 1$) and

$$\mathcal{F}(t) = \lambda(1 - \exp(-\chi t))$$

with

$$\lambda = \frac{18\alpha\kappa\mu}{(7\mu + 3\kappa)(7\mu(1-\alpha) + 3\kappa)}, \quad \chi = \frac{\rho(7\mu(1-\alpha) + 3\kappa)}{7\mu + 3\kappa}$$

for the medium with purely elastic behaviour under volume compression ($\beta = 0$).

Remark 2.2. A similar result holds for the plane problem in $\mathbb{R}_+^2$. In this case the best value of $\mathcal{K}(t)$ in the inequality analogous to (2.5) is

$$\mathcal{K}(t) = \mathcal{F}(t) + \frac{2}{\gamma}(1 + \gamma)E\left(\frac{2\sqrt{\gamma}}{1 + \gamma}\right)$$

$$= \mathcal{F}(t) + 1 + \frac{1}{2^2}\gamma^2 + \frac{1}{2^2 4^2}\gamma^4 + \ldots + \left[\frac{(2m - 3)!!}{2^m m!}\right]^2\gamma^{2m} + \ldots,$$

where γ and $\mathcal{F}(t)$ are the same as in the preceding theorem.

Remark 2.3. The value $\mathcal{K}(t)$ in the two above formulas is greater than one which implies that the classical maximum modulus principle is not valid for the viscoelastic medium described by the three parametric model.

2.5. On the best constant in the inequality of the Miranda-Agmon type for solutions of elliptic equations

Let Ω be a domain in $\mathbb{R}^n$ with smooth boundary $\partial\Omega$ and compact closure $\overline{\Omega}$. Consider the elliptic operator

$$P(D) = \sum_{|\alpha| \leq 2m} a_\alpha D^\alpha$$

with constant complex coefficients, where $D^\alpha = \partial^{|\alpha|}/\partial x_1^{\alpha_1}\ldots\partial x_n^{\alpha_n}$, and $\alpha = (\alpha_1,\ldots,\alpha_n)$ is the multiindex of order $|\alpha| = \alpha_1 + \ldots + \alpha_n$. By $P_0(\xi)$ we denote the principal homogeneous part of the polynomial $P(\xi)$. For $n = 2$ we assume additionally that all ξ_2-roots of the polynomial $P_0(\xi)$ are distributed equally in the upper and lower parts of the plane for all $\xi_1 \in \mathbb{R}^1\backslash\{0\}$.

Let $\mathbb{R}^n_+(\nu) = \{x \in \mathbb{R}^n : (x,\nu) > 0\}$, where ν is a unit vector and let $K(\nu)$ be the best constant in the Miranda-Agmon inequality ([1, 2, 43])

$$\sup_{\overline{\mathbb{R}^n_+(\nu)}} |\nabla_{m-1}u| \leq \mathcal{K} \sup_{\partial\mathbb{R}^n_+(\nu)} |\nabla_{m-1}u|.$$

Here

$$|\nabla_{m-1}u| = \left(\sum_{|\alpha|=m-1} \frac{(m-1)!}{\alpha!}|D^\alpha u|^2 \right)^{1/2}$$

and u is an arbitrary solution of the equation $P_0(D)u = 0$, smooth in $\overline{\mathbb{R}^n_+(\nu)}$ and such that $u(x) = O(|x|^{m-1})$ for large $|x|$.

The following result obtained in [27] gives the best constant in a weak form of the Agmon-Miranda inequality.

Theorem 2.4. *For any solution of the equation $P(D)u = 0$, smooth on $\overline{\Omega}$, the inequality*

$$\max_{\overline{\Omega}} |\nabla_{m-1}u| \leq \left(\sup_{\nu\in S^{n-1}} K(\nu) + \varepsilon \right) \max_{\partial\Omega} |\nabla_{m-1}u| + c(\varepsilon)\|u\|_{L(\Omega)}$$

is valid, where ε is any positive number and $c(\varepsilon)$ is a positive constant independent of u.

The next theorem (see [27, 39]) contains the best constant K in the Miranda inequality ([42])

$$\sup_{\overline{\mathbb{R}^n_+}} |\nabla u| \leq \mathcal{K} \sup_{\partial\mathbb{R}^n_+} |\nabla u|,$$

where u is a solution of the biharmonic equation in $\mathbb{R}^n_+$ from $C^\infty(\overline{\mathbb{R}^n_+})$ with $\nabla u(x_1,\ldots,x_{n-1},0)$ being a bounded vector-valued function on $\partial\mathbb{R}^n_+$ and $u = O(|x|)$ for large $|x|$.

Theorem 2.5. *The best constant K is*

$$K = \frac{2\Gamma(\frac{n}{2})}{\sqrt{\pi}\,\Gamma(\frac{n-1}{2})} \int_0^{\pi/2} \left[4 + n(n-4)\cos^2\vartheta\right]^{1/2} \sin^{n-2}\vartheta\, d\vartheta.$$

In particular, $K = 4/\pi$ for $n = 2$, $K = 1/2 + 2\pi\sqrt{3}/9$ for $n = 3$ and $K = 2$ for $n = 4$.

3. Maximum modulus principle for elliptic systems

This section contains criteria for validity of the maximum modulus principle for second order strongly elliptic systems [28, 39].

3.1. Introduction

The maximum modulus principle for elliptic second order systems have been widely studied (cf. e.g. [7, 39, 44, 46, 48, 60]). The first articles aiming at the study and applications of the classical maximum modulus principle for solutions of elliptic second order systems concerned systems with scalar coefficients of the first and second derivatives of the unknown vector-function ([7, 46, 60]). Later C. Miranda [44] showed that the classical maximum modulus principle is valid not only for weakly coupled systems. More precisely, he considered elliptic second order systems with a scalar principal part and with arbitrary coefficients of derivatives of order less than two and found a sufficient condition for validity of the classical maximum modulus principle. This condition has a form of an algebraic inequality for coefficients of the system. Various modifications of the maximum modulus principle were considered in [22, 49].

In this section we give criteria for validity of the maximum modulus principle for solutions of the strongly elliptic system

$$(3.1) \qquad \sum_{j,k=1}^{n} \mathcal{A}_{jk}(x)\frac{\partial^2 u}{\partial x_j \partial x_k} - \sum_{j=1}^{n} \mathcal{A}_j(x)\frac{\partial u}{\partial x_j} - \mathcal{A}_0(x)u = 0$$

with real or complex coefficients. Here $\mathcal{A}_{jk}, \mathcal{A}_j, \mathcal{A}_0$ are $(m \times m)$-matrix-valued functions and u is a m-component vector-valued function. Without loss of generality we assume that $\mathcal{A}_{jk} = \mathcal{A}_{kj}$.

3.2. Model systems

I begin with the simple case of the homogeneous operator with the constant coefficients.

3.2.1. The case of real coefficients

We introduce the operator

$$\mathfrak{A}_0(\partial/\partial x) = \sum_{j,k=1}^{n} \mathcal{A}_{jk} \frac{\partial^2}{\partial x_j \partial x_k},$$

where $\mathcal{A}_{jk}$ are constant real $(m \times m)$-matrices. Assume that the operator $\mathfrak{A}_0$ is strongly elliptic, i.e. that for all $\zeta = (\zeta_1, \dots, \zeta_m) \in \mathbb{R}^m$ and $\sigma = (\sigma_1, \dots, \sigma_n) \in \mathbb{R}^n$, with ζ, $\sigma \neq 0$, we have the inequality

$$\left(\sum_{j,k=1}^{n} \mathcal{A}_{jk}\sigma_j\sigma_k\zeta,\ \zeta \right) > 0.$$

Let $\mathcal{D}$ be a domain in $\mathbb{R}^n$ with boundary $\partial\mathcal{D}$ and closure $\overline{\mathcal{D}}$. Let $C(\overline{\mathcal{D}})$ denote the space of bounded m-component vector-valued functions which are continuous in $\overline{\mathcal{D}}$. The norm on $C(\overline{\mathcal{D}})$ is $\|u\| = \sup\{|u(x)| : x \in \overline{\mathcal{D}}\}$. The notation $C(\partial\mathcal{D})$ has a similar meaning. By $C^2(\mathcal{D})$ we denote the space of m-component vector-valued functions with continuous derivatives up to the second order in $\mathcal{D}$.

Consider the strongly elliptic system

$$(3.2) \qquad \sum_{j,k=1}^{n} \mathcal{A}_{jk} \frac{\partial^2 u}{\partial x_j \partial x_k} = 0, \quad x \in \mathcal{D},$$

where u is a m-component vector-valued function. Let

$$(3.3) \qquad \mathcal{K}(\mathcal{D}) = \sup \frac{\|u\|_{C(\overline{\mathcal{D}})}}{\|u\|_{C(\partial\mathcal{D})}},$$

where the supremum is taken over all vector-valued functions in the class $C^2(\mathcal{D}) \cap C(\overline{\mathcal{D}})$ satisfyng the system $\mathfrak{A}_0(\partial/\partial x)u = 0$. Clearly, $\mathcal{K}(\mathcal{D})$ is the best constant in the inequality

$$|u(x)| \leq \mathcal{K} \sup\{|u(x')| : x' \in \partial\mathcal{D}\},$$

where $x \in \mathcal{D}$ and u is a solution of (3.2) in the class $C^2(\mathcal{D}) \cap C(\overline{\mathcal{D}})$. If $\mathcal{K}(\mathcal{D}) = 1$, then the maximum modulus principle holds for system (3.2).

According to [3, 55, 57], there exists a bounded solution of the problem

$$\mathfrak{A}_0(\partial/\partial x)u = 0 \text{ in } \mathbb{R}_+^n, \quad u = f \text{ on } \partial\mathbb{R}_+^n,$$

with $f \in C(\partial\mathbb{R}_+^n)$, such that u is continuous up to $\partial\mathbb{R}_+^n$, and can be represented in the form

$$(3.4) \qquad u(x) = \int_{\partial\mathbb{R}_+^n} U\left(\frac{y-x}{|y-x|}\right) \frac{x_n}{|y-x|^n} f(y')dy'.$$

Here $y = (y', 0)$, $y' = (y_1, \ldots, y_{n-1})$, and U is a continuous $(m \times m)$-matrix-valued function on the closure of the hemisphere $S_-^{n-1} = \{x \in \mathbb{R}^n : |x| = 1, x_n < 0\}$ such that the integral of U on S^{n-1} is the identity matrix.

The uniqueness of the solution of the Dirichlet problem in the class $C^2(\mathbb{R}_+^n) \cap C(\overline{\mathbb{R}_+^n})$ can be derived by means of standard arguments from (3.4) and from local estimates of the derivatives of solutions to elliptic systems (see [3, 57]). The following four theorems are borrowed from [39].

Theorem 3.1. *The formula*

$$\mathcal{K}(\mathbb{R}_+^n) = \sup_{|z|=1} \int_{S_-^{n-1}} |U^*(\sigma)z|\,d\sigma,$$

*is valid, where the * denotes passage to the transposed matrix.*

Theorem 3.2. *The equality $\mathcal{K}(\mathbb{R}_+^n) = 1$ is valid if and only if*

$$(3.5) \qquad \mathfrak{A}_0(\partial/\partial x) = \mathcal{A} \sum_{j,k=1}^{n} a_{jk} \frac{\partial^2}{\partial x_j \partial x_k},$$

where $\mathcal{A}$ is a constant $(m \times m)$-matrix with positive-definite symmetric part, and $((a_{jk}))$ is a positive-definite $(n \times n)$-matrix.

Theorem 3.3. *Let $\mathcal{D}$ be a domain in $\mathbb{R}^n$ with compact closure and C^1-boundary. Then*

$$\mathcal{K}(\mathcal{D}) \geq \sup\{\mathcal{K}(\mathbb{R}_+^n(\nu)) : \nu \in S^{n-1}\},$$

where $\mathbb{R}_+^n(\nu)$ is a half-space with inward normal ν.

Theorem 3.4. *Let $\mathcal{D}$ be a domain in $\mathbb{R}^n$ with compact closure and C^1-boundary. The equality $\mathcal{K}(\mathcal{D}) = 1$ holds if and only if the operator $\mathfrak{A}_0(\partial/\partial x)$ has the form (3.5).*

Remark 3.1. Does the quantity $\mathcal{K}(\mathcal{D})$ depend on the domain $\mathcal{D}$ in general? And can the inequality sign be replaced by equality in the estimate given in Theorem 3.3?

We answer the first of these questions in the positive and the second in the negative by considering the Lamé system in a three-dimensional ball B and using the solution constructed by G. Polya [47].

3.2.2. The case of complex coefficients

We introduce the operator

$$\mathfrak{A}_0(\partial/\partial x) = \sum_{j,k=1}^{n} \mathcal{A}_{jk} \frac{\partial^2}{\partial x_j \partial x_k},$$

where $\mathcal{A}_{jk}$ are constant complex $(m \times m)$-matrices. Assume that the operator $\mathfrak{A}_0$ is strongly elliptic, i.e. that

$$\mathrm{Re}\langle \sum_{j,k=1}^{n} \mathcal{A}_{jk}\sigma_j\sigma_k\zeta, \ \zeta \rangle > 0$$

for all $\zeta = (\zeta_1, \ldots, \zeta_m) \in \mathbb{C}^m$ and $\sigma = (\sigma_1, \ldots, \sigma_n) \in \mathbb{R}^n$, with $\zeta, \ \sigma \neq 0$. Here $\mathbb{C}^m$ is a complex linear m-dimensional space with elements $a + ib$, where $a, b \in \mathbb{R}^m$. The scalar product in $\mathbb{C}^m$ is $\langle c, d \rangle = c_1\overline{d_1} + \ldots + c_m\overline{d_m}, \ c = (c_1, \ldots, c_m), \ d = (d_1, \ldots, d_m)$. The length of the vector d in $\mathbb{C}^m$ is $|d| = \langle d, d \rangle^{1/2}$.

Let $\mathcal{R}_{jk}$ and $\mathcal{H}_{jk}$ be constant real $(m \times m)$-matrices such that $\mathcal{A}_{jk} = \mathcal{R}_{jk} + i\mathcal{H}_{jk}$. Define

$$\mathfrak{R}_0(\partial/\partial x) = \sum_{j,k=1}^{n} \mathcal{R}_{jk}\frac{\partial^2}{\partial x_j \partial x_k}, \quad \mathfrak{H}_0(\partial/\partial x) = \sum_{j,k=1}^{n} \mathcal{H}_{jk}\frac{\partial^2}{\partial x_j \partial x_k}.$$

Separating the real and imaginary parts of the system $\mathfrak{A}_0(\partial/\partial x)g = 0$, where $g = v + iw$, we get a system with real coefficients

$$\mathfrak{R}_0(\partial/\partial x)v - \mathfrak{H}_0(\partial/\partial x)w = 0, \quad \mathfrak{H}_0(\partial/\partial x)v + \mathfrak{R}_0(\partial/\partial x)w = 0,$$

which, like the original system, is strongly elliptic.

We introduce the matrix

$$\mathfrak{K}_0(\partial/\partial x) = \begin{pmatrix} \mathfrak{R}_0(\partial/\partial x) & -\mathfrak{H}_0(\partial/\partial x) \\ \mathfrak{H}_0(\partial/\partial x) & \mathfrak{R}_0(\partial/\partial x) \end{pmatrix}.$$

Let $\mathbf{C}(\mathcal{D})$ be the space of m-component complex vector-valued functions $g = v + iw$ which are bounded and continuous on $\mathcal{D} \subset \mathbb{R}^n$. The norm on $\mathbf{C}(\mathcal{D})$ is $\|g\| = \sup\{(|v(x)|^2 + |w(x)|^2)^{1/2} : x \in \mathcal{D}\}$. The notation $\mathbf{C}(\partial\mathcal{D})$ has a similar meaning. By $\mathbf{C}^2(\mathcal{D})$ we denote the space of m-component complex vector-valued functions with continuous derivatives up to the second order in $\mathcal{D}$.

By analogy with the definition of $\mathcal{K}(\mathcal{D})$, let

$$\mathcal{K}'(\mathcal{D}) = \sup \frac{\|g\|_{\mathbf{C}(\overline{\mathcal{D}})}}{\|g\|_{\mathbf{C}(\partial\mathcal{D})}},$$

where the supremum is over all vector-valued functions in the class $\mathbf{C}^2(\mathcal{D}) \cap \mathbf{C}(\overline{\mathcal{D}})$ satisfyng the system $\mathfrak{A}_0(\partial/\partial x)g = 0$ in $\mathcal{D}$.

It is clear that the constant $\mathcal{K}'(\mathcal{D})$ for the system $\mathfrak{A}_0(\partial/\partial x)g = 0$ with complex coefficients coincides with the constant $\mathcal{K}(\mathcal{D})$ for the system $\mathfrak{K}_0(\partial/\partial x)\{v, w\} = 0$ with real coefficients if we replace m by $2m$, $\mathfrak{A}_0(\partial/\partial x)$ by $\mathfrak{K}_0(\partial/\partial x)$, and u by $\{v, w\}$ in definition (3.3). Therefore, all assertions about $\mathcal{K}'(\mathcal{D})$ are direct consequences of the analogous assertions about $\mathcal{K}(\mathcal{D})$. Using this fact, from Theorems 3.1–3.4 we get the assertions marked below by primes [39].

Theorem 3.1′. *The formula*

$$\mathcal{K}'(\mathbb{R}^n_+) = \sup_{|z|=1} \int_{S^{n-1}_-} |\mathcal{U}^*(\sigma)z|\, d\sigma$$

is valid, where $\mathcal{U}$ is the $(2m \times 2m)$-matrix-valued function on S^{n-1} appearing in the integral representation for a solution of the Dirichlet problem in $\mathbb{R}^n_+$ for the system $\mathfrak{K}_0(\partial/\partial x)\{v, w\} = 0$ (analogous to the representation (3.4)) and $z \in \mathbb{R}^{2m}$.

Theorem 3.2′. *The equality $\mathcal{K}'(\mathbb{R}^n_+) = 1$ is valid if and only if*

$$(3.6) \qquad \mathfrak{A}_0(\partial/\partial x) = A \sum_{j,k=1}^{n} a_{jk} \frac{\partial^2}{\partial x_j \partial x_k},$$

where A is a constant complex-valued $(m \times m)$-matrix such that $\mathrm{Re}\langle A\zeta, \zeta \rangle > 0$ for all $\zeta \in \mathbb{C}^m$, $\zeta \neq 0$, and $((a_{jk}))$ is a real positive-definite $(n \times n)$-matrix.

Theorem 3.3′. *Let $\mathcal{D}$ be a domain in $\mathbb{R}^n$ with compact closure and C^1-boundary. Then*

$$\mathcal{K}'(\mathcal{D}) \geq \sup\{\mathcal{K}'(\mathbb{R}^n_+(\nu)) : \nu \in S^{n-1}\},$$

where $\mathbb{R}^n_+(\nu)$ is a half-space with inward normal ν.

Theorem 3.4′. *Let $\mathcal{D}$ be a domain in $\mathbb{R}^n$ with compact closure and C^1-boundary. The equality $\mathcal{K}'(\mathcal{D}) = 1$ holds if and only if the operator $\mathfrak{A}_0(\partial/\partial x)$ has the form (3.6).*

3.3. General systems

Now I turn to elliptic systems with lower order terms and variable coefficients [28].

3.3.1. The case of real coefficients

Let Ω be a domain in $\mathbb{R}^n$ with compact closure $\overline{\Omega}$ and with boundary $\partial\Omega$ of the class $C^{2,\alpha}, 0 < \alpha \leq 1$. The space of $(m \times m)$-matrix-valued functions whose elements have continuous derivatives up to order k and satisfy the Hölder condition with exponent $\alpha, 0 < \alpha \leq 1$, in $\overline{\Omega}$ is denoted by $C^{k,\alpha}_m(\overline{\Omega})$.

We introduce the differential operator

$$\mathfrak{A}(x, \partial/\partial x) = \sum_{j,k=1}^{n} A_{jk}(x) \frac{\partial^2}{\partial x_j \partial x_k} - \sum_{j=1}^{n} A_j(x) \frac{\partial}{\partial x_j} - A_0(x),$$

where A_{jk}, A_j, A_0 are real $(m \times m)$-matrix-valued functions in the spaces $C^{2,\alpha}_m(\overline{\Omega})$, $C^{1,\alpha}_m(\overline{\Omega})$, $C^{0,\alpha}_m(\overline{\Omega})$, respectively. When the coefficients of the operator $\mathfrak{A}(x, \partial/\partial x)$

do not depend on x we use the notation $\mathfrak{A}(\partial/\partial x)$. Let the principal homogeneous part of the operator $\mathfrak{A}(x,\partial/\partial x)$ be denoted by $\mathfrak{A}_0(x,\partial/\partial x)$.

We assume that $\mathfrak{A}(x,\partial/\partial x)$ is strongly elliptic in $\overline{\Omega}$ that is for all $x \in \overline{\Omega}$, $\zeta = (\zeta_1,\ldots,\zeta_m) \in \mathbb{R}^m$, $\sigma = (\sigma_1,\ldots,\sigma_n) \in \mathbb{R}^n$, with ζ, $\sigma \neq 0$, the inequality

$$\Big(\sum_{j,k=1}^{n} \mathcal{A}_{jk}(x)\sigma_j\sigma_k\zeta,\ \zeta \Big) > 0$$

is valid.

Theorem 3.5. *The classical maximum modulus principle*

$$(3.7) \qquad \|u\|_{C(\overline{\omega})} - \|u|_{\partial\omega}\|_{C(\partial\omega)},$$

is valid for solutions of the system $\mathfrak{A}(x,\partial/\partial x)u = 0$ in arbitrary domain $\omega \subset \Omega$ with boundary $\partial\omega$ of the class $C^{2,\alpha}$ if and only if:

(i) for all $x \in \overline{\Omega}$ the equalities

$$\mathcal{A}_{jk}(x) = \mathcal{A}(x)a_{jk}(x),\ \ 1 \leq j,\ k \leq n,$$

hold, where $\mathcal{A}$ and $((a_{jk}))$ are positive-definite in $\overline{\Omega}$ $(m \times m)$- and $(n \times n)$-matrix-valued functions, respectively;

(ii) for all $x \in \Omega$ and for any ξ_j, $\zeta \in \mathbb{R}^m$, $j = 1,\ldots,n$, with $(\xi_j,\zeta) = 0$, the inequality

$$\sum_{j,k=1}^{n} a_{jk}(x)(\xi_j,\xi_k) + \sum_{j=1}^{n}(\mathcal{A}^{-1}(x)\mathcal{A}_j(x)\xi_j,\zeta) + (\mathcal{A}^{-1}(x)\mathcal{A}_0(x)\zeta,\zeta) \geq 0$$

is valid.

The next assertion immediately follows from this theorem.

Corollary 3.1. *The classical maximum modulus principle holds for solutions of the system $\mathfrak{A}(x,\partial/\partial x)u = 0$ in an arbitrary domain $\omega \subset \Omega$ with boundary $\partial\omega$ of the class $C^{2,\alpha}$ if and only if condition (i) of Theorem 3.5 is satisfied and*

(ii′) for all $x \in \Omega$ and for any $\zeta \in \mathbb{R}^m\backslash\{0\}$ the inequality

$$\sum_{i,j=1}^{n} b_{ij}(x)\big[|\zeta|^{-2}(\mathcal{A}^{-1}(x)\mathcal{A}_i(x)\zeta,\zeta)(\mathcal{A}^{-1}(x)\mathcal{A}_j(x)\zeta,\zeta)$$

$$-(\mathcal{A}_i^*(x)(\mathcal{A}^*(x))^{-1}\zeta,\ \mathcal{A}_j^*(x)(\mathcal{A}^*(x))^{-1}\zeta)\big] + 4((\mathcal{A}^*(x))^{-1}\mathcal{A}_0(x)\zeta,\zeta) \geq 0$$

is valid. Here $((b_{ij}))$ is the inverse matrix of $((a_{ij}))$ and $*$ means passage to the transposed matrix.

Corollary 3.2. *The classical maximum modulus principle (3.7) holds for solutions of the system*

$$\sum_{j,k=1}^{n} \mathcal{A}_{jk}(x)\frac{\partial^2 u}{\partial x_j \partial x_k} - \sum_{j=1}^{n} \mathcal{A}_j(x)\frac{\partial u}{\partial x_j} = 0$$

in an arbitrary domain $\omega \subset \Omega$ with boundary $\partial\omega$ of the class $C^{2,\alpha}$ if and only if

$$\mathcal{A}_{jk}(x) = \mathcal{A}(x)a_{jk}(x), \quad \mathcal{A}_j(x) = \mathcal{A}(x)a_j(x), \ 1 \le j, \ k \le n.$$

Here $\mathcal{A}$ and $((a_{jk}))$ are positive-definite in $\overline{\Omega}$ $(m \times m)$- and $(n \times n)$-matrix-valued functions, respectively, and a_j are scalar functions.

Remark 3.2. In [28] we show by an example that the possibility to represent the principal part of the system $\mathfrak{A}(x, \partial/\partial x)u = 0$ in the form

$$\mathcal{A}(x)\sum_{j,k=1}^{n} a_{jk}(x)\frac{\partial^2 u}{\partial x_j \partial x_k}$$

everywhere in $\overline{\Omega}$, when the coefficients $\mathcal{A}_{jk}$ depend on x, is not necessary for validity of the classical maximum modulus principle (3.7), where u is a solution of the system $\mathfrak{A}(x, \partial/\partial x)u = 0$ in Ω which belongs to $C^2(\Omega) \cap C(\overline{\Omega})$.

3.3.2. The case of complex coefficients

For the spaces of vector-valued and matrix-valued functions with complex components we retain the same notation as in the case of real components but use bold letter.

We introduce the differential operator

$$\mathfrak{A}(x, \partial/\partial x) = \sum_{j,k=1}^{n} \mathcal{A}_{jk}(x)\frac{\partial^2}{\partial x_j \partial x_k} - \sum_{j=1}^{n} \mathcal{A}_j(x)\frac{\partial}{\partial x_j} - \mathcal{A}_0(x),$$

where $\mathcal{A}_{jk}, \mathcal{A}_j, \mathcal{A}_0$ are complex $(m \times m)$-matrix-valued functions in the spaces $\mathbf{C}_m^{2,\alpha}(\overline{\Omega}), \mathbf{C}_m^{1,\alpha}(\overline{\Omega}), \mathbf{C}_m^{0,\alpha}(\overline{\Omega})$, respectively. Suppose that the operator $\mathfrak{A}(x, \partial/\partial x)$ is strongly elliptic in $\overline{\Omega}$, that is for all $x \in \overline{\Omega}$, $\eta = (\eta_1, \ldots, \eta_m) \in \mathbb{C}^m$, $\sigma = (\sigma_1, \ldots, \sigma_n) \in \mathbb{R}^n$, with η, $\sigma \neq 0$, the inequality

$$\mathrm{Re}\langle \sum_{j,k=1}^{n} \mathcal{A}_{jk}(x)\sigma_j\sigma_k\zeta, \ \zeta \rangle > 0$$

is valid.

Let $\mathcal{R}_{jk}, \mathcal{H}_{jk}, \mathcal{R}_j, \mathcal{H}_j, \mathcal{R}_0, \mathcal{H}_0$ be real $(m \times m)$-matrix-valued functions such that

$$\mathcal{A}_{jk} = \mathcal{R}_{jk} + i\mathcal{H}_{jk}, \quad \mathcal{A}_j = \mathcal{R}_j + i\mathcal{H}_j, \quad \mathcal{A}_0 = \mathcal{R}_0 + i\mathcal{H}_0.$$

We use the notation

$$\mathfrak{R}(x, \partial/\partial x) = \sum_{j,k=1}^{n} \mathcal{R}_{jk}(x)\frac{\partial^2}{\partial x_j \partial x_k} - \sum_{j=1}^{n} \mathcal{R}_j(x)\frac{\partial}{\partial x_j} - \mathcal{R}_0(x),$$

$$\mathfrak{H}(x, \partial/\partial x) = \sum_{j,k=1}^{n} \mathcal{H}_{jk}(x)\frac{\partial^2}{\partial x_j \partial x_k} - \sum_{j=1}^{n} \mathcal{H}_j(x)\frac{\partial}{\partial x_j} - \mathcal{H}_0(x).$$

Separating the real and imaginary parts of the system $\mathfrak{A}(x, \partial/\partial x)u = 0$, where $u = v + iw$, we get a system with real coefficients,

$$\mathfrak{R}(x, \partial/\partial x)v - \mathfrak{H}(x, \partial/\partial x)w = 0, \quad \mathfrak{H}(x, \partial/\partial x)v + \mathfrak{R}(x, \partial/\partial x)w = 0,$$

which, like the original system, is strongly elliptic.

All the facts given below and prowed in [28] are resulted by application of corresponding assertions on the maximum modulus for system with real coefficients to the systems obtained after the separation of real and imaginary parts.

Theorem 3.5′. *The classical maximum modulus principle* (3.7) *is valid for solutions of the system* $\mathfrak{A}(x, \partial/\partial x)u = 0$ *in an arbitrary domain* $\omega \subset \Omega$ *with boundary* $\partial\omega$ *of the class* $C^{2,\alpha}$ *if and only if:*

(i) for all $x \in \overline{\Omega}$ *the equalities*

$$\mathcal{A}_{jk}(x) = \mathcal{A}(x)a_{jk}(x), \; 1 \le j, \; k \le n,$$

hold, where $\mathcal{A}$ *is a complex* $(m \times m)$*-matrix-valued function such that* $\mathrm{Re}\langle\mathcal{A}(x)\eta, \eta\rangle > 0$ *for all* $x \in \overline{\Omega}$, $\eta \in \mathbb{C}^m\backslash\{0\}$ *and* $((a_{jk}))$ *is a real positive-definite* $(n \times n)$*-matrix-valued function in* $\overline{\Omega}$;

(ii) for all $x \in \Omega$ *and for any* $\xi_j, \; \zeta \in \mathbb{C}^m$, $j = 1,\ldots,n$, *with* $\mathrm{Re}\langle\xi_j, \zeta\rangle = 0$, *the inequality*

$$\mathrm{Re}\{\sum_{j,k=1}^{n} a_{jk}(x)\langle\xi_j, \xi_k\rangle + \sum_{j=1}^{n}\langle\mathcal{A}^{-1}(x)\mathcal{A}_j(x)\xi_j, \zeta\rangle + \langle\mathcal{A}^{-1}(x)\mathcal{A}_0(x)\zeta, \zeta\rangle\} \ge 0$$

is valid.

Corollary 3.1′. *The classical maximum modulus principle holds for solutions of the system* $\mathfrak{A}(x, \partial/\partial x)u = 0$ *in an arbitrary domain* $\omega \subset \Omega$ *with boundary* $\partial\omega$ *of the class* $C^{2,\alpha}$ *if and only if condition* (i) *of Theorem 3.5′ is satisfied and*

(ii′) for all $x \in \Omega$ *and for any* $\zeta \in \mathbb{C}^m\backslash\{0\}$ *the inequality*

$$\sum_{i,j=1}^{n} b_{ij}(x)\big[|\zeta|^{-2}\mathrm{Re}\langle\mathcal{A}^{-1}(x)\mathcal{A}_i(x)\zeta, \zeta\rangle\langle\mathcal{A}^{-1}(x)\mathcal{A}_j(x)\zeta, \zeta\rangle$$

$$-\mathrm{Re}\langle\mathcal{A}_i^*(x)(\mathcal{A}^*(x))^{-1}\zeta, \; \mathcal{A}_j^*(x)(\mathcal{A}^*(x))^{-1}\zeta\rangle\big] + 4\mathrm{Re}\langle(\mathcal{A}^*(x))^{-1}\mathcal{A}_0(x)\zeta, \zeta\rangle \ge 0$$

is valid. Here $((b_{ij}))$ *is the inverse matrix of* $((a_{ij}))$ *and* * *means passage to the transposed matrix.*

Corollary 3.2′. *The classical maximum modulus principle* (3.7) *holds for solutions of the system*

$$\sum_{j,k=1}^{n} \mathcal{A}_{jk}(x)\frac{\partial^2 u}{\partial x_j \partial x_k} - \sum_{j=1}^{n} \mathcal{A}_j(x)\frac{\partial u}{\partial x_j} = 0$$

in an arbitrary domain $\omega \subset \Omega$ *with boundary* $\partial\omega$ *of the class* $C^{2,\alpha}$ *if and only if*

$$\mathcal{A}_{jk}(x) = \mathcal{A}(x)a_{jk}(x), \;\; \mathcal{A}_j(x) = \mathcal{A}(x)a_j(x), \; 1 \le j, \; k \le n.$$

Here $\mathcal{A}$ *and* $((a_{ij}))$ *are matrix-valued functions defined in Theorem* 3.5′ *and* a_j *are scalar functions.*

For scalar uniformly elliptic equation with complex coefficients of the general form

$$(3.8) \qquad \sum_{j,k=1}^{n} c_{jk}(x)\frac{\partial^2 u}{\partial x_j \partial x_k} - \sum_{j=1}^{n} c_j(x)\frac{\partial u}{\partial x_j} - c_0(x)u = 0,$$

Theorem 3.5′ and Corollary 3.1′ imply

Corollary 3.3′. *The classical maximum modulus principle* (3.7) *is valid for solutions of the equation* (3.8) *in an arbitrary domain* $\omega \subset \Omega$ *with boundary* $\partial\omega$ *of the class* $C^{2,\alpha}$ *if and only if for all* $x \in \overline{\Omega}$

$$(i) \; c_{jk}(x) = c(x)a_{jk}(x), \; 1 \le j, \; k \le n,$$

where $\operatorname{Re} c(x) > 0$ *for all* $x \in \overline{\Omega}$, *and* $((a_{jk}))$ *is a real positive-definite* $(n \times n)$-*matrix-valued function in* $\overline{\Omega}$;
  (ii) *the inequality*

$$4\operatorname{Re}\frac{c_0(x)}{c(x)} \ge \sum_{j,k=1}^{n} b_{jk}(x)\operatorname{Im}\frac{c_j(x)}{c(x)}\operatorname{Im}\frac{c_k(x)}{c(x)}$$

holds, where $((b_{jk}))$ *is the* $(n \times n)$-*matrix inverse of* $((a_{jk}))$.

4. Maximum modulus principle for parabolic systems

The present section contains criteria for validity of the maximum modulus principle for second order partial differential equations, parabolic in the sense of Petrovskii [30]. Besides, it is stated that the criterion for validity of the maximum modulus principle remains valid for zero boundary values as well [34].

4.1. Introduction

It is well known that solutions of parabolic second order equations with real coefficients in the cylinder

$$Q_T = \{(x,t) : x \in \Omega,\ 0 < t \le T\},\ \Omega \subset \mathbb{R}^n,$$

satisfy the maximum modulus principle. Namely, for any solution of the equation

$$\frac{\partial u}{\partial t} - \sum_{j,k=1}^{n} a_{jk}(x,t)\frac{\partial^2 u}{\partial x_j \partial x_k} + \sum_{j=1}^{n} a_j(x,t)\frac{\partial u}{\partial x_j} + a_0(x,t)u = 0,$$

where $((a_{ij}))$ is a positive-definite $(n \times n)$-matrix-valued function and $a_0 \ge 0$, the inequality

$$|u(x,t)| \le \sup\{|u(y,\tau)| : (y,\tau) \in \partial Q_T,\ \tau < T\}$$

holds. This classical fact was extended to parabolic second order systems with scalar coefficients of the first and second derivatives in [59], where a sufficient condition for validity of the classical maximum modulus principle was proved. Maximum principles for weakly coupled parabolic systems are discussed in the books [48, 62].

In this section we give criteria for validity of the classical maximum modulus principle for solutions of the uniformly parabolic system in the sense of Petrovskii

$$(4.1) \qquad \frac{\partial u}{\partial t} - \sum_{j,k=1}^{n} \mathcal{A}_{jk}(x,t)\frac{\partial^2 u}{\partial x_j \partial x_k} + \sum_{j=1}^{n} \mathcal{A}_j(x,t)\frac{\partial u}{\partial x_j} + \mathcal{A}_0(x,t)u = 0$$

with real or complex coefficients. Here $\mathcal{A}_{jk}, \mathcal{A}_j, \mathcal{A}_0$ are $(m \times m)$-matrix-valued functions and u is a m-component vector-valued function. Without loss of generality we may assume that $\mathcal{A}_{jk} = \mathcal{A}_{kj}$.

4.2. The case of real coefficients

Below we use the following notations. Let F be a closed set in the Euclidean space. By $C(F)$ we denote the space of continuous and bounded m-component vector-valued functions on F with the norm

$$\|u\| = \sup\ \{|u(q)| : q \in F\}.$$

Let $\Pi_T = D \times (0,T]$, where D is either a bounded domain Ω in $\mathbb{R}^n$ or $D = \mathbb{R}^n$ and $0 < T < \infty$. By $C^{(2,1)}(\Pi_T)$ we mean the space of m-component vector-valued functions on Π_T whose derivatives with respect to x up to the second order and first derivative with respect to t are continuous. Let $\mathbb{R}_T^{n+1} = \mathbb{R}^n \times (0,T]$ and $Q_T = \Omega \times (0,T]$. By $C_m^{k+\alpha,\alpha/2}(\overline{\mathbb{R}_T^{n+1}})$ we denote the space of $(m \times m)$-matrix-valued functions with derivatives up to order k with respect to x which are bounded in $\overline{\mathbb{R}_T^{n+1}}$ and satisfy the uniform Hölder condition on $\overline{\mathbb{R}_T^{n+1}}$ with exponent α, $0 < \alpha \leq 1$, with respect to the parabolic distance $(|x - x'|^2 + |t - t'|)^{1/2}$ between points (x, t) and (x', t') in $\overline{\mathbb{R}_T^{n+1}}$.

Throughout the subsection we make the following assumptions:

(A) For any point $(x, t) \in \overline{\mathbb{R}_T^{n+1}}$, the real parts of λ-roots of the equation

$$\det \Big(\sum_{j,k=1}^{n} \mathcal{A}_{jk}(x, t)\sigma_j \sigma_k + \lambda I \Big) = 0$$

satisfy the inequality $\mathrm{Re}\, \lambda(x, t, \sigma) \leq -\delta|\sigma|^2$, where $\delta = \mathrm{const} > 0$, for any $\sigma = (\sigma_1, \dots, \sigma_n) \in \mathbb{R}^n$, I is the identity matrix of order m, and $|\cdot|$ is the Euclidean length of a vector.

(B) $\mathcal{A}_{jk} \in C_m^{2+\alpha, \alpha/2}(\overline{\mathbb{R}_T^{n+1}})$, $\mathcal{A}_j \in C_m^{1+\alpha, \alpha/2}(\overline{\mathbb{R}_T^{n+1}})$, $\mathcal{A}_0 \in C_m^{\alpha, \alpha/2}(\overline{\mathbb{R}_T^{n+1}})$, where $1 \leq j, k \leq n$.

We put

$$\mathcal{K}(\mathbb{R}^n, T) = \sup \frac{\|u\|_{C(\overline{\mathbb{R}_T^{n+1}})}}{\|u|_{t=0}\|_{C(\mathbb{R}^n)}},$$

where the supremum is taken over all functions in $C^{(2,1)}(\mathbb{R}_T^{n+1}) \cap C(\overline{\mathbb{R}_T^{n+1}})$ satisfying system (4.1).

We introduce one more constant

$$\mathcal{K}(\Omega, T) = \sup \frac{\|u\|_{C(\overline{Q}_T)}}{\|u|_{\overline{\Gamma}_T}\|_{C(\overline{\Gamma}_T)}},$$

where $\Gamma_T = \{(x, t) \in \partial Q_T : t < T\}$ and the supremum is taken over all solutions of system (4.1) in $C^{(2,1)}(Q_T) \cap C(\overline{Q}_T)$.

According to [13], there exists one and only one function in $C^{(2,1)}(\mathbb{R}_T^{n+1}) \cap C(\overline{\mathbb{R}_T^{n+1}})$, which solves the Cauchy problem

$$\frac{\partial u}{\partial t} - \sum_{j,k=1}^{n} \mathcal{A}_{jk}(x, t)\frac{\partial^2 u}{\partial x_j \partial x_k} + \sum_{j=1}^{n} \mathcal{A}_j(x, t)\frac{\partial u}{\partial x_j} + \mathcal{A}_0(x, t)u = 0 \quad \text{in} \quad \mathbb{R}_T^{n+1}, \quad u|_{t=0} = \psi,$$

with $\psi \in C(\mathbb{R}^n)$. This solution can be represented in the form

$$u(x, t) = \int_{\mathbb{R}^n} G(t, 0, x, \eta)\psi(\eta)d\eta,$$

where $G(t, \tau, x, \eta)$ is the Green matrix for system (4.1).

The following results are obtained in [30].

Theorem 4.1. *The formula*

$$\mathcal{K}(\mathbb{R}^n, T) = \sup_{x \in \mathbb{R}^n} \ \sup_{0 < t \le T} \ \sup_{|z|=1} \int_{\mathbb{R}^n} |G^*(t, 0, x, \eta) z| d\eta$$

*is valid, where the * denotes passage to the transposed matrix.*

We have the following necessary condition for validity of the maximum modulus principle (i.e. $\mathcal{K}(D, T) = 1$) for solutions of system (4.1).

Theorem 4.2. *Let the classical maximum modulus principle be valid for solutions of system (4.1) in $C^{(2,1)}(\Pi_T) \cap C(\overline{\Pi}_T)$. Then:*
(i) *for all $x \in D$ the equalities*

$$A_{jk}(x, 0) = a_{jk}(x)I, \ 1 \le j, k \le n,$$

hold, where $((a_{jk}))$ is a positive-definite $(n \times n)$-matrix-valued function in $\overline{D}$;
(ii) *for all $x \in D$ and for all ξ_j, $\zeta \in \mathbb{R}^m$, $j = 1, \ldots, n$, with $(\xi_j, \zeta) = 0$, the inequality*

$$\sum_{j,k=1}^n a_{jk}(x)(\xi_j, \xi_k) + \sum_{j=1}^n (\mathcal{A}_j(x, 0)\xi_j, \zeta) + (\mathcal{A}_0(x, 0)\zeta, \zeta) \ge 0$$

is valid.

Next we present a theorem on a sufficient condition for validity of the classical maximum modulus principle for second order systems with scalar principal part.

Theorem 4.3. *Let the coefficients of the system*

$$\frac{\partial u}{\partial t} - \sum_{j,k=1}^n a_{jk}(x, t)\frac{\partial^2 u}{\partial x_j \partial x_k} + \sum_{j=1}^n \mathcal{A}_j(x, t)\frac{\partial u}{\partial x_j} + \mathcal{A}_0(x, t)u = 0$$

satisfy the condition:
(iii) *for all $(x, t) \in \Pi_T$ and for all ξ_j, $\zeta \in \mathbb{R}^m$, $j = 1, \ldots, n$, with $(\xi_j, \zeta) = 0$, the inequality*

$$\sum_{j,k=1}^n a_{jk}(x, t)(\xi_j, \xi_k) + \sum_{j=1}^n (\mathcal{A}_j(x, t)\xi_j, \zeta) + (\mathcal{A}_0(x, t)\zeta, \zeta) \ge 0$$

holds. Then $\mathcal{K}(D, T) = 1$.

If the coefficients of system (4.1) do not depend on t, then the above mentioned necessary and sufficient conditions coincide. More precisely, the following statement concerning the system

$$(4.2) \qquad \frac{\partial u}{\partial t} - \sum_{j,k=1}^n \mathcal{A}_{jk}(x)\frac{\partial^2 u}{\partial x_j \partial x_k} + \sum_{j=1}^n \mathcal{A}_j(x)\frac{\partial u}{\partial x_j} + \mathcal{A}_0(x)u = 0$$

holds.

Theorem 4.4. *The classical maximum modulus principle is valid for solutions of system (4.2) in $C^{(2,1)}(\Pi_T) \cap C(\overline{\Pi}_T)$ if and only if:*
(i) for all $x \in D$ the equalities

$$\mathcal{A}_{jk}(x) = a_{jk}(x)I, \ 1 \le j, \ k \le n,$$

hold, where $((a_{jk}))$ is a positive-definite $(n \times n)$-matrix-valued function in $\overline{D}$;
(ii) for all $x \in D$ and for all ξ_j, $\zeta \in \mathbb{R}^m$, $j = 1, \ldots, n$, with $(\xi_j, \zeta) = 0$, the inequality

$$\sum_{j,k=1}^{n} a_{jk}(x)(\xi_j, \xi_k) + \sum_{j=1}^{n}(\mathcal{A}_j(x)\xi_j, \zeta) + (\mathcal{A}_0(x)\zeta, \zeta) \ge 0$$

is valid.

Theorem 4.4 implies

Corollary 4.1. *The classical maximum modulus principle is valid for solutions of the system*

$$\frac{\partial u}{\partial t} - \sum_{j,k=1}^{n} \mathcal{A}_{jk}(x)\frac{\partial^2 u}{\partial x_j \partial x_k} + \sum_{j=1}^{n} \mathcal{A}_j(x)\frac{\partial u}{\partial x_j} = 0$$

in $C^{(2,1)}(\Pi_T) \cap C(\overline{\Pi}_T)$ if and only if for all $x \in D$ the equalities

$$\mathcal{A}_{jk}(x) = a_{jk}(x)I, \ \ \mathcal{A}_j(x) = a_j(x)I$$

hold, where $((a_{jk}))$ is a positive-definite $(n \times n)$-matrix-valued function and a_j are scalar functions in $\overline{D}$.
   In particular, the classical maximum modulus principle is valid for solutions of the system

$$\frac{\partial u}{\partial t} - \sum_{j,k=1}^{n} \mathcal{A}_{jk}(x)\frac{\partial^2 u}{\partial x_j \partial x_k} = 0$$

in $C^{(2,1)}(\Pi_T) \cap C(\overline{\Pi}_T)$ if and only if for all $x \in D$ the equalities

$$\mathcal{A}_{jk}(x) = a_{jk}(x)I$$

hold, where $((a_{jk}))$ is a positive-definite $(n \times n)$-matrix-valued function in $\overline{D}$.

The next assertion immediately follows from Theorem 4.4.

Corollary 4.2. *The classical maximum modulus principle holds for solutions of system (4.2) in $C^{(2,1)}(\Pi_T) \cap C(\overline{\Pi}_T)$ if and only if condition (i) of Theorem 4.4 is satisfied and* (ii′) *for all $x \in D$ and for any $\zeta \in \mathbb{R}^m$, $|\zeta| = 1$ the inequality*

$$\sum_{i,j=1}^{n} b_{ij}(x)\big[(\mathcal{A}_i(x)\zeta, \zeta)(\mathcal{A}_j(x)\zeta, \zeta) - (\mathcal{A}_i^*(x)\zeta,\ \mathcal{A}_j^*(x)\zeta)\big] + 4(\mathcal{A}_0(x)\zeta, \zeta) \geq 0$$

holds, where $((b_{ij}))$ is the $(n \times n)$-matrix-valued function inverse of $((a_{ij}))$ and $\mathcal{A}_j^*(x)$ is the matrix transposed of $\mathcal{A}_j(x)$.

Remark 4.1. We give an example of a system whose principal part is not a scalar differential operator in the whole domain and for which the classical maximum modulus principle is valid in $\mathbb{R}_T^{n+1}$.

Consider the parabolic system

$$(4.3) \qquad \frac{\partial v}{\partial t} - \sum_{j,k=1}^{n} \mathcal{A}_{jk}(x,t)\frac{\partial^2 v}{\partial x_j \partial x_k} + \sum_{j=1}^{n} \mathcal{A}_j(x,t)\frac{\partial v}{\partial x_j} + \mathcal{A}_0(x,t)v + \lambda^2 v = 0$$

in $\mathbb{R}_T^{n+1}$, where $\mathcal{A}_{ij} \in C_m^{2+\alpha,\alpha/2}(\overline{\mathbb{R}_T^{n+1}})$, $\mathcal{A}_j \in C_m^{1+\alpha,\alpha/2}(\overline{\mathbb{R}_T^{n+1}})$, $\mathcal{A}_0 \in C_m^{\alpha,\alpha/2}(\overline{\mathbb{R}_T^{n+1}})$ and λ=const. Suppose the coefficients of system (4.3) do not depend on t in the layer $\mathbb{R}_\delta^{n+1}$, $0 < \delta < T$. Let the matrix-valued functions $\mathcal{A}_{jk}(x,0)$, $\mathcal{A}_j(x,0)$, $\mathcal{A}_0(x,0)$, denoted by $\mathcal{A}_{jk}(x)$, $\mathcal{A}_j(x)$, $\mathcal{A}_0(x)$, respectively, satisty the conditions (i), (ii) of Theorem 4.4. Suppose further that $|v(x,0)| \leq 1$. Then $|v(x,t)| \leq 1$ in $\overline{\mathbb{R}_\delta^{n+1}}$. By M we denote the value $\mathcal{K}(\mathbb{R}^n, T)$ for system (4.1). Since $u(x,t) = v(x,t)\exp(\lambda^2 t)$ is the solution of (4.1), then

$$\sup\left\{|v(x,t)| : (x,t) \in \overline{\mathbb{R}_T^{n+1}}\backslash\overline{\mathbb{R}_\delta^{n+1}}\right\} \leq M\exp(-\lambda^2\delta).$$

Thus, the solution of the Cauchy problem for system (4.3) satisfies the classical maximum modulus principle for sufficiently large values of λ.

Remark 4.2. (*On the maximum modulus principle for the parabolic systems with zero boundary data*). Consider system (4.2) in the cylinder $Q_T = \Omega \times (0,T]$ with zero Dirichlet data on $S_T = \partial\Omega \times [0,T]$. Suppose that Ω is a subdomain of $\mathbb{R}^n$ with compact closure and boundary $\partial\Omega$ of the class $C^{2+\alpha}$, $\alpha \in (0,1)$.

By $\mathring{C}(\overline{Q}_T)$ we denote the space of vector-valued functions from $C(\overline{Q}_T)$ vanishing on the cylindric surface S_T. Let $\mathring{\mathcal{K}}(\Omega,T)$ denote the best constant in the inequality

$$|u(x,t)| \leq \mathcal{K}\sup\{|u(y,0)| : y \in \overline{\Omega}\},$$

where $(x,t) \in Q_T$ and u is a solution of (4.2) in $C^{(2,1)}(Q_T) \cap \mathring{C}(\overline{Q}_T)$.

Since $1 \leq \mathring{\mathcal{K}}(\Omega,T)$, the conditions of Theorem 4.4 are sufficient for $\mathring{\mathcal{K}}(\Omega,T) = 1$. Are they also necessary for $\mathring{\mathcal{K}}(\Omega,T) = 1$? In [34] it is shown that the answer is positive. For instance, in the case of real coefficients we have the following assertion.

Theorem 4.4. *The maximum modulus principle is valid for solutions of system* (4.2) *in* $C^{(2,1)}(Q_T) \cap \mathring{C}(\overline{Q}_T)$ *if and only if:*

(i) *for all* $x \in \Omega$ *the equalities*

$$\mathcal{A}_{jk}(x) = a_{jk}(x)I, \ \ 1 \leq j, k \leq n,$$

hold, where $((a_{jk}(x)))$ *is a positive-definite* $(n \times n)$-*matrix-valued function in* $\overline{\Omega}$;

(ii) *for all* $x \in \Omega$ *and for all* $\xi_j, \ \zeta \in \mathbb{R}^m, \ j = 1, \ldots, n,$ *with* $(\xi_j, \zeta) = 0,$ *the inequality*

$$\sum_{j,k=1}^{n} a_{jk}(x)(\xi_j, \xi_k) + \sum_{j=1}^{n}(\mathcal{A}_j(x)\xi_j, \zeta) + (\mathcal{A}_0(x)\zeta, \zeta) \geq 0$$

is valid.

We note that the last Theorem can be interpreted as a criterion for the $C(\overline{\Omega})$-contractivity of a semigroup generated by the operator

$$\mathfrak{A}(x, \partial/\partial x) = \sum_{j,k=1}^{n} \mathcal{A}_{jk}(x)\frac{\partial^2}{\partial x_j \partial x_k} - \sum_{j=1}^{n}\mathcal{A}_j(x)\frac{\partial}{\partial x_j} - \mathcal{A}_0(x)$$

defined on vector-valued functions u satisfying the boundary condition $u\big|_{\partial\Omega} = 0.$

4.3. The case of complex coefficients

We can extend the results of the Subsection 4.2 to systems (4.1), (4.2) with complex coefficients and solutions $u = v + iw$, where v and w are m-component vector-valued functions with real components. The results are obtained by application of corresponding assertions on the maximum modulus for the real case to systems obtained by the separation of real and imaginary parts. Thus we can formulate theorems and corollaries similar to those in Subsection 4.2. By $\mathbb{C}^m$ we denote the complex linear m-dimensional space with the inner product $\langle \cdot, \cdot \rangle$.

Let F be a closed set in the Euclidean space. By $\mathbf{C}(F)$ we denote the space of continuous and bounded on F vector-valued functions $u = v + iw$ with m complex componenets. The norm on $\mathbf{C}(F)$ is

$$||u|| = \sup \left\{ (|v(x)|^2 + |w(x)|^2)^{1/2} \ : \ x \in F \right\}.$$

Analogously, for others spaces of vector-valued and matrix-valued functions with complex components we retain the same notation as in the case of real components but use bold letter.

Throughout the subsection we assume that for coefficients of system (4.1) the condition (A) from subsection 4.2 is satisfied and $\mathcal{A}_{jk} \in \mathbf{C}_m^{2+\alpha, \alpha/2}(\overline{\mathbb{R}_T^{n+1}})$, $\mathcal{A}_j \in \mathbf{C}_m^{1+\alpha, \alpha/2}(\overline{\mathbb{R}_T^{n+1}})$, $\mathcal{A}_0 \in \mathbf{C}_m^{\alpha, \alpha/2}(\overline{\mathbb{R}_T^{n+1}})$, $1 \leq j, k \leq n.$

As in the definition of $\mathcal{K}(\mathbb{R}^n, T)$ we put

$$\mathcal{K}'(\mathbb{R}^n, T) = \sup \frac{\|u\|_{\mathbf{C}(\overline{\mathbb{R}_T^{n+1}})}}{\|u|_{t=0}\|_{\mathbf{C}(\mathbb{R}^n)}},$$

where the supremum is taken over all functions $u = v + iw$ in the class $\mathbf{C}^{(2,1)}(\mathbb{R}_T^{n+1}) \cap \mathbf{C}(\overline{\mathbb{R}_T^{n+1}})$ satisfying system (4.1) with complex coefficients.

By analogy with the definition of the constant $\mathcal{K}(\Omega, T)$ let

$$\mathcal{K}'(\Omega, T) = \sup \frac{\|u\|_{\mathbf{C}(\overline{Q}_T)}}{\left\|u\big|_{\overline{\Gamma}_T}\right\|_{\mathbf{C}(\overline{\Gamma}_T)}},$$

where the supremum is taken over all vector-valued functions $u - v + iw$ in the class $\mathbf{C}^{(2,1)}(Q_T) \cap \mathbf{C}(\overline{Q}_T)$ which satisfy system (4.1) with complex coefficients.

If $\mathcal{K}'(D, T) = 1$ then the classical maximum modulus principle is valid for system (4.1) with complex coefficients.

Theorem 4.2′. *Let the classical maximum modulus principle be valid for system (4.1) in $\mathbf{C}^{(2,1)}(\Pi_T) \cap \mathbf{C}(\overline{\Pi}_T)$. Then:*
(i) for all $x \in D$ the equalities

$$\mathcal{A}_{jk}(x, 0) = a_{jk}(x)I, \ 1 \leq j, k \leq n,$$

hold, where $((a_{jk}))$ is a real positive-definite $(n \times n)$-matrix-valued function in $\overline{D}$;
(ii) for all $x \in D$ and for all ζ_j, $\zeta \in \mathbb{C}^m$, $j = 1, \ldots, n$, with $\mathrm{Re}\,\langle \xi_j,\ \zeta \rangle = 0$, the inequality

$$\mathrm{Re} \left\{ \sum_{j,k=1}^{n} a_{jk}(x)\langle \xi_j, \xi_k \rangle + \sum_{j=1}^{n} \langle \mathcal{A}_j(x, 0)\xi_j,\ \zeta \rangle + \langle \mathcal{A}_0(x, 0)\zeta, \zeta \rangle \right\} \geq 0$$

is valid.

Theorem 4.4′. *The classical maximum modulus principle is valid for solutions of system (4.2) in $\mathbf{C}^{(2,1)}(\Pi_T) \cap \mathbf{C}(\overline{\Pi}_T)$ if and only if:*
(i) for all $x \in D$ the equalities

$$\mathcal{A}_{jk}(x) = a_{jk}(x)I, \ 1 \leq j, k \leq n,$$

hold, where $((a_{jk}))$ is a real positive-definite $(n \times n)$-matrix-valued function in $\overline{D}$;
(ii) for all $x \in D$ and for all ξ_j, $\zeta \in \mathbb{C}^m$, $j = 1, \ldots, n$, with $\mathrm{Re}\,\langle \xi_j,\ \zeta \rangle = 0$, the inequality

$$\mathrm{Re} \left\{ \sum_{j,k=1}^{n} a_{jk}(x)\langle \xi_j, \xi_k \rangle + \sum_{j=1}^{n} \langle \mathcal{A}_j(x)\xi_j,\ \zeta \rangle + \langle \mathcal{A}_0(x)\zeta, \zeta \rangle \right\} \geq 0$$

is valid.

286   Kresin

Corollary 4.1′. *The classical maximum modulus principle is valid for solutions of the system*

$$\frac{\partial u}{\partial t} - \sum_{j,k=1}^{n} \mathcal{A}_{jk}(x)\frac{\partial^2 u}{\partial x_j \partial x_k} + \sum_{j=1}^{n} \mathcal{A}_j(x)\frac{\partial u}{\partial x_j} = 0$$

in $\mathbf{C}^{(2,1)}(\Pi_T) \cap \mathbf{C}(\overline{\Pi}_T)$ *if and only if for all* $x \in D$ *the equalities*

$$\mathcal{A}_{jk}(x) = a_{jk}(x)I, \ \mathcal{A}_j(x) = a_j(x)I, \ 1 \le j,k \le n,$$

hold, where $((a_{jk}))$ *is a real positive-definite* $(n \times n)$*-matrix-valued function and* a_j *are real scalar functions in* $\overline{D}$*.*

In particular, the classical maximum modulus principle is valid for solutions of the system

$$\frac{\partial u}{\partial t} - \sum_{j,k=1}^{n} \mathcal{A}_{jk}(x)\frac{\partial^2 u}{\partial x_j \partial x_k} = 0$$

in $\mathbf{C}^{(2,1)}(\Pi_T) \cap \mathbf{C}(\overline{\Pi}_T)$ *if and only if for all* $x \in D$ *the equalities*

$$\mathcal{A}_{jk}(x) = a_{jk}(x)I,$$

hold, where $((a_{jk}))$ *is a real positive-definite* $(n \times n)$*-matrix-valued function in* $\overline{D}$*.*

Corollary 4.2′. *The classical maximum modulus principle is valid for solutions of system* (4.2) *in* $\mathbf{C}^{(2,1)}(\Pi_T) \cap \mathbf{C}(\overline{\Pi}_T)$ *if and only if the condition* (i) *of Theorem 4.4′ is satisfied and*
(ii′) *for all* $x \in D$ *and for any* $\zeta \in \mathbb{C}^m$, $|\zeta| = 1$ *the inequality*

$$\sum_{i,j=1}^{n} b_{ij}(x)\big[\mathrm{Re}\,\langle \mathcal{A}_i(x)\zeta, \zeta\rangle\mathrm{Re}\langle \mathcal{A}_j(x)\zeta,\zeta\rangle - \langle \mathcal{A}_i^*(x)\zeta,\ \mathcal{A}_j^*(x)\zeta\rangle\big] + 4\mathrm{Re}\langle \mathcal{A}_0(x)\zeta,\zeta\rangle \ge 0$$

holds. Here $((b_{ij}(x)))$ *is the* $(n \times n)$*-matrix inverse of* $((a_{ij}(x)))$ *and* $\mathcal{A}_j^*(x)$ *is the adjoint matrix of* $\mathcal{A}_j(x)$*.*

We remark that the sum

$$\sum_{i,j=1}^{n} b_{ij}(x)\langle \mathcal{A}_i^*(x)\zeta,\ \mathcal{A}_j^*(x)\zeta\rangle$$

is real by the symmetry of the matrix $((b_{ij}(x)))$.

The next assertion follows from Corollary 4.2′ and concerns the scalar parabolic equation with complex coefficients

$$(4.4) \qquad \frac{\partial u}{\partial t} - \sum_{j,k=1}^{n} a_{jk}(x)\frac{\partial^2 u}{\partial x_j \partial x_k} + \sum_{j=1}^{n} a_j(x)\frac{\partial u}{\partial x_j} + a_0(x)u = 0.$$

Corollary 4.3'. *The classical maximum modulus principle is valid for equation* (4.4) *in* $\mathbf{C}^{(2,1)}(\Pi_T) \cap \mathbf{C}(\overline{\Pi}_T)$ *if and only if:*

(i) the $(n \times n)$-matrix-valued function $((a_{jk}(x)))$ is real and positive-definite in $\overline{D}$;

(ii) for all $x \in D$ the inequality

$$4\mathrm{Re}\ a_0(x) \geq \sum_{j,k=1}^{n} b_{jk}(x)\mathrm{Im}\ a_j(x)\ \mathrm{Im}\ a_k(x)$$

holds, where $((b_{jk}))$ is the $(n \times n)$-matrix inverse of $((a_{jk}))$.

5. Maximum norm principle for parabolic systems

This section contains a generalization of the maximum modulus principle for parabolic systems. Here the modulus is replaced by a generalized norm, that is by the Minkowski functional of a compact body.

5.1. Introduction

In the present section we consider parabolic systems of the form

$$(5.1) \qquad \frac{\partial u}{\partial t} - \sum_{j,k=1}^{n} \mathcal{A}_{jk}(x,t)\frac{\partial^2 u}{\partial x_j \partial x_k} + \sum_{j=1}^{n} \mathcal{A}_j(x,t)\frac{\partial u}{\partial x_j} + \mathcal{A}_0(x,t)u = 0,$$

where u is a m-component vector-valued function and $\mathcal{A}_{jk}, \mathcal{A}_j, \mathcal{A}_0$ are real $(m \times m)$-matrix-valued functions.

The case $m = 1$ in (5.1) is classical. It is well known that solutions of the second order uniformly parabolic equation

$$(5.2) \qquad \frac{\partial u}{\partial t} - \sum_{j,k=1}^{n} a_{jk}(x,t)\frac{\partial^2 u}{\partial x_j \partial x_k} + \sum_{j=1}^{n} a_j(x,t)\frac{\partial u}{\partial x_j} + a_0(x,t)u = 0$$

with $a_0(x,t) \geq 0$ and with real coefficients in the cylinder

$$Q_T = \{(x,t) : x \in \Omega,\ 0 < t \leq T\},\ \Omega \subset \mathbb{R}^n,$$

satisfy the maximum principle. In other words, for any solution of (5.2), where $a_0(x,t) \geq 0$, which is continuous in $\overline{Q}_T$ together with the first derivative with respect to t and derivatives with respect to x up to order two in Q_T, the inequality

$$(5.3) \qquad \min\left\{0,\ \min_{(y,\tau)\in\overline{\Gamma}_T} u(y,\tau)\right\} \leq u(x,t) \leq \max\left\{0,\ \max_{(y,\tau)\in\overline{\Gamma}_T} u(y,\tau)\right\}$$

is valid. Here $(x,t) \in \overline{Q}_T$ and $\Gamma_T = \{(y,\tau) \in \partial Q_T : 0 \le \tau < T\}$. This means that if the initial-boundary conditions belong to the interval $[\alpha, \beta]$ containing the origin then the same is valid for a solution of (5.2), i.e. any closed interval containing the origin is an invariant set of equation (5.2).

In the case of the uniformly parabolic equation of the form

$$(5.4) \qquad \frac{\partial u}{\partial t} - \sum_{j,k=1}^{n} a_{jk}(x,t)\frac{\partial^2 u}{\partial x_j \partial x_k} = 0$$

with real coefficients in the cylinder Q_T the following inequality

$$\min_{(y,\tau)\in\overline{\Gamma}_T} u(y,\tau) \le u(x,t) \le \max_{(y,\tau)\in\overline{\Gamma}_T} u(y,\tau)$$

is valid instead of (5.3). Here $(x,t) \in \overline{Q}_T$. This means that if the initial-boundary conditions belong to the interval $[\alpha, \beta]$ the same is valid for a solution of (5.4), i.e. any closed interval is an invariant set of (5.4).

In the present section we give criteria for validity of the so called *maximum norm principle* for solutions of system (5.1) which is uniformly parabolic in the sense of Petrovskii. The norm $|\cdot|$ is understood in a generalized sense, namely as the Minkowski functional of a compact convex body in $\mathbb{R}^m$ containing the origin. The geometric aspect of the obtained criteria consists of establishing a correspondence between a compact convex body $B \subset \mathbb{R}^m$ containing the origin and a subclass of systems (5.1) for which B is invariant.

We shall consider system (5.1) in the layer $\mathbb{R}_T^{n+1} = \mathbb{R}^n \times (0, T]$ and in the cylinder $Q_T = \Omega \times (0, T]$, where Ω is a bounded subdomain of $\mathbb{R}^n$. By $C_m^{k+\alpha,\alpha/2}(\overline{\mathbb{R}_T^{n+1}})$ we denote the space of $(m \times m)$-matrix-valued functions defined on $\overline{\mathbb{R}_T^{n+1}}$ with bounded derivatives in x up to order k which satisfy the uniform Hölder condition on $\overline{\mathbb{R}_T^{n+1}}$ with exponent $\alpha, 0 < \alpha \le 1$, with respect to the parabolic distance $(|x - x'|^2 + |t - t'|)^{1/2}$ between points (x,t) and (x',t') in $\overline{\mathbb{R}_T^{n+1}}$.

Throughout the section we make the following assumptions:

(A) For any point $(x,t) \in \overline{\mathbb{R}_T^{n+1}}$, the real parts of the λ-roots of the equation

$$\det\left(\sum_{j,k=1}^{n} \mathcal{A}_{jk}(x,t)\sigma_j\sigma_k + \lambda I \right) = 0$$

satisfy the inequality $\operatorname{Re} \lambda(x,t,\sigma) \le -\delta|\sigma|^2$ for any $\sigma = (\sigma_1,\ldots,\sigma_n) \in \mathbb{R}^n$, where $\delta=\text{const} > 0$, I is the identity matrix of order m, and $|\cdot|$ is the Euclidean length of a vector;

(B) $\mathcal{A}_{jk} \in C_m^{2+\alpha,\alpha/2}(\overline{\mathbb{R}_T^{n+1}})$, $\quad \mathcal{A}_j \in C_m^{1+\alpha,\alpha/2}(\overline{\mathbb{R}_T^{n+1}})$, $\quad \mathcal{A}_0 \in C_m^{\alpha,\alpha/2}(\overline{\mathbb{R}_T^{n+1}})$, $1 \le j,k \le n$.

By definition ([41, 58]), the norm in the Minkowski sense (following [58] we shall call it the norm) is a function $|\cdot| : \mathbb{R}^m \to \mathbb{R}$ with the following properties

(a) $\quad |\zeta| \geq 0$ for all $\zeta \in \mathbb{R}^m$,

(b) $\quad |\zeta| = 0$ if and only if $\zeta = 0$,

(c) $\quad |\lambda\zeta| = \lambda|\zeta|$ for all $\zeta \in \mathbb{R}^m$ and $\lambda \geq 0$,

(d) $\quad |\zeta_1 + \zeta_2| \leq |\zeta_1| + |\zeta_2|$ for all $\zeta_1, \zeta_2 \in \mathbb{R}^m$.

The unit ball $\mathcal{B} = \{\zeta \in \mathbb{R}^m : |\zeta| \leq 1\}$ of an arbitrary norm in $\mathbb{R}^m$ is a compact convex body which contains the origin in its interior. Conversely, for every compact convex body $B \subset \mathbb{R}^m$, which contains the origin in its interior, there is precisely one norm $|\cdot|$ such that $B = \mathcal{B}$, namely, $|\zeta| = \inf \{r : r^{-1}\zeta \in B, \ r > 0\}$. In other words, the compact convex bodies which contain the origin in their interior are in 1-1 correspondence with the norms in $\mathbb{R}^m$.

Given the norm $|\cdot|$, one can introduce a new function $|\cdot|_d : \mathbb{R}^m \to \mathbb{R}$ by the equality

$$|z|_d = \sup \{(z, \zeta) : |\zeta| \leq 1\}.$$

The function $|\cdot|_d$, similarly to $|\cdot|$, satisfies (a)–(d), and is called the dual norm.

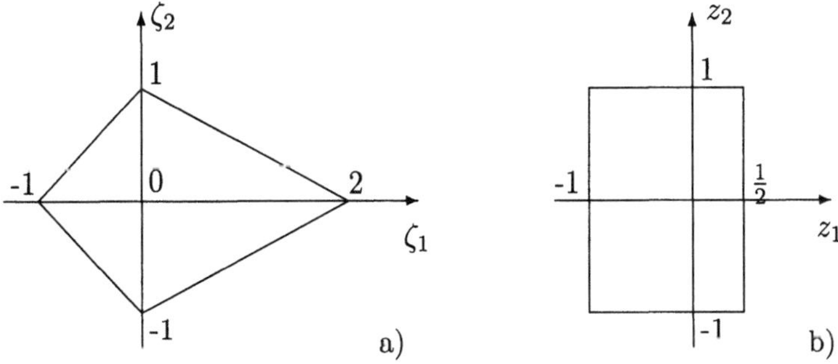

Figure 1: Example: a) The unit ball of the norm $|\zeta| = \alpha|\zeta_1| + |\zeta_2|$ with $\alpha = 1/2$ for $\zeta_1 \geq 0$, and $\alpha = 1$ for $\zeta_1 < 0$; b) The unit ball of the corresponding dual norm $|z|_d = \max\{\beta|z_1|, |z_2|\}$, where $\beta = 2$ for $z_1 \geq 0$, and $\beta = 1$ for $z_1 < 0$.

The symbol $(\cdot, \cdot)$ will be used as the standard notation of the scalar product in $\mathbb{R}^m$. The original and dual norms satisfy the following duality relation

$$|\zeta| = \sup \{(z, \zeta) : |z|_d \leq 1\}$$

which is a corollary of the equality $|\cdot|_{dd} = |\cdot|$.

Let F be a closed set in the Euclidean space of dimension n or $n+1$. By $C(F)$ we denote the space of continuous and bounded m-component vector-valued functions on F with the norm

$$||u|| = \sup \{ |u(q)| : q \in F \}.$$

Let $\Pi_T = D \times (0, T]$, where D is either a bounded domain Ω in $\mathbb{R}^n$ or $D = \mathbb{R}^n$ and $0 < T < \infty$. By $C^{(2,1)}(\Pi_T)$ we mean the space of m-component vector-valued functions on Π_T whose derivatives with respect to x up to the second order and the first derivative with respect to t are continuous.

By $\mathcal{K}(\mathbb{R}^n, T)$ we denote the best constant in the inequality

$$|u(x, t)| \leq \mathcal{K} \sup \{|u(y, 0)| : y \in \mathbb{R}^n\},$$

where $(x, t) \in \mathbb{R}_T^{n+1}$ and u is a solution of system (5.1) in $C^{(2,1)}(\mathbb{R}_T^{n+1}) \cap C(\overline{\mathbb{R}_T^{n+1}})$.

Besides the constant $\mathcal{K}(\mathbb{R}^n, T)$, we define the best constant $\mathcal{K}(\Omega, T)$ in the inequality

$$|u(x, t)| \leq \mathcal{K} \sup \{|u(y, \tau)| : (y, \tau) \in \overline{\Gamma}_T\},$$

where $(x, t) \in Q_T$ and u is a solution of (5.1) in $C^{(2,1)}(Q_T) \cap C(\overline{Q_T})$.

We give (for detailed exposition see [32]) separate necessary and sufficient conditions for validity of the *maximum norm principle* (i.e. $\mathcal{K}(\Omega, T) = 1$, $\mathcal{K}(\mathbb{R}^n, T) = 1$) for solutions of the system

$$(5.5) \qquad \frac{\partial u}{\partial t} - \sum_{j,k=1}^{n} \mathcal{A}_{jk}(x, t) \frac{\partial^2 u}{\partial x_j \partial x_k} = 0.$$

If the coefficients of system (5.5) do not depend on t, i.e.

$$(5.6) \qquad \frac{\partial u}{\partial t} - \sum_{j,k=1}^{n} \mathcal{A}_{jk}(x) \frac{\partial^2 u}{\partial x_j \partial x_k} = 0,$$

then the above mentioned necessary and sufficient conditions coincide. The criteria for validity of the maximum norm principle in this case are formulated as a number of equivalent algebraic conditions describing the relation between the geometry of the unit sphere of the given norm $|\cdot|$ in $\mathbb{R}^m$ and coefficients of the system under consideration. Simpler formulated criteria are given for certain classes of norms: for differentiable norms, p-norms $(1 \leq p \leq \infty)$ in $\mathbb{R}^m$, as well as for norms whose unit balls are m-pyramids, m-bipyramids, cylindrical bodies, m-parallelepipeds. The case $m = 2$ is studied separately.

Assuming that the norm $|\cdot|$ is twice continuously differentiable in $\mathbb{R}^m \backslash \{0\}$, we give (see [33]) separate necessary and sufficient conditions for validity of the *maximum norm principle* for solutions of (5.1). If the coefficients of system (5.1) do not depend on t, i.e. the system under consideration is

$$(5.7) \qquad \frac{\partial u}{\partial t} - \sum_{j,k=1}^{n} \mathcal{A}_{jk}(x) \frac{\partial^2 u}{\partial x_j \partial x_k} + \sum_{j=1}^{n} \mathcal{A}_j(x) \frac{\partial u}{\partial x_j} + \mathcal{A}_0(x) u = 0,$$

then the above mentioned necessary and sufficient conditions coincide. The criterion for the validity of the maximum norm principle in this case is that the principal part of the system is scalar and that the coefficients of the system satisfy a certain inequality, describing the relation between the geometry of the sphere of the given norm $|\cdot|$ and coefficients of the system. Simpler formulated criteria are given for certain classes of norms, in particular, for p-norms $(2 < p < \infty)$ in $\mathbb{R}^m$.

5.2. Representation of the constant $\mathcal{K}(\mathbb{R}^n, T)$

We introduce the operators

$$\mathfrak{A}(x, t, \partial/\partial x) = \sum_{j,k=1}^{n} \mathcal{A}_{jk}(x, t)\frac{\partial^2}{\partial x_j \partial x_k} - \sum_{j=1}^{n} \mathcal{A}_j(x, t)\frac{\partial}{\partial x_j} - \mathcal{A}_0(x, t),$$

$$\mathfrak{A}_0(x, t, \partial/\partial x) = \sum_{j,k=1}^{n} \mathcal{A}_{jk}(x, t)\frac{\partial^2}{\partial x_j \partial x_k},$$

where $\mathcal{A}_{jk}$, $\mathcal{A}_j$, $\mathcal{A}_0$ are real $(m \times m)$-matrix-valued functions satisfying the conditions (A) and (B) in Sec. 5.1. Without loss of generality one may assume that $\mathcal{A}_{jk} = \mathcal{A}_{kj}$, $1 \leq j$, $k \leq n$.

The unique solution $u(x, t)$ in the class $C^{(2,1)}(\mathbb{R}_T^{n+1}) \cap C(\overline{\mathbb{R}_T^{n+1}})$ of the Cauchy problem

$$\partial u/\partial t - \mathfrak{A}(x, t, \partial/\partial x)u = 0 \ \text{ in } \ \mathbb{R}_T^{n+1}, \ u\Big|_{t=0} = \psi$$

with $\psi \in C(\mathbb{R}^n)$ can be represented in the form

$$u(x, t) = \int_{\mathbb{R}^n} G(t, 0, x, \eta)\psi(\eta)d\eta,$$

where $G(t, \tau, x, \eta)$ is the Green matrix of this problem.

In what follows * means passage to the transposed matrix.

Theorem 5.1. *The formula*

$$\mathcal{K}(\mathbb{R}^n, T) = \sup_{x \in \mathbb{R}^n} \ \sup_{0 < t \leq T} \ \sup_{|z|_d=1} \int_{\mathbb{R}^n} |G^*(t, 0, x, \eta)z|_d \, d\eta$$

is valid.

By $\mathcal{K}(y)$ we denote the constant $\mathcal{K}(\mathbb{R}^n, T)$ for the system

$$\frac{\partial u}{\partial t} - \sum_{j,k=1}^{n} \mathcal{A}_{jk}(y, 0)\frac{\partial^2 u}{\partial x_j \partial x_k} = 0$$

in which $y \in \mathbb{R}^n$ serves as a parameter.

The relation between the constants $\mathcal{K}(D,T)$ and $\mathcal{K}(y)$ is given by the following assertion.

Lemma 5.1. *The inequality*

$$\mathcal{K}(D,T) \geq \sup\{\mathcal{K}(y): \ y \in D\}$$

is valid, where D is either a bounded domain $\Omega \subset R^n$, or $D = \mathbb{R}^n$.

5.3.   Maximum norm principle for the system $\partial u/\partial t - \mathfrak{A}_0(x,t,\partial/\partial x)u = 0$

This section contains criteria for validity of the maximum norm principle for parabolic systems with derivatives in x only of the second order [32].

5.3.1.   Necessary and sufficient conditions for validity of the maximum norm principle for the system $\partial u/\partial t - \mathfrak{A}_0(x,t,\partial/\partial x)u = 0$

Given the norm $|\cdot|$, by $\mathbf{S}_\nu^{m-1}$ we denote the set of points ζ of the unit sphere $\mathbf{S}^{m-1} = \{\zeta \in \mathbb{R}^m : |\zeta| = 1\}$ for which there exists the unit outward normal to $\mathbf{S}^{m-1}$. By $\nu(\zeta)$ we denote the vector obtained by parallel translation of the unit outward normal to $\mathbf{S}^{m-1}$ at the point $\zeta \in \mathbf{S}_\nu^{m-1}$ into the origin $0 \in \mathbb{R}^m$.

The next assertion, crucial in what follows, contains a necessary condition for validity of the maximum norm principle for parabolic systems of the form (5.5) with constant coefficients in $\mathbb{R}_T^{n+1}$.

Proposition 5.1. *Let the maximum norm principle hold for solutions of the system*

$$\frac{\partial u}{\partial t} - \sum_{j,k=1}^{n} \mathcal{A}_{jk}\frac{\partial^2 u}{\partial x_j \partial x_k} = 0$$

with constant coefficients in $\mathbb{R}_T^{n+1}$. Then there exist functions $\mu_{jk} : \mathbf{S}_\nu^{m-1} \to \mathbb{R}$, $1 \leq j,k \leq n$, such that

$$\mathcal{A}_{jk}^*\nu(\zeta) = \mu_{jk}(\zeta)\nu(\zeta), \ \ 1 \leq j,k \leq n,$$

for any $\zeta \in \mathbf{S}_\nu^{m-1}$.

The following assertion results from Proposition 5.1 and Lemma 5.1.

Proposition 5.2. *Let the maximum norm principle be valid for solutions of system (5.5) in Π_T. Then there exist functions $\mu_{jk} : D \times \mathbf{S}_\nu^{m-1} \to \mathbb{R}$, $1 \leq j,k \leq n$, such that for any $\zeta \in \mathbf{S}_\nu^{m-1}$ and for all $x \in D$*

$$\mathcal{A}_{jk}^*(x,0)\nu(\zeta) = \mu_{jk}(x,\zeta)\nu(\zeta), \ 1 \leq j,k \leq n.$$

We now give a sufficient condition for validity of the maximum norm principle.

Proposition 5.3. *Let*

$$\mathcal{A}^*_{jk}(x,t)\nu(\zeta) = \mu_{jk}(x,t,\zeta)\nu(\zeta),\ 1 \le j,k \le n,$$

for any $\zeta \in \mathbf{S}^{m-1}_\nu$ and for all $(x,t) \in \Pi_T$ with $\mu_{jk} : \Pi_T \times \mathbf{S}^{m-1}_\nu \to \mathbb{R}$, $1 \le j,k \le n$. Then the maximum norm principle holds for solutions of system (5.5) in Π_T.

5.3.2. Criteria for validity of the maximum norm principle for the system $\partial u/\partial t - \mathfrak{A}_0(x, \partial/\partial x)u = 0$

To formulate the main theorem on a series of equivalent criteria for validity of the maximum norm principle for solutions of system (5.6) with coefficients which depend only on x we introduce the following notation.

Let $\mathcal{N} = \{\nu(\zeta) : \zeta \in \mathbf{S}^{m-1}_\nu\}$. By $[v_1, \dots, v_m]$ we mean the $(m \times m)$-matrix whose columns are m-component vectors $v_1, \dots, v_m$. Further, let

$$A[D] = \{\mathcal{A}_{jk}(x) : 1 \le j,k \le n,\ \ x \in D\}.$$

The symbol * means passage to the transposed matrix.

Theorem 5.2. *The maximum norm principle is valid for solutions of system (5.6) in $\Pi_T = D \times (0,T]$ if and only if one of the following equivalent conditions holds:*

(i) for any matrix $\mathcal{A} \in A[D]$ there exists a function $\mu : \mathbf{S}^{m-1}_\nu \to \mathbb{R}$ such that

$$\mathcal{A}^*\nu(\zeta) = \mu(\zeta)\nu(\zeta)$$

for all $\zeta \in \mathbf{S}^{m-1}_\nu$;

(ii) the family of eigenspaces $\mathcal{L}_1, \dots, \mathcal{L}_s$ of each matrix $\mathcal{A}^*$, where $\mathcal{A} \in A[D]$, is such that

$$\cup^s_{i=1}\mathcal{L}_i \supset \mathcal{N};$$

(iii) the family of matrices $A[D]$ is simultaneously diagonable by any matrix of the form $([\nu_1, \dots, \nu_m]^*)^{-1}$, where $\nu_1, \dots, \nu_m$ is a linear independent system of normals to $\mathbf{S}^{m-1}$;

(iv) any plane $T'_\zeta \mathbf{S}^{m-1}$, passing through the origin of $\mathbb{R}^m$ and parallel to the plane $T_\zeta \mathbf{S}^{m-1}$, tangent to the sphere $\mathbf{S}^{m-1}$ at $\zeta \in \mathbf{S}^{m-1}_\nu$, is an invariant subspace of any matrix $\mathcal{A} \in A[D]$.

Remark 5.1. A condition similar to (i) of Theorem 5.2 was obtained earlier in [10] as a criterion of positively invariant regions for systems of nonlinear diffusion equations

$$\frac{\partial u}{\partial t} = \mathcal{D}(x,u) \sum_{j,k=1}^n a_{jk}(x)\, \frac{\partial^2 u}{\partial x_j \partial x_k} + F(u).$$

The next assertion is also a criterion for validity of the maximum norm principle for solutions of system (5.6). However, we formulate it separately as a theorem on a representation of the matrix coefficients $\mathcal{A}_{jk}(x)$.

In the following theorem, the spaces $L_1, \ldots, L_s \subset \mathbb{R}^m$, $1 \leq s \leq m$, form a family such that

$$(L) \qquad\qquad \mathbb{R}^m = \oplus_{i=1}^s L_i, \quad \cup_{i=1}^s L_i \supset \mathcal{N}.$$

Let s_{max} denote the maximum value of s for which (L) holds for the given norm. Further, let $\nu_1, \ldots, \nu_m$ be a fixed linear independent system of normals to $\mathbf{S}^{m-1}$.

Theorem 5.3. *The maximum norm principle is valid for solutions of system (5.6) in $\Pi_T = D \times (0, T]$ if and only if:*

(a) *for all $x \in D$,*

$$\mathcal{A}_{jk}(x) = ([\nu_1, \ldots, \nu_m]^*)^{-1} \operatorname{diag} \{a_{1,jk}(x), \ldots, a_{m,jk}(x)\} [\nu_1, \ldots, \nu_m]^*,$$

with $1 \leq j, k \leq n$;

(b) *to any matrix $\mathcal{A}_{jk}(x)$ there corresponds a partition of the set $\{\nu_1, \ldots, \nu_m\}$ into subsets $\mathcal{N}_1, \ldots, \mathcal{N}_s$, $1 \leq s \leq m$, such that*

1) *subspaces $L_1 = \operatorname{Span} \mathcal{N}_1, \ldots, L_s = \operatorname{Span} \mathcal{N}_s$ satisfy*

$$\cup_{i=1}^s L_i \supset \mathcal{N}$$

and

2) *if the vectors ν_p and ν_q, $1 \leq p, q \leq m$, $p \neq q$, belong to the same set $\mathcal{N}_i$, $1 \leq i \leq s$, then the eigenvalues $a_{p,jk}(x)$ and $a_{q,jk}(x)$ of the matrix $\mathcal{A}_{jk}(x)$ are equal.*

Remark 5.2. It is clear that $\operatorname{Span} \mathcal{N} = \mathbb{R}^m$. Therefore for $s = 1$ the condition (L) holds if $L_1 = \operatorname{Span} \mathcal{N}$. Moreover, $s = 1$ can be the only value of s for which (L) holds for the given norm. It is shown that (L) is valid for a smooth norm only for $s = 1$. In other cases (L) can hold for several values of s. For instance, if the unit ball of the norm is a cylinder and its lateral surface has a smooth directrix, then (L) holds for $s = 1$ and for $s = 2$. If $\mathbf{S}^{m-1}$ is the surface of an m-dimensional parallelepiped, then (L) holds for $s = 1, \ldots, m$. Here, it is easily seen, that the choice of subspaces $L_1, \ldots, L_s$ for a fixed s can be done in different ways (for instance, this is the case with a parallelepiped).

Remark 5.3. Let B be a convex compact body in $\mathbb{R}^m$, $0 \overline{\in} \operatorname{int} B$. We are looking for an answer to the question if B is the invariant set of system (5.5).

Let u_0 denote an m-component vector directed from 0 to $u_0 \in \operatorname{int} B$. If a vector-valued function u is a solution of system (5.5), then $v = u - u_0$ is also a solution. Let $B' = \{v : v = u - u_0, \ u \in B\}$. Since $u \in B \Leftrightarrow v \in B'$ and $B' \ni 0$, then the conditions of invariance of the set B for system (5.5) in the case int $B \overline{\ni} 0$ are the same as in the case int $B \ni 0$ (cf. Theorem 5.2).

5.3.3. Certain particular cases and examples

In this section we consider particular cases of Theorem 5.3 on necessary and sufficient condition for validity of the maximum norm principle for solutions of system (5.6). We give geometric description of certain norms for which $s_{\max} = 1$, $s_{\max} = 2$, $s_{\max} = m$. Special attention is paid for the case $m = 2$. At the end we give criterion for validity of the maximum p-norm principle, $1 \le p \le \infty$, in $\mathbb{R}^m$ for solutions of system (5.6).

   a) The case $m = 2$

The following assertion contains a complete description of the relationship between the structure of system (5.6) with two unknown functions and the geometry of the unit ball of the corresponding norm which follows from the maximum norm principle.

Corollary 5.1. *Let $m = 2$. The maximum norm principle is valid for solutions of system (5.6) in Π_T if and only if, for all $x \in D$:*

$$\text{(i)} \quad \mathcal{A}_{jk}(x) = a_{jk}(x)I, \quad 1 \le j, k \le n,$$

in the case when the unit ball of the norm is not a parallelogram. Here I is the unit (2×2)-matrix.

$$\text{(ii)} \quad \mathcal{A}_{jk}(x) = ([\nu_1, \nu_2]^*)^{-1} \operatorname{diag} \{a_{1,jk}(x)a_{2,jk}(x)\}[\nu_1, \nu_2]^*,$$

$1 \le j, k \le n$, in the case when the unit ball of the norm is a parallelogram. Here ν_1, ν_2 are linear independent normals to the sides of the parallelogram and $a_{1,jk}(x)$, $a_{2,jk}(x)$ are real-valued functions.

   In particular, if $\nu_1 = e_1$, $\nu_2 = e_2$, then

$$\mathcal{A}_{jk}(x) = \operatorname{diag} \{a_{1,jk}(x), \, a_{2,jk}(x)\}.$$

   b) The case $m > 2$, $s_{\max} = 1$

Corollary 5.2. *Let the condition (L) hold for the given norm for $s_{\max} = 1$. Then the maximum norm principle is valid for solutions of system (5.6) in Π_T if and only if, for all $x \in D$, the equalities*

$$\mathcal{A}_{jk}(x) = a_{jk}(x)I, \quad 1 \le j, k \le n,$$

hold.

According to Corollary 5.2, the maximum norm principle for these classes of norms is valid only for scalar systems.

   It is shown in [32] that $s_{\max} = 1$ for differentiable norms, for the norms whose unit balls are non-simple m-polytopes, m-simplexes, m-pyramids, m-bipyramids.

We remind that by definition (see, for instance, [40]), m-polytope is called simple if its vertices belong to exactly m facets and m-bipyramid is a convex body composed of two m-pyramids with the general base.

We note that for a particular case of the differentiable norm, namely, for the Euclidean norm in $\mathbb{R}^m$, criterion for validity of the maximum modulus principle for solutions of system (5.6) was given in Section 4.

c) The case $m > 2$, $s_{\max} = 2$

Now, we give a criterion for validity of the maximum norm principle for solutions of the system

$$\frac{\partial u}{\partial t} - \sum_{j,k=1}^{n} \mathcal{A}_{jk}(x) \frac{\partial^2 u}{\partial x_j \partial x_k} = 0$$

under the assumption that the unit ball of the norm is a cylindric body for which the condition (L) holds with $s_{\max} = 2$.

Corollary 5.3. *Let the unit ball of the given norm be a cylindric body. Suppose there exists a family of normals $\nu_0, \nu_1, \ldots, \nu_{m-1}$ to its lateral surface such that each proper subset of the family forms a linear independent system. Further, let ν_m denote the normal to one of cylinder's basements.*

The maximum norm principle for solutions of system (5.6) is valid in Π_T if and only if, for all $x \in D$,

$$\mathcal{A}_{jk}(x) = ([\nu_1, \ldots, \nu_m]^*)^{-1} \Lambda_{jk}(x)[\nu_1, \ldots, \nu_m]^*, \quad 1 \leq j, k \leq n,$$

where the $(m \times m)$-matrix-valued function $\Lambda_{jk}(x)$ has the form

$$\Lambda_{jk}(x) = \operatorname{diag}\{a_{1,jk}(x), \ldots, a_{1,jk}(x), \ a_{2,jk}(x)\}.$$

In particular, if the body is a right cylinder and $\nu_m = e_m$, then

$$\mathcal{A}_{jk}(x) = \operatorname{diag}\{a_{1,jk}(x), \ldots, a_{1,jk}(x), \ a_{2,jk}(x)\}.$$

For example, the conditions of the last corollary hold if the unit ball of the norm is a cylindric body with a smooth lateral surface.

d) The case $m > 2$, $s_{\max} = m$

Obviously, the condition (L) holds for a norm with $s_{max} = m$ if and only if the unit ball of this norm is a parallelepiped. The next assertion follows immediately from Theorem 5.3.

Corollary 5.4. *Let the unit ball of the norm be a parallelepiped and let $\nu_1, \ldots, \nu_m$ denote the normals to its facets which have a common vertex.*

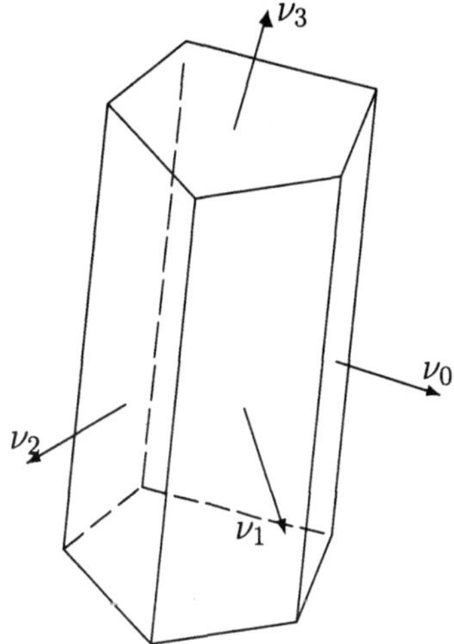

Figure 2: The unit ball of a norm in Corollary 5.3 ($s_{max} = 2$)

Then the maximum norm principle is valid for solutions of system (5.6) *in* Π_T *if and only if, for all* $x \in D$,

$$\mathcal{A}_{jk}(x) = ([\nu_1, \ldots, \nu_m]^*)^{-1} \Lambda_{jk}(x)[\nu_1, \ldots, \nu_m]^{-1}, \quad 1 \le j, k \le n,$$

where

$$\Lambda_{jk}(x) = \mathrm{diag}\,\{a_{1,jk}(x), \ldots, a_{m,jk}(x)\}.$$

In particular, if $\nu_1 = e_1, \ldots, \nu_m = e_m$, *then*

$$\mathcal{A}_{jk}(x) = \mathrm{diag}\,\{a_{1,jk}(x), \ldots, a_{m,jk}(x)\}.$$

e) Example: criterion for validity of the maximum p-norm principle, $1 \le p \le \infty$

Now, we give applications of above results to p-norms in $\mathbb{R}^m$, that is to the norms

(5.8)
$$\|\zeta\|_p = \left(\sum_{j=1}^m |\zeta_j|^p\right)^{1/p},$$

$1 \le p < \infty$, and

$$\|\zeta\|_\infty = \max\{|\zeta_j| : 1 \le j \le m\},$$

where $\zeta = (\zeta_1, \ldots, \zeta_m) \in \mathbb{R}^m$.

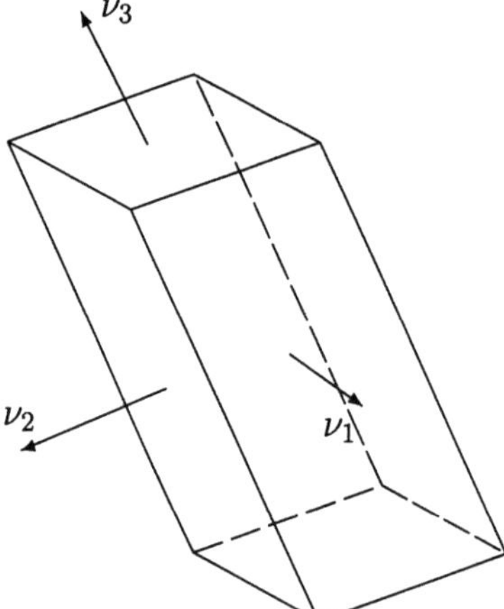

Figure 3: The unit ball of a norm subject to the condition $s_{max} = m$

Corollary 5.5. *The maximum p-norm principle is valid for solutions of system* (5.6) *in* Π_T *if and only if, for all* $x \in D$,

$$\text{(i)} \; A_{jk}(x) = \frac{1}{2} \begin{pmatrix} \alpha_{jk}(x) & \beta_{jk}(x) \\ \beta_{jk}(x) & \alpha_{jk}(x) \end{pmatrix}, \quad 1 \le j,k \le n,$$

for $p = 1$, $m = 2$, *where*

$$\begin{aligned} \alpha_{jk}(x) &= a_{1,jk}(x) + a_{2,jk}(x), \\ \beta_{jk}(x) &= a_{1,jk}(x) - a_{2,jk}(x); \end{aligned}$$

$$\text{(ii)} \; A_{jk}(x) = a_{jk}(x)I,$$

for $p = 1$, $m > 2$ *and for* $1 < p < \infty$,

$$\text{(iii)} \; A_{jk}(x) = \text{diag}\,\{a_{1,jk}(x), \ldots, a_{m,jk}(x)\}$$

for $p = \infty$.

5.4. Maximum norm principle with respect to smooth norms for the system $\partial u/\partial t - \mathfrak{A}(x,t,\partial/\partial x)u = 0$

This section contains criteria for validity of the maximum norm principle for general linear parabolic systems [33].

5.4.1. Necessary and sufficient conditions for validity of the maximum norm principle for the system $\partial u/\partial t - \mathfrak{A}(x,t,\partial/\partial x)u = 0$

In what follows we suppose that the norm $|\cdot|$ is twice continuously differentiable in $\mathbb{R}^m \backslash \{0\}$.

By $\mathcal{H}(|\zeta|)$ we denote the Hesse matrix $((\partial^2 |\zeta|/\partial \zeta_j \partial \zeta_k))$ of the norm $|\cdot|$.

Proposition 5.4. *Let the maximum norm principle hold for solutions of system* (5.1) *in the class* $C^{(2,1)}(\Pi_T) \cap C(\overline{\Pi}_T)$. *Then:*

(i) *for all* $x \in D$ *the equalities*

$$\mathcal{A}_{jk}(x,0) = a_{jk}(x)I, \quad 1 \leq j,k \leq n,$$

hold, where $((a_{jk}(x)))$ *is a positive-definite* $(n \times n)$-*matrix-valued function in* $\overline{D}$;

(ii) *for all* $x \in D$ *and for all* $\xi_j \in \mathbb{R}^m, \zeta \in \mathbb{R}^m \backslash \{0\}$, $j = 1,\ldots,n$, *with* $(\xi_j, \nabla|\zeta|) = 0$, *the inequalitiy*

$$\sum_{j,k=1}^{n} a_{jk}(x)(\mathcal{H}(|\zeta|)\xi_j, \xi_k) + \sum_{j=1}^{n}(\mathcal{A}_j(x,0)\xi_j, \nabla|\zeta|) + (A_0(x,0)\zeta, \nabla|\zeta|) \geq 0$$

is valid.

Next we formulate a statement on a sufficient condition for validity of the maximum norm principle for second order systems with scalar principal part.

Proposition 5.5. *Let the coefficients of the parabolic system*

$$\frac{\partial u}{\partial t} - \sum_{j,k=1}^{n} a_{jk}(x,t)\frac{\partial^2 u}{\partial x_j \partial x_k} + \sum_{j=1}^{n} \mathcal{A}_j(x,t)\frac{\partial u}{\partial x_j} + A_0(x,t)u = 0$$

satisfy the condition:

(iii) *for all* $(x,t) \in \Pi_T$ *and for all* $\xi_j \in \mathbb{R}^m$, $\zeta \in \mathbb{R}^m \backslash \{0\}$, $j = 1,\ldots,n$, *with* $(\xi_j, \nabla|\zeta|) = 0$, *the inequalitiy*

$$\sum_{j,k=1}^{n} a_{jk}(x,t)(\mathcal{H}(|\zeta|)\xi_j, \xi_k) + \sum_{j=1}^{n}(\mathcal{A}_j(x,t)\xi_j, \nabla|\zeta|) + (A_0(x,t)\zeta, \nabla|\zeta|) \geq 0$$

is valid. Then $\mathcal{K}(D,T) = 1$.

5.4.2. Criteria for validity of the maximum norm principle for the system $\partial u/\partial t - \mathfrak{A}(x, \partial/\partial x)u = 0$

The necessary (Proposition 5.4) and sufficient (Proposition 5.5) conditions coincide if the coefficients of system (5.1) do not depend on t. In this case we obtain the following criterion for validity of the maximum norm principle.

Theorem 5.4. *The maximum norm principle is valid for solutions of system* (5.7) *in the class* $C^{(2,1)}(\Pi_T) \cap C(\overline{\Pi}_T)$ *if and only if:*

(i) *for all* $x \in D$ *the equalities*

$$\mathcal{A}_{jk}(x) = a_{jk}(x)I, \quad 1 \leq j,k \leq n,$$

hold, where $((a_{jk}(x)))$ *is a positive-definite* $(n \times n)$-*matrix-valued function in* $\overline{D}$;

(ii) *for all $x \in D$ and for all $\xi_j \in \mathbb{R}^m, \zeta \in \mathbb{R}^m \backslash \{0\}$, $j = 1, \ldots, n$, with $(\xi_j, \nabla|\zeta|) = 0$, the inequalitiy*

$$\sum_{j,k=1}^{n} a_{jk}(x)(\mathcal{H}(|\zeta|)\xi_j, \xi_k) + \sum_{j=1}^{n}(\mathcal{A}_j(x)\xi_j, \nabla|\zeta|) + (A_0(x)\zeta, \nabla|\zeta|) \geq 0$$

is valid.

We note that a special case of this Theorem for the Euclidean norm in $\mathbb{R}^m$ was given in Section 4.

5.4.3. Certain particular cases

Let $\mathcal{T}_\zeta = \{\xi \in \mathbb{R}^m : (\nabla|\zeta|, \xi) = 0\}$ be the space of vectors lying in the tangent hyperplane to the sphere of the norm at the point $\zeta \in \mathbb{R}^m \backslash \{0\}$. Further, let $\mathbb{R}_*^m = \{\zeta \in \mathbb{R}^m : \mathcal{H}(|\zeta|)\mathcal{T}_\zeta = 0\}$ and $\mathcal{N}_* = \{\nu(\zeta) : \zeta \in \mathbf{S}^{m-1} \cap \mathbb{R}_*^m\}$.

Theorem 5.4 implies the following assertion.

Corollary 5.6. *Let $\mathcal{N}_* \supset \{\nu_1, \ldots, \nu_m\}$ and suppose that the set $\{\nu_1, \ldots, \nu_m\}$ forms a linear independent system. Then the maximum norm principle is valid for solutions of system (5.7) in the class $C^{(2,1)}(\Pi_T) \cap C(\overline{\Pi}_T)$ if and only if:*

(i) *for all $x \in D$ the equalities*

$$\mathcal{A}_{jk}(x) = a_{jk}(x)I,$$

$$\mathcal{A}_j(x) = ([\nu_1, \ldots, \nu_m]^*)^{-1} \operatorname{diag}\{a_{1,j}(x), \ldots, a_{m,j}(x)\}[\nu_1, \ldots, \nu_m]^*,$$

$1 \leq j, k \leq n$, *hold, where $((a_{jk}(x)))$ is a positive-definite $(n \times n)$-matrix-valued function and $a_{1,j}(x), \ldots, a_{m,j}(x)$ are scalar functions in $\overline{D}$;*

(ii) *for all $x \in D$ and for all $\xi_j \in \mathbb{R}^m, \zeta \in \mathbb{R}^m \backslash \{0\}$, $j = 1, \ldots, n$, with $(\xi_j, \nabla|\zeta|) = 0$, the inequalitiy*

$$\sum_{j,k=1}^{n} a_{jk}(x)(\mathcal{H}(|\zeta|)\xi_j, \xi_k) + \sum_{j=1}^{n}(\mathcal{A}_j(x)\xi_j, \nabla|\zeta|) + (A_0(x)\zeta, \nabla|\zeta|) \geq 0$$

is valid.

The following assertion follows from Corollary 5.6.

Corollary 5.7. *Let $\mathcal{N}_* \supset \{\nu_1, \ldots, \nu_{m+1}\}$ and suppose that any proper subset of $\{\nu_1, \ldots, \nu_{m+1}\}$ forms a linear independent system. Then the maximum norm principle is valid for solutions of system (5.7) in the class $C^{(2,1)}(\Pi_T) \cap C(\overline{\Pi}_T)$ if and only if:*

(i) *for all $x \in D$ the equalities*

$$\mathcal{A}_{jk}(x) = a_{jk}(x)I, \quad \mathcal{A}_j(x) = a_j(x)I, \quad 1 \leq j, k \leq n,$$

hold, where $((a_{jk}(x)))$ is a positive-definite $(n \times n)$-matrix-valued function and $a_1(x), \ldots, a_m(x)$ are scalar functions in $\overline{D}$;

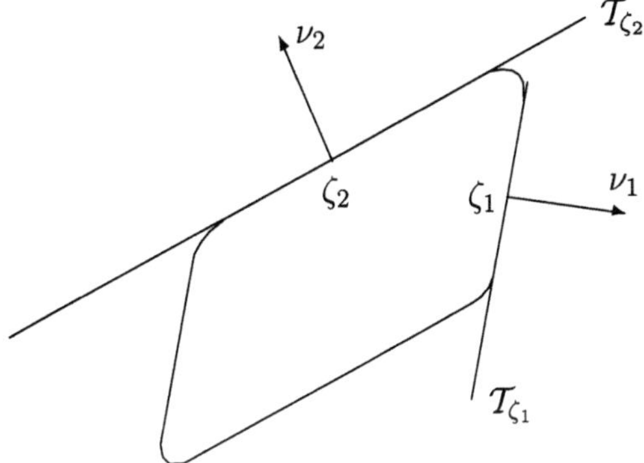

Figure 4: The unit ball of a norm in Corollary 5.6, $\mathcal{N}_* \supset \{\nu_1, \nu_2\}$

(ii) *for all $x \in D$ and for any $\zeta \in \mathbb{R}^m \backslash \{0\}$ the inequalitiy*

$$(A_0(x)\zeta, \nabla|\zeta|) \geq 0$$

is valid.

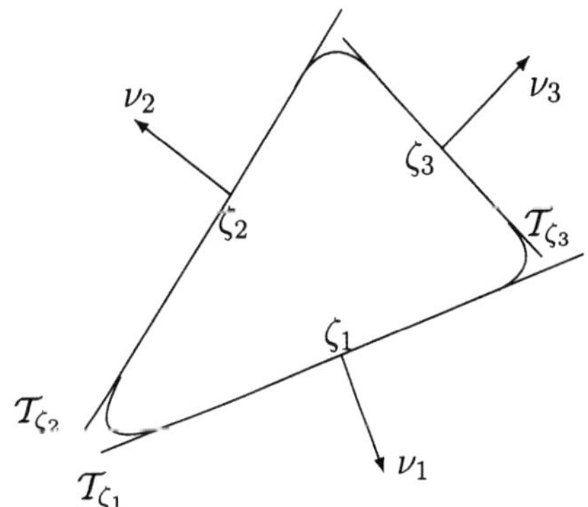

Figure 5: The unit ball of a norm subject to the conditions of Corollary 5.7, $\mathcal{N}_* \supset \{\nu_1, \nu_2, \nu_3\}$

Example: criterion for validity of the maximum p-norm principle, $2 < p < \infty$

Now, we give application of Corollary 5.6 to the p-norm (5.8) in $\mathbb{R}^m$, $2 < p < \infty$. In this case the p-norm is twice continuously differentiable. We have

$$\nabla\|\zeta\|_p = \|\zeta\|_p^{1-p}\left(\zeta_1|\zeta_1|^{p-2}e_1 + \ldots + \zeta_m|\zeta_m|^{p-2}e_m\right),$$

$$\mathcal{H}_{jk}(|\zeta|) = (p-1)\left(\delta_{jk}\|\zeta\|_p^{1-p}|\zeta_j|^{p-2} - \|\zeta\|_p^{-1}\frac{\zeta_j|\zeta_j|^{p-2}}{\|\zeta\|_p^{p-1}} \cdot \frac{\zeta_k|\zeta_k|^{p-2}}{\|\zeta\|_p^{p-1}}\right),$$

where $1 \leq j, k \leq m$. Since $\mathcal{N}_* = \{e_1, -e_1, \ldots, e_m, -e_m\}$, we put $\nu_1 = e_1, \ldots, \nu_m = e_m$. Therefore, $[\nu_1, \ldots, \nu_m] = I$. Let $\mathcal{Z}_p = \mathrm{diag}\,\{\,|\zeta_1|^{p-2}, \ldots, |\zeta_m|^{p-2}\,\}$.

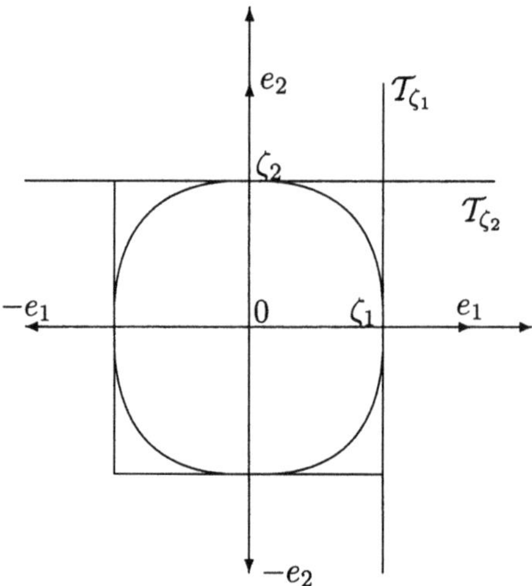

Figure 6: The unit ball of the p-norm for $m = 2$, $2 < p < \infty$, $\mathcal{N}_* = \{e_1, -e_1, e_2, -e_2\}$

Applying the Corollary 5.6, we obtain

Corollary 5.8. *The maximum p-norm principle $(2 < p < \infty)$ is valid for solutions of system (5.7) in the class $C^{(2,1)}(\Pi_T) \cap C(\overline{\Pi}_T)$ if and only if:*

(i) *for all $x \in D$ the equalities*

$$\mathcal{A}_{jk}(x) = a_{jk}(x)I, \quad \mathcal{A}_j(x) = \mathrm{diag}\,\{a_{1,j}(x), \ldots, a_{m,j}(x)\}, \quad 1 \leq j, k \leq n,$$

hold, where $((a_{jk}(x)))$ is a positive-definite $(n \times n)$-matrix-valued function and $a_{1,j}(x), \ldots, a_{m,j}(x)$ are scalar functions in $\overline{D}$;

(ii) *for all $x \in D$ and for any ξ_j, $\zeta \in \mathbb{R}^m$, with $(\xi_j, \mathcal{Z}_p\, \zeta) = 0$, $j = 1, \ldots, n$, the inequalitiy*

$$(p-1) \sum_{j,k=1}^{n} a_{jk}(x)\,(\mathcal{Z}_p\xi_j, \xi_k) + \sum_{j=1}^{n} (\mathcal{A}_j(x)\xi_j, \mathcal{Z}_p\, \zeta) + (A_0(x)\zeta, \mathcal{Z}_p\, \zeta) \geq 0$$

is valid.

Acknowledgements

The research of the author was supported in part by the Ministry of Science and by the Center of Scientific Absorption of the Ministry of Absorption, State of Israel.

References

[1] AGMON, S., DOUGLIS, A. and NIRENBERG, L., Estimates near the boundary for solutions of elliptic partial differential equations satisfying general boundary conditions. I, Comm. Pure Appl. Math. **12** (1959), 623–737.

[2] AGMON, S., Maximum theorems for solutions of higher order elliptic equations, Bull. Amer. Math. Soc. **66** (1960), 77–80.

[3] AGMON, S., DOUGLIS, A. and NIRENBERG, L., Estimates near the boundary for solutions of elliptic partial differential equations satisfying general boundary conditions. II, Comm. Pure Appl. Math. **17** (1964), 35–92.

[4] AMANN, H., Invariant sets and existence theorems for semilinear parabolic and elliptic systems, J. Math. Anal. Appl. **65** (1978), 432–467.

[5] ANGELL, T.S., KLEINMAN, R.E. and KRÁL, J., Layer potentials on boundaries with corners and edges, Časopis pro pěstování matematiky **113**(4) (1988), 387–402.

[6] BEBERNES, J. and SMITT, K., Invariant sets and the Hukuhara-Kneser property for systems of parabolic partial differential equations, Rocky Mountain J. Math. **7** (1977), 557–567.

[7] BITSADZE, A.V., On elliptic systems of second order partial differential equations, Dokl. Akad. Nauk SSSR **112** (1957), 983–986 (Russian).

[8] BURAGO, JU.D. and MAZ'YA, V.G., Multidimensional potential theory and the solution of the boundary value problems for regions with irregular boundaries, Zap. Nauchn. Sem. Leningr. Otd. Mat. Inst. Steklov (LOMI) **3**, Part 1 (1967), 1–86; English translation: Sem. V.A. Steklov Math. Inst., Leningrad **3** (1967), 3–39.

[9] CARLEMAN, T., Über das Neumann-Poincaresche Problem für ein Gebiet mit Ecken, Inaugural Dissertation, Tr. Almqvist and Wiksell, Uppsala, 1916.

[10] CHUEH, K.N., CONLEY C.C. and SMOLLER, J.A., Positively invariant regions for systems of nonlinear diffusion equations, Indiana Univ. Math. J. **26** (1977), 373–391.

[11] COSNER, C. and SCHAEFFER, P.W., On the development of functionals which satisfy a maximum principle, Appl. Anal. **26** (1987), 45–60.

[12] COSTABEL, M., Boundary integral operators on Lipschitz domains: Elementary results, SIAM J. Math. Anal. **19** (1988), 613–626.

[13] EIDEL'MAN, S.D., Parabolic Systems, North-Holland, Amsterdam, 1969.

[14] ELSCHNER, J., The double layer potential operator over polyhedral domains. I: Solvability in weighted Sobolev spaces, Appl. Analysis **45** (1992), 117–134.

[15] FABES, E., Layer potential methods for boundary value problems on Lipschitz domains, in: Potential theory. Survey and problems, Proceedings, Prague 1987 J. Král, J. Lukeš, I. Netuka and J. Veselý (eds.) Lecture Notes in Mathematics, **1344**, Springer, Berlin-Heidelberg-New York, 1988.

[16] FEDERER, H., Geometric Measure Theory, Springer, Berlin-Heidelberg-New York, 1969.

[17] GERMAIN, P., Course de Mécanique des Milieux Continus, t. I. Théorie Générale, Masson et C^{je}, Éditeurs, Paris, 1973.

[18] GOHBERG, I. and MARCUS, A., Some remarks on topologically equivalent norms, Izvestia Mold. Fil. Akad. Nauk SSSR **10**(76) (1960), 91–94.

[19] GRACHEV, N.V., Representation and estimates for inverse operators of the potential theory integral equations in a polyhedron, in: Potential Theory, Proc Int. Conf. on Potential Theory, Nagoya, 1990, (Masanori Kishi, ed.), 201–206.

[20] GRACHEV, N.V. and MAZ'YA, V.G., On the Fredholm radius for operators of the double layer potential type on piecewise smooth boundaries, Vestnik Leningr. Univ. **19**(4) (1986), 60–64 (in Russian).

[21] GRACHEV, N.V. and MAZ'YA, V.G., Solvability of a boundary integral equation on a polyhedron, Report LiTH-MAT-R-91-50, Linköping Univ., Sweden.

[22] HILE, G. and PROTTER, M.H., Maximum principles for a class of first-order elliptical systems, J. Diff. Equations **24** (1977), 136–151.

[23] KRÁL, J., The Fredholm method in potential theory, Trans. Amer. Math. Soc. **125**(3) (1966), 511–547.

[24] KRÁL, J. and WENDLAND, W.L., Some examples concerning applicability of the Fredholm-Radon method in potential theory, Aplikace Matematiky **31** (1986), 293–298.

[25] KRÁL, J. and WENDLAND, W.L., On the applicability of the Fredholm-Radon method in potential theory and the panel method, in: Panel methods in fluid mechanics with emphasis in aero-dynamics, (J. Ballmann, R. Appler, W. Hackbush, eds.), Notes on numerical Fluid Mechanics **21**, Vieweg, Braunschweig, 1988, 120–136.

[26] KRESIN, G.I. and MAZ'YA, V.G., On the maximum of displacements in the visco-elastic half-space (three parametrical model), Vestnik Leningr. Univ. **22** (1985), 47–51 (Russian).

[27] KRESIN, G.I. and MAZ'YA, V.G., On the exact constant in the inequality of Miranda-Agmon type for solutions of elliptic equations, Izvestia Vys. Uch. Zaved., Matem. Ser. **5** (1988), 41–50 (Russian).

[28] KRESIN, G.I. and MAZ'YA, V.G., Criteria for validity of the maximum modulus principle for solutions of linear strongly elliptic systems, Potential Analysis **2** (1993), 73–99.

[29] KRESIN, G.I., On the exact constant in the inequality for a norm of solutions of the Lamé system in a half-space, Functional Differential Equations. Israel Seminar. The College of Judea and Samaria **1** (1993), 132–135.

[30] KRESIN, G.I. and MAZ'YA, V.G., Criteria for validity of the maximum modulus principle for solutions of linear parabolic systems, Ark. Math. **32** (1994), 121–155.

[31] KRESIN, G.I. and MAZ'YA, V.G., The norm and the essential norm of the double layer elastic and hydrodynamics potentials in the space of continuous functions, Math. Meth. in Appl. Sc. **18** (1995), 1095–1131.

[32] KRESIN, G.I. and MAZ'YA, V.G., Criteria for validity of the maximum norm principle for parabolic systems, Potential Analysis, to appear.

[33] KRESIN, G.I. and MAZ'YA, V.G., On the maximum principle with respect to smooth norms for linear strongly coupled parabolic systems, Functional Differential Equations, to appear.

[34] KRESIN, G.I. and MAZ'YA, V.G., On the maximum modulus principle for linear parabolic systems with zero boundary data, Functional Differential Equations **5** (1998), 165–181.

[35] KUPRADZE, V.D., GEGELIA, T.G., BASHELEISHVILI, M.O. and BURCHULADZE T.V., Three-dimensional problems of the mathematical theory of elasticity and thermoelasticity, 2nd ed., Nauka, Moscow, 1976; English transl.: North-Holland, 1979.

[36] LADYZHENSKAYA, O.A., Mathematical questions in the hydrodynamics of a viscous incompressible fluid, 2nd ed., Nauka, Moscow, 1970; English transl. of 1st ed., The mathematical theory of viscous incompressible flow, Gordon and Breach, New York, 1963.

[37] MAZ'YA, V.G., Boundary integral equations of elasticity in domains with piecewise smooth boundaries, in: EQUADIFF 6 Proc., Brno Univ., 1985, 235–242.

[38] MAZ'YA, V.G., Boundary Integral Equations, in: Encyclopedia of Math. Sc., **13**, Springer, Berlin-Heidelberg-New York, 1991.

[39] MAZ'YA, V.G. and KRESIN, G.I., On the maximum principle for strongly elliptic and parabolic second order systems with constant coefficients, Mat. Sb. **125** (1984), 458–480 (Russian). English transl.: Math. USSR-Sb. **53** (1986), 457–479.

[40] McMULLEN, P. and SHEPPARD, G.C., Convex Polytopes and the Upper Bound Conjecture, London Math. Soc. Lect. Notes Series, **3**, Cambridge Univ. Press, 1971.

[41] MINKOWSKI, H., Theorie der konvexen Körper, insbesondere Begründung ihres Oberflächenbegriffs. Posthum. Gesammelte Abhandlungen von Herman Minkowski (ed. D. Hilbert), vol. 2, Leipzig: Teubner, 1911.

[42] MIRANDA, C., Formule di maggiorazione e teorema di esistenza per le funzioni biarmoniche di due variabili, Giorn. Mat. Bataglini **78** (1948–1949), 97–118.

[43] MIRANDA, C., Teorema del massimo modulo e teorema di esistenza e di unicità per il problema di Dirichlet relativo alle equazioni ellitiche in due variabili, Ann. Mat. Pura Appl., ser. 4 **46** (1958), 265–311.

[44] MIRANDA, C., Sul teorema del massimo modulo per una classe di sistemi ellitici di equazioni del secondo ordine e per le equazioni a coefficienti complessi, Istit. Lombardo Accad. Sci. Lett. Rend. A **104** (1970), 736–745.

[45] PARTON, V.Z. and PERLIN P.I., Methods in the mathematical theory of elasticity, Nauka, Moscow, 1981 (Russian).

[46] PINI, B., Sui sistemi equazioni lineari a derivate parziali del secondo ordine dei tipi ellittico e parabolico, Rend. del Sem. Mat. della Univ. di Padova **22** (1953), 265–280.

[47] POLYA, G., Liegt die Stelle der größten Beanspruchungen an der Oberfläche?, Z. Angew. Math. Mech., **10** (1930), 353–360.

[48] PROTTER, M.H. and WEINBERGER, H.F., Maximum Principles in Differential Equations, Prentice-Hall, Englewood Cliffs, N.J., 1967; Springer, Berlin-Heidelberg-New York, 1985.

[49] PROTTER, M.H., Maximum principles, Pittman Research Notes, Math. ser. **175** (1988), 1–14.

[50] RADON, J., Über lineare Functionaltransformationen und Functionalgleichungen, Sitzungsber. Akad. Wiss. Wien, Abt. 2a, **128**, H. 7, (1919), 1083–1121.

[51] RADON, J., Über die Randwertaufgaben beim logarithmischen Potential, Sitzungsber. Akad. Wiss. Wien, Abt. 2a, **128**, H. 7, (1919), 1123–1167.

[52] RATHSFELD, A., The invertibility of the double layer potential operator in the space of continuous functions defined on a polyhedron: The panel method, Appl. Analysis **45** (1992), 135–177.

[53] REDHEFFER, R. and WALTER, W., Invariant sets for systems of partial differential equations. I. Parabolic equations, Arch. Rational Mech. Anal. **67** (1978), 41–52.

[54] REMPEL, S. and SCHMIDT, G., Eigenvalues for spherical domains with corners via boundary integral equations, Int. Equat. Oper. Theory **14** (1991), 229–250.

[55] SHAPIRO, Z.YA., The first boundary value problem for an elliptic system of differential equations, Mat. Sb. **28**(70) (1951), 55–78 (Russian).

[56] SHELEPOV, V.YU., Investigations of matrix integral equations in the space of continuous functions by Ya. B. Lopatinskii's method, in: General theory of boundary value problems. Collect. Sci. Works, Naukova dumka, Kiev, 1983, 220–226 (Russian).

[57] SOLONNIKOV, V.A., On general boundary value problems for systems elliptic in the Douglis-Nirenberg sense, Izv. Akad. Nauk SSSR, ser. Mat. **28** (1964), 665–706; English transl. in: Amer. Math. Soc. Transl.(2) **56** (1966).

[58] STOER, J. and WITZGALL, C., Convexity and Optimization in Finite Dimensions I, Springer, Berlin-Heidelberg-New York, 1970.

[59] STYS, T., On the unique solvability of the first Fourier problem for a parabolic system of second order linear differential equations, Comment. Math. Prace Mat. **9** (1965), 282–283 (Russian).

[60] SZEPTYCKI, P., Existence theorem for the first boundary value problem for a quasilinear elliptic system, Bulletin Acad. Polon. des Sciences **7** (1959), 419–424.

[61] VERCHOTA, G.C., Layer potentials and regularity for the Dirichlet problem for Laplace's equation in Lipschitz domains, J. Funct. Anal. **59** (1984), 419–424.

[62] WALTER, W., Differential and Integral Inequalities, Springer, Berlin-Heidelberg-New York, 1970.

[63] WASOWSKI, J., Maximum principle for a certain strongly elliptic system of linear equations of second order, Bull. Acad. Polon. Sci., **18** (1970), 741–745.

[64] WEINBERGER, H.F., Invariant sets for weakly coupled parabolic and elliptic systems, Rend. Mat. **8** (1975), 295–310.

The Research Institute,
The Colledge of Judea and Samaria,
44837 Ariel,Israel,
kresin@research.yosh.ac.il

1991 Mathematics Subject Classification: Primary 31A10, 31B10, 35B45, 35B50, 35Q30, 35Q72; Secondary 31A30, 31B30, 35J40, 35J55, 35K45, 35K50, 73F15, 76D07, 76C02

Submitted: 07.02.1999

Operator Theory:
Advances and Applications, Vol. 109
© 1999 Birkhäuser Verlag Basel/Switzerland

L^p-contractivity of semigroups generated by parabolic matrix differential operators

Mikael Langer

This paper is devoted to the study of contraction semigroups generated by second order linear partial differential operators and is written in collaboration with V. Maz'ya. The paper begins with a brief survey of the results obtained earlier by the author and V. Maz'ya. A class of weakly coupled systems is studied and a close relationship between the generation of $(L^p)^N$-contractive semigroups of the corresponding operators and $(L^2)^N$-contractivity of the semigroups generated by some associated operators is obtained. More precisely it is seen that $(L^2)^N$-dissipativity of an associated operator implies $(L^p)^N$-dissipativity of the original operator, whereas the converse holds for some subclass of operators, which includes the scalar operators.

For operators associated with systems uniformly parabolic in the sense of Petrovskiĭ, necessary and sufficient conditions in algebraic terms are presented in order for the generation of contraction semigroups on $(L^p)^N$ for all $p \in [1,\infty]$ simultaneously. As a consequence, a necessary and sufficient condition is obtained for the Schrödinger operator with magnetic field to generate contraction semigroups on $L^p(\Omega)$ for all $p \in [1,\infty]$ simultaneously.

1. Introduction

It is well known that general elliptic operators generate analytic semigroups on L^p, $1 < p < \infty$. For an elliptic operator, the contraction property in L^2 is equivalent to dissipativity and therefore, a large class of strongly elliptic higher order scalar and vector operators are generators of contraction semigroups. This is however not true if $p \neq 2$. In Langer and Maz'ya [9], it is shown that higher order differential operators fail to generate contraction semigroups. More specifically, for differential operators having $L^1_{\mathrm{loc}}(\Omega)$-elements in their coefficient matrices, the following theorem holds:

Theorem *If $\Omega \subset \mathbf{R}^n$ is open and $1 \leq p < \infty$, $p \neq 2$, $N \geq 1$, no linear partial differential operator of order higher than two which contains $(C_0^\infty(\Omega))^N$ in its domain of definition can generate a contraction semigroup on $(L^p(\Omega))^N$.*

In the same paper it was also proven that if $p > 1$ and the L^p-dissipativity criterion is restricted to the cone of nonnegative functions for differential operators with real-valued coefficients, the criterion still fails for operators of order higher than two, except for some fourth order operators if $\frac{3}{2} \leq p \leq 3$. A class of such fourth order operators was also presented.

Kresin and Maz'ya [10] proved that arbitrary higher order differential operators fail to generate contraction semigroups on $(L^\infty(\Omega))^N$, where $\Omega \subset \mathbf{R}^n$ and the norm is given by

$$(1.1) \qquad \left\| \left(\sum_{i=1}^N |u_i|^2 \right)^{1/2} \right\|_{L^\infty(\Omega)}.$$

In the same paper, they also gave necessary and sufficient conditions, stated in algebraic form, for L^∞-contractivity for second order strongly elliptic systems.

In L^p, $1 < p < \infty$, it is since long known that second order elliptic operators with real coefficients, and also some weakly coupled second order elliptic systems, generate contraction semigroups under certain restrictions on their coefficients; see for example Maz'ya and Sobolevskiĭ [11]. This is also the case for $p = 1$: sufficient conditions for a second order elliptic scalar operator to generate a contraction semigroup on L^1 are given in Brezis and Strauss [2]. In Amann [1], necessary and sufficient conditions are given for a second order elliptic scalar operator to simultaneously generate contraction semigroups on all L^p-spaces, $p \in [1, \infty]$.

1.1. Notation

The function spaces appearing in this paper will, unless otherwise stated, all have complex scalar fields. If X is a function space, X^+ will denote the set of nonnegative functions in X. Analogously, $\mathbf{R}^+$ denotes the set of nonnegative real numbers. Let Ω be an open subset of $\mathbf{R}^n$, let $p \in [1, \infty]$ and suppose that k is a nonnegative integer.

The set of all complex-valued functions defined on Ω, having continuous partial derivatives of all orders up to and including k, is denoted by $C^k(\Omega)$. By $C^k(\overline{\Omega})$, we will mean the collection of restrictions of functions in $C^k(\mathbf{R}^n)$ to the set Ω. The set $C^\infty(\Omega)$ is the intersection of all $C^m(\Omega)$, $m \in \mathbf{N}$, and $C_0^\infty(\Omega)$ consists of the functions in $C^\infty(\Omega)$ having compact support. If Ω is a bounded set, $C(\overline{\Omega})$ will denote the space obtained by letting the set $C^0(\overline{\Omega})$ inherit the norm of the space $L^\infty(\Omega)$. The subspace $C_0(\overline{\Omega})$ of $C(\overline{\Omega})$ consists of the functions that are zero on the boundary of Ω.

In this text, a mollifier φ is a function fulfilling $\varphi \in (C_0^\infty(\mathbf{R}^n))^+$ and normalized as $\int \varphi \, dm_n = 1$, where the measure m_n is understood to be the n-dimensional Lebesgue measure on $\mathbf{R}^n$.

The usual Hölder space is denoted by $C^{k,s}(\overline{\Omega})$, that is, if $0 < s \leq 1$, then $C^{k,s}(\overline{\Omega})$ consists of the functions f in $C^k(\Omega)$ for which there exists a constant C such that

$$|\partial^\alpha f(x) - \partial^\alpha f(y)| \leq C|x - y|^s$$

for all $x, y \in \Omega$ and all multi-indices α with $|\alpha| \leq k$.

Let $\langle \cdot, \cdot \rangle$ denote the standard scalar product on $\mathbf{C}^N$. For $f \in (L^p(\Omega))^N$, define $|f| = \langle f, f \rangle^{1/2}$ pointwise and equip $(L^p(\Omega))^N$ with the norm

$$\|f\|_{(L^p(\Omega))^N} = \left\| |f| \right\|_p.$$

This norm will henceforth also be denoted by $\|\cdot\|_p$, as the context will show which norm is meant.

We will also use the ordinary Sobolev space $W^{k,p}(\Omega)$, consisting of the functions $f \in L^p(\Omega)$ whose distributional derivatives $\partial^\alpha f$, of order less than or equal to k, all belong to $L^p(\Omega)$. The norm is given by

$$\|f\|_{k,p} = \left(\sum_{|\alpha| \leq k} \|\partial^\alpha f\|_p^p \right)^{1/p}.$$

$W_0^{k,p}(\Omega)$ is defined as the closure of $C_0^\infty(\Omega)$ in the norm of $W^{k,p}(\Omega)$. The spaces $W^{k,2}(\Omega)$ and $W_0^{k,2}(\Omega)$ are denoted by $H^k(\Omega)$ and $H_0^k(\Omega)$, respectively.

1.2.   Description of results

Second order weakly coupled elliptic systems are studied in Section 3, where it is shown that $(L^p)^N$-dissipativity of the corresponding operator under certain conditions can be expressed in terms of $(L^2)^N$-dissipativity of an associated operator. In fact, if

$$\begin{aligned}
\mathcal{A}_p u &= \partial_i(a_{ij}\partial_j u) + a_i\partial_i u + Au, \quad u \in (W^{2,p}(\Omega) \cap W_0^{1,p}(\Omega))^N, \\
\mathcal{A}u &= \frac{4}{pq}\partial_i(a_{ij}\partial_j u) + \frac{1}{2p}\bigl(p(A + A^*) - 2\partial_i a_i I\bigr)u, \quad u \in (H^2(\Omega) \cap H_0^1(\Omega))^N,
\end{aligned}$$

where a_{ij}, a_i are real-valued, A is an $N \times N$-matrix with complex-valued entries and q is the conjugate exponent to p, then the following theorem holds:

Theorem 3.7 *If $\mathcal{A}$ generates a contraction semigroup on $(L^2(\Omega))^N$, then $\mathcal{A}_p$ generates a contraction semigroup on $(L^p(\Omega))^N$. Conversely, if there is a basis of constant eigenvectors to $A + A^*$, then $\mathcal{A}$ generates a contraction semigroup on $(L^2(\Omega))^N$ if $\mathcal{A}_p$ generates a contraction semigroup on $(L^p(\Omega))^N$. In particular, the converse holds in the scalar case.*

We also show that there is an operator $\mathcal{A}_p$ which generates a contraction semigroup on $(L^p(\Omega))^2$, but the corresponding operator $\mathcal{A}$ fails to do so on the space $(L^2(\Omega))^2$.

Section 4 is devoted to the initial boundary value problem with zero Dirichlet boundary conditions for linear second order systems which are uniformly parabolic in the sense of Petrovskiĭ and not necessarily weakly coupled. By using the results by Kresin and Maz'ya [10], we give algebraic criteria for the L^p-maximum principle to hold simultaneously for all $p \in [1,\infty]$. This generalizes a result obtained by Amann [1], who treated the scalar case. Let A_{ij}, A_i and A be $N \times N$-matrices with complex-valued entries. If for $p \in [1,\infty]$, $\mathcal{A}_p$ is the differential operator

$$u \mapsto \partial_i(A_{ij}\partial_j u) + A_i\partial_i u + Au$$

with a certain domain $D(\mathcal{A}_p) \subset (L^p(\Omega))^N$ (see page 326), then the result of Section 4 is:

Theorem 4.5 *The operators $\mathcal{A}_p$ generate contraction semigroups on $(L^p(\Omega))^N$ for all $p \in [1, \infty)$ and on $(C_0(\overline{\Omega}))^N$ for $p = \infty$ simultaneously if and only if*

(a) *there are real-valued scalar functions a_{ij} on $\overline{\Omega}$ such that for every i, j, $A_{ij} = a_{ij}I$ and the $n \times n$-matrix $[a_{ij}]$ is positive definite,*

(b) *for all $\eta_i, \zeta \in \mathbf{C}^N$, $i = 1, \ldots, n$, with $\Re\langle \eta_i, \zeta \rangle = 0$, the inequalities*

$$\Re\left\{ a_{ij}\langle \eta_i, \eta_j \rangle - \langle A_i \eta_i, \zeta \rangle - \langle A\zeta, \zeta \rangle \right\} \geq 0,$$
$$\Re\left\{ a_{ij}\langle \eta_i, \eta_j \rangle + \langle A_i \zeta, \eta_i \rangle - \langle (A - \partial_i A_i)\zeta, \zeta \rangle \right\} \geq 0$$

hold on Ω.

As a corollary, we get that the Schrödinger operator with magnetic field

$$-(i\nabla + m)^t(i\nabla + m) - V$$

generates contraction semigroups on $L^p(\Omega)$ for all $p \in [1, \infty]$ simultaneously if and only if $\Re V \geq 0$ on Ω, where V is complex-valued and m is an $\mathbf{R}^n$-valued function on Ω.

2. Preliminaries

We remind that, according to Lumer-Phillips (see e.g. Goldstein [5]), the generation of strongly continuous contraction semigroups is characterized by the following statement: if the operator A generates a contraction semigroup, then $(0, \infty) \subset \rho(A)$, A is densely defined and A is dissipative with respect to any duality section. Conversely, if $\rho(A) \cap (0, \infty) \neq \emptyset$, A is densely defined and dissipative, then A generates a contraction semigroup. The theorem of Lumer and Phillips was also proved independently, in a slightly weaker form (the norm was assumed to be Gateaux-differentiable), by Maz'ya and Sobolevskiĭ in 1960 [11].

In the Banach space $(L^p(\Omega))^N$, where $\Omega \subset \mathbf{R}^n$ is open, $1 \leq p < \infty$ and N is a positive integer, the dissipativity criterion takes the following form. Let A be an operator on $(L^p(\Omega))^N$, where $p \in [1, \infty)$. If

$$(2.1) \qquad\qquad \Re \int_\Omega \langle Au, u \rangle |u|^{p-2} \, dm_n \leq 0$$

for all $u \in D(A)$, then A is dissipative (with respect to any duality section). Conversely, if $p > 1$ and A is dissipative, then (2.1) holds for all $u \in D(A)$. If $p = 1$ and (2.1) fails to hold on $D(A)$, then A is not dissipative with respect to all duality sections.

In the inequality (2.1), the combination $|u|^{p-2}u$ in the integrand is taken to be zero where u vanishes — even in the case $p = 1$.

3. Weakly coupled systems

In this section, we will consider second order linear elliptic partial differential operators acting on vector-valued functions, with the property that the coefficients of the derivatives of positive order are real-valued scalar functions. It will be shown that the generation of a contraction semigroup on $(L^p)^N$ for an operator is closely connected with the generation of a contraction semigroup on $(L^2)^N$ for an another operator.

Indices will appear extensively and where not otherwise indicated, the letters i and j will be used for indices ranging over $\{1,\ldots,n\}$ and the letters k and ℓ for indices ranging over $\{1,\ldots,N\}$, adopting the convention to sum over repeated indices within terms of an expression.

Let Ω be an open and bounded subset of $\mathbf{R}^n$, having a $C^{2,\alpha}$-boundary for some $\alpha \in (0,1]$, let $1 < p < \infty$ and let $\mathcal{A}_p$ be the elliptic operator in $(L^p(\Omega))^N$ given by

$$\mathcal{A}_p u = \partial_i(a_{ij}\partial_j u) + a_i \partial_i u + A u,$$

the domain being $D(\mathcal{A}_p) = (W^{2,p}(\Omega) \cap W_0^{1,p}(\Omega))^N$, where the functions a_{ij}, a_i in $C^1(\overline{\Omega})$ are real-valued and the $N \times N$-matrix A has $C(\overline{\Omega})$-functions as entries. Without loss of generality, the matrix $[a_{ij}]$ is assumed to be pointwise symmetric. We will also use the operator

$$\mathcal{A}u = \frac{4}{pq}\partial_i(a_{ij}\partial_j u) + \frac{1}{2p}\left(p(A + A^*) - 2\partial_i a_i I\right)u,$$

with domain $D(\mathcal{A}) = (H^2(\Omega) \cap H_0^1(\Omega))^N$, where q is the conjugate exponent to p and A^* is the complex conjugate transpose of A.

The first lemma will be used later to characterize L^p-dissipativity of $\mathcal{A}_p$ in terms of L^2-dissipativity of $\mathcal{A}$. It does not require ellipticity, but the results in the rest of this section do need the ellipticity assumption.

Lemma 3.1. *The operator $\mathcal{A}_p$ is dissipative in $(L^p(\Omega))^N$ if and only if*

$$(3.1) \qquad \Re \int_\Omega \left(a_{ij}\langle \partial_i v, \partial_j v\rangle + \langle(p^{-1}\partial_i a_i I - A)v, v\rangle\right) dm_n$$

$$- \frac{(p-2)^2}{p^2}\int_{\{v \neq 0\}} a_{ij}\Re\langle\partial_i v, v\rangle\Re\langle\partial_j v, v\rangle|v|^{-2}\, dm_n \geq 0$$

for all $v \in (H_0^1(\Omega))^N$.

Proof. Let X denote all functions $h \in (C^2(\overline{\Omega}))^N$ with $h|_{\partial\Omega} = 0$, suppose that $u \in X$, set $E_r = \{x \in \Omega : |u(x)| > r\}$ for every $r \geq 0$ and let $\mathcal{X}_r$ denote the characteristic function of the set E_r. Furthermore, set $\varphi_\varepsilon(t) = \max(t,\varepsilon)$ for $t \geq 0$, $\varepsilon > 0$ and define $f_\varepsilon = \varphi_\varepsilon \circ |u|$. Since φ_ε is a Lipschitz function for each ε, f_ε is absolutely continuous along all lines parallel to the coordinate axes and

$\partial_i f_\varepsilon = f_\varepsilon^{-1} \Re \langle \partial_i u, u \rangle \chi_\varepsilon$ almost everywhere. Integrating by parts and using that u is zero on the boundary gives

$$
(3.2) \quad -\int_\Omega \langle \partial_i (a_{ij} \partial_j u), u \rangle f_\varepsilon^{p-2}\, dm_n
$$

$$
= \int_\Omega a_{ij} \langle \partial_j u, \partial_i u \rangle f_\varepsilon^{p-2}\, dm_n
$$

$$
+ (p-2) \int_{E_\varepsilon} a_{ij} \langle \partial_j u, u \rangle \Re \langle \partial_i u, u \rangle f_\varepsilon^{p-4}\, dm_n.
$$

We observe that $|\partial_i u|^2 |u|^{p-2} \in L^1$ for every i if $p \geq 2$. If $1 < p < 2$, let for the moment $a_{ij} = \delta_{ij}$, $N = 1$ and suppose that u is real-valued. Then equality (3.2) transforms to

$$
-\int_\Omega \Delta u\, u\, f_\varepsilon^{p-2}\, dm_n = \int_\Omega |\nabla u|^2 f_\varepsilon^{p-2}\, dm_n + (p-2) \int_{E_\varepsilon} |\nabla u|^2 f_\varepsilon^{p-2}\, dm_n
$$

$$
(3.3) \qquad\qquad = (p-1) \int_{E_\varepsilon} |\nabla u|^2 |u|^{p-2}\, dm_n
$$

$$
+ \varepsilon^{p-2} \int_{\Omega \setminus E_\varepsilon} |\nabla u|^2\, dm_n.
$$

Since the left-hand side of (3.3) converges as $\varepsilon \to 0$ by dominated convergence, the right-hand side is bounded from above. Both integrands on the right are positive so monotone convergence implies that $|\nabla u|^2 |u|^{p-2} \chi_0 \in L^1$. Note that the second term on the right is bounded for every $p > 1$, implying that

$$
\lim_{\varepsilon \to 0} \varepsilon^{p-2} \int_{\Omega \setminus E_\varepsilon} |\nabla u|^2\, dm_n = 0.
$$

If u is complex-valued and N is arbitrary, this discussion can be applied to the real and imaginary parts of any of the components of u, showing that the function $|\partial_i u|^2 |u|^{p-2} \chi_0 \in L^1$. As a consequence of the Cauchy inequality, it follows from this that also $|\partial_i u||\partial_j u||u|^{p-2} \chi_0 \in L^1$. Moreover,

$$
(3.4) \qquad\qquad \lim_{\varepsilon \to 0} \varepsilon^r \int_{\Omega \setminus E_\varepsilon} |\partial_i u|^2\, dm_n = 0
$$

for every i and any $r > -1$.

Set $v_\varepsilon = f_\varepsilon^{p/2 - 1} u$. From the relation

$$
\partial_i v_\varepsilon = f_\varepsilon^{p/2 - 1} \partial_i u + (p/2 - 1)(\Re \langle \partial_i u, u \rangle) f_\varepsilon^{p/2 - 3} \chi_\varepsilon u,
$$

it follows that

$$
\Re \langle \partial_i v_\varepsilon, \partial_j v_\varepsilon \rangle = \Re \langle \partial_i u, \partial_j u \rangle f_\varepsilon^{p-2} + (p^2/4 - 1) \Re \langle \partial_i u, u \rangle \Re \langle \partial_j u, u \rangle f_\varepsilon^{p-4} \chi_\varepsilon
$$

and that

$$\Re\langle\partial_i v_\varepsilon, v_\varepsilon\rangle\Re\langle\partial_j v_\varepsilon, v_\varepsilon\rangle|v_\varepsilon|^{-2} = \begin{cases} p^2 4^{-1}\Re\langle\partial_i u, u\rangle\Re\langle\partial_j u, u\rangle|u|^{p-4}, & \text{on } E_\varepsilon, \\ \Re\langle\partial_i u, u\rangle\Re\langle\partial_j u, u\rangle\varepsilon^{p-2}|u|^{-2}, & \text{on } E_0 \setminus E_\varepsilon. \end{cases}$$

From these identities, we immediately obtain

$$\Re\int_\Omega a_{ij}\langle\partial_i v_\varepsilon, \partial_j v_\varepsilon\rangle\, dm_n - \frac{(p-2)^2}{p^2}\int_{E_0} a_{ij}\Re\langle\partial_i v_\varepsilon, v_\varepsilon\rangle\Re\langle\partial_j v_\varepsilon, v_\varepsilon\rangle|v_\varepsilon|^{-2}\, dm_n$$

$$(3.5) \qquad = \ \Re\int_\Omega a_{ij}\langle\partial_i u, \partial_j u\rangle f_\varepsilon^{p-2}\, dm_n$$

$$+ (p-2)\int_{E_\varepsilon} a_{ij}\Re\langle\partial_j u, u\rangle\Re\langle\partial_i u, u\rangle f_\varepsilon^{p-4}\, dm_n - \frac{(p-2)^2}{p^2}R(\varepsilon),$$

where

$$|R(\varepsilon)| \ = \ \left|\int_{E_0\setminus E_\varepsilon} a_{ij}\Re\langle\partial_i u, u\rangle\Re\langle\partial_j u, u\rangle\varepsilon^{p-2}|u|^{-2}\, dm_n\right|$$

$$\leq \ c\varepsilon^{p-2}\int_{E_0\setminus E_\varepsilon} |\partial_i u||\partial_j u|\, dm_n.$$

By (3.4) combined with the Cauchy inequality, $R(\varepsilon)$ tends to zero as ε tends to zero so dominated convergence implies that the left-hand side of expression (3.5) tends to the real part of the limit of the left-hand side of (3.2). We continue with the equalities

$$\Re\int_\Omega a_i\langle\partial_i u, u\rangle f_\varepsilon^{p-2}\, dm_n = (2-p)\Re\int_{E_\varepsilon} a_i\langle\partial_i u, u\rangle f_\varepsilon^{p-2}\, dm_n$$

$$(3.6) \qquad - \Re\int_\Omega a_i\langle\partial_i u, u\rangle f_\varepsilon^{p-2}\, dm_n - \int_\Omega \partial_i a_i\, |u|^2 f_\varepsilon^{p-2}\, dm_n$$

$$= \ -\frac{1}{p}\int_\Omega \partial_i a_i\, |u|^2 f_\varepsilon^{p-2}\, dm_n + \frac{p-2}{p}\int_{\Omega\setminus E_\varepsilon} a_i\Re\langle\partial_i u, u\rangle f_\varepsilon^{p-2}\, dm_n$$

$$= \ -\frac{1}{p}\int_\Omega \partial_i a_i\, |v_\varepsilon|^2\, dm_n + \frac{p-2}{p}\varepsilon^{p-2}\int_{\Omega\setminus E_\varepsilon} a_i\Re\langle\partial_i u, u\rangle\, dm_n,$$

obtained by integration by parts. The last term tends to zero as ε tends to zero since $|\langle\partial_i u, u\rangle|$ is bounded by $C\varepsilon$ on $\Omega \setminus E_\varepsilon$. By combining the relation pointed out above between the left-hand sides of the equations (3.2) and (3.5), adding the left-hand side of (3.6), subtracting the integrand $\langle Au, u\rangle f_\varepsilon^{p-2}$ and finally passing to the limit, we see that

$$(3.7) \quad -\Re\int_\Omega \langle\partial_i(a_{ij}\partial_j u) + a_i\partial_i u + Au, u\rangle|u|^{p-2}\, dm_n$$

314 Langer

$$= \int_{E_0} a_{ij}\big(\Re\langle\partial_i u,\partial_j u\rangle|u|^{p-2} + (p-2)\Re\langle\partial_i u,u\rangle\Re\langle\partial_j u,u\rangle|u|^{p-4}\big)\,dm_n$$

$$(3.8) \qquad + \int_{E_0} \langle(p^{-1}\partial_i a_i I - A)u,u\rangle|u|^{p-2}\,dm_n$$

$$= \lim_{\varepsilon\to 0}\Big(\Re\int_{\Omega}\big(a_{ij}\langle\partial_i v_\varepsilon,\partial_j v_\varepsilon\rangle + \langle(p^{-1}\partial_i a_i I - A)v_\varepsilon,v_\varepsilon\rangle\big)\,dm_n$$

$$(3.9) \qquad -\frac{(p-2)^2}{p^2}\int_{E_0} a_{ij}\Re\langle\partial_i v_\varepsilon,v_\varepsilon\rangle\Re\langle\partial_j v_\varepsilon,v_\varepsilon\rangle|v_\varepsilon|^{-2}\,dm_n\Big)$$

holds.

Suppose that (3.1) is true for all $v \in (H_0^1(\Omega))^N$. We will now show that the given operator is dissipative. Set $Y = (W^{2,p}(\Omega)\cap W_0^{1,p}(\Omega))^N$ and note that $X \subset Y$ due to the boundedness of Ω. Let $u \in X$ be arbitrary and let v_ε be as above. Then $v_\varepsilon \in (H_0^1(\Omega))^N$ for every ε, as f_ε is absolutely continuous along all lines parallel to the coordinate axes. Thus, (3.9) is nonnegative so we conclude that (3.7) is nonnegative for every $u \in X$. The fact that X is dense in the space Y, together with Lemma 3.3 in Langer and Maz'ya [9], implies that (3.7) is nonnegative for every $u \in Y$ and hence, the dissipativity of $\mathcal{A}_p$ follows.

To prove the converse statement, let $v \in X$ be arbitrary, define the functions $g_\varepsilon = (|v|^2 + \varepsilon)^{1/2}$ for $\varepsilon > 0$, set $u_\varepsilon = g_\varepsilon^{2/p-1}v$ and let $E = \{x \in \Omega : v(x) \neq 0\}$. Then

$$\partial_i u_\varepsilon = g_\varepsilon^{2/p-1}\partial_i v + (2/p-1)g_\varepsilon^{2/p-3}(\Re\langle\partial_i v,v\rangle)v$$

so, for any continuous matrix function B, we have

$$(3.10)\quad \Re\langle\partial_i u_\varepsilon,\partial_j u_\varepsilon\rangle|u_\varepsilon|^{p-2} + (p-2)\Re\langle\partial_i u_\varepsilon,u_\varepsilon\rangle\Re\langle\partial_j u_\varepsilon,u_\varepsilon\rangle|u_\varepsilon|^{p-4}$$

$$= \Re\langle\partial_i v,\partial_j v\rangle|v|^{p-2}g_\varepsilon^{2-p} + \Re\langle\partial_i v,v\rangle\Re\langle\partial_j v,v\rangle\Big((p-2)|v|^{p-4}g_\varepsilon^{2-p}$$

$$+ (p-1)(4/p-2)|v|^{p-2}g_\varepsilon^{-p} + (p-1)(2/p-1)^2|v|^p g_\varepsilon^{-p-2}\Big),$$

$$(3.11)\quad \langle Bu_\varepsilon,u_\varepsilon\rangle|u_\varepsilon|^{p-2} = \langle Bv,v\rangle|v|^{p-2}g_\varepsilon^{2-p}$$

on the set E. Since

$$|v|^\alpha g_\varepsilon^\beta \leq \begin{cases} |v|^{\alpha+\beta}, & \beta < 0, \\ c|v|^\alpha, & \beta \geq 0 \end{cases}$$

on E, (3.10) is majorized by the L^1-function

$$C|\partial_i v||\partial_j v|\big(1 + |v|^{p-2}\big)$$

and (3.11) is uniformly bounded with respect to ε. Furthermore, g_ε tends to $|v|$ pointwise as $\varepsilon \to 0$, so by replacing u by u_ε in (3.7) and (3.8) and taking limits, dominated convergence gives

$$(3.12)\quad -\lim_{\varepsilon\to 0}\Re\int_{\Omega}\langle\partial_i(a_{ij}\partial_j u_\varepsilon) + a_i\partial_i u_\varepsilon + Au_\varepsilon,u_\varepsilon\rangle|u_\varepsilon|^{p-2}\,dm_n$$

$$= \Re \int_{E_0} \left(a_{ij}\langle \partial_i v, \partial_j v\rangle + \langle (p^{-1}\partial_i a_i I - A)v, v\rangle \right) dm_n$$

$$- \frac{(p-2)^2}{p^2} \int_{E_0} a_{ij}\Re\langle \partial_i v, v\rangle \Re\langle \partial_j v, v\rangle |v|^{-2}\, dm_n.$$

In view of (3.4), the domain of integration of the first integral on the right of (3.12) can be replaced by Ω and hence, the dissipativity of $\mathcal{A}_p$ implies that (3.1) holds for every $v \in X$, since $u_\varepsilon \in X$ for every ε. By definition, X is dense in $(H_0^1(\Omega))^N$ and an analogous argument as used in the proof of Lemma 3.3 in Langer and Maz'ya [9] shows that (3.1) holds for every $v \in (H_0^1(\Omega))^N$. $\qquad\square$

In some of the lemmas and theorems below, we will restrict ourselves to operators $\mathcal{A}_p$ whose principal part $\mathcal{P}$ is positive, that is, the principal part fulfills $\mathcal{P}(x,\xi) > 0$ for all $x \in \overline{\Omega}$ and all $\xi \in \mathbf{R}^n \setminus \{0\}$. That this is no restriction when dealing with elliptic operators follows from the following lemma.

Lemma 3.2. *If $\mathcal{A}$ or $\mathcal{A}_p$ is dissipative, then the principal part $\mathcal{P}$ of $\mathcal{A}_p$ is positive.*

Proof. Let $\mathcal{A}_p$ be dissipative and assume, without loss of generality, that $0 \in \Omega$ and let $u \in C_0^\infty(\mathbf{R}^n)$ be nonzero and have support in $U \subset \Omega$. Write u_i for $\partial_i u$ and define for every $\varepsilon > 0$ the function v_ε by $v_\varepsilon(x) = (u(\varepsilon^{-1}x), 0, \ldots, 0)$. For sufficiently small ε, we have $v_\varepsilon|_\Omega \in (H_0^1(\Omega))^N$. Replacing v by v_ε in (3.1) and multiplying by ε^{2-n}, Lemma 3.1 gives

$$0 \leq \varepsilon^{2-n}\int_\Omega \left(\varepsilon^{-2}a_{ij}(x)(u_i u_j)(\varepsilon^{-1}x) + b(x)u^2(\varepsilon^{-1}x) \right) dx$$

$$- \frac{(p-2)^2}{p^2}\varepsilon^{2-n}\int_\Omega \varepsilon^{-2}a_{ij}(x)(u_i u_j)(\varepsilon^{-1}x)\, dx$$

$$= \int_U \left(ca_{ij}(\varepsilon x)u_i(x)u_j(x) + \varepsilon^2 b(\varepsilon x)u^2(x) \right) dx$$

$$\to c\int_U a_{ij}(0)u_i u_j\, dm_n = c\int_U \mathcal{P}(0, \nabla u)\, dm_n, \quad \varepsilon \to 0,$$

where $b \in C(\overline{\Omega})$ and $c = 4/(pq)$. Since $\mathcal{P}(0,\xi) \neq 0$ for every $\xi \neq 0$ and ∇u is not identically zero, it follows that $\mathcal{P}(0,\xi)$ is positive at some point. But ellipticity and continuity implies then that $\mathcal{P}(0,\xi)$ is positive for every $\xi \neq 0$. A simple translation argument shows that $\mathcal{P}(x,\xi) > 0$ for every $x \in \Omega$ and $0 \neq \xi \in \mathbf{R}^n$. Finally, the continuity of the coefficients and the ellipticity imply positivity for every $x \in \overline{\Omega}$.

The proof that the dissipativity of $\mathcal{A}$ implies positivity of the principal part of $\mathcal{A}_p$ is completely analogous but instead of using Lemma 3.1, the L^2-dissipativity criterion can be used directly. $\qquad\square$

In order to investigate the relation between the dissipativity of $\mathcal{A}_p$ and the dissipativity of $\mathcal{A}$, we now continue by introducing some functionals and defining two constants associated with $\mathcal{A}$ and $\mathcal{A}_p$:

$$J(v) \;=\; \int_\Omega \left(\frac{4}{pq} a_{ij}\langle \partial_i v, \partial_j v\rangle + \Re\langle (p^{-1}\partial_i a_i I - A)v, v\rangle \right) dm_n,$$

$$J_p(v) \;=\; \int_\Omega \left(a_{ij}\langle \partial_i v, \partial_j v\rangle + \Re\langle (p^{-1}\partial_i a_i I - A)v, v\rangle \right) dm_n$$
$$\qquad - \frac{(p-2)^2}{p^2} \int_{\{v\neq 0\}} a_{ij}\Re\langle \partial_i v, v\rangle \Re\langle \partial_j v, v\rangle |v|^{-2}\, dm_n,$$

$$\mu \;=\; \inf\{ J(v) : v \in (H^1_0(\Omega))^N, \|v\|_2 = 1\},$$

$$\mu_p \;=\; \inf\{ J_p(v) : v \in (H^1_0(\Omega))^N, \|v\|_2 = 1\}.$$

Note that the expression for J_p comes from the statement of Lemma 3.1 and that the expression for J comes from the same lemma if it is used on the operator $\mathcal{A}$ with $p = 2$ instead of on $\mathcal{A}_p$ (since $2\Re\langle Av, v\rangle = \langle (A + A^*)v, v\rangle$).

It is clear that J and J_p are real-valued since $[a_{ij}]$ is symmetric. Assuming that the principal part of $\mathcal{A}_p$ is positive, it follows immediately that J is bounded from below on the set $\{v \in (H^1_0)^N : \|v\|_2 = 1\}$. Hence, μ is finite. This is also the case with μ_p, but the proof is postponed to Lemma 3.4. However, assuming that μ_p is finite, the study of μ and μ_p is justified by the following lemma.

Lemma 3.3. *Let $1 < p < \infty$ and suppose that the principal part of $\mathcal{A}_p$ is positive. Then $\mathcal{A}$ and $\mathcal{A}_p$ generate the semigroups T on $(L^2(\Omega))^N$ and T_p on $(L^p(\Omega))^N$, respectively, fulfilling the inequalities*

$$(3.13) \qquad \|T(t)\| \le e^{-\mu t}, \quad \|T_p(t)\| \le e^{-\mu_p t}, \quad t \ge 0.$$

The constants μ and μ_p are the best possible.

Proof. By temporarily replacing $\mathcal{A}_p$ by $\mathcal{A}_p + \mu_p I$, we see that the infimum of J_p is zero. According to Lemma 3.1, this implies that the operator $\mathcal{A}_p + \mu_p I$ is dissipative. In Grisvard [4] it is proven that $\mathcal{A}_p - \lambda I$ is invertible with a bounded inverse (defined on $(L^p(\Omega))^N$) as soon as $\mathcal{A}_p - \lambda I$ is one to one. This is the case when $\lambda > -\mu_p$ since the dissipativity of $\mathcal{A}_p + \mu_p I$ implies that

$$-\Re \int_\Omega \langle (\mathcal{A}_p - \lambda I)v, v\rangle |v|^{p-2}\, dm_n = -\Re \int_\Omega \langle (\mathcal{A}_p + \mu_p I)v, v\rangle |v|^{p-2}\, dm_n$$
$$+ (\mu_p + \lambda)\|v\|_p^p \ge (\mu_p + \lambda)\|v\|_p^p, \quad v \in D(\mathcal{A}_p).$$

The theorem by Lumer-Phillips now shows that $\mathcal{A}_p + \mu_p I$ generates a contraction semigroup Q_p on $(L^p(\Omega))^N$. Set $T_p(t) = e^{-\mu_p t}Q_p(t)$ for $t \ge 0$ and it follows immediately that T_p is a semigroup on $(L^p(\Omega))^N$ with $\|T_p(t)\| \le e^{-\mu_p t}$ for $t \ge 0$. Furthermore, from the definition of a generator, it follows that $\mathcal{A}_p$ is the generator of T_p. By the definition of J_p and Lemma 3.1, μ_p is the smallest number making $\mathcal{A}_p + \mu_p I$ dissipative, hence μ_p is the best constant in (3.13).

The proof of $\mathcal{A}$ generating the semigroup T with the indicated properties is completely analogous — just replace $\mathcal{A}_p$, J_p and L^p, by $\mathcal{A}$, J and L^2, respectively.

$\square$

3.1. The relation between μ and μ_p

Note that it follows from Lemma 3.3 that if $\mu = \mu_p$, then $\mathcal{A}_p$ generates a contraction semigroup on $(L^p)^N$ if and only if $\mathcal{A}$ generates a contraction semigroup on $(L^2)^N$. We will therefore take an interest in the relation between μ and μ_p.

Lemma 3.4. *Suppose that $1 < p < \infty$ and that the principal part of $\mathcal{A}_p$ is positive. Then $\mu \leq \mu_p$.*

Proof. Let $v \in (H_0^1(\Omega))^N$ and choose $x \subset \Omega$ such that x is a Lebesgue point to all components of v and $\partial_i v$. Let $\lambda_1, \ldots, \lambda_n$ be the eigenvalues of $[a_{ij}(x)]$ and let $T : \mathbf{R}^n \to \mathbf{R}^n$ be a linear transformation that takes the corresponding set of orthonormal eigenvectors to the coordinate axes. By introducing the function $u = v \circ T^{-1}$ and denoting the kth component of v and u by v_k and u_k, respectively, it follows that $\nabla v_k = T^{-1} \circ \nabla u_k \circ T$. Set $y = Tx$ and the relation

$$(3.14) \qquad \big(a_{ij}\langle \partial_i v, \partial_j v\rangle\big)(x) = (\nabla v_k(x))^t [a_{ij}(x)]\overline{\nabla v_k(x)}$$

$$= \sum_k \lambda_i |\partial_i u_k(y)|^2 = \lambda_i |\partial_i u(y)|^2$$

is obtained. Suppose for the moment that the components of v are real-valued, let $\tilde{v}$ be a function that fulfills the same properties as v and set $\tilde{u} = \tilde{v} \circ T^{-1}$. Then

$$(3.15) \qquad \big(a_{ij}\langle \partial_i v, v\rangle\langle \partial_j \tilde{v}, \tilde{v}\rangle\big)(x) = \big(v_k \tilde{v}_\ell (\nabla v_k)^t [a_{ij}]\nabla \tilde{v}_\ell\big)(x)$$

$$= \lambda_i (u_k \tilde{u}_\ell \partial_i u_k \, \partial_i \tilde{u}_\ell)(y)$$

$$= \lambda_i \big(\langle \partial_i u, u\rangle\langle \partial_i \tilde{u}, \tilde{u}\rangle\big)(y).$$

Since in general

$$\Re\langle \partial_i v, v\rangle = \langle \partial_i \Re v, \Re v\rangle + \langle \partial_i \Im v, \Im v\rangle,$$

expression (3.15) can be used on each term of $\Re\langle \partial_i v, v\rangle\Re\langle \partial_j v, v\rangle$. If $v(x) \neq 0$, this gives

$$(3.16) \quad \big(a_{ij}\Re\langle \partial_i v, v\rangle\Re\langle \partial_j v, v\rangle|v|^{-2}\big)(x) = \lambda_i \big(\Re\langle \partial_i u(y), u(y)\rangle\big)^2 |u(y)|^{-2}$$

$$\leq \lambda_i |\partial_i u(y)|^2 = \big(a_{ij}\langle \partial_i v, \partial_j v\rangle\big)(x),$$

the last equality coming from (3.14) and the inequality arising from the Cauchy inequality since the eigenvalues λ_i are positive due to the positivity of the principal part of $\mathcal{A}_p$. The arbitrariness of x now implies that

$$\int_\Omega a_{ij}\langle \partial_i v, \partial_j v\rangle \, dm_n - \int_{\{v \neq 0\}} a_{ij}\Re\langle \partial_i v, v\rangle\Re\langle \partial_j v, v\rangle|v|^{-2} \, dm_n \geq 0.$$

Defining the functional $v \mapsto K(v)$ by the left-hand side of the inequality above, we immediately get the functional equality

$$(3.17) \qquad\qquad J + \frac{(p-2)^2}{p^2} K = J_p.$$

Thus $J \leq J_p$, implying that $\mu \leq \mu_p$. $\qquad\qquad\qquad\qquad\qquad\qquad\qquad\qquad$ $\square$

Lemma 3.5. *Suppose that $1 < p < \infty$ and that the principal part of $\mathcal{A}_p$ is positive. Then $\mu = \mu_p$ if and only if at least one of the nonzero generalized solutions of the equation*

$$(3.18) \quad -\frac{4}{pq}\partial_i(a_{ij}\partial_j v) + \frac{1}{2p}(2\partial_i a_i I - p(A + A^*))v = \mu v, \quad v \in (H_0^1(\Omega))^N$$

is of the form $v = fc$ for some real-valued scalar function f and some $c \in \mathbf{C}^N$. Moreover, μ is the least eigenvalue of the left-hand side of the equation.

Proof. We have $2\Re\langle Av, v\rangle = \langle (A + A^*)v, v\rangle$ for all complex v where $A + A^*$ is self-adjoint. It follows from the theory of compact, self-adjoint operators, see e.g. Ladyžhenskaya [7, Ch. II] for the scalar case, that the infimum μ is the least eigenvalue of the operator defined by the right-hand side of (3.18) (whose spectrum only consists of countably many real eigenvalues with $+\infty$ as only possible accumulation point). Furthermore, the infimum is attained by the normalized eigenfunctions corresponding to μ and by no other functions. By general elliptic theory, see Grisvard [4], it follows from the regularity of the coefficients that solutions of (3.18) belong to $(C^1(\overline{\Omega}) \cap H^2(\Omega))^N$.

We go back and adopt the notation used in the proof of Lemma 3.4. Equality in (3.16) holds if and only if $\partial_i u(y) = b_i u(y)$ for every i and some real constants b_i (depending on y), or equivalently, if and only if $\partial_i v(x) = c_i v(x)$, c_i still being real. This shows that $K(v) = 0$ for a $C^1(\Omega)$-function v exactly when $\partial_i v = c_i v$ on $E_v = \{x \in \Omega : v(x) \neq 0\}$ for a collection of real-valued functions $\{c_i\}$. Thus, $K(v) = 0$ and $v \in C^1(\Omega)$ implies that

$$\begin{aligned}
\partial_i\big(|v|^{-1}v\big) &= |v|^{-1}\partial_i v - |v|^{-3}(\Re\langle\partial_i v, v\rangle)v \\
&= c_i|v|^{-1}v - |v|^{-3}c_i|v|^2 v = 0
\end{aligned}$$

on E_v, from which it follows that $v = |v|c$ for some $c \in \mathbf{C}^N$ with unit length on each component of E_v. Conversely, a straightforward verification shows that for C^1-functions v of this form, $K(v) = 0$. Equation (3.17) gives that $\mu = \mu_p$ if and only if $K(w) = 0$ for one of the minimizers w of J which, by the discussion above, is equivalent to that w is of the form $w = |w|c$ on each component of E_w.

Assume that one of the minimizers w is of this form and suppose that E_w consists of at least two components. Denote one of the components by F and let g be the real or imaginary part of any of the N components of w. Then $g \in C^1(\overline{\Omega})$ and g

vanishes on the boundary of Ω. Define

$$f_\varepsilon(t) = \min(t + \varepsilon, 0) + \max(t - \varepsilon, 0), \quad t \in \mathbf{R}, \ \varepsilon > 0$$

and set $g_\varepsilon = f_\varepsilon \circ g$. Then $g_\varepsilon|_F \in H_0^1(F)$ since f_ε is Lipschitz and the support of $g_\varepsilon|_F$ is a compact subset of F. It is easy to see that $g_\varepsilon|_F \to g|_F$ in $H^1(F)$ as $\varepsilon \to 0$. Hence, $g|_F \in H_0^1(F)$ and we conclude that $g\chi_F \in H_0^1(\Omega)$. Set $w_1 = \chi_F w$ and $w_2 = w - w_1$ and it follows from the previous discussion that $w_1, w_2 \in (H_0^1(\Omega))^N$. Moreover,

$$\mu = J(w_1) + J(w_2) \geq \|w_1\|_2^2 \mu + \|w_2\|_2^2 \mu = \mu,$$

so $J(w_1/\|w_1\|_2) = \mu$. Hence, $w_1/\|w_1\|_2$ is also a minimizer of J and consequently fulfills equation (3.18). Since w_1 is zero on an open subset of Ω, the Aronzajn-Cordes uniqueness theorem, see Hörmander [6], implies that w_1 is identically zero on Ω. This contradiction shows that E_w only consists of one component and we can conclude that $w = fc$ for some nonnegative $f \in C^1$ and $c \in \mathbf{C}^N$. $\square$

From Lemma 3.5, we obtain a necessary condition in algebraic terms for the equality $\mu = \mu_p$ to hold:

Corollary 3.6. *Suppose that the principal part of A_p is positive. If $\mu = \mu_p$, then there is a constant eigenvector to $A + A^*$ on $\overline{\Omega}$.*

Proof. By defining the matrix E as

$$E = -\frac{pq}{4}((p^{-1}\partial_i a_i - \mu)I - 2^{-1}(A + A^*)),$$

equation (3.18) becomes

(3.19) $$\partial_i(a_{ij}\partial_j v) + Ev = 0.$$

Assume that $\mu = \mu_p$ and let $v = fc$ be a solution of equation (3.19) given by Lemma 3.5. Without loss of generality, assume that $|c| = 1$. Consider the problem

$$\inf\{J(gc) : g \in H_0^1(\Omega), \|g\|_2 = 1\}.$$

The infimum is of course attained for $g = f$ and is μ and since

$$J(gc) = \frac{4}{pq} \int_\Omega \left(a_{ij}\partial_i g\, \partial_j g - \langle Ec, c\rangle g^2\right) dm_n + \mu\|g\|_2^2,$$

the same theory as used in the proof of Lemma 3.5, or the simple fact that we have $\frac{d}{d\varepsilon}J((f + \varepsilon\varphi)c)|_{\varepsilon=0} = 0$ for all $\varphi \in C_0^\infty(\Omega)$, implies that f satisfies the Euler equation

(3.20) $$\partial_i(a_{ij}\partial_j f) + \langle Ec, c\rangle f = 0.$$

Substituting $v = fc$ into (3.19), we also have the equation

(3.21) $$\partial_i(a_{ij}\partial_j f)c + fEc = 0.$$

By multiplying the scalar equation (3.20) with c and subtracting (3.21), it follows that $fEc = f\langle Ec, c\rangle c$. Suppose that $Ec \neq \langle Ec, c\rangle c$ at some point in $\overline{\Omega}$. Due to the continuity of E, this would imply that f vanishes on some subset of Ω with nonempty interior. As in the proof of Theorem 3.5, the Aronzajn-Cordes uniqueness theorem then would imply that f is identically zero on Ω. This is not the case since v is nonzero. Hence, $Ec = \langle Ec, c\rangle c$ on $\overline{\Omega}$, which by the definition of E implies that c is a constant eigenvector to $A + A^*$ on $\overline{\Omega}$. $\square$

3.2. Dissipativity of $\mathcal{A}$ and $\mathcal{A}_p$

We are now in the position to state the main results of this section.

Theorem 3.7. *If $\mathcal{A}$ generates a contraction semigroup on $(L^2(\Omega))^N$, then $\mathcal{A}_p$ generates a contraction semigroup on $(L^p(\Omega))^N$. Conversely, if there is a basis of constant eigenvectors to $A + A^*$, then $\mathcal{A}$ generates a contraction semigroup on $(L^2(\Omega))^N$ if $\mathcal{A}_p$ generates a contraction semigroup on $(L^p(\Omega))^N$. In particular, the converse holds in the scalar case.*

Proof. If $\mathcal{A}$ or $\mathcal{A}_p$ generate a contraction semigroup, Lemma 3.2 shows that the principal part of $\mathcal{A}_p$ is positive.

If $\mathcal{A}$ generates a contraction semigroup, Lemma 3.3 implies that $\mu \geq 0$ so by Lemma 3.4, it follows that $\mu_p \geq 0$. According to Lemma 3.3, $\mathcal{A}_p$ generates a contraction semigroup.

The converse is proved by first treating the scalar case and then extending the result to when $A + A^*$ has a basis of constant eigenvectors. Suppose that $N = 1$ and that $\mathcal{A}_p$ generates a contraction semigroup. Then $\mu_p \geq 0$ so $J_p(v) \geq 0$ for every $v \in H_0^1(\Omega)$. Consider the functional K defined in the proof of Lemma 3.4. We have for real-valued $v \in H_0^1(\Omega)$

$$K(v) = \int_\Omega a_{ij}\partial_i v\, \partial_j v\, dm_n - \int_{\{v \neq 0\}} a_{ij}\partial_i v\, \partial_j v\, dm_n = 0,$$

since $|\nabla v| = 0$ almost everywhere on the set $\{x \in \Omega : v(x) = 0\}$. In view of (3.17) and the nonnegativity of J_p, this gives that $J(v) \geq 0$ for all real-valued v. From the equality $J(u + iv) = J(u) + J(v)$, holding for all real-valued functions u and v, it follows that $\mu \geq 0$ and we conclude from Lemma 3.3 that $\mathcal{A}$ generates a contraction semigroup.

Let N be arbitrary, suppose that $\mathcal{A}_p$ generates a contraction semigroup and let T be a linear transformation that diagonalizes $A + A^*$ in such a way that T is an isometry. Let $f \in D(\mathcal{A}_p)$ and set $g = Tf$. Since T is an isometry,

$$\Re \int_\Omega \langle \mathcal{A}_p g, g\rangle |g|^{p-2}\, dm_n = \Re \int_\Omega \langle T^{-1}\mathcal{A}_p Tf, f\rangle |f|^{p-2}\, dm_n,$$

showing that dissipativity of $\mathcal{A}_p$ is equivalent to dissipativity of $f \mapsto T^{-1}\mathcal{A}_p Tf$. This equivalence together with the fact that

$$\Re\langle \mathcal{A}_p g, g \rangle = \Re\langle (\mathcal{A}_p - A + 2^{-1}(A + A^*))g, g \rangle$$

for all $g \in D(\mathcal{A}_p)$, allows us to assume without loss of generality that A is a diagonal matrix. Let $u \in W^{2,p} \cap W_0^{1,p}$, fix some $k \in \{1, \ldots, N\}$ and set $w = ue_k$ where e_k is the unit vector in $\mathbf{C}^N$ directed along the kth coordinate axis. Finally, define the scalar operator $\mathcal{A}_p^k$ as the "kth row" of $\mathcal{A}_p$, that is,

$$\mathcal{A}_p^k \varphi = \mathcal{A}_p(\varphi e_k), \quad \varphi \in W^{2,p} \cap W_0^{1,p}.$$

Define $\mathcal{A}^k$ analogously. The dissipativity of $\mathcal{A}_p$ implies that

$$0 \geq \Re \int_\Omega \langle \mathcal{A}_p w, w \rangle |w|^{p-2}\, dm_n = \Re \int_\Omega (\mathcal{A}_p^k u)\bar{u}|u|^{p-2}\, dm_n,$$

so $\mathcal{A}_p^k$ is dissipative. From the case $N = 1$ treated above, the dissipativity of $\mathcal{A}^k$ follows for every k. Finally, let $v = (v_1, \ldots, v_N) \in D(\mathcal{A})$. Since

$$\int_\Omega \langle \mathcal{A}v, v \rangle\, dm_n = \sum_k \int_\Omega (\mathcal{A}^k v_k)\bar{v}_k\, dm_n,$$

the dissipativity of each $\mathcal{A}^k$ implies the dissipativity of $\mathcal{A}$, so $\mu \geq 0$ and we conclude that $\mathcal{A}$ generates a contraction semigroup. $\qquad\square$

Example 3.8. In the scalar case $N = 1$, the equality $\mu = \mu_p$ always holds according to Lemma 3.5. This is however not true in general: consider an operator $\mathcal{A}_p$ in $(L^p(\Omega))^2$ with a positive principal part and suppose that the matrix coefficient A of $\mathcal{A}_p$ is given as

$$A(x) = \begin{pmatrix} 1 & |x| \\ |x| & -1 \end{pmatrix}, \quad x \in \overline{\Omega}.$$

Since $\mathbf{C}^2$ is spanned by the eigenvector

$$\begin{pmatrix} 1 \\ |x|/(1 + \sqrt{1 + |x|^2}) \end{pmatrix}$$

of A and one eigenvector orthogonal to it, A has no constant eigenvectors. According to Corollary 3.6, μ is strictly smaller than μ_p. In particular, we can add a suitable multiple of the identity operator to $\mathcal{A}_p$ to make $\mathcal{A}_p$ generate a contraction semigroup on $(L^p(\Omega))^2$, while $\mathcal{A}$ does *not* generate a contraction semigroup on $(L^2(\Omega))^2$.

4. Coupled systems

We will now treat the initial boundary value problem with zero Dirichlet boundary conditions for linear second order systems which are uniformly parabolic in the sense of Petrovskiĭ. These systems will not necessarily be weakly coupled as in the previous section. Our goal is to give criteria for the L^p-maximum principle to hold for all $p \in [1, \infty]$ simultaneously and then formulate the result in terms of the generation of contraction semigroups.

First we will study a parabolic system and use one of the results in Kresin and Maz'ya [10] to get necessary and sufficient conditions for the solution of the parabolic system to not increase in maximum norm as the time increases. Using duality and interpolation, we get criteria for this maximum principle to hold for all $p \in [1, \infty]$ simultaneously. After that, the results will be written in semigroup language. This will require some work to determine the domain of the operators for different p. But first, some notation.

Let $\alpha \in (0, 1)$, let Ω be a bounded domain in $\mathbf{R}^n$ with $C^{2,\alpha}$-boundary and let for $T > 0$ the cylinder $\Omega \times (0, T)$ be denoted by Q_T. Furthermore, let $C_0^{2,\alpha}$ be the subset of $(C^{2,\alpha}(\overline{\Omega}))^N$ consisting of the functions that vanish on the boundary $\partial\Omega$ of Ω.

If f is a function defined on Q_T, $\partial_i f$ will as before denote the partial derivative of f with respect to the i:th space coordinate, whereas $\partial_t f$ will denote the time derivative $\partial_{n+1} f$. Let $\mathcal{A}$ be the differential operator

$$\mathcal{A}u = \partial_i(A_{ij}\partial_j u) + A_i \partial_i u + Au$$

and let $\mathcal{A}^*$ denote the formally adjoint operator of $\mathcal{A}$, i.e.

$$\mathcal{A}^*v = \partial_i(A_{ij}^*\partial_j v) - A_i^*\partial_i v + (A^* - \partial_i A_i^*)v.$$

Here, the coefficients are $N \times N$-matrix-valued functions on Ω and the adjoint A^* of A is the matrix obtained by taking the complex conjugate transpose of A. It is assumed that the elements of A_{ij} belong to $C^{2,\alpha}(\overline{\Omega})$, the elements of A_i belong to $C^{1,\alpha}(\overline{\Omega})$ and the elements of of A belong to $C^{0,\alpha}(\overline{\Omega})$. We will furthermore assume that $A_{ij} = A_{ji}$ and that there exists a $\delta > 0$ such that for every $x \in \overline{\Omega}$ and every $\xi = (\xi_1, \ldots, \xi_N) \in \mathbf{R}^N$, the zeros of the polynomial

$$\lambda \mapsto \det(\xi_i\xi_j A_{ij} + \lambda I)$$

satisfy the inequality $\Re\lambda \leq -\delta|\xi|^2$, I being the identity matrix of order N.

4.1. The parabolic PDE setting

We introduce the initial boundary value problem

$$(4.1) \qquad \begin{cases} \partial_t u - \mathcal{A}u = 0, & \text{on } Q_T, \\ u(\cdot, 0) = \varphi, & \text{on } \Omega, \\ u|_{\partial\Omega \times [0,T]} = 0, \end{cases}$$

where $\varphi \in C_0^{2,\alpha}$. The requirements in the beginning of the section on the coefficients of $\mathcal{A}$ imply that the system in (4.1) is uniformly parabolic in the sense of Petrovskiĭ. According to Theorem 10.1 in Ladyženskaya, Solonnikov and Ural'ceva [8], there exists a unique classical solution to (4.1) with Hölder continuous second derivatives in the space variables and first derivative in time (in fact, the regularity assumptions on the coefficients and the boundary can be relaxed and were imposed only for the simplicity of the presentation).

In Kresin and Maz'ya [10], necessary and sufficient conditions for the validity of the maximum modulus principle for the system in (4.1) with nonzero boundary data were found. A similar proof yields the following result.

Theorem 4.1. *Let u be the solution of* (4.1). *In order that*

$$(4.2) \qquad \|u(\,\cdot\,,t)\|_\infty \le \|\varphi\|_\infty, \quad \forall t \in [0,T],$$

for all $\varphi \in C_0^{2,\alpha}$, it is necessary and sufficient that

(a) *there are real-valued scalar functions a_{ij} on $\overline{\Omega}$ such that for every i,j, $A_{ij} = a_{ij}I$ and the $n \times n$-matrix $[a_{ij}]$ is positive definite,*

(b) *for all $\eta_i, \zeta \in \mathbf{C}^N$, $i = 1,\ldots,n$, with $\Re\langle \eta_i, \zeta \rangle = 0$, the inequality*

$$\Re\left\{ a_{ij}\langle \eta_i, \eta_j \rangle - \langle A_i \eta_i, \zeta \rangle - \langle A\zeta, \zeta \rangle \right\} \ge 0$$

holds on Ω.

Remark 4.2. As pointed out in Kresin and Maz'ya [10], condition (b) is in the scalar case $n = 1$ reduced to the requirement that the inequality

$$-4\Re A \ge b_{ij}\Im A_i \Im A_j$$

holds on Ω, where $[b_{ij}] = [a_{ij}]^{-1}$.

By duality, an inequality of the type (4.2) will be obtained in the L^1-norm. In the course of obtaining this similar inequality, the following auxiliary initial boundary value problem will be needed:

$$(4.3) \qquad \begin{cases} \partial_t v + \mathcal{A}^* v = 0, & \text{on } Q_T, \\ v(\,\cdot\,,T) = \psi, & \text{on } \Omega, \\ v|_{\partial\Omega \times [0,T]} = 0, \end{cases}$$

where $\psi \in C_0^{2,\alpha}$. As previously for the problem (4.1), the problem (4.3) has a unique classical solution. By setting $v(\,\cdot\,,t) = u(\,\cdot\,,T-t)$ for $t \in [0,T]$ in (4.3) and then applying Theorem 4.1 to the resulting system instead of on the system (4.1), we get the following corollary.

Corollary 4.3. *Let v be the solution of* (4.3). *In order that*

$$\|v(\,\cdot\,,t)\|_\infty \le \|\psi\|_\infty, \quad \forall t \in [0,T],$$

for all $\psi \in C_0^{2,\alpha}$, it is necessary and sufficient that

(a) *there are real-valued scalar functions a_{ij} on $\overline{\Omega}$ such that for every i, j, $A_{ij} = a_{ij}I$ and the $n \times n$-matrix $[a_{ij}]$ is positive definite,*

(b) *for all $\eta_i, \zeta \in \mathbf{C}^N$, $i = 1, \ldots, n$, with $\Re\langle \eta_i, \zeta \rangle = 0$, the inequality*

$$\Re\left\{ a_{ij}\langle \eta_i, \eta_j \rangle + \langle A_i\zeta, \eta_i \rangle - \langle (A - \partial_i A_i)\zeta, \zeta \rangle \right\} \geq 0$$

holds on Ω.

We now state and prove the result of this subsection.

Theorem 4.4. *Let u be the solution of (4.1). In order that*

$$(4.4) \qquad \|u(\,\cdot\,,t)\|_p \leq \|\varphi\|_p, \quad \forall t \in [0, T],$$

for all $\varphi \in C_0^{2,\alpha}$ and for all $p \in [1, \infty]$, it is necessary and sufficient that

(a) *there are real-valued scalar functions a_{ij} on $\overline{\Omega}$ such that for every i, j, $A_{ij} = a_{ij}I$ and the $n \times n$-matrix $[a_{ij}]$ is positive definite,*

(b) *for all $\eta_i, \zeta \in \mathbf{C}^N$, $i = 1, \ldots, n$, with $\Re\langle \eta_i, \zeta \rangle = 0$, the inequalities*

$$\begin{aligned} \Re\left\{ a_{ij}\langle \eta_i, \eta_j \rangle - \langle A_i\eta_i, \zeta \rangle - \langle A\zeta, \zeta \rangle \right\} &\geq 0, \\ \Re\left\{ a_{ij}\langle \eta_i, \eta_j \rangle + \langle A_i\zeta, \eta_i \rangle - \langle (A - \partial_i A_i)\zeta, \zeta \rangle \right\} &\geq 0 \end{aligned}$$

hold on Ω.

Proof. Let u and v be the solutions of (4.1) and (4.3), respectively. Let $\tau \in [0, T)$ and set $\tilde{v}(\,\cdot\,,t) = v(\,\cdot\,,t + \tau)$ for $t \in [0, T - \tau]$. Integrating by parts and applying the equality $\langle \mathcal{A}u, \tilde{v} \rangle = \langle u, \mathcal{A}^*\tilde{v} \rangle$ gives

$$\begin{aligned} 0 &= \int_0^{T-\tau}\!\!\int_\Omega \langle \partial_t u - \mathcal{A}u, \tilde{v} \rangle \, dm_n \, dt \\ &= \int_\Omega \left(\langle u(\,\cdot\,,T-\tau), \tilde{v}(\,\cdot\,,T-\tau) \rangle - \langle u(\,\cdot\,,0), \tilde{v}(\,\cdot\,,0) \rangle \right) dm_n \\ &\quad - \int_0^{T-\tau}\!\!\int_\Omega \langle u, \partial_t \tilde{v} + \mathcal{A}^*\tilde{v} \rangle \, dm_n \, dt. \end{aligned}$$

Since $\tilde{v}$ fulfills the same differential equation as v, the last integral vanishes and we obtain the identity

$$(4.5) \qquad \int_\Omega \langle u(\,\cdot\,,T-\tau), \psi \rangle \, dm_n = \int_\Omega \langle \varphi, v(\,\cdot\,,\tau) \rangle \, dm_n$$

We need only prove the necessity of the second inequality in (b), since the necessity of (a) and the necessity of the first inequality in (b) result from Theorem 4.1

along with inequality (4.4) where $p = \infty$. Equality (4.5), together with the Cauchy and Hölder inequalities, imply that

$$\left| \int_\Omega \langle \varphi, v(\,\cdot\,, \tau) \rangle \, dm_n \right| \leq \|\psi\|_\infty \|u(\,\cdot\,, T - \tau)\|_1 \leq \|\psi\|_\infty \|\varphi\|_1,$$

where (4.4) with $p = 1$ has been used in the second inequality. The functional on $(L^1)^N$ defined by the integral in the left-hand side of the inequalities above has norm $\|v(\,\cdot\,, \tau)\|_\infty$. By the arbitrariness of φ and the denseness of $C_0^{2,\alpha}$ in $(L^1)^N$,

$$\|v(\,\cdot\,, \tau)\|_\infty \leq \|\psi\|_\infty$$

for all ψ. Since τ is arbitrary, this is equivalent to the maximum modulus principle for the problem (4.3). Hence, Corollary 4.3 gives the necessity of the second inequality in condition (b).

We now turn to the sufficiency part. Set $t = T - \tau$. By (4.5) together with the sufficiency part of Corollary 4.3, we have the following inequality:

$$\left| \int_\Omega \langle u(\,\cdot\,, t), \psi \rangle \, dm_n \right| \leq \|\varphi\|_1 \|v(\,\cdot\,, \tau)\|_\infty \leq \|\varphi\|_1 \|\psi\|_\infty.$$

As in the L^1-case, the integral on the left in the inequality above defines a functional on $(C_0(\overline{\Omega}))^N$. By using the Cauchy and Hölder inequalities and by applying the functional to $(C_0(\overline{\Omega}))^N$-approximations of the function $|u(\,\cdot\,, t)|^{-1} u(\,\cdot\,, t)$, it is easily shown that the norm of the functional is $\|u(\,\cdot\,, t)\|_1$. Since φ is arbitrary, we conclude that

$$(4.6) \qquad \|u(\,\cdot\,, t)\|_1 \leq \|\varphi\|_1.$$

Take some $t \in (0, T]$ and define the operator G on $C_0^{2,\alpha}$ by

$$G\varphi = u(\,\cdot\,, t),$$

where u as before is the solution to (4.1). Due to (4.6), G can be extended by continuity to $(L^1)^N$ and

$$(4.7) \qquad \|G\|_{\mathcal{B}((L^1)^N)} \leq 1.$$

By the sufficiency part of Theorem 4.1, we also have the inequality

$$\|G\varphi\|_\infty \leq \|\varphi\|_\infty, \quad \varphi \in C_0^{2,\alpha}.$$

Let $f \in (L^\infty)^N$. By multiplying f with the characteristic function of a suitable precompact set in Ω and then convolve the result with a suitable mollifier, it follows that f can be arbitrarily well approximated in $(L^1)^N$ by a function $h \in (C_0^\infty(\Omega))^N$ in such a way that $\|h\|_\infty \leq \|f\|_\infty$. Thus, take a sequence $\{f_k\}_k$ in $(C_0^\infty(\Omega))^N$ with $\|f_k - f\|_1 \to 0$ as $k \to \infty$ and with $\|f_k\|_\infty \leq \|f\|_\infty$ for every k. Since

$$\|Gf_k - Gf\|_1 \leq \|f_k - f\|_1 \to 0, \quad k \to \infty,$$

there is a subsequence of $\{Gf_k\}_k$ such that the subsequence converges pointwise to Gf almost everywhere on Ω. Rename the subsequence to the original sequence and it follows that

$$|(Gf)(x)| = \lim_{k\to\infty} |(Gf_k)(x)| \leq \lim_{k\to\infty} \|Gf_k\|_\infty \leq \lim_{k\to\infty} \|f_k\|_\infty \leq \|f\|_\infty$$

for almost every x in Ω. Consequently,

$$(4.8) \qquad\qquad \|G\|_{\mathcal{B}((L^\infty)^N)} \leq 1.$$

The norm estimates in (4.7) and (4.8) now enables us to use the Riesz-Thorin interpolation theorem on G and we get

$$\|G\|_{\mathcal{B}((L^p)^N)} \leq 1, \quad p \in [1, \infty].$$

The arbitrariness of t finally gives the desired result. $\qquad\qquad\square$

4.2.   The semigroup setting

With Theorem 4.4 in mind, we will now define extensions of the operator $\mathcal{A}$ with domain $C_0^{2,\alpha}$ to operators in the various $(L^p)^N$-spaces. Thus, let $\mathcal{A}_p$ be the operator $\mathcal{A}$ with domain

$$D(\mathcal{A}_p) = (W^{2,p}(\Omega) \cap W_0^{1,p}(\Omega))^N, \quad 1 < p < \infty$$

and let $\mathcal{A}_1$ be the operator obtained by taking the closure in $(L^1)^N$ of the operator $\mathcal{A}$ with domain $C_0^{2,\alpha}$.

Every $\mathcal{A}_p$ is thus a densely defined operator in $(L^p)^N$. In the case $p = \infty$, we have to proceed in another manner since $C_0^{2,\alpha}$ is not dense in $(L^\infty)^N$. Instead, the natural function space in which to define the operator $\mathcal{A}_\infty$, is in the closure of $C_0^{2,\alpha}$ in the L^∞-norm; that is, the space $(C_0(\overline{\Omega}))^N$. By Sobolev's embedding theorem, $W^{2,r}(\Omega)$ is a subset of $C(\overline{\Omega})$ for all $r > n/2$. Take some $s > n/2$ and let $\mathcal{A}_\infty$ be the operator $\mathcal{A}$ with domain

$$D(\mathcal{A}_\infty) = \left\{ u \in (W^{2,s}(\Omega) \cap W_0^{1,s}(\Omega))^N : \mathcal{A}u \in (C_0(\overline{\Omega}))^N \right\}.$$

and regard $\mathcal{A}_\infty$ as an operator in $(C_0(\overline{\Omega}))^N$.

Using the results obtained on the parabolic systems in the previous subsection, we can prove the following theorem about the generation of contraction semigroups for all the operators $\mathcal{A}_p$.

Theorem 4.5. *The operators $\mathcal{A}_p$ generate contraction semigroups on $(L^p(\Omega))^N$ for all $p \in [1, \infty)$ and on $(C_0(\overline{\Omega}))^N$ for $p = \infty$ simultaneously if and only if*

(a) *there are real-valued scalar functions a_{ij} on $\overline{\Omega}$ such that for every i, j, $A_{ij} = a_{ij}I$ and the $n \times n$-matrix $[a_{ij}]$ is positive definite,*

(b) *for all $\eta_i, \zeta \in \mathbf{C}^N$, $i = 1, \ldots, n$, with $\Re\langle \eta_i, \zeta \rangle = 0$, the inequalities*

$$\Re\left\{ a_{ij}\langle \eta_i, \eta_j \rangle - \langle A_i \eta_i, \zeta \rangle - \langle A\zeta, \zeta \rangle \right\} \geq 0,$$
$$\Re\left\{ a_{ij}\langle \eta_i, \eta_j \rangle + \langle A_i\zeta, \eta_i \rangle - \langle (A - \partial_i A_i)\zeta, \zeta \rangle \right\} \geq 0$$

hold on Ω.

Proof. The necessity of (a) and (b) is easy to show. Let $p \in [1, \infty]$. The contraction semigroup generated by $\mathcal{A}_p$ gives the solutions to the abstract Cauchy problem associated with the operator $\mathcal{A}_p$. Since $C_0^{2,\alpha}$ is contained in the function space whose elements the semigroup acts on and the solutions to the parabolic system (4.1) are unique, inequality (4.4) holds for the chosen p due to the contractivity of the semigroup. Since p is arbitrary, Theorem 4.4 shows that (a) and (b) follow.

We next show the sufficiency of (a) and (b). This will be achieved by constructing a contraction semigroup for every p and show that its generator is precisely $\mathcal{A}_p$: define the map $t \mapsto Q(t)$, where $t \in [0, T]$ and $Q(t) : C_0^{2,\alpha} \to C_0^{2,\alpha}$, by

$$Q(t)\varphi = u(\,\cdot\,, t)$$

where u is the solution to the parabolic system (4.1) with initial data φ. Defining $Q(t) = Q(T)Q(t - T)$ recursively, Q is extended to $\mathbf{R}^+$. It is easily seen that $Q(s + t) = Q(s)Q(t)$ and that $Q(0)$ is the identity operator on $C_0^{2,\alpha}$. Theorem 4.4 shows that

$$\|Q(t)\varphi\|_p \leq \|\varphi\|_p, \quad t \in \mathbf{R}^{\,|}, \ p \in [1, \infty], \ \varphi \in C_0^{2,\alpha}.$$

$C_0^{2,\alpha}$ is dense in $(L^p)^N$ for $1 \leq p < \infty$ and in $(C_0(\overline{\Omega}))^N$ so extend for each $p \in [1, \infty)$ and $t \in \mathbf{R}^+$ the operator $Q(t)$ by continuity to $Q_p(t) \in \mathcal{B}((L^p(\Omega))^N)$ and to $Q_\infty(t) \in \mathcal{B}((C_0(\overline{\Omega}))^N)$, respectively. Then $\|Q_p(t)\| \leq 1$ and the algebraic properties of Q are preserved. Thus Q_p fulfills the semigroup properties every p if the continuity property can be established.

Let $\varphi \in C_0^{2,\alpha}$ and let u be the corresponding solution of (4.1). The solution is classical so, if $N = 1$ and if u is real-valued, the mean value theorem gives

$$(4.9) \qquad \begin{aligned} \left|\left(t^{-1}(Q(t)\varphi - \varphi) - \mathcal{A}\varphi\right)(x)\right| \\ = \ & |t^{-1}(u(x, t) - u(x, 0)) - \partial_t u(x, 0)| \\ = \ & |\partial_t u(x, \xi(x, t)) - \partial_t u(x, 0)| \leq C|\xi(x, t)|^\gamma \\ \leq \ & Ct^\gamma, \quad x \in \Omega \end{aligned}$$

for some $C > 0$, $\gamma \in (0, 1)$, coming from the Hölder continuity of $\partial_t u$. If u is complex-valued and N is arbitrary, the inequality between the first and last row in (4.9) still holds, since it can be used on each of the components and real/imaginary parts of u together with the triangle inequality. Since Ω is bounded, (4.9) especially implies that $t \mapsto Q_p(t)\varphi$ is continuous at 0 in the norm of $(L^p(\Omega))^N$ for every

$\varphi \in C_0^{2,\alpha}$ and $p \in [1,\infty)$. The same is true for $p = \infty$ since $Q_\infty(t)\varphi$ is zero on the boundary of Ω for every t. Let p be arbitrary, take $f \in (L^p)^N$ or $f \in (C_0(\overline{\Omega}))^N$ if $p = \infty$, let $\varepsilon > 0$ and choose $\varphi \in C_0^{2,\alpha}$ with $\|f - \varphi\|_p < \varepsilon$. Then

$$\begin{aligned} \|Q_p(t)f - f\|_p &\leq \|Q_p(t)\|\|f - \varphi\|_p + \|Q_p(t)\varphi - \varphi\|_p + \|\varphi - f\|_p \\ &\leq 2\varepsilon + \|Q_p(t)\varphi - \varphi\|_p, \end{aligned}$$

showing that the continuity of $t \mapsto Q_p(t)f$ at 0 follows from the continuity on $C_0^{2,\alpha}$. This is actually equivalent to Q_p being a strongly continuous contraction semigroup on $(L^p)^N$ for every $p \in [1,\infty)$ and on $(C_0(\overline{\Omega}))^N$ for $p = \infty$.

Let for each p the operator B_p be the generator of Q_p. Inequality (4.9) shows that the $(L^p)^N$-limit

$$\lim_{t \to 0+} \|t^{-1}(Q_p(t) - I)\varphi - A\varphi\|_p = 0, \quad \varphi \in C_0^{2,\alpha}, \; p \in [1,\infty),$$

so B_p is an extension of the operator $(A, C_0^{2,\alpha})$ for every $p < \infty$. If $p = \infty$, we must in addition require that $A\varphi$ is zero on the boundary of Ω in order for the limit to be zero.

Let $p \in (1,\infty)$. Since B_p is dissipative, so is the operator $(A, C_0^{2,\alpha})$. Since $C_0^{2,\alpha}$ is dense in the domain of A_p, Lemma 3.3 in Langer and Maz'ya [9] implies that the extension A_p of $(A, C_0^{2,\alpha})$ is dissipative. As in the proof of Lemma 3.3, the injectivity of $A_p - \lambda I$ for $\lambda > 0$ follows from the dissipativity of A_p and according to Grisvard [4], the injectivity yields the invertibility of $A_p - \lambda I$ for $\lambda > 0$. By the theorem of Lumer-Phillips, A_p generates a contraction semigroup on $(L^p)^N$.

The next step is to prove that A_1 generates a contraction semigroup in $(L^1)^N$. Since B_1 is a closed extension of $(A, C_0^{2,\alpha})$ and A_1 is the smallest closed extension of the same operator, A_1 is a restriction of B_1. $B_1 - I$ is injective since B_1 is a semigroup generator so if we can show that $A_1 - I$ maps $D(A_1)$ onto $(L^1)^N$, it follows that $A_1 = B_1$ and thus A_1 generates a contraction semigroup.

Denote the space $(W^{2,2} \cap W_0^{1,2})^N$ by F and take $u \in F$. Then there is a sequence $\{u_k\}$ in $C_0^{2,\alpha}$ with limit u in F. We get

$$\|A_1 u_k\|_1 \leq C\|u_k\|_{2,1} \leq C_1\|u_k\|_{2,2},$$

where the first inequality follows from the smoothness of the coefficients of A and the second follows from the boundedness of Ω. This inequality shows that $\{A_1 u_k\}$ is a Cauchy sequence in $(L^1)^N$ and thus convergent, implying that $u \in D(A_1)$ due to the fact that A_1 is closed. Hence $F \subset D(A_1)$. Since A_2 generates a contraction semigroup, $(A_1 - I)(F) = (A_2 - I)(F) = (L^2)^N$, so the range of $A_1 - I$ includes $(L^2)^N$. Take $f \in (L^1)^N$, let $\{f_k\} \subset (L^2)^N$ have limit f in $(L^1)^N$ and set $u_k = (A_2 - I)^{-1} f_k$. By using that A_r generates a contraction semigroup and that u_k and f_k belong to $(L^r)^N$ for $r \in [1,2]$, the relation

$$\begin{aligned} \|u_k\|_1 &= \lim_{r \to 1+} \|u_k\|_r = \lim_{r \to 1+} \|(A_2 - I)^{-1} f_k\|_r \\ &= \lim_{r \to 1+} \|(A_r - I)^{-1} f_k\|_r \leq \lim_{r \to 1+} \|f_k\|_r = \|f_k\|_1 \end{aligned}$$

is obtained. It shows that $\{u_k\}$ converges in $(L^1)^N$ to a limit function u. Since $\mathcal{A}_1$ is closed and

$$\lim_{k\to\infty} \|(\mathcal{A}_1 - I)u_k - f\|_1 = \lim_{k\to\infty} \|f_k - f\|_1 = 0,$$

we conclude that $u \in D(\mathcal{A}_1)$ and $(\mathcal{A}_1 - I)u = f$. Thus $\mathcal{A}_1 - I$ maps $D(\mathcal{A}_1)$ onto $(L^1)^N$.

Finally, we treat the case $p = \infty$. Since $\mathcal{A}_\infty$ is a restriction of $\mathcal{A}_s$ and $\mathcal{A}_s$ generates a contraction semigroup, $\mathcal{A}_\infty - \lambda I$ is injective for $\lambda > 0$. From the definition of $D(\mathcal{A}_\infty)$, it follows that the range of $\mathcal{A}_\infty - \lambda I$ is $(L^s \cap C_0(\overline{\Omega}))^N$ so $\mathcal{A}_\infty - \lambda I$ maps $D(\mathcal{A}_\infty)$ onto $(C_0(\overline{\Omega}))^N$ due to the boundedness of Ω. Let $\lambda > 0$, take $f \in (C_0(\overline{\Omega}))^N$ and set $u = (\mathcal{A}_\infty - \lambda I)^{-1}f$. The operators $\mathcal{A}_r$, $r \in [s, \infty)$, generate contraction semigroups so

$$\begin{aligned}
\|u\|_\infty &= \lim_{r\to\infty} \|u\|_r = \lim_{r\to\infty} \left\|(\mathcal{A}_s - \lambda I)^{-1}f\right\|_r \\
&= \lim_{r\to\infty} \left\|(\mathcal{A}_r - \lambda I)^{-1}f\right\|_r \leq \lim_{r\to\infty} \lambda^{-1}\|f\|_r = \lambda^{-1}\|f\|_\infty,
\end{aligned}$$

implying that $\|(\mathcal{A}_\infty - \lambda I)^{-1}\| \leq \lambda^{-1}$. The operator $\mathcal{A}_\infty$ is densely defined since $(C_0^\infty(\Omega))^N$ is contained in $D(\mathcal{A}_\infty)$ and $C_0^\infty(\Omega)$ is dense in $C_0(\overline{\Omega})$. By the theorem of Hille-Yosida (see e.g. Goldstein [5]), the operator $\mathcal{A}_\infty$ generates a contraction semigroup on the space $(C_0(\overline{\Omega}))^N$. $\qquad\square$

As an example, consider the Schrödinger operator with magnetic field, see e.g. Simon [12] or Cycon et al. [3],

$$-(i\nabla + m)^t(i\nabla + m) - V,$$

i.e. the scalar operator

$$\mathcal{A} = \Delta - 2im \cdot \nabla - i(\nabla \cdot m) - |m|^2 - V,$$

where m is an $\mathbf{R}^n$-valued function on Ω, the function V is complex-valued and the domain Ω and the functions m, V fulfill the smoothness assumptions in the beginning of this section. Using the inequality in the remark following Theorem 4.1 and its dual counterpart, Theorem 4.5 gives necessary and sufficient conditions for $\mathcal{A}$ to generate contraction semigroups on all L^p-spaces simultaneously as

$$-4\Re\mathcal{A} \geq \sum_{j=1}^{n}(\Im A_j)^2, \qquad -4\Re(\overline{A - \partial_j A_j}) \geq \sum_{j=1}^{n}(-\Im\overline{A_j})^2.$$

A simple verification shows that these two equations are equivalent to the condition $\Re V \geq 0$ on Ω.

References

[1] AMANN, H., Dual Semigroups and Second Order Linear Elliptic Boundary Value Problems, *Isr. J. Math.* **45** (1983), 225–254

[2] BREZIS, H. and W. A. STRAUSS, Semi-linear Second-order Elliptic Equations in L^1, *J. Math. Soc. Japan* **25** (1973), 565–590

[3] CYCON, H. L., FROESE, R. G., KIRSCH, W. and B. SIMON, *Schrödinger Operators*, Springer-Verlag Berlin Heidelberg, 1987

[4] GRISVARD, P., *Elliptic Problems in Nonsmooth Domains*, Pitman Publishing Ltd., London, 1985

[5] GOLDSTEIN, J. A., *Semigroups of Linear Operators and Applications*, Oxford University Press, Oxford, 1985

[6] HÖRMANDER, L., *The Analysis of Linear Partial Differential Operators III*, Springer-Verlag, Berlin Heidelberg, 1985

[7] LADYŽHENSKAYA, O. A., *The Boundary Value Problems of Mathematical Physics*, Springer-Verlag New York, 1985

[8] LADYŽHENSKAYA, O. A., SOLONNIKOV, V. A. and N. N. URAL'CEVA, *Linear and Quasilinear Equations of Parabolic Type*, Amer. Math. Soc. Providence, R. I., 1968

[9] LANGER, M. and V. G. MAZ'YA, On L^p-contractivity of Semigroups Generated by Linear Partial Differential Operators, *J. Func. Analysis*, To appear

[10] KRESIN, G. I. and V. G. MAZ'YA, Criteria for Validity of the Maximum Modulus Principle for Solutions of Linear Parabolic Systems, *Ark. Mat.* **32** (1994), 121–155

[11] MAZ'YA, V. G. and P. E. SOBOLEVSKIĬ, On Generating Operators of Semigroups (Russian), *Uspekhi Mat. Nauk* **17:6** (1962), 151–154

[12] SIMON, B., Schrödinger Semigroups, *Bull. Amer. Math. Soc.* **7** (1982), 447–526

Dept. of Mathematics,
Linköping University,
SE-581 83 Linköping, Sweden,
milan@mai.liu.se

1991 Mathematics Subject Classification: Primary 47D06

Submitted: 15.02.1999

Operator Theory:
Advances and Applications, Vol. 109

Curriculum vitae of Vladimir Maz'ya

Birth: 1937, December 31, Leningrad (USSR)

Education:

- Graduated from the Leningrad University in 1960

- Candidate (Ph. D.) degree in 1962 from the University of Moscow

- D. Sc. degree in 1965 from the University of Leningrad

Employment:

- 1960–1964 junior researcher, Institute of Mathematics and Mechanics of Leningrad University

- 1964–1986 senior researcher at the same Institute

- 1968–1972 lecturer, Leningrad Shipbuilding Institute

- 1986–1990 head of Laboratory for Mathematical Modeling in Mechanics, Institute of Problems in Mechanical Engineering (the USSR Acad. Sci., Leningrad)

- 1990–March 1993 högskolelektor, Mathematical Department of Linköping University

- March 1993– professor, Mathematical Department of Linköping University

Awards, honours:

- 1962 Prize from the Leningrad Mathematical Society to a junior mathematician

- 1971 title of professor in Applied Mathematics awarded by the Higher Sertification Commission of the USSR (VAK)

- 1990 Doctor honoris causa of the University of Rostock

- January 1993 Conference in honour of Vladimir Maz'ya "Sobolev Spaces and their related fields", Kyoto

- May 26–29 1998, Minisymposium dedicated to V. Maz'ya "Mathematical Aspects of Boundary Element Methods" at the IABEM International Symposium on Boundary Element Methods, École Polytechnique, Palaiseau (Paris), France

- August 31–Sept 4, 1998 A conference in honour of Vladimir Maz'ya, Functional Analysis, Partial Differential Equations and Applications, Rostock, Germany

- Eidus, D., Khvoles, A., Kresin, G., Merzbach, E., Prössdorf, S., Shaposhnikova, T., Sobolevskii, P., Mathematical work of Vladimir Maz'ya (on the occasion of his 60th birthday), Funct. Differ. Equ. 4 (1997), no. 1–2, 3–11 (1998).

Additional information:

- Organizer and Chairman of two International Conferences on Applied and Industrial Mathematics in Linköping, December 1991 and June 1994

- Visiting position of Maître des Recherches, École Polytechnique, Palaiseau, 1994

Selected addresses:

- International Congress of Mathematicians, Moscow, 1966

- Mathematical Congress of GDR, 1975 (plenary speaker)

- International Conference on Potential Theory, Amersfoort, 1991 (plenary speaker)

- Colloque Équations aux Dérivées Partielles, Saint Jean de Monts, 1992 (plenary speaker)

- Wiener Centennial Symposium, Cambridge, Massachusetts, MIT, 1994 (plenary speaker)

- IABEM Symposium on Boundary Integral Methods for Nonlinear Problems, Siena, 1995 (closing lecture)

- International Symposium Problemi Attuali dell'Analisi e della Fisica Matematica in memory of G. Fichera, Taormina, 1998 (opening lecture)

Editorial Boards and Committees:

- Zeitschrift für Analysis and ihre Anwendungen, Heldermann Verlag, Berlin

- Potential Analysis, Kluwer Academic Publishers, The Netherlands

- Asymptotic Analysis, North-Holland, Amsterdam-London-New York-Tokyo

- Functional Differential Equations, The College of Judea and Samaria, Kedumim-Ariel, Israel

- Nonlinear Mathematical Physics, Luleå

- Applicable Analysis, Overseas Publishers Association, The Netherlands

- A member of The Friends of Mittag-Leffler Institute, 1999-2000

Degrees completed under V. Maz'ya's supervison at Linköping University:

- Doctor, 1995 – G. Bastay

- Doctor, 1996 – J. Björn

- Doctor, 1997 – M. Langer

- Doctor, 1997 – T. Ivanov

- Licentiate, 1995 – M. Langer, H. Åkermark

- Licentiate, 1996 – T. Ivanov

- Licentiate, 1997 – S. Eilertsen

- Licentiate, 1999 – J. Åslund

- Master, 1995 – J. Åslund, S. Eilertsen

- Master, 1998 – J. Östensson

Publications of Vladimir Maz'ya

Books

1. Burago, Yu. D.; Maz'ya, V. G. *Certain Questions of Potential Theory and Function Theory for Regions with Irregular Boundaries.* (Russian) Zap. Naučn. Sem. Leningrad. Otdel. Mat. Inst. Steklov. (LOMI) **3** 1967 152 pp; English translation: Potential Theory and Function Theory for Irregular Regions. Translated from Russian. Seminars in Mathematics, V. A. Steklov Mathematical Institute, Leningrad, Vol. 3 Consultants Bureau, New York 1969 vii+68 pp.

2. Mazja, W. *Einbettungssätze für Sobolewsche Räume.* Teil 1. (German) [Embedding Theorems for Sobolev Spaces. Part 1] Translated from the Russian by J. Nagel. With English, French and Russian summaries. Teubner-Texte zur Mathematik. [Teubner Texts on Mathematics] BSB B. G. Teubner Verlagsgesellschaft, Leipzig, 1979. 204 pp.; *Einbettungssätze für Sobolewsche Räume.* Teil 2. (German) [Embedding Theorems for Sobolev Spaces. Part 2] With English, French and Russian summaries. Teubner-Texte zur Mathematik [Teubner Texts in Mathematics], 28. BSB B. G. Teubner Verlagsgesellschaft, Leipzig, 1980. 188 pp.

3. Maz'ya, V. G.; Nazarov, S. A.; Plamenevskiĭ , B. A. *Asymptotic Behavior of Solutions of Elliptic Boundary Value Problems under Singular Perturbations of the Domain.* (Russian) With Georgian and English summaries. Tbilis. Gos. Univ., Inst. Prikl. Mat., Tbilisi, 1981. 207 pp.

4. Gelman, I. W.; Mazja, W. G. *Abschätzungen für Differentialoperatoren im Halbraum.* (German) [Estimates for differential operators in the half-space] Translated from the Russian by Ehrhard Herbst and Werner Plischke. Edited by G. Wildenhain. Mathematische Lehrbücher und Monographien, II. Abteilung: Mathematische Monographien [Mathematical Textbooks and Monographs, Part II: Mathematical Monographs], 54. Akademie-Verlag, Berlin, 1981. 221 pp.; Birkhäuser Verlag, Basel-Boston, Mass., 1982.

5. Mazja, W. *Zur Theorie Sobolewscher Räume.* (German) [On the theory of Sobolev spaces] Translated from the Russian by J. Nagel. With English, French and Russian summaries. Teubner-Texte zur Mathematik [Teubner Texts in Mathematics], 38. BSB B. G. Teubner Verlagsgesellschaft, Leipzig, 1981. 170 pp.

6. Maz'ya, V. G.; Morozov, N. F.; Plamenevskii; B. A.; Stupialis, L.,*Elliptic Boundary Value Problems,* Amer. Math. Soc. Transl. (1984), **123**, Ser. 2, 268 pp.

7. Maz'ya, V. G. *Sobolev Spaces.* (Russian) Leningrad. Univ., Leningrad, 1985. 416 pp.; English translation from the Russian by T. O. Shaposhnikova. Springer-Verlag, Berlin-New York, 1985. xix+486 pp.

8. Maz'ya, V. G.; Shaposhnikova, T. O. *Theory of Multipliers in Spaces of Differentiable Function.* Monographs and Studies in Mathematics, 23. Pitman, Boston, Mass.-London, 1985. xiii+344 pp.; Russian edition by Leningrad. Univ., Leningrad, 1986. 404 pp.

9. Mazja, W. G.; Nasarow, S. A.; Plamenewski, B. A. *Asymptotische Theorie elliptischer Randwertaufgaben in singulär gestörten Gebieten.* I. (German) [Asymptotic theory of elliptic boundary value problems in singularly perturbed domains. I.] *Störungen isolierter Randsingularitäten.* [Perturbations of isolated boundary singularities] Mathematische Lehrbücher und Monographien, II. Abteilung: Mathematische Monographien [Mathematical Textbooks and Monographs, Part II: Mathematical Monographs], 82. Akademie-Verlag, Berlin, 1991. 432 pp.; II. (German) [Asymptotic theory of elliptic boundary value problems in singularly perturbed domains. II] Nichtlokale Störungen. [Nonlocal perturbations] Mathematische Lehrbücher und Monographien, II. Abteilung: Mathematische Monographien [Mathematical Textbooks and Monographs, Part II: Mathematical Monographs], 83. Akademie-Verlag, Berlin, 1991. 319 pp.

10. Kozlov, V.; Maz'ya, V., *Theory of a Higher-order Sturm-Liouville Equation.* Lecture Notes in Mathematics, 1659. Springer-Verlag, Berlin, 1997. xii+140 pp.

11. Kozlov, V. A.; Maz'ya, V. G.; Rossmann, J. *Elliptic Boundary Value Problems in Domains with Point Singularities.* Mathematical Surveys and Monographs, 52. American Mathematical Society, Providence, RI, 1997, x+414 pp.

12. Maz'ya, V.; Poborchi, S. *Differentiable Functions on Bad Domains.* World Scientific, 1997, XVIII+481 pp.

13. Maz'ya, V.; Shaposhnikova, T. *Jacques Hadamard, a Universal Mathematician.* History of Mathematics, 14. American Mathematical Society, Providence, RI; London Mathematical Society, London, 1998. xxviii+574 pp.

14. Kozlov, V.; Maz'ya, V. *Differential Equations with Operator Coefficients.* Springer-Verlag, 1999, XV+441.

15. Kozlov, V.; Maz'ya, V.; A. Movchan, A. *Asymptotic Analysis of Fields in Multistructures.* Oxford University Press, to appear.

16. Kuznetsov, N.; Maz'ya, V.; Vainberg, B. *Linear Time-harmonic Water Waves. A Mathematical Approach.* Cambridge University Press, to appear.

Papers

1959

1. Maz'ya, V. G. *Solution of Dirichlet's problem for an equation of elliptic type.* (Russian) Dokl. Akad. Nauk SSSR **129**, 257–260.

1960

2. Maz'ya, V. G. *Classes of domains and embedding theorems for function spaces.* Soviet Math. Dokl. **1**, 882–885.

1961

3. Maz'ya, V. G. *Some estimates of solutions of second-order elliptic equations.* (Russian) Dokl. Akad. Nauk SSSR **137**, 1057–1059.

4. Maz'ya, V. G. *On p-conductivity and theorems on embedding certain functional spaces into a C-space.* (Russian) Dokl. Akad. Nauk SSSR **140**, 299–302.

1962

5. Maz'ya, V. G. *The negative spectrum of the higher-dimensional Schrödinger operator.* (Russian) Dokl. Akad. Nauk SSSR **144**, 721–722.

6. Maz'ya, V. G. *On the solvability of the Neumann problem.* (Russian) Dokl. Akad. Nauk SSSR **147**, 294–296.

7. Burago, Yu. D.; Maz'ya, V. G.; Sapožnikova, V. D. *On the double layer potential for non-regular domains.* (Russian) Dokl. Akad. Nauk SSSR **147**, 523–525.

8. Maz'ya, V. G.; Sobolevskiĭ, P. E. *On generating operators of semi-groups.* (Russian) Uspehi Mat. Nauk **17**:6 (108), 151–154.

9. Maz'ya, V. G. *Embedding theorems for arbitrary sets.* (Russian) Uspehi Mat. Nauk **17**:1, 247–248.

1963

10. Maz'ya, V. G. *The Dirichlet problem for elliptic equations of arbitrary order in unbounded domains.* (Russian) Dokl. Akad. Nauk SSSR **150**, 1221–1224.

11. Maz'ya, V. G. *On the boundary regularity of solutions of elliptic equations and of a conformal mapping.* (Russian) Dokl. Akad. Nauk SSSR **152**, 1297–1300.

1964

12. Maz'ya, V. G.; Sapožnikova, V. D. *A remark on the regularization of a singular system in the isotropic theory of elasticity.* (Russian) Vestnik Leningrad. Univ. Ser. Mat. Meh. Astronom. **19**:2, 165–167. Erratum in: Vestnik Leningrad. Univ. Ser. Mat. Meh. Astronom. **9**:4 (1977), 160.

13. Maz'ya, V. G. *On the theory of the higher-dimensional Schrödinger operator.* (Russian) Izv. Akad. Nauk SSSR Ser. Mat. **28**, 1145–1172.

14. Maz'ya, V. G. *The solvability in $\overset{\circ}{W}{}_2^2$ of the Dirichlet problem for a region with a smooth irregular boundary.* (Russian) Vestnik Leningrad. Univ. **19**:7 163–165.

15. Maz'ya, V. G.; Sapožnikova, V. D. *Solution of the Dirichlet and Neumann problems for irregular domains by potential-theoretic methods.* (Russian) Dokl. Akad. Nauk SSSR **159**, 1221–1223.

1965

16. Maz'ya, V. G.; Plamenevskiĭ, B. A. *On singular equations with a vanishing symbol.* (Russian) Dokl. Akad. Nauk SSSR **160**, 1250–1253.

17. Maz'ya, V. G. *Polyharmonic capacity in the theory of the first boundary-value problem.* (Russian) Sibirsk. Mat. Ž. **6**, 127–148.

18. Maz'ya, V. G.; Plamenevskiĭ, B. A. *The Cauchy problem for hyperbolic singular integral equations of convolution type.* (Russian) Vestnik Leningrad. Univ. **20**:19, 161–163.

19. Maz'ya, V. G. *On the theory of the multi-dimensional Schrödinger operator.* (Russian) Vestnik Leningrad. Univ. **20**:1, 135–137.

1966

20. Maz'ya, V. G. *On the modulus of continuity of a solution of the Dirichlet problem near an irregular boundary.* (Russian) 1966 Problems Math. Anal. Boundary Value Problems Integr. Equations (Russian) pp. 45–58 Izdat. Leningrad. Univ., Leningrad.

21. Burago, Ju. D.; Maz'ya, V. G.; Sapožnikova, V. D. *On the theory of potentials of a double and a simple layer for regions with irregular boundaries.* (Russian) 1966 Problems Math. Anal. Boundary Value Problems Integr. Equations (Russian) 3–34 Izdat. Leningrad. Univ., Leningrad.

22. Mazz'ja, V. G. *Boundary value problems in domains with irregular boundaries.* Reports of the Internat. Congress of Math., Section 7, Moscow, 1966, 42.

1967

23. Maz'ya, V. G. *Solvability in $\overset{\circ}{W}{}_2^2$ of the Dirichlet problem in a region with a smooth irregular boundary.* (Russian) Vestnik Leningrad. Univ. **22**:7, 87–95.

24. Maz'ya, V. G.; Mihlin, S. G. *The Cosserat spectrum of the equations of elasticity theory.* (Russian) Vestnik Leningrad. Univ. **22**:13, 58–63.

25. Maz'ya, V. G. *The behavior near the boundary of the solution of the Dirichlet problem for an elliptic equation of the second order in divergence form.* (Russian) Mat. Zametki **2**, 209–220.

26. Maz'ya, V. G.; Havin, V. P. *Approximation in the mean by harmonic functions.* (Russian) Zap. Naučn. Sem. Leningrad. Otdel. Mat. Inst. Steklov. (LOMI) **5**, 196–200.

27. Veržbinskiǐ, G. M.; Maz'ya, V. G. *The asymptotics of solutions of the Dirichlet problem near a non-regular frontier.* (Russian) Dokl. Akad. Nauk SSSR **176**, 498–501.

28. Maz'ya, V. G. *Closure in the metric of the generalized Dirichlet integral.* (Russian) Zap. Naučn. Sem. Leningrad. Otdel. Mat. Inst. Steklov. (LOMI)**5** 192–195.

1968

29. Babič, V. M.; Bakel'man, I. Ja.; Košelev, A. I.; Maz'ya, V. G. *Solomon Grigor'evič Mihlin: On the sixtieth anniversary of his birth.* (Russian) Uspehi Mat. Nauk **23**:4 (142), 269–272.

30. Maz'ya, V. G.; Havin, V. P. *Approximation in the mean by analytic functions.* (Russian) Vestnik Leningrad. Univ. **23**:13, 62–74.

31. Maz'ya, V. G.; Havin, V. P. *The Cauchy problem for Laplace's equation.* (Russian) Vestnik Leningrad. Univ. **23**:7, 146–147.

32. Maz'ya, V. G. *Examples of nonregular solutions of quasilinear elliptic equations with analytic coefficients.* (Russian) Funkcional. Anal. i Priložen. **2**:3, 53–57.

33. Maz'ya, V. G. *The Neumann problem in regions with nonregular boundaries.* (Russian) Sibirsk. Mat. Ž. **9**, 1322–1350.

1969

34. Maz'ya, V. G. *The boundedness of the first derivatives of the solution of the Dirichlet problem in a region with smooth nonregular boundary.* (Russian) Vestnik Leningrad. Univ. **24**:1, 72–79.

35. Maz'ya, V. G.; Panejah, B. P. *Degenerate elliptic pseudo-differential operators on a smooth manifold without boundary.* (Russian) Funkcional. Anal. i Priložen. **3**:2, 91–92.

36. Maz'ya, V. G. *Weak solutions of the Dirichlet and Neumann problems.* (Russian) Trudy Moskov. Mat. Obšč. **20**, 137–172.

37. Maz'ya, V. G.; Havin, V. P. *On the uniqueness theorem of L. Carleson for analytic functions with finite Dirichlet integral.* (Russian) 1969 Problems of Math. Anal., no. 2: Linear Operators and Operator Equations (Russian) pp. 153–156 Izdat. Leningrad. Univ., Leningrad.

38. Krol', I. N.; Maz'ya, V. G. *The lack of continuity and Hölder continuity of solutions of a certain quasilinear equation.* (Russian) Zap. Naučn. Sem. Leningrad. Otdel. Mat. Inst. Steklov. (LOMI) **14** 89–91.

39. Maz'ya, V. G.; Haĭkin, Ju. E. *A remark on the continuity in L_2 of a singular integral operator.* (Russian) Vestnik Leningrad. Univ. **24**:19, 156–159.

1970

40. Maz'ya, V. G. *The degenerate problem with oblique derivative.* (Russian) Uspehi Mat. Nauk **25**:2 (152), 275–276.

41. Maz'ya, V. G.; Panejah, B. P. *Degenerate elliptic pseudodifferential operators with simple complex characteristics.* (Russian) Uspehi Mat. Nauk**25**:1 (151), 193–194.

42. Maz'ya, V. G.; Havin, V. P. *A nonlinear analogue of the Newtonian potential, and metric properties of (p, l)-capacity.* (Russian) Dokl. Akad. Nauk SSSR **194**, 770–773.

43. Maz'ya, V. G. *The continuity at a boundary point of the solutions of quasi-linear elliptic equations.* (Russian) Vestnik Leningrad. Univ. **25**:13, 42–55; erratum: Vestnik Leningrad. Univ. **27**:1, 160; English translation: Vestnik Leningrad. Univ. Math. **3** (1976), 225–242.

44. Maz'ya, V. G., *Some questions from the theory of general differential operators.* In the book: Mikhlin, S. G. Mathematical physics, an advanced course. Translated from the Russian. North-Holland Series in Applied Mathematics and Mechanics, Vol. 11 North-Holland Publishing Co., Amsterdam-London; American Elsevier Publishing Co., Inc., New York 1970 xv+561 pp.; German translation: Maz'ya, W.G. Einige Fragen der Theorie allgemeiner Differentialoperatoren, In the book: Michlin, S. G. Lehrgang der mathematischen Physik. (German) Übersetzung aus dem Russischen: Mathematische Lehrbücher und Monographien. I. Abteilung: Mathematische Lehrbücher, Band XV. Akademie-Verlag, Berlin, 1972. xiv+475 pp.

45. Maz'ya, V. G. *Classes of sets and measures that are connected with embedding theorems.* (Russian) Embedding theorems and their applications (Proc. Sympos., Baku, 1966) (Russian), pp. 142–159. Izdat. "Nauka", Moscow, 1970.

46. Maz'ya, V. G. The selfadjointness of the Laplace operator. (Russian) Embedding theorems and their applications (Proc. Sympos., Baku, 1966) (Russian), pp. 160–162, 246. Izdat. "Nauka", Moscow, 1970.

47. Maz'ya, V. G.; Panejah, B. P. *Coercive estimates and regularity of solutions of degenerate elliptic pseudodifferential equations.* (Russian) Funkcional. Anal. i Prilozen. **4**:4, 41–56.

1971

48. Maz'ya, V. G.; Plamenevskiĭ, B. A. *The asymptotics of the solutions of differential equations with operator coefficients.* (Russian) Dokl. Akad. Nauk SSSR **196**, 512–515.

49. Maz'ya, V. G.; Plamenevskiĭ, B. A. *The oblique derivative problem in a domain with a piecewise smooth boundary.* (Russian) Funkcional. Anal. i Priložen **5**:3, 102–103.

50. Veržbinskiĭ, G. M.; Maz'ya, V. G. *Asymptotic behavior of the solutions of second order elliptic equations near the boundary.* I. (Russian) Sibirsk. Mat. Ž. **12**, 1217–1249.

1972

51. Gel'man, I. V.; Maz'ya, V. G. *Estimates for differential operators with constant coefficients in a half-space.* (Russian) Dokl. Akad. Nauk SSSR **202**, 751–754.

52. Maz'ya, V. G. *The Neumann problem for elliptic operators of arbitrary order in domains with nonregular boundaries.* (Russian) Vestnik Leningrad. Univ. no. 1, 26–33.

53. Maz'ya, V. G. *The degenerate problem with an oblique derivativ.* (Russian) Mat. Sb. (N.S.) **87** (129), 417–454.

54. Maz'ya, V. G. *Applications of certain integral inequalities to the theory of quasilinear elliptic equations.* (Russian) Comment. Math. Univ. Carolinae **13**, 535–552.

55. Maz'ya, V. G. *The removable singularities of bounded solutions of quasilinear elliptic equations of arbitrary order.* (Russian) Boundary value problems of mathematical physics and related questions in the theory of functions, 6. Zap. Naučn. Sem. Leningrad. Otdel. Mat. Inst. Steklov. (LOMI) **27**, 116–130.

56. Maz'ya, V. G.; Havin, V. P. *Approximation in the mean by harmonic functions.* (Russian) Investigations on linear operators and the theory of functions, III. Zap. Naučn. Sem. Leningrad. Otdel. Mat. Inst. Steklov. (LOMI) **30**, 91–105.

57. Veržbinskiĭ, G. M.; Maz'ya, V. G. *Asymptotic behavior of the solutions of second order elliptic equations near the boundary.* II. (Russian) Sibirsk. Mat. Ž. **13**, 1239–1271.

58. Maz'ya, V. G. *On Beurling's theorem on the minimum principle for positive harmonic functions.* (Russian) Investigations on linear operators and the theory of functions, III. Zap. Naučn. Sem. Leningrad. Otdel. Mat. Inst. Steklov. (LOMI) **30**, 76–90.

59. Maz'ya, V. G.; Havin, V. P. *On the theory of nonlinear potentials and (p, l)-capacity.* (Russian) Vestnik Leningrad. Univ. Mat. Meh. Astronom. **13**:3, 46–51.

60. Maz'ya, V. G. *Certain integral inequalities for functions of several variables.* (Russian) Problems of mathematical analysis, no. 3: Integral and differential operators, Differential equations (Russian), pp. 33–68. Izdat. Leningrad. Univ., Leningrad.

61. Maz'ya, W.G. *Einige Fragen der Theorie allgemeiner Differentialoperatoren.* In the book: Michlin, S. G. Lehrgang der mathematischen Physik. (German) Übersetzung aus dem Russischen: Mathematische Lehrbücher und Monographien. I. Abteilung: Mathematische Lehrbücher, Band XV. Akademie-Verlag, Berlin, 1972. xiv+475 pp.; extended version in: Mitteilungen der Math. Gesellschaft der DDR (1975) H.1, 92–116.

62. Maz'ya, V. G.; Plamenevskiĭ, B. A. *The asymptotic behavior of solutions of differential equations in Hilbert space.* (Russian) Izv. Akad. Nauk SSSR Ser. Mat. **36**, 1080–1133; erratum, ibid. **37** (1973), 709–710.

63. Maz'ya, V. G.; Plamenevskiĭ, B. A.*A certain class of manifolds with singularities.* (Russian) Izv. Vysš. Učebn. Zaved. Matematika, no. 11(126), 46–52.

64. Krol', I. N.; Maz'ya, V. G. *The absence of the continuity and Hölder continuity of the solutions of quasilinear elliptic equations near a nonregular boundary.* (Russian) Trudy Moskov. Mat. Obšč. **26**, 75–94.

65. Vaĭnberg, B. R.; Maz'ya, V. G. *Certain stationary problems of the linear theory of surface waves.* (Russian) Dokl. Akad. Nauk SSSR **205**, 310–313.

66. Maz'ya, V. G.; Havin, V. P. *A nonlinear potential theory.* (Russian) Uspehi Mat. Nauk **27**:6, 67–138.

1973

67. Maz'ya, V. G. *A certain embedding operator, and set functions of the type of (p, l)-capacity.* (Russian) Comment. Math. Univ. Carolinae **14**, 155–175.

68. Maz'ya, V. G.; Plamenevskiĭ, B. A. *Elliptic boundary value problems with discontinuous coefficients on manifolds with singularities.* (Russian) Dokl. Akad. Nauk SSSR **210**, 529–532.

69. Maz'ya, V. G. *The coercivity of the Dirichlet problem in a domain with irregular boundary.* (Russian) Izv. Vysš. Učebn. Zaved. Matematika, no. 4(131), 64–76.

70. Maz'ya, V. G.; Plamenevskiĭ, B. A. *The asymptotic behavior of solutions of the Navier-Stokes equations near the edges.* (Russian) Dokl. Akad. Nauk SSSR **210**, 803–806.

71. Maz'ya, V. G. *The oblique derivative problem in a domain with edges of various dimensions.* (Russian) Vestnik Leningrad. Univ. **7** Mat. Meh. Astronom. Vyp. 2 (1973), 34–39.

72. Maz'ya, V. G. *The (p, l)-capacity, embedding theorems and the spectrum of a selfadjoint elliptic operator.* (Russian) Izv. Akad. Nauk SSSR Ser. Mat. **37**, 356–385.

73. Maz'ya, V. G. *The continuity and boundedness of functions in S. L. Sobolev spaces.* (Russian) Problems of mathematical analysis, no. 4: Integral and differential operators, Differential equations , pp. 46–77. Izdat. Leningrad. Univ., Leningrad.

74. Maz'ya, V. G.; Havin, V. P.*Application of the (p, l)-capacity to certain problems of the theory of exceptional sets.* (Russian) Mat. Sb. (N.S.) **90** (132) (1973), 558–591.

75. Maz'ya, V. G.; Plamenevskiĭ, B. P. *The behavior of the solutions of quasilinear elliptic boundary value problems in the neighborhood of a conical point.* (Russian) Boundary value problems of mathematical physics and related questions in the theory of functions, 7. Zap. Naučn. Sem. Leningrad. Otdel. Mat. Inst. Steklov. (LOMI) **38**, 94–97.

76. Maz'ya, V. G.; Plamenevskiĭ, B. A.*Elliptic boundary value problems in a domain with a piecewise smooth boundary.* (Russian) Proceedings of the Symposium on Continuum Mechanics and Related Problems of Analysis (Tbilisi, 1971), Vol. 1 , pp. 171–181. Izdat. "Mecniereba", Tbilisi.

77. Maz'ya, V. G. *The degenerate oblique derivative problem.* (Russian) Proceedings of the Symposium on Continuum Mechanics and Related Problems of Analysis (Tbilisi, 1971), Vol. 1 (Russian), pp. 165–170. Izdat. "Mecniereba", Tbilisi.

78. Vaĭnberg, B. R.; Maz'ya, V. G. *On the plane problem of the motion of a solid immersed in a fluid.* Trudy Moskov. Mat. Obšč. **28**, 35–56.

79. Vaĭnberg, B. R.; Maz'ya, V. G. *On the problem of the steady oscillations of a layer of fluid of variable depth.* (Russian) Trudy Moskov. Mat. Obšč. **28**, 57–74.

1974

80. Maz'ya, V. G.; Plamenevskiĭ, B. A. *The fundamental solutions of elliptic boundary value problems, and the Miranda-Agmon maximum principle in domains with conical points.* (Russian) Sakharth. SSSR Mecn. Akad. Moambe **73**, 277–280.

81. Gel'man, I. V.; Maz'ya, V. G. *Estimates on the boundary for differential operators with constant coefficients in a half-space.* (Russian) Izv. Akad. Nauk SSSR Ser. Mat. **38**, 663–720.

82. Kuznecov, N. G.; Maz'ya, V. G. *On the problem of the steady-state oscillations of a layer of fluid in the presence of an obstacle.* (Russian) Dokl. Akad. Nauk SSSR **216**, 759–762.

83. Veržbinskiĭ, G. M.; Maz'ya, V. G. *The closure in L_p of the operator of the Dirichlet problem in a domain with conical points.* (Russian) Izv. Vysš. Učebn. Zaved. Matematika, no. 6(145), 8–19.

84. Maz'ya, V. G. *The connection between two forms of capacity.* (Russian) Vestnik Leningrad. Univ. Mat. Mech. Astronom. **7**:2, 33–40.

85. Maz'ya, V. G.; Plamenevskiĭ, B. A. *The coefficients in the asymptotic expansion of the solutions of elliptic boundary value problems to near conical points.* (Russian) Dokl. Akad. Nauk SSSR **219**, 286–289.

86. Maz'ya, V. G.; Paneyah, B. *Degenerate elliptic pseudo-differential operators and the problem with oblique derivative.* (Russian) Collection of articles dedicated to the memory of Ivan Georgievč Petrovskiĭ. Trudy Moskov. Mat. Obšč. **31**, 237–295.

87. Maz'ya, V. G.; Havin, V. P. *The solutions of the Cauchy problem for the Laplace equation (uniqueness, normality, approximation).* (Russian) Trudy Moskov. Mat. Obšč. **30**, 61–114.

1975

88. Maz'ya, V. G.; Plamenevskiĭ, B. A. *Boundary value problems for a second order elliptic equation in a domain with ribs.* (Russian) Collection of articles dedicated to the memory of Academician V. I. Smirnov. Vestnik Leningrad. Univ. Mat. Meh. Astronom. **1**:1, 102–108.

89. Gel'man, I. V.; Maz'ya, V. G. *The domination of differential operators with constant coefficients in a half-space.* (Russian) Dokl. Akad. Nauk SSSR **221**:3, 528–531.

90. Maz'ya, V. G.; Plamenevskiĭ, B. A. *The coefficients in the asymptotic expansion of the solutions of elliptic boundary value problems in a cone.* (Russian) Boundary value problems of mathematical physics and related questions of the theory of functions, 8. Zap. Naučn. Sem. Leningrad. Otdel. Mat. Inst. Steklov. (LOMI) **52**, 110–127.

91. Maz'ya, V. G. *The index of closure of the operator of the Dirichlet problem in a domain with a nonregular boundary.* (Russian) Problems of mathematical analysis, no. 5: Linear and nonlinear differential equations, Differential operators, pp. 98–121. Izdat. Leningrad. Univ., Leningrad.

92. Maz'ya, V. G.; Gel'man, I. V. *Estimates for differential operators with constant coefficients in a half-space.* (Russian) Mat. Sb. (N.S.) **96** (138), 240–275.

93. Maz'ya, V. G. *The summability of functions belonging to Sobolev spaces.* (Russian) Problems of mathematical analysis, no. 5: Linear and nonlinear differential equations, Differential operators, pp. 66–98. Izdat. Leningrad. Univ., Leningrad.

94. Maz'ya, V. G. *Einige Richtungen und Probleme der Theorie elliptischer Gleichungen.* Mitteilungen Gesellschaft der DDR, H. 1, 26–91.

95. Maz'ya, W.G. *Einige Fragen der Theorie allgemeiner Differentialoperatoren.* Mitteilungen der Math. Gesellschaft der DDR, H.1, 92–116.

96. Maz'ya, V. G.; Plamenevskiĭ, B. A. L_p *estimates, and the asymptotic behavior of the solutions of elliptic boundary value problems in domains with edges.* (Russian) Conference on Differential Equations and Applications (Ruse, 1975). Godišnik Visš. Učebn. Zaved. Priložna Mat. **11**:2, 113–123.

1976

98. Maz'ya, V. G.; Plamenevskiĭ, B. A. *The coefficients in the asymptotic expansion of the solutions of elliptic boundary value problems near an edge.* (Russian) Dokl. Akad. Nauk SSSR **229**:1, 33–36.

99. Maz'ya, V. G.; Haĭkin, Ju. E. *The continuity of singular integral operators in normed spaces.* (Russian) Vestnik Leningrad. Univ.Mat. Meh. Astronom. **1**:1, 28–34.

1977

100. Maz'ya, V. G. *The connection between Martin's and Euclid's topologies.* (Russian) Dokl. Akad. Nauk SSSR **233**:1, 27–30.

101. Maz'ya, V. G.; Otelbaev, M. *Embedding theorems and the spectrum of a certain pseudodifferential operator.* (Russian) Sibirsk. Mat. Ž. **18**:5, 1073–1087.

102. Maz'ya, V. G.; Plamenevskiĭ, B. A. *The asymptotic behavior of the solution of the Dirichlet problem near an isolated singularity on the boundary.* (Russian) Vestnik Leningrad. Univ. Mat. Meh. Astronom, **13**:3, 60–66.

103. Maz'ya, V. G.; Plamenevskiĭ, B. A.; Haĭkin, Ju. E. *A well-posed problem for singular integral equations with a symbol that goes to zero.* (Russian) Differencial'nye Uravnenija **13**:8, 1479–1486.

104. Fichera, G.; Maz'ya, V. G. *In honour of Professor Solomon G. Mikhlin on the occasion of his seventieth birthday.* Applicable Anal. **7**:3, 167–170.

105. Dudučava, R. V.; Maz'ya, V. G. *A uniqueness theorem for the integral equation of a thin rectangular airfoil.* (Russian) Sakharth. SSR Mecn. Akad. Moambe **87**:1, 53–56.

106. Maz'ya, V. G. *Strong capacity-estimates for "fractional" norms.* (Russian) Numerical methods and questions on organization of computations. Zap. Naučn. Sem. Leningrad. Otdel. Mat. Inst. Steklov. (LOMI) **70**, 161–168.

107. Maz'ya, V. G.; Plamenevskiĭ, B. A. *Elliptic boundary value problems on manifolds with singularities.* (Russian) Problems in mathematical analysis, no. 6: Spectral theory, boundary value problems, pp. 85–142. Izdat. Leningrad. Univ., Leningrad.

108. Maz'ya, V. G.; Plamenevskiĭ, B. A. *The pseudo-analyticity of the solutions of elliptic equations in the space $\mathbf{R}^n$.* (Russian) Sakharth. SSR Mecn. Akad. Moambe **85**:1, 37–40.

109. Maz'ya, V. G.; Plamenevskiĭ, B. A. *The coefficients in the asymptotics of solutions of elliptic boundary value problems with conical points.* (Russian) Math. Nachr. **76**, 29–60.

110. Maz'ya, V. G. *Solvability of the problem of oscillations of a fluid in the presence of an immersed body.* (Russian) Boundary value problems of mathematical physics and related questions in the theory of functions, 10. Zap. Naučn. Sem. Leningrad. Otdel. Mat. Inst. Steklov. (LOMI) **69**, 124–128.

111. Maz'ya, V. G. *On a stationary problem on small oscillations of a fluid in the presence of an embedded body.* (Russian) Partial differential equations, 57–79, Proc. Sobolev Sem., no. 2, Akad. Nauk SSSR Sibirsk. Otdel., Inst. Mat., Novosibirsk.

112. Maz'ya, V. G. *Behaviour of solutions to the Dirichlet problem for the biharmonic operator at a boundary point.* Dokl. (Russian) Acad. Nauk USSR **235**:6, 1263–1266.

113. Maz'ya, V. G. *Local square summability of a convolution.* (Russian) Investigations on linear operators and the theory of functions, VIII. Zap. Nauchn. Sem. Leningrad. Otdel. Mat. Inst. Steklov. (LOMI) **73**, 211–216. Erratum in: Zap. Nauchn. Sem. Leningrad. Otdel. Mat. Inst. Steklov. (LOMI) **113** (1981).

1978

114. Maz'ya, V. G.; Plamenevskiĭ, B. A. *Estimates in L_p and in Hölder classes, and the Miranda-Agmon maximum principle for the solutions of elliptic boundary value problems in domains with singular points on the boundary.* (Russian) Math. Nachr. **81**, 25–82.

115. Masja, W.; Nagel, J. *Über äquivalente Normierung der anisotropen Funktionalräume $H^\mu(\mathbf{R}^n)$.* (German) Beiträge Anal. no. 12, 7–17.

116. Maz'ya, V. G.; Morozov, N. F.; Plamenevskiĭ, B. A. *The stressed-strained state in a neighborhood of a crack apex at nonlinear bending of a plate.* (Russian) Dokl. Akad. Nauk SSSR **243**:4, 889–892.

117. Maz'ya, V. G.; Plamenevskiĭ, B. A. *Estimates of the Green functions and Schauder estimates of the solutions of elliptic boundary value problems in a two-sided corner.* (Russian) Sibirsk. Mat. Zh. **19**:5, 1065–1082.

118. Maz'ya, V. G.; Plamenevskiĭ, B. A. *L_p-estimates of solutions of elliptic boundary value problems in domains with ribs.* (Russian) Trudy Moskov. Mat. Obshch. **37**, 49–93.

119. Maz'ya, V. G.; Plamenevskiĭ, B. A. *Schauder estimates for the solutions of elliptic boundary value problems in domains with edges on the boundary.* (Russian) Partial differential equations, pp. 69–102, Trudy Sem. S. L. Soboleva, no. 2, Akad. Nauk SSSR Sibirsk. Otdel., Inst. Mat., Novosibirsk.

120. Maz'ya, V. G.; Plamenevskiĭ, B. A. *Weighted spaces with inhomogeneous norms, and boundary value problems in domains with conical points.* (Russian) Elliptische Differentialgleichungen (Meeting, Rostock, 1977), pp. 161–190, Wilhelm-Pieck-Univ., Rostock.

121. Maz'ya, V. G. *An integral inequality.* (Russian) Sem. Inst. Prikl. Mat. Dokl. no. 12-13, 33–36.

122. Maz'ya, V. G. *On regularity of a boundary point for elliptic equations.* (Russian) Zap. Nauchn. Sem. Leningrad. Otdel. Mat. Inst. Steklov. (LOMI) **81**, 197–199.

123. Maz'ya, V. G. *Multipliers in Sobolev spaces.* (Russian) In the book: Application of function theory and functional analysis methods to problems of mathematical physics. Pjatoe Sovetso-Čehoslovackoe Soveščanie, 1976, 181–189; Novosibirsk.

1979

124. Kresin, G. I.; Maz'ya, V. G. *The essential norm of an operator of the double-layer potential type in the space C_m.* (Russian) Dokl. Akad. Nauk SSSR **246**:2, 272–275.

125. Maz'ya, V. G. *Behaviour of solutions to the Dirichlet problem for the biharmonic operator at a boundary point.* Equadiff IV (Proc. Czechoslovak Conf. Differential Equations and their Applications, Prague, 1977), pp. 250–262, Lecture Notes in Math., **703**, Springer, Berlin.

126. Maz'ya, V. G.; Šapošnikova, T. O. *Multipliers in spaces of functions with fractional derivatives.* (Russian) Dokl. Akad. Nauk SSSR **244**:5, 1065–1067.

127. Maz'ya, V. G.; Šapošnikova, T. O. *Traces and extensions of the multipliers in the space W_p^l.* (Russian) Uspckhi Mat. Nauk **34**:2 (206), 205–206.

128. Maz'ya, V. G.; Nazarov, S. A.; Plamenevskiĭ , B. A. *Asymptotic behavior of the solutions of elliptic boundary value problems in the case of variation of the domain near conic points.* (Russian) Dokl. Akad. Nauk SSSR **249**:1, 94–96.

129. Maz'ya, V. G.; Šapošnikova, T. O. *Multipliers in S. L. Sobolev spaces.* (Russian) Vestnik Leningrad. Univ. Mat. Mekh. Astronom. no. 2, 33–40.

130. Maz'ya, V. G. *Summability, with respect to an arbitrary measure, of functions from S. L. Sobolev L. N. Slobodeckiĭ spaces.* (Russian) Investigations on linear operators and the theory of functions, IX. Zap. Nauchn. Sem. Leningrad. Otdel. Mat. Inst. Steklov. (LOMI) **92**, 192–202.

131. Maz'ya, V. G.; Plamenevskiĭ , B. A. *Asymptotic behavior of the fundamental solutions of elliptic boundary value problems in domains with conical points.* (Russian) Boundary value problems. Spectral theory, pp. 100–145, Probl. Mat. Anal., 7, Leningrad. Univ., Leningrad.

132. Maz'ya, V. G. *A new integral representation of differentiable functions and its applications.* (Russian) Soobshch. Akad. Nauk Gruzin. SSR **95**:3, 537–540.

133. Maz'ya, V.; Plamenevskiĭ , B.; Stupjalis, L. *The three-dimensional problem of the steady-state motion of a fluid with a free surface.* (Russian) Differentsial'nye Uravneniya i Primenen.—Trudy Sem. Protsessy Optimal. Upravleniya I Sektsiya no. 23, 157 pp.

134. Maz'ya, V. G.; Plamenevskiĭ , B. A. *Pseudoanalyticity of the solutions of a perturbed polyharmonic equation in* $\mathbf{R}^n$. (Russian) Scattering theory. Theory of oscillations, pp. 75–91, Probl. Mat. Fiz., **9**, Leningrad. Univ., Leningrad.

135. Maz'ya, V. G.; Šapošnikova, T. O. *Multipliers in spaces of differentiable functions.* (Russian) Theory of cubature formulas and the application of functional analysis to problems of mathematical physics (Proc. Sem. S. L. Sobolev, no. 1, 1979) (Russian), pp. 37–90, Trudy Sem. S. L. Soboleva, no. 1, 1979, Akad. Nauk SSSR Sibirsk. Otdel., Inst. Mat., Novosibirsk.

136. Maz'ya, V. G.; Morozov, N. F.; Plamenevskiĭ , B. A. *Nonlinear bending of a plate with a crack.* (Russian) Differential and integral equations. Boundary value problems, pp. 145–163, Tbilis. Gos. Univ., Tbilisi.

137. Maz'ya, V. G. *The spectrum of V. A. Steklov's problem for a second-order equation with nonnegative characteristic form.* (Russian) Soobshch. Akad. Nauk Gruzin. SSR **95**:1, 41–44.

138. Maz'ya, V. G.; Gel'man, I. V. *Estimates for the maximal operator in a half-space.* I. (Russian) Beiträge Anal. no. 14, 7–24.

1980

139. Maz'ya, V. G. *An integral representation of functions, satisfying homogeneous boundary conditions, and its applications.* (Russian) Izv. Vyssh. Uchebn. Zaved. Mat. no. 2, 34–44.

140. Vaĭnberg, B. R.; Maz'ya, V. G. *A characteristic Cauchy problem for a hyperbolic equation.* (Russian) Uspekhi Mat. Nauk **35**:1 (211), 193–194.

141. Maz'ya, V. G.; Plamenevskiĭ , B. A. *A problem of the motion of a fluid with a free surface in a faceted vessel.* (Russian) Dokl. Akad. Nauk SSSR **250**:6, 1315–1317.

142. Maz'ya, V. G.; Šapošnikova, T. O. *On conditions for the boundary in the L_p-theory of elliptic boundary value problems.* (Russian) Dokl. Akad. Nauk SSSR **251**:5, 1055–1059.

143. Maz'ya, V. G.; Plamenevskiĭ , B. A. *On the first boundary value problem for the equations of hydrodynamics in a domain with a piecewise smooth boundary.* (Russian) Boundary value problems of mathematical physics and related questions in the theory of functions, 12. Zap. Nauchn. Sem. Leningrad. Otdel. Mat. Inst. Steklov. (LOMI) **96**, 179–186.

144. Gel'man, I. V.; Maz'ya, V. G. *The domination of differential operators with constant coefficients in a half-space.* (Russian) Math. Nachr. **95**, 47–78.

145. Maz'ya, V. G.; Šapošnikova, T. O. *A coercive estimate for solutions of elliptic equations in spaces of multipliers.* (Russian) Vestnik Leningrad. Univ. Mat. Mekh. Astronom. no. 1, 41–47.

146. Maz'ya, V. G.; Šapošnikova, T. O. *Theory of multipliers in spaces of differentiable functions and their applications.* (Russian) Theory of cubature formulas and numerical mathematics (Proc. Conf., Novosibirsk, 1978), pp. 225–233, "Nauka" Sibirsk. Otdel., Novosibirsk.

147. Maz'ya, V. G. *An embedding theorem and multipliers in pairs of S. L. Sobolev spaces.* (Russian) Theory of analytic functions and harmonic analysis. Akad. Nauk Gruzin. SSR Trudy Tbiliss. Mat. Inst. Razmadze **66**, 59–69.

148. Maz'ya, V. G.; Hvoles, A. A. *Embedding of the space $\overset{\circ}{L}{}_p^l(\Omega)$ into a space of generalized functions.* (Russian) Theory of analytic functions and harmonic analysis. Akad. Nauk Gruzin. SSR Trudy Tbiliss. Mat. Inst. Razmadze **66**, 70–83.

149. Maz'ya, V. G.; Shaposhnikova, T. O. *Multipliers of S. L. Sobolev spaces in a domain.* (Russian) Math. Nachr. **99**, 165–183.

150. Maz'ya, V. G.; Shaposhnikova, T. O. *Multipliers in pairs of spaces of potentials.* (Russian) Math. Nachr. **99**, 363–379.

151. Maz'ya, V. G.; Shaposhnikova, T. O. *On the regularity of the boundary in the L_p-theory of elliptic boundary value problems.* I. (Russian) Partial differential equations, pp. 39–56, Trudy Sem. S. L. Soboleva, no. 2, Akad. Nauk SSSR Sibirsk. Otdel., Inst. Mat., Novosibirsk.

152. Maz'ya, V. G.; Preobrazhenskiĭ , S. P. *Some estimates of (l, p)-capacities and their application to embedding theorems.* (Russian) Soobshch. Akad. Nauk Gruzin. SSR **100**:1, 25–28.

153. Maz'ya, V. G.; Gel'man, I. V. *Estimates for the maximal operator in a half-space.* II. (Russian) Beiträge Anal. no. 15, 7–25.

154. Maz'ya, V.; Nazarov, S.; Plamenevskiĭ, B. *Asymptotic behavior of the solutions of a quasilinear equation in nonregular perturbed domains.* (Russian) Differentsial'nye Uravneniya i Primenen.—Trudy Sem. Protsessy Optimal. Upravleniya I Sektsiya no.**27**, 17–50.

1981

155. Maz'ya, Y. G.; Šapošnikova, T. O. *Multipliers on the spaces W_p^m, and their applications.* (Russian) Vestnik Leningrad. Univ. Mat. Mekh. Astronom. no. 1, 42–47.

156. Maz'ya, V. G. *On the extension of functions belonging to S. L. Sobolev spaces.* (Russian) Investigations on linear operators and the theory of functions, XI. Zap. Nauchn. Sem. Leningrad. Otdel. Mat. Inst. Steklov. (LOMI) **113**, 231–236.

157. Maz'ya, V. G.; Shaposhnikova, T. O. *Sufficient conditions for belonging to classes of multipliers*. (Russian) Math. Nachr. **100**, 151–162.

158. Maz'ya, V. G.; Nazarov, S. A.; Plamenevskiĭ , B. A. *On the asymptotic behavior of the solutions to the Dirichlet problem in a three-dimensional domain with a cut-out thin body*. (Russian) Dokl. Akad. Nauk SSSR **256**1, 37–39.

159. Maz'ya, V. G.; Shaposhnikova, T. O. *Multipliers in pairs of spaces of differentiable functions*. (Russian) Trudy Moskov. Mat. Obshch. **43**, 37–80.

160. Maz'ya, V. G.; Plamenevskiĭ , B. A. *Properties of solutions of three-dimensional problems of the theory of elasticity and hydrodynamics in domains with isolated singular points*. (Russian) Comm. Math. Phys. **82**:2, 99–120.

161. Maz'ya, V. G.; Nazarov, S. A.; Plamenevskiĭ , B. A. *Asymptotics of the solutions of the Dirichlet problem in a domain with an excluded thin tube*. (Russian) Uspekhi Mat. Nauk **36**:5 (221), 183–184.

162. Maz'ya, V. G.; Nazarov, S. A.; Plamenevskiĭ , B. A. *The asymptotic behavior of solutions of the Dirichlet problem in a domain with a cut out thin tube*. (Russian) Mat. Sb. (N.S.) **116**:2, 187–217.

163. Maz'ya, V. G.; Shaposhnikova, T. O. *Change of variables as an operator on a pair of S. L. Sobolev spaces*. (Russian) Vestnik Leningrad. Univ. Mat. Mekh. Astronom. no. 1, 43–48.

164. Kresin, G. I.; Maz'ya, V. G. *On the maximum principle for Lamé and Stokes systems in a half-space*. (Russian) Akad. Nauk Armyan. SSR Dokl. **73**:1, 46–50.

165. Maz'ya, V. G.; Shaposhnikova, T. O. *On the regularity of the boundary in the L_p-theory of elliptic boundary value problems*. II. (Russian) Theory of cubature formulas and the application of functional analysis to problems of mathematical physics, pp. 57–102, Trudy Sem. S. L. Soboleva, no. 1, Akad. Nauk SSSR Sibirsk. Otdel., Inst. Mat., Novosibirsk.

166. Vaĭnberg, B. R.; Maz'ya, V. G. *The characteristic Cauchy problem for a hyperbolic equation*. (Russian) Trudy Sem. Petrovsk. no. 7, 101–117.

167. Maz'ya, V. G.; Plamenevskiĭ , B. A. *On the maximum principle for the biharmonic equation in a domain with conical points*. (Russian) Izv. Vyssh. Uchebn. Zaved. Mat. no. 2, 52–59.

168. Maz'ya, V. G.; Nazarov, S. A.; Plamenevskiĭ , B. A. *On the asymptotic behavior of solutions of elliptic boundary value problems with irregular perturbations of the domain*. (Russian) Probl. Mat. Anal., 8, pp. 72–153, Leningrad. Univ., Leningrad.

169. Maz'ya, V. G. *On the influence of boundary conditions on embedding theorems*. (Russian) In the book: Boundary value problems of mathematical physics, 66–72, Kiev.

170. Kresin, G. I.; Maz'ya, V. G. *The essential norm of an operator of the double-layer potential type in the space C_m*. (Russian) Funkzion. Anal. i Vychisl. Matem. "Nauka", Alma-Ata, 131–165.

1982

171. Maz'ya, V. G.; Nazarov, S. A.; Plamenevskiĭ , B. A. *Homogeneous solutions of the Dirichlet problem in the exterior of a thin cone*. (Russian) Dokl. Akad. Nauk SSSR **266**:2, 281–284.

172. Maz'ya, V. G. *Theory of multipliers in spaces of differentiable functions and its applications*. Nonlinear analysis, function spaces and applications, Vol. 2 (P'ısek, 1982), pp. 150–190, Teubner-Texte zur Math. **49**, Teubner, Leipzig.

173. Kerimov, T. M.; Maz'ya, V. G.; Novruzov, A. A. *An analogue of Wiener's criterion for the Zaremba problem in a cylindrical domain*. (Russian) Funktsional. Anal. i Prilozhen. **16**:4, 70–71.

174. Maz'ya, V. G.; Nazarov, S. A.; Plamenevskiĭ , B. A. *Asymptotic behavior of the solution of the Dirichlet problem in a domain with a thin bridge*. (Russian) Funktsional. Anal. i Prilozhen. **16**:2, 39–46.

175. Maz'ya, V. G.; Nazarov, S. A.; Plamenevskiĭ , B. A. *Absence of a De Giorgi-type theorem for strongly elliptic equations with complex coefficients*. (Russian) Boundary value problems of mathematical physics and related questions in the theory of functions, 14. Zap. Nauchn. Sem. Leningrad. Otdel. Mat. Inst. Steklov. (LOMI) **115**, 156–168.

1983

176. Maz'ya, V. G. *Functions with a finite Dirichlet integral in a domain with a cusp at the boundary*. (Russian) Investigations on linear operators and the theory of functions, XII. Zap. Nauchn. Sem. Leningrad. Otdel. Mat. Inst. Steklov. (LOMI) **126**, 117–137.

177. Maz'ya, V. G.; Nazarov, S. A.; Plamenevskiĭ , B. A. *Calculation of the asymptotics of "coefficients of intensity" in the coming together of corner or conic points*. (Russian) Zh. Vychisl. Mat. i Mat. Fiz. **23**:2, 333–346.

178. Zargaryan, S. S.; Maz'ya, V. G. *Singularities of solutions of a system of equations of potential theory for Zaremba's problem.* (Russian) Vestnik Leningrad. Univ. Mat. Mekh. Astronom. no. 1, 43–48.

179. Maz'ya, V. G.; Donchev, T. *Regularity in the sense of Wiener of a boundary point for a polyharmonic operator*. (Russian) C. R. Acad. Bulgare Sci. **36**:2, 177–179.

180. Maz'ya, V. G.; Shaposhnikova, T. O. *Theory of multipliers in spaces of differentiable functions*. (Russian) Uspekhi Mat. Nauk **38**:3 (231), 23–86.

181. Kresin, G. I.; Maz'ya, V. G. *The maximum principle for second-order elliptic and parabolic systems*. (Russian) Dokl. Akad. Nauk SSSR **273**:1, 38–41.

182. Maz'ya, V. G.; Nazarov, S. A.; Plamenevskiĭ, B. A. *Singularities of solutions of the Dirichlet problem in the exterior of a thin cone.* (Russian) Mat. Sb. (N.S.) **122** (164):4, 435–457.

183. Maz'ya, V. G.; Nazarov, S. A.; Plamenevskiĭ, B. A. *Bending of a near-polygonal plate with a free open boundary.* (Russian) Izv. Vyssh. Uchebn. Zaved. Mat. no. 8, 34–40.

184. Maz'ya, V. G.; Plamenevskiĭ, B. A. *The first boundary value problem for classical equations of mathematical physics in domains with piecewise-smooth boundaries. I.* (Russian) Z. Anal. Anwendungen **2**:4, 335–359.

185. Maz'ya, V. G.; Plamenevskiĭ, B. A. *The first boundary value problem for classical equations of mathematical physics in domains with piecewise smooth boundaries. II.* (Russian) Z. Anal. Anwendungen **2**:6, 523–551.

1984

186. Maz'ya, V. G.; Preobrazenskii, S. P. *Estimates for capacities and traces of potentials.* Internat. J. Math. Math. Sci. **7**:1, 41–63.

187. Maz'ya, V. G. *The modulus of continuity of a harmonic function at a boundary point.* (Russian) Investigations on linear operators and the theory of functions, XIII. Zap. Nauchn. Sem. Leningrad. Otdel. Mat. Inst. Steklov. (LOMI) **135**, 87–95.

188. Maz'ya, V. G.; Poborchiĭ, S. V. *Extension of functions from S. L. Sobolev spaces to the exterior and interior of a small domain.* (Russian) Vestnik Leningrad. Univ. Mat. Mekh. Astronom. no. 2, 27–32.

189. Maz'ya, V. G.; Nazarov, S. A.; Plamenevskiĭ, B. A. *Asymptotic expansions of eigenvalues of boundary value problems for the Laplace operator in domains with small openings.* (Russian) Izv. Akad. Nauk SSSR Ser. Mat. **48**:2, 347–371.

190. Maz'ya, V.G.; Kresin, G.I. *The maximum principle for second-order strongly elliptic and parabolic systems with constant coefficients.* (Russian) Mat. Sb. (N.S.) **125**(167):4, 458–480.

191. Maz'ya, V. G.; Nazarov, S. A.; Plamenevskiĭ, B. A. *The Dirichlet problem in domains with thin cross connections.* (Russian) Sibirsk. Mat. Zh. **25**:2, 161–179.

192. Zargaryan, S. S.; Maz'ya, V. G. *The asymptotic form of the solutions of integral equations of potential theory in the neighbourhood of the corner points of a contour.* (Russian) Prikl. Mat. Mekh. **48**:1, 169–174; translation in J. Appl. Math. Mech. **48**:1, 120–124 (1985).

193. Maz'ya, V. G.; Poborchiĭ, S. V. *Extension of functions belonging to S. L. Sobolev spaces into the exterior of a domain with a cusp on the boundary.* (Russian) Dokl. Akad. Nauk SSSR **275**:5, 1066–1069.

194. Maz'ya, V. G.; Nazarov, S. A.; Plamenevskiĭ, B. A. *Elliptic boundary value problems in domains of the type of the exterior of a cusp.* (Russian) Linear and nonlinear partial differential equations. Spectral asymptotic behavior, 105–148, Probl. Mat. Anal., **9**, Leningrad. Univ., Leningrad.

195. Maz'ya, V. G.; Nazarov, S. A. *On the Sapondjan-Babuška paradox for problems of the theory of thin plates.* (Russian) Dokl. Acad. Nauk Arm. SSR **48**:3, 127–130.

196. Maz'ya, V. G. *Über die Regularität eines Randpunktes für elliptische Differentialgleichungen.* (German) Linear and complex analysis problem book, 199 research problems, Lecture Notes in Math. **1043**, 507–514.

197. Maz'ya, V. G. Rossman, J. *Über die Lösbarkeit und Asymptotik der Lösungen elliptischer Randwertaufgaben in Gebieten mit Kanten.* (German) Akad. der Wiss. der DDR, Inst. für Math. I Preprint P-Math. 07/84, s. 1–50; II Preprint P-Math. 30/84, s. 1-50; III Preprint P-Math. 31/84, s. 1–50.

1985

198. Kresin, G. I.; Maz'ya, V. G. *On the maximum of displacements in a viscoelastic half-space (a three-dimensional model).* (Russian) Vestnik Leningrad. Univ. Mat. Mekh. Astronom. no. 4, 47–51.

199. Kozlov, V. A.; Maz'ya, V. G.; Parton, V. Z. *Asymptotic form of the stress-intensity coefficients in quasistatic temperature problems for a domain with a cut.* (Russian) Prikl. Mat. Mekh. **49**:4, 627–636; translation in J. Appl. Math. Mech. **49**:4, 482–489 (1986).

200. Gel'man, I. V.; Maz'ya, V. G. *Estimates in a half-space for systems of differential operators with constant coefficients.* (Russian) Qualitative analysis of solutions of partial differential equations, 70–99, Akad. Nauk SSSR Sibirsk. Otdel., Inst. Mat., Novosibirsk.

201. Kuznetsov, N. G., Maz'ya, V. G. *Asymptotic expansions for transient surface waves due to short-period oscillating disturbances.* (Russian) Proc. Leningrad Shipbuild. Inst./ Math. Modelling and Automated Design in Shipbuilding, 57–64.

1986

202. Maz'ya, V. G.; Nazarov, S. A. *The apex of a cone can be irregular in Wiener's sense for a fourth-order elliptic equation.* (Russian) Mat. Zametki **39**:1, 24–28.

203. Maz'ya, V. G.; Kufner, A. *Variations on the theme of the inequality $(f')^2 \leq 2f \sup |f''|$.* Manuscripta Math. **56**:1, 89–104.

204. Kuznetsov, N. G.; Maz'ya, V. G. *Asymptotic expansions for surface waves caused by short-term disturbances.* (Russian) Asymptotic methods, 103–138, "Nauka" Sibirsk. Otdel., Novosibirsk.

205. Kozlov, V. A.; Maz'ya, V. G. *Estimates of the L_p-means and asymptotic behavior of the solutions of elliptic boundary value problems in a cone. I. The case of the model operator.* (Russian) Seminar Analysis, 55–91, Akad. Wiss. DDR, Berlin.

206. Grachev, N. V.; Maz'ya, V. G. *The Fredholm radius of operators of double layer potential type on piecewise smooth surfaces.* (Russian) Vestnik Leningrad. Univ. Mat. Mekh. Astronom. no. 4, 60–64.

207. Maz'ya, V. G. *Boundary integral equations of elasticity in domains with piecewise smooth boundaries.* Equadiff 6 (Brno, 1985), 235–242, Lecture Notes in Math., **1192**, Springer, Berlin-New York.

208. Maz'ya, V. G.; Poborchiĭ, S. V. *Extension of functions in S. L. Sobolev classes to the exterior of a domain with the vertex of a peak on the boundary. I.* (Russian) Czechoslovak Math. J. **36** (111):4, 634–661.

209. Maz'ya, V. G.; Nazarov, S. A. *Paradoxes of the passage to the limit in solutions of boundary value problems for the approximation of smooth domains by polygons.* (Russian) Izv. Akad. Nauk SSSR Ser. Mat. **50**:6, 1156–1177.

210. Maz'ya, V. G. *On potential theory for the Lamé system in a domain with a piecewise smooth boundary.* (Russian) Partial differential equations and their applications (Tbilisi, 1982), 123–129, Tbilis. Gos. Univ., Tbilisi.

211. Maz'ya, V. G., Slutskii, A. S., Fomin, V. A. *Asymptotic behaviour of the stress function near the vertex of a crack in the problem of torsion under steady-state creep.* (Russian) Mech. Tverd. Tela **4**, 170–176.

1987

212. Maz'ya, V. G.; Sulimov, M. G. *Asymptotic behavior of solutions of one-dimensional difference equations with constant operator coefficients.* (Russian) Mat. Sb. (N.S.) **132** (174):4, 451–469.

213. Maz'ya, V. G.; Poborchiĭ, S. V. *Extension of functions in S. L. Sobolev classes to the exterior of a domain with the vertex of a peak on the boundary. II.* (Russian) Czechoslovak Math. J. *37* (112):1, 128–150.

214. Maz'ya, V. G.; Slutskiĭ, A. S. *Averaging of differential equations on a fine grid.* (Russian) Dokl. Akad. Nauk SSSR **293**:4, 792–796.

215. Kozlov, V. A.; Maz'ya, V. G. *Singularities of solutions of the first boundary value problem for the heat equation in domains with conical points. I.* (Russian) Izv. Vyssh. Uchebn. Zaved. Mat. no. 2, 38–46.

216. Kozlov, V. A.; Maz'ya, V. G. *Singularities of solutions of the first boundary value problem for the heat equation in domains with conical points. II.* (Russian) Izv. Vyssh. Uchebn. Zaved. Mat. no. 3, 37–44.

217. Maz'ya, V. G.; Nazarov, S. A. *Asymptotic behavior of energy integrals under small perturbations of the boundary near corner and conic points.* (Russian) Trudy Moskov. Mat. Obshch. **50**, 79–129.

218. Maz'ya, V. G.; Slutskiĭ, A. S. *Averaging of a differential operator on a fine periodic curvilinear net.* (Russian) Math. Nachr. **133**, 107–133.

219. Kuznetsov, N. G.; Maz'ya, V. G. *Asymptotic expansions for surface waves that can be induced by rapidly oscillating or accelerating perturbations.* (Russian) Asymptotic methods, 136–175, "Nauka" Sibirsk. Otdel., Novosibirsk.

220. Maz'ya, V. G.; Slutskiĭ, A. S. *Averaging of difference equations with rapidly oscillating coefficients.* (Russian) Seminar Analysis (Berlin, 1986/87), 63–92, Akad. Wiss. DDR, Berlin.

221. Maz'ya, V. G. *A boundary integral equation of the Dirichlet problem in a plane domain with a cusp at the boundary.* (Russian) Current problems in mathematical physics, Vol. II (Tbilisi, 1987), 263–270, Tbilis. Gos. Univ., Tbilisi.

1988

222. Maz'ya, V. G.; Solov'ev, A. A. *Solvability of an integral equation of the Dirichlet problem in a plane domain with cusps on the boundary.* (Russian) Dokl. Akad. Nauk SSSR **298**:6, 1312–1315; translation in Soviet Math. Dokl. **37**:1, 255–258.

223. Maz'ya, V. G. *Classes of domains, measures and capacities in the theory of spaces of differentiable functions.* (Russian) Current problems in mathematics. Fundamental directions, Vol. 26 (Russian), 159–228, Itogi Nauki i Tekhniki, Akad. Nauk SSSR, Vsesoyuz. Inst. Nauchn. i Tekhn. Inform., Moscow; English translation: Analysis, III, 141–211, Encyclopaedia Math. Sci., **26**, Springer, Berlin, 1991.

224. Maz'ya, V. G. *Boundary integral equations.* (Russian) Current problems in mathematics. Fundamental directions, Vol 27 (Russian), 131–228, Itogi Nauki i Tekhniki, Akad. Nauk SSSR, Vsesoyuz. Inst. Nauchn. i Tekhn. Inform., Moscow; English translation: Analysis, IV, 127–222, Encyclopaedia Math. Sci., **27**, Springer, Berlin, 1991.

225. Maz'ya, V. G.; Solov'ev, A. A. *Asymptotic behavior of the solution of an integral equation of the Neumann problem in a plane domain with cusps at the boundary.* (Russian) Soobshch. Akad. Nauk Gruzin. SSR **130**:1, 17–20.

226. Maz'ya, V. G.; Rossmann, J. *Über die Asymptotik der Lösungen elliptischer Randwertaufgaben in der Umgebung von Kanten.* (German) [On the asymptotics of solutions of elliptic boundary value problems in the neighborhood of edges] Math. Nachr. **138**, 27–53.

227. Kozlov, V. A.; Maz'ya, V. G. *Estimates of the L_p-means and asymptotic behavior of the solutions of elliptic boundary value problems in a cone. II. Operators with variable coefficients.* (Russian) Math. Nachr. **137**, 113–139.

228. Kozlov, V. A.; Maz'ya, V. G. *Spectral properties of operator pencils generated by elliptic boundary value problems in a cone.* (Russian) Funktsional. Anal. i Prilozhen. **22**:2, 38–46, 96; translation in Functional Anal. Appl. **22**: 2, 114–121.

229. Kerimov, T. M.; Maz'ya, V. G.; Novruzov, A. A. *A criterion for the regularity of the infinitely distant point for the Zaremba problem in a half-cylinder.* (Russian) Z. Anal. Anwendungen **7**:2, 113–125.

230. Kresin, G. I.; Maz'ya, V. G. *A sharp constant in a Miranda-Agmon-type inequality for solutions of elliptic equations.* (Russian) Izv. Vyssh. Uchebn. Zaved. Mat. no. 5, 41–50; translation in Soviet Math. (Iz. VUZ) **32**:5, 49–59.

231. Kozlov, V. A.; Maz'ya, V. G.; Parton, V. Z. *Thermal shock in a region with a crack.* (Russian) Prikl. Mat. Mekh. **52**:2, 318–326; translation in J. Appl. Math. Mech. **52**:2, 250–256.

232. Kuznetsov, N. G.; Maz'ya, V. G. *Unique solvability of a plane stationary problem connected with the motion of a body submerged in a fluid.* (Russian) Differentsial'nye Uravneniya **24**:11, 1928–1940; translation in Differential Equations **24**:11, 1291–1301.

233. Kozlov, V. A.; Maz'ya, V. G. *An asymptotic formula for eigenfunctions of the Dirichlet problem in a domain with a conic point.* (Russian) Vestnik Leningrad. Univ. Mat. Mekh. Astronom. no. 4, 30–33,; translation in Vestnik Leningrad Univ. Math. **21**:4, 36–40.

234. Grachev, N. V.; Maz'ya, V. G. *Representations and estimates for inverse operators of integral equations of potential theory for surfaces with conic points.* (Russian) Soobshch. Akad. Nauk Gruzin. SSR **32**:1, 21–24.

235. Maz'ya, V. G. *Inversion formulas for boundary integral equations and their applications.* (Russian) Functional and numerical methods in mathematical physics, 127–131, "Naukova Dumka", Kiev.

236. Maz'ya, V. G., Poborchiĭ, S. V. *Traces of functions in Sobolev spaces on the boundary of a domain with a peak.* Preprint, MD 87.91, VGM SVP, TR 88-01, University of Maryland.

237. Kuznetsov, N.G.; Maz'ya, V.G. *Unique solvability of the plane Neumann-Kelvin problem.* (Russian) Mat. Sb. (N.S.) **135** (177):4, 440–462; translation in Math. USSR-Sb. 63 (1989), no. 2, 425–446.

1989

238. Maz'ya, V. G.; Poborchiĭ, S. V. *Traces of functions with a summable gradient in a domain with a cusp at the boundary.* (Russian) Mat. Zametki **45**:1, 57–65, 140; translation in Math. Notes **45**:1-2, 39–44.

239. Maz'ya, V. G.; Poborchiĭ, S. V. *Traces of functions from S. L. Sobolev spaces on small and large components of the boundary.* (Russian) Mat. Zametki **45**:4, 69–77, 126; translation in Math. Notes **45**:3-4, 312–317.

240. Maz'ya, V. G.; Nazarov, S. A. *Singularities of solutions of the Neumann problem at a conic point.* (Russian) Sibirsk. Mat. Zh. **30**:3, 52–63, 218; translation in Siberian Math. J. **30**:3, 387–396.

241. Maz'ya, V. G.; Solov'ev, A. A. *An integral equation of the Dirichlet problem in a plane domain with cusps on the boundary.* (Russian) Mat. Sb. **180**:9, 1211–1233.

242. Kozlov, V. A.; Maz'ya, V. G.; Parton, V. Z. *Thermal shock in a thin plate with a crack in the presence of heat exchange with the surrounding medium.* (Russian) Izv. Akad. Nauk Armyan. SSR Ser. Mekh. **42**:2, 41–49.

243. Kresin, G. I.; Maz'ya, V. G. *On the maximum modulus principle for solutions of linear parabolic systems.* Seminar Analysis (Berlin, 1988/1989), 41–50, Akad. Wiss. DDR, Berlin.

244. Levin, A. V.; Maz'ya, V. G. *Asymptotics of the densities of harmonic potentials near the apex of a cone.* (Russian) Z. Anal. Anwendungen **8**:6, 501–514.

245. Maz'ya, V. G.; Poborchiĭ, S. V. *Traces of functions in Sobolev spaces on a boundary of a domain with a cusp.* (Russian) Trudy Inst. Mat. (Novosibirsk) **14**, Sovrem. Probl. Geom. Analiz., 182–208; translation in Siberian Advances in Mathematics **1**:3 (1991), 75–107.

246. Kozlov, V. A.; Maz'ya, V. G. *Iterative procedures for solving ill-posed boundary value problems that preserve the differential equations.* (Russian) Algebra i Analiz **1**:5, 144–170; translation in Leningrad Math. J. **1** (1990), no. 5, 1207–1228.

247. Kozlov, V. A.; Kondrat'ev, V. A.; Maz'ya, V. G. *On sign variability and the absence of "strong" zeros of solutions of elliptic equations.* (Russian) Izv. Akad. Nauk SSSR Ser. Mat. **53**:2, 328–344; translation in Math. USSR-Izv. **34** (1990), no. 2, 337–353.

1990

248. Grachev, N. V.; Maz'ya, V. G. *The Fredholm radius of integral operators of potential theory.* (Russian) Nonlinear equations and variational inequalities. Linear operators and spectral theory, 109–133, Probl. Mat. Anal., **11**, Leningrad. Univ., Leningrad.

249. Maz'ya, V. G.; Solov'ev, A. A. *On a boundary integral equation for the Neumann problem for a domain with a peak.* (Russian) Trudy Leningrad. Mat. Obshch. **1**, 109–134.

250. Maz'ya, V. G. *A mathematical algorithm for reconstruction of the optic distribution of the power dencity under the sondage of a laser beam.* Preprint 35, Leningrad Department of the Institute for Engineering Studies Acad. Nauk SSSR, 1–45.

251. Maz'ya, V. G.; Morozov, N. F.; Nazarov, S. A. *On the elastic strain energy release due to the variation of the domain near the angular stress concentrator.* Preprint, LiTH-MAT-R-90-21, Linköping University.

252. Kozlov, V. A.; Maz'ya, V. G.; Parton, V. Z. *Some mathematical problems of thermoelasticity. Problems of the long-term strength of power equipment.* Proc. of the Polzunov Central Research Institute of power equipment, Leningrad, no. 260, 55–67.

253. Maz'ya, V. G.; Tashchiyan, G. M. *On the behavior of the gradient of the solution of the Dirichlet problem for the biharmonic equation near a boundary point of a three-dimensional domain.* (Russian) Sibirsk. Mat. Zh. **31**:6, 113–126; translation in Siberian Math. J. **31**:6, 970–983 (1991).

1991

254. Kozlov, V. A.; Maz'ya, V. G.; Fomin, A. V. *An iterative method for solving the Cauchy problem for elliptic equations.* (Russian) Zh. Vychisl. Mat. i Mat. Fiz. **31**:1, 64–74.

255. Kozlov, V. A.; Maz'ya, V. G. *On stress singularities near the boundary of a polygonal crack.* Proc. Roy. Soc. Edinburgh Sect. A **117**:1-2, 31–37.

256. Maz'ya, V. G.; Rossmann, J. *On the Agmon-Miranda maximum principle for solutions of elliptic equations in polyhedral and polygonal domains.* Ann. Global Anal. Geom. **9**:3, 253–303.

257. Grachev, N. V.; Maz'ya, V. G. *A contact problem for the Laplace equation in the exterior of the boundary of a dihedral angle.* (Russian) Math. Nachr. **151**, 207–231.

258. Kozlov, V. A.; Maz'ya, V. G. *On the spectrum of an operator pencil generated by the Dirichlet problem in a cone.* (Russian) Mat. Sb. **182**:5, 638–660.

259. Maz'ya, V. G.; Poborchi, S. V. *Boundary traces of functions from Sobolev spaces on a domain with a cusp* [translation of Trudy Inst. Mat. (Novosibirsk) **14** (1989), Sovrem. Probl. Geom. Analiz., 182–208; Siberian Advances in Mathematics **1** (1991), no. 3, 75–107.

260. Kozlov, V. A.; Maz'ya, V. G. *On the spectrum of an operator pencil generated by the Neumann problem in a cone.* (Russian) Algebra i Analiz **3**:2, 111–131; translation in St. Petersburg Math. J. **3** (1992), no. 2, 333–353.

261. Grachev, N.; Maz'ya, V. G. *Estimates for kernels of the inverse operators of the integral equations of elasticity on surfaces with conic points.* Preprint LiTH-MAT-R-91-07, Linköping University.

262. Grachev, N.; Maz'ya, V. G. *Invertibility of the boundary integral operators of elasticity on surfaces with conic points.* Preprint LiTH-MAT-R-91-08, Linköping University.

263. Grachev, N.; Maz'ya, V. G. *Estimates for fundamental solutions of the Neumann problem in a polyhedron.* Preprint, LiTH-MAT-R-91-28, Linköping University.

264. Kozlov, V. A.; Maz'ya, V. G. *On the asymptotic behaviour of solutions of ordinary differential equations with operator coefficients* I, Preprint, LiTH-MAT-R-91-47, Linköping University.

265. Grachev, N.; Maz'ya, V. G. *Solvability of boundary integral equations in a polyhedron.* Preprint, LiTH-MAT-R-91-50, Linköping University.

266. Maz'ya, V. G. *A new approximation method and its applications to the calculation of volume potentials. Boundary point method.* DFG-Kolloquium des DFG-Forschungsschwerpunktes "Randelementmethoden", 30 September-5 October, Schloss Reisenburg (1991), 8.

1992

267. Kozlov, V. A.; Maz'ya, V. G.; Schwab, C. *On singularities of solutions of the displacement problem of linear elasticity near the vertex of a cone.* Arch. Rational Mech. Anal. **119**:3, 197–227.

268. Maz'ya, V. G.; Rossmann, J. *On the Agmon-Miranda maximum principle for solutions of strongly elliptic equations in domains of* $\mathbf{R}^n$ *with conical points.* Ann. Global Anal. Geom. **10**:2, 125–150.

269. Kozlov, V.; Maz'ya, V. *Solvability and asymptotic behaviour of solutions of ordinary differential equations with variable operator coefficients.* Journées "Équations aux Dérivées Partielles" (Saint-Jean-de-Monts, 1992), Exp. no. V, 12 pp., École Polytech., Palaiseau.

270. Maz'ya, V. G.; Rossmann, J. *Stable asymptotics of the solution to the Dirichlet problem for elliptic equations of second order in domains with angular points or edges.* Operator calculus and spectral theory (Lambrecht, 1991), 215–224, Oper. Theory Adv. Appl., **57**, Birkhäuser, Basel.

271. Maz'ya, V. G.; Mahnke, R. *Asymptotics of the solution of a boundary integral equation under a small perturbation of a corner.* Z. Anal. Anwendungen **11**:2, 173–182.

272. Maz'ya, V. G.; Poborchii, S. V. *Embedding theorems for Sobolev spaces in domains with cusps.* Preprint, LiTH-MAT-R-92-14, Linköping University.

273. Maz'ya, V. G.; Slutskii, A. S. *An asymptotic solution of a non-linear Dirichlet problem with strong singularity at the corner point* I. Preprint, LiTH-MAT-R-92-15, Linköping University.

274. Kozlov, V. A.; Maz'ya, V. G. *On the asymptotic behaviour of solutions of ordinary differential equations with operator coefficients* I, Preprint, LiTH-MAT-R-92-18, Linköping University.

275. Kozlov, V. A.; Maz'ya, V. G. *On the asymptotic behaviour of solutions of ordinary differential equations with operator coefficients* III, Preprint, LiTH-MAT-R-92-29, Linköping University.

276. Maz'ya, V.; Rossmann, J. *On a problem of Babuška (stable asymptotics of the solution to the Dirichlet problem for elliptic equations of second order in domains with angular points).* Math. Nachr. **155**, 199–220.

1993

277. Maz'ya, V. G.; Hänler, M. *Approximation of solutions of the Neumann problem in disintegrating domains.* Math. Nachr. **162**, 261–278.

278. Maz'ya, V. G.; Vaĭnberg, B. R. *On ship waves.* Wave Motion **18**:1, 31–50.

279. Maz'ya, V.; Sulimov, M. *Asymptotics of solutions of difference equations with variable coefficients.* Math. Nachr. **161**, 155–170.

280. Kresin, G. I.; Maz'ya, V. G. *Criteria for validity of the maximum modulus principle for solutions of linear strongly elliptic second order systems.* Potential Anal. **2**:1, 73–99.

281. Maz'ya, V. *Solvability and asymptotic behavior of solutions of ordinary differential equations with operator coefficients.* Second International Conference on Mathematical and Numerical Aspects of Wave Propagation (Newark, DE, 1993), 354–362, SIAM, Philadelphia, PA.

282. Maz'ya, V. G.; Slodichka, M. *Some time-marching algorithms for semilinear parabolic equations based upon approximate approximations.* Preprint, LiTH-MAT-R-93-38, Linköping University.

1994

283. Kozlov, V.; Maz'ya, V.; Rozin, L., *On certain hybrid iterative methods for solving boundary value problems.* SIAM J. Numer. Anal. **31**:1, 101–110.

284. Kozlov, V.; Maz'ya, V.; Fomin, A., *The inverse problem of coupled thermoelasticity.* Inverse Problems **10**:1, 153–160.

285. Maz'ya, V. G.; Rossmann, J. *On the behaviour of solutions to the Dirichlet problem for second order elliptic equations near edges and polyhedral vertices with critical angles.* Z. Anal. Anwendungen **13**:1, 19–47.

286. Kresin, G. I.; Maz'ya, V. G. *Criteria for validity of the maximum modulus principle for solutions of linear parabolic systems.* Ark. Mat. **32**:1, 121–155.

287. Kozlov, V. A.; Maz'ya, V. G.; Parton, V. Z. *Asymptotics of the intensity factors for stresses induced by heat sources.* J. Thermal Stresses **17**:3, 309–320.

288. Maz'ya, V. *Approximate approximations.* The mathematics of finite elements and applications (Uxbridge, 1993), 77–104, Wiley, Chichester.

289. Kozlov, V. A.; Maz'ya, V. G.; Schwab, C. *On singularities of solutions to the Dirichlet problem of hydrodynamics near the vertex of a cone.* J. Reine Angew. Math. **456**, 65–97.

290. Kozlov, V. A.; Maz'ya, V. G.; Movchan, A. B. *Asymptotic analysis of a mixed boundary value problem in a multi-structure.* Asymptotic Anal. **8**:2, 105–143.

291. Carlsson, A.; Maz'ya, V., *On approximation in weighted Sobolev spaces and self-adjointness.* Math. Scand. **74**:1, 111–124.

292. Maz'ya, V.; Karlin, V., *Semi-analytic time-marching algorithms for semilinear parabolic equations.* BIT **34**:1, 129–147.

1995

293. Maz'ya, V. G.; Poborchiĭ, S. V. *On traces of functions in S. L. Sobolev spaces on the boundary of a thin cylinder.* (Russian) Trudy Tbiliss. Mat. Inst. Razmadze Akad. Nauk Gruzii **99**, 17–36.

294. Maz'ya, V.; Netrusov, Y., *Some counterexamples for the theory of Sobolev spaces on bad domains.* Potential Anal. **4**:1, 47–65.

295. Maz'ya, V. G.; Verbitsky, I. E. *Capacitary inequalities for fractional integrals, with applications to partial differential equations and Sobolev multipliers.* Ark. Mat. **33**:1, 81–115.

296. Kresin, G. I.; Maz'ya, V. G. *The norm and the essential norm of the double layer elastic and hydrodynamic potentials in the space of continuous functions.* Math. Methods Appl. Sci. **18**:14, 1095–1131.

297. Maz'ya, V.; Schmidt, G., *"Approximate approximations" and the cubature of potentials.* Atti Accad. Naz. Lincei Cl. Sci. Fis. Mat. Natur. Rend. Lincei (9) Mat. Appl. **6**:3, 161–184.

298. Kozlov, V. A.; Maz'ya, V. G.; Movchan, A. B. *Asymptotic representation of elastic fields in a multi-structure.* Asymptotic Anal. **11**:4, 343–415.

299. Karlin, V.; Maz'ya, V., *Time-marching algorithms for initial-boundary value problems based upon "approximate approximations".* BIT **35**:4, 548–560.

1996

300. Maz'ya, V.; Schmidt, G., *On approximate approximations using Gaussian kernels.* IMA J. Numer. Anal. **16**:1, 13–29.

301. Maz'ya, V. G.; Poborchi, S. V. *Extension of functions in Sobolev spaces on parameter dependent domains.* Math. Nachr. **178**, 5–41.

302. Maz'ya, V.; Soloviev, A. A. *Boundary integral equations of the logarithmic potential theory for domains with peaks.* Atti Accad. Naz. Lincei Cl. Sci. Fis. Mat. Natur. Rend. Lincei (9) Mat. Appl. **6** (1995), no. 4, 211–236 (1996).

303. Kozlov, V. A.; Maz'ya, V. G. *Singularities in solutions to mathematical physics problems in non-smooth domains.* Partial differential equations and functional analysis, 174–206, Progr. Nonlinear Differential Equations Appl., **22**, Birkhäuser, Boston, MA.

304. Kozlov, V. A.; Maz'ya, V. G. *On "power-logarithmic" solutions of the Dirichlet problem for elliptic systems in $K_d \times \mathbf{R}^{n-d}$, where K_d is a d-dimensional cone.* Atti Accad. Naz. Lincei Cl. Sci. Fis. Mat. Natur. Rend. Lincei (9) Mat. Appl. **7**:1, 17–30.

1997

305. Kozlov, V. A.; Maz'ya, V. G. *On "power-logarithmic" solutions to the Dirichlet problem for the Stokes system in a dihedral angle.* Math. Methods Appl. Sci. **20**:4, 315–346.

306. Karlin, V.; Maz'ya, V., *Time-marching algorithms for nonlocal evolution equations based upon "approximate approximations".* SIAM J. Sci. Comput. **18**:3, 736–752.

307. Livshits, M.; Maz'ya, V., *Solvability of the two-dimensional Kelvin-Neumann problem for a submerged circular cylinder.* Appl. Anal. **64**:1-2, 1–5.

308. Maz'ya, V., *Unsolved problems connected with the Wiener criterion.* The Legacy of Norbert Wiener: A Centennial Symposium (Cambridge, MA, 1994), 199–208, Proc. Sympos. Pure Math., **60**, Amer. Math. Soc., Providence, RI.

309. Maz'ya, V.; Soloviev, A. *L_p-theory of a boundary integral equation on a cuspidal contour.* Appl. Anal.**65**:3-4, 289–305.

310. Kozlov, V. A.; Maz'ya, V. G. *On "power-logarithmic" solutions to the Dirichlet problem for the Stokes system in a dihedral angle.* Math. Methods Appl. Sci. **20**:4, 315–346.

311. Maz'ya, V. *Asymptotic theory of operator differential equations and its applications.* Modern mathematical methods in diffraction theory and its applications in engineering (Freudenstadt, 1996), 163–173, Methoden Verfahren Math. Phys., **42**, Lang, Frankfurt am Main.

312. Kuznetsov, N.; Maz'ya, V. *Asymptotic analysis of surface waves due to high-frequency disturbances.* Atti Accad. Naz. Lincei Cl. Sci. Fis. Mat. Natur. Rend. Lincei (9) Mat. Appl. **8**:1, 5–29.

313. Kozlov, V.; Maz'ya, V. G.; Rossmann, J. *Spectral properties of operator pencils generated by elliptic boundary value problems for the Lamé system.* Rostock. Math. Kolloq. no. 51, 5–24.

314. Maz'ya, V. G. *Boundary integral equations on a contour with peaks.* IABEM Symposium on Boundary Integral Methods for Nonlinear Problems (Pontignano, 1995), 145–153, Kluwer Acad. Publ., Dordrecht.

1998

315. Maz' ya, V.; Soloviev, A. *L_p-theory of boundary integral equations on a contour with inward peak.* Z. Anal. Anwendungen **17**:3, 641–673.

316. Maz' ya, V., Soloviev A. *L_p-theory of boundary integral equations on a contour with outward peak.* Integral Equations Operator Theory **32**:1, 75–100.

317. Kresin, G.I., Maz'ya, V.G. *On the maximum modulus principle for linear parabolic systems with zero boundary data.* Functional Differential Equations **5**:1–2, 165–181.

318. Maz'ya, V. *From Warschawski's conformal mapping theorem to higher order multi-dimensional elliptic equations.* Analysis, numerics and applications of differential and integral equations. Pitman Res. Notes Math. Ser., **379**, Longman, Harlow, 137–142.

319. Kozlov, V.; Maz'ya, V. *Comparison principles for nonlinear operator differential equations in Banach spaces.* Birman's 70-th Anniversary Collection, American Mathematical Society, Translations 2, **189**.

320. Maz' ya, V., Soloviev A. *Integral equations of logarithmic potential theory in Hölder spaces, on contours with peak.* (Russian) Algebra i Analiz **10**:5, 85-142.

321. Kozlov, V.A., Maz'ya, V., Rossmann, J. *Conic singularities of solutions to problems in hydrodynamics of a viscous fluid with a free surface.* Math. Scand. **83**, 103–141.

1999

322. Maz'ya, V.; Netrusov, Y.; Poborchi, S. *Boundary values of Sobolev functions on non-Lipschitz domains bounded by Lipschitz surfaces.* Algebra i Analiz (Russian), **11**:1.

323. Langer, M.; Maz'ya, V. *On L^p-contractivity of semigroups generated by linear partial differential operators.* J. Functional Analysis, to appear.

324. Maz'ya, V.; Shaposhnikova, T. *On pointwise interpolation inequalities for derivatives.* Mathematica Bohemica, to appear.

325. Maz'ya, V.; Shaposhnikova, T. *Traces and extensions of multipliers in pairs of Sobolev spaces.* Complex Analysis, Operator Theory, and Related Topics: S.A. Vinogradov - In Memoriam, Birkhäuser, to appear.

326. Karlin, V.; Maz'ya, V.; Movchan, A.; Willis, J.; Bullough, R. *Numerical analysis of nonlinear hypersingular integral equations of the Peierls type in dislocation theory.* SIAM J. Appl. Anal., to appear.

327. Kozlov, V.; Maz'ya, V. *Angle singularities of solutions to the Neumann problem for the two-dimensional Riccati equation.* Asympt. Anal., to appear.

328. Kresin, G.I., Maz'ya, V.G. *Criteria for validity of the maximum norm principle for parabolic systems.* Potential Analysis, to appear.

329. Kresin, G.I., Maz'ya, V.G. *On the maximum norm principle with respect to smooth norms for linear strongly coupled parabolic systems.* Functional Differential Equations, to appear.

330. Hansson, K.; Maz'ya, V.; Verbitsky, I. *Criteria of solvability for multidimensional Riccati's equations.* Ark. för Mat. **37**:1, 87–220.

331. Maz'ya, V. *On the Wiener-type regularity of a boundary point for the polyharmonic operator.* Applicable Analysis, to appear.

332. Björn J., Maz'ya, V. *Capacity estimates for solutions of the Dirichlet problem for second order elliptic equations in divergence form.* Potential Analysis, to appear.

333. Maz'ya, V., Schmidt, G. *Approximate wavelets and the approximation of pseudodifferential operators.* Appl. Comp. Harm. Anal., to appear.

334. Maz'ya, V. *On Wiener-type regularity of a boundary point for higher-order elliptic equations.* Summer School on Functional Spaces, Prague, September, 1998, to appear.

335. Ivanov, T., Maz'ya, V., Schmidt, G. *Boundary layer approximate approximations and cubature of potentials in domains.* Advances in Comp. Math., to appear.